Student Solutions Manual

General Chemistry

and

General Chemistry

with Qualitative Analysis

Whitten • Davis • Peck

Yi-Noo Tang
Texas A&M University

Wendy Keeney-Kennicutt
Texas A&M University

Saunders College Publishing
Harcourt Brace College Publishers

Fort Worth   Philadelphia   San Diego   New York   Orlando
San Antonio   Austin   Montreal   London   Sydney   Tokyo

Printed in the United States of America.

Keeney-Kennicult & Tang; Student Solutions Manual to
accompany General Chemistry and General Chemistry with
Qualitative Analysis, 5E. Whitten, Davis and Peck

ISBN 0-03-015694-7

567 018 987654321

Forward to the Students

This Solutions Manual supplements the textbook, *General Chemistry*, fifth edition, by Kenneth W. Whitten, Raymond E. Davis and M. Larry Peck. The solutions of the 1148 even-numbered problems at the end of the chapters have been worked out in a detailed, step-by-step fashion.

Your learning of chemistry serves two purposes: (1) to accumulate fundamental knowledge in chemistry which you will use to understand the world around you, and (2) to enhance your ability to make logical deductions in science. This ability comes when you know how to reason in a scientific way and how to perform the mathematical manipulations necessary for solving certain problems. The excellent textbook by Whitten, Davis, and Peck provides you with a wealth of chemical knowledge, accompanied by good solid examples of logical scientific deductive reasoning. The problems at the end of the chapters are a review, a practice and, in some cases, a challenge to your scientific problem-solving abilities. It is the fundamental spirit of this Solution Manual to help you to understand the scientific deductive process involved in each problem.

In this manual, we provide you with a solution and an answer to the numerical problems, but our emphasis lies on providing the step-by-step reasoning behind the mathematical manipulations. In some cases, we present as many as three different approaches to solve the same problem, since we understand that each of you has your own unique learning style. In stoichiometry as well as in many other types of calculations, the "unit factor" method is universally emphasized in general chemistry textbooks. We think that the over-emphasis of this method may train you to regard chemistry problems as being simply mathematical manipulations in which the only objective is to cancel units and get the answer. Our goal is for you to understand the principles behind the calculations and hopefully to visualize with your mind's eye the chemical processes and the experimental techniques occurring as the problem is being worked out on paper. And so we have dissected the "unit factor" method for you and introduced chemical meaning into each of the steps.

We gratefully acknowledge the tremendous help provided by Frank Kolar in the preparation of this manuscript.

Yi-Noo Tang and
Wendy L. Keeney-Kennicutt

Department of Chemistry
Texas A&M University

Table of Contents

1 The Foundations of Chemistry

1-2.   *Refer to Sections 1-1 and 1-7, and the Key Terms for Chapter 1.*

(a) Weight is a measure of the gravitational attraction of the earth for a body. Although the mass of an object remains constant, its weight will vary depending on its distance from the center of the earth. One kilogram of mass at sea level weighs about 2.2 pounds (9.8 newtons), but that same one kilogram of mass weighs less at the top of Mt. Everest. In more general terms, it is a measure of the gravitational attraction of one body for another. The weight of an object on the moon is about 1/7th that of the same object on the earth..

(b) Potential energy is the energy that matter possesses by virtue of its position, condition, or composition. Your chemistry book lying on a table has potential energy due to its position. Energy is released if it falls from the table.

(c) Kinetic energy is the energy that matter possesses by virtue of its motion. The kinetic energy belonging to a moving train is easily transferred to a stalled car on the tracks.

(d) An endothermic process is a process that absorbs heat energy. The boiling of water is a physical process that requires heat and therefore is endothermic.

1-4.   *Refer to Section 1-2.*

Solids: are rigid and have definite shapes; they occupy a fixed volume and are thus very difficult to compress; the hardness of a solid is related to the strength of the forces holding the particles of a solid together; the stronger the forces, the harder is the solid object.

Liquids: occupy essentially constant volume but have variable shape; they are difficult to compress; particles can pass freely over each other; their boiling points increase with increasing forces of attraction among the particles.

Gases: expand to fill the entire volume of their containers; they are very compressible with relatively large separations between particles.

The three states are alike in that they all exhibit definite mass and volume under a given set of conditions. All consist of some combination of atoms, molecules or ions. The differences are stated above. Additional differences occur in their relative densities:

$$\text{gases} < < < \text{liquids} < \text{solids}.$$

1-6.   *Refer to Section 1-3.*

(a) Baking powder releasing bubbles of carbon dioxide when added to water is a chemical property since a new substance is being formed and a change in composition is occurring.

(b) The composition of steel is a physical property. It can be determined without a composition change.

(c) The density of gold is a physical property, since it can be observed without any change in the composition of the gold.

(d) The ability of iron to dissolve in hydrochloric acid with the evolution of hydrogen gas is a chemical property of iron, since during the reaction, its composition is changing and a new substance is being formed.

(e) The ability of fine steel wool to burn in air is a chemical property of steel wool since a compositional change in the steel wool occurs.

(a) A wet towel drying in the sun is a physical change since the water evaporates from the towel.

(b) When lemon juice is added to tea, a color change occurs. This is a chemical change; the lemon is reacting with some component of the tea.

(c) Hot air rising over a radiator is a physical change. The heat causes the air to expand and its density to decrease; the air then rises.

(d) When coffee is brewed, the hot water dissolves the soluble components in the ground coffee. Therefore, this is a physical change. (Note: there are some chemists who consider dissolution a chemical change.)

1-10. *Refer to Sections 1-1 and 1-4, and the Key Terms for Chapter 1.*

(a) Combustion is an exothermic process in which a chemical reaction releases heat.

(b) The freezing of water is an exothermic process. Heat must be removed from the molecules in the liquid state to cause solidification.

(c) The melting of ice is an endothermic process. The system requires heat to break the attractive forces that hold solid water together.

(d) The boiling of water is an endothermic process. Molecules of liquid water must absorb energy to break away from the attractive forces that hold liquid water together in order to form gaseous molecules.

(e) The condensing of steam is an exothermic process. The heat stored in water vapor must be removed for the vapor to liquefy. The condensation process is the opposite of boiling.

1-12. *Refer to Section 1-5 and the Key Terms for Chapter 1.*

(a) A substance is a kind of matter in which all samples have identical chemical composition and physical properties, e.g., iron (Fe) and water (H_2O).

(b) A mixture is a sample of matter composed of two or more substances in variable composition, each substance retaining its identity and properties, e.g., soil (minerals, water, organic matter, living organisms, etc.) and seawater (water, different salts, dissolved gases, organic compounds, living organisms, etc.).

(c) An element is a substance that cannot be decomposed into simpler substances by chemical means, e.g., nickel (Ni) and nitrogen (N).

(d) A compound is a substance composed of two or more elements in fixed proportions. Compounds can be decomposed into their constituent elements by chemical means. Examples include water (H_2O) and sodium chloride (NaCl).

1-14. *Refer to Section 1-5.*

(a) Coffee is an aqueous, homogeneous mixture of organic compounds (tannins, caffeine, etc.) found in the beans of coffee plants.

(b) Silver is an element, a pure substance that cannot be decomposed into simpler substances by chemical means.

(c) Calcium carbonate is a compound, $CaCO_3$, consisting of the elements Ca, C and O in the fixed atomic ratio, 1:1:3.

(d) Ink from a ballpoint pen is a homogeneous mixture of solvent, water and dyes.

(e) Toothpaste is a heterogeneous mixture of water, organic and inorganic compounds to aid in cleaning, whitening and preventing cavities, which can include dicalcium phosphate, glycerin sorbitol, cellulose gum and sodium fluoride.

1-16. *Refer to Section 1-5 and the Key Terms for Chapter 1.*

A heterogeneous mixture is a mixture that does not have a uniform composition and properties throughout the mixture.

(a) A blend of salt (a compound) and sulfur (an element) is a heterogeneous mixture.

(b) Milk (raw and unhomogenized) is a heterogeneous mixture of fat, water, vitamins, etc. Homogenized milk is a heterogeneous mixture on the microscopic level.

(c) A sample of clean air is a homogeneous mixture of gases (nitrogen, oxygen, argon, carbon dioxide and water vapor).

(d) Gasoline is a homogeneous mixture of various hydrocarbons (compounds containing C and H), anti-knock additives and detergents.

(e) A chocolate chip cookie is a heterogeneous mixture of flour, water, baking soda, baking powder, butter and, of course, chocolate chips.

1-18. *Refer to Section 1-8.*

(a) 52600

(b) 0.00000410

(c) 1600.

(d) 0.08206

(e) 9346

(f) 0.009346

1-20. *Refer to Section 1-8.*

(a) Density is a measured quantity. The density of platinum, 21.45 g/cm^3, has 4 significant figures.

(b) The given average speed, 114.277 mi/hr, is a measured quantity. This value has 6 significant figures.

(c) The mile, defined as 5280 feet, is an exact number and has an infinite number of significant digits.

(d) The curie, defined as a unit of radioactivity, is an exact number and also has an infinite number of significant digits.

1-22. *Refer to Section 1-8.*

$$? \text{ pounds} = 12 \text{ cars} \times \frac{1532.5 \text{ pounds}}{1 \text{ car}} = 1.8390 \times 10^4 \text{ pounds}$$

(The answer has 5 significant figures since "12 cars" is an exact number.)

1-24. *Refer to Section 1-8, and Examples 1-1 and 1-2.*

(a) 423.1 in + 0.256 in - 116 in = 307.356 in = **307 in** (Answer must have only 3 significant figures.)

(b) volume of a sphere = $\frac{4}{3} \pi r^3 = \frac{4}{3} (3.141593)(3.31 \text{ in})^3 = 151.9052 \text{ in}^3 = \textbf{152 in}^3$

(Answer must be rounded to 3 significant figures.)

(c) $(6.057 \times 10^3 \text{ m}) - 9.35 \text{ m} = 6057 \text{ m} - 9.35 \text{ m} = 6047.65 \text{ m} = \textbf{6048 m}$
(Answer is limited to 4 significant figures)

(d) $(8.54 \times 10^5 \text{ mi})/(22 \text{ days}) = 3.8818 \times 10^4 \text{ mi/day} = \textbf{3.9} \times \textbf{10}^4 \textbf{ mi/day}$
(Answer is limited to 2 significant figures)

1-26. *Refer to Section 1-9, the conversion factors from Table 1-7, and Example 1-3.*

(a) $? \text{ km} = 16.3 \text{ m} \times \dfrac{1 \text{ km}}{1000 \text{ m}} = \textbf{1.63} \times \textbf{10}^{-2} \textbf{ km}$

(b) $? \text{ m} = 16.3 \text{ km} \times \dfrac{1000 \text{ m}}{1 \text{ km}} = \textbf{1.63} \times \textbf{10}^4 \textbf{ m}$

(c) $? \text{ g} = 247 \text{ kg} \times \dfrac{1000 \text{ g}}{1 \text{ kg}} = \textbf{2.47} \times \textbf{10}^5 \textbf{ g}$

(d) $? \text{ mL} = 4.32 \text{ L} \times \dfrac{1000 \text{ mL}}{1 \text{ L}} = \textbf{4.32} \times \textbf{10}^3 \textbf{ mL}$

(e) $? \text{ L} = 85.9 \text{ dL} \times \dfrac{1 \text{ L}}{10 \text{ dL}} = \textbf{8.59 L}$

(f) $? \text{ cm}^3 = 7654 \text{ L} \times \dfrac{1000 \text{ cm}^3}{1 \text{ L}} = \textbf{7.654} \times \textbf{10}^6 \textbf{ cm}^3$ (Note: $1 \text{ cm}^3 = 1 \text{ mL}$)

1-28. *Refer to Section 1-9, the conversion factors from Table 1-7, and Example 1-3.*

These answers are precise to 2 significant figures since the initial value, 95 mile/hr, has only 2 significant figures.

(a) $? \dfrac{\text{km}}{\text{hr}} = 95 \dfrac{\text{miles}}{\text{hr}} \times \dfrac{1.609 \text{ km}}{1 \text{ mile}} = \textbf{1.5} \times \textbf{10}^2 \dfrac{\textbf{km}}{\textbf{hr}}$

(b) $? \dfrac{\text{ft}}{\text{s}} = 95 \dfrac{\text{miles}}{\text{hr}} \times \dfrac{5280 \text{ ft}}{1 \text{ mile}} \times \dfrac{1 \text{ hr}}{60 \text{ min}} \times \dfrac{1 \text{ min}}{60 \text{ s}} = \textbf{1.4} \times \textbf{10}^2 \dfrac{\textbf{ft}}{\textbf{s}}$

1-30. *Refer to Section 1-9, the conversion factors from Table 1-7, and Example 1-4.*

To determine which quantity is larger, convert one quantity into the units of the other and compare.

(a) $? \text{ cg} = 24.0 \text{ mg} \times \dfrac{1 \text{ g}}{1000 \text{ mg}} \times \dfrac{100 \text{ cg}}{1 \text{ g}} = 2.40 \text{ cg}$; therefore, **24.0 cg** is larger.

(b) $? \text{ m} = 250 \text{ cm} \times \dfrac{1 \text{ m}}{100 \text{ cm}} = 2.5 \text{ m}$; therefore, **250 cm** is larger.

(c) $? \text{ Å} = 0.8 \text{ nm} \times \dfrac{1 \text{ m}}{10^9 \text{ nm}} \times \dfrac{1 \text{ Å}}{10^{-10} \text{ m}} = 8 \text{ Å}$; therefore, **0.8 nm is equivalent to 8 Å**.

(d) $? \text{ m}^3 = 10 \text{ L} \times \dfrac{1000 \text{ mL}}{1 \text{ L}} \times \dfrac{1 \text{ cm}^3}{1 \text{ mL}} \times \dfrac{(1 \text{ m})^3}{(100 \text{ cm})^3} = 1 \times 10^{-2} \text{ m}^3$; therefore, **6.4 m³** is larger.

1-32. *Refer to Section 1-9 and Example 1-10.*

(a) ? tons ore = 8.40 tons hematite $\times \dfrac{100 \text{ tons ore}}{9.24 \text{ tons hematite}}$ = **90.9 tons ore**

(b) ? kg ore = 8.40 kg hematite $\times \dfrac{100 \text{ kg ore}}{9.24 \text{ kg hematite}}$ = **90.9 kg ore**

1-34. *Refer to Section 1-9.*

? kg salt = 157 gallons water $\times \dfrac{3.67 \text{ kg water}}{1 \text{ gallon water}} \times \dfrac{11 \text{ kg salt}}{10^6 \text{ kg water}}$ = **6.3 $\times$ 10^{-3} kg salt** removed daily

---

1-36. *Refer to Section 1-9, the conversion factors listed in Table 1-7, and Example 1-9.*

? cents/L = $\dfrac{\$1.159}{1 \text{ gal}} \times \dfrac{1 \text{ gal}}{4 \text{ qt}} \times \dfrac{1.057 \text{ qt}}{1 \text{ L}} \times \dfrac{100 \text{ cents}}{\$1}$ = **30.63 cents/L**

1-38. *Refer to Section 1-9, Tables 1-5 and 1-7 for the conversion factors, and Example 1-5.*

(a) $?\,\dfrac{\text{cm}^3}{\text{drop}} = \dfrac{1.0 \text{ mL}}{22 \text{ drops}} \times \dfrac{1 \text{ cm}^3}{1\text{mL}}$ = **0.045 cm³/drop**

$?\,\dfrac{\mu\text{L}}{\text{drop}} = \dfrac{1.0 \text{ mL}}{22 \text{ drops}} \times \dfrac{1 \text{ L}}{1000 \text{ mL}} \times \dfrac{1 \times 10^6 \,\mu\text{L}}{1 \text{ L}}$ = **45 µL/drop**

(b) Plan: (1) Determine the drop volume in mm³.

(2) Calculate the drop radius using $V = \frac{4}{3}\pi r^3$.

(3) Calculate the diameter, $D = 2r$.

(1) $?\,\dfrac{\text{mm}^3}{\text{drop}} = \dfrac{1.0 \text{ mL}}{22 \text{ drops}} \times \dfrac{(1 \text{ cm})^3}{1 \text{ mL}} \times \dfrac{(10 \text{ mm})^3}{(1 \text{ cm})^3}$ = 45 mm³/drop

(2) Solving for r, $r = \sqrt[3]{\dfrac{3V}{4\pi}} = \sqrt[3]{\dfrac{3(45 \text{ mm}^3)}{4(3.14159)}}$ = 2.2 mm

(3) drop diameter = $2r$ = **4.4 mm**

1-40. *Refer to Section 1-9, Tables 1-5 and 1-7 for conversion factors, and Example 1-9.*

(a) ? pounds = 326 kg $\times \dfrac{1000 \text{ g}}{1 \text{ kg}} \times \dfrac{1 \text{ lb}}{453.6 \text{ g}}$ = **719 lb**

(b) ? kg = 326 pounds $\times \dfrac{453.6 \text{ g}}{1 \text{ lb}} \times \dfrac{1 \text{ kg}}{1000 \text{ g}}$ = **148 kg**

(c) ? cg = 326 ounces $\times \dfrac{1 \text{ g}}{0.03527 \text{ oz}} \times \dfrac{1 \text{ cg}}{10^{-2} \text{ g}}$ = **9.24 $\times$ 10^5 cg**

1-42. *Refer to Section 1-9, Table 1-7 for conversion factors, and Example 1-4.*

Each cesium atom has a diameter $= 2 \times 2.62\ \text{Å} = 5.24\ \text{Å}$

$$? \text{ Cs atoms} = 1.00 \text{ inch} \times \frac{2.54\ \text{cm}}{1\ \text{in}} \times \frac{1\ \text{m}}{100\ \text{cm}} \times \frac{1\ \text{Å}}{10^{-10}\ \text{m}} \times \frac{1\ \text{atom}}{5.24\ \text{Å}} = \mathbf{4.85 \times 10^7 \text{ atoms}}$$

1-44. *Refer to Section 1-11 and Example 1-11.*

$$\text{Density} \left(\text{units} = \frac{\text{g}}{\text{mL or cm}^3} \right) = \frac{m}{V} = \frac{\text{mass (g)}}{\text{volume (mL)}} = \frac{50.6\ \text{g}}{21.72\ \text{mL}} = \mathbf{2.33\ \frac{\text{g}}{\text{mL}}}$$

1-46. *Refer to Section 1-11 and Example 1-11.*

$$\text{Density (mg/mm}^3) = \frac{m}{V} = \frac{5.536\ \text{mg}}{(2.20\ \text{mm} \times 1.36\ \text{mm} \times 1.12\ \text{mm})} = 1.65\ \text{mg/mm}^3$$

$$\text{Density (g/cm}^3) = \frac{1.65\ \text{mg}}{1\ \text{mm}^3} \times \frac{1\ \text{g}}{1000\ \text{mg}} \times \frac{(10\ \text{mm})^3}{(1\ \text{cm})^3} = \mathbf{1.65\ \text{g/cm}^3}$$

1-48. *Refer to Section 1-11, and Example 1-11.*

$$\text{Volume, } V, \text{ of a neutron} = \frac{4}{3}\pi r^3 = \frac{4}{3}(3.14159)(1.5 \times 10^{-15}\ \text{m})^3 = 1.4 \times 10^{-44}\ \text{m}^3$$

$$\text{Density (g/cm}^3) = \frac{m\ (\text{g})}{V\ (\text{cm}^3)} = \frac{1.675 \times 10^{-24}\ \text{g}}{1.4 \times 10^{-44}\ \text{m}^3} \times \frac{(1\ \text{m})^3}{(100\ \text{cm})^3} = \mathbf{1.2 \times 10^{14}\ \text{g/cm}^3}$$

1-50. *Refer to Section 1-11.*

(a) We know that $\text{Density} \left(\frac{\text{g}}{\text{cm}^3} \right) = \frac{m}{V} = \frac{\text{mass of substance (g)}}{\text{volume of substance (cm}^3)}$

Therefore, volume of container = volume of water $= \dfrac{\text{mass of water (g)}}{\text{density of water (g/cm}^3)}$

$$= \frac{\text{total mass(container + water) - mass of container}}{1.0000\ \text{g/cm}^3}$$

$$= \frac{(99.646\ \text{g - 77.664 g})}{1.0000\ \text{g/cm}^3}$$

$$= \mathbf{21.982\ \text{cm}^3}$$

(b) mass of metal = total mass (container + metal) - mass of container = 85.308 g - 77.664 g = **7.644 g**

(c) mass of water added = total mass (container + metal + water) - mass (container + metal)

$$= 106.442\ \text{g - 85.308 g}$$
$$= \mathbf{21.134\ \text{g}}$$

(d) Since $D = \dfrac{m}{V}$, then $V\ (\text{cm}^3) = \dfrac{m\ (\text{g})}{D\ (\text{g/cm}^3)} = \dfrac{21.134\ \text{g}}{1.0000\ \text{g/cm}^3} = \mathbf{21.134\ \text{cm}^3}$

(e) volume of metal = total volume of container (from (a)) - volume of water (from (d))

$$= 21.982 \text{ cm}^3 - 21.134 \text{ cm}^3$$
$$= \textbf{0.848 cm}^3$$

(f) Therefore, the density of the metal $= \dfrac{\text{mass (g)}}{\text{volume (cm}^3)} = \dfrac{7.644 \text{ g}}{0.848 \text{ cm}^3} = \textbf{9.014 g/cm}^3$

1-52.   *Refer to Section 1-11.*

The specific gravity of a substance is the ratio of its density to the density of water. Since the density of water = 1.00 g/mL in the temperature range of 0°C to 25°C, the specific gravity is numerically equal to the density and is unitless.

(a) Method 1: $D = \dfrac{m}{V}$; $V \text{(cm}^3) = \dfrac{m \text{ (g)}}{D \text{ (g/cm}^3)} = \dfrac{765 \text{ g}}{10.5 \text{ g/cm}^3} = \textbf{72.9 cm}^3$ since 0.765 kg ≡ 765 g

Method 2:   Dimensional Analysis

$$? \text{ cm}^3 \text{ silver} = 0.765 \text{ kg} \times \dfrac{1000 \text{ g}}{1 \text{ kg}} \times \dfrac{1 \text{ cm}^3}{10.5 \text{ g}} = \textbf{72.9 cm}^3$$

(b) length of each edge (cm) $= \sqrt[3]{V} = \sqrt[3]{72.9 \text{ cm}^3} = \textbf{4.18 cm}$

(c) length of each edge (in) $= 4.18 \text{ cm} \times \dfrac{1 \text{ in}}{2.54 \text{ cm}} = \textbf{1.65 in}$

1-54.   *Refer to Section 1-12.*

(a) A Celsius degree must be larger than a Fahrenheit degree, since it takes only 100 Celsius degrees (100°C - 0°C) to cover the same temperature interval (boiling point of pure water to the freezing point of pure water) as does 180 Fahrenheit degrees (212°F - 32°F).

(b) Since the size of a kelvin is the same as a Celsius degree, a kelvin must also be larger than a Fahrenheit degree.

1-56.   *Refer to Section 1-12, and Examples 1-16 and 1-17.*

In determining the correct number of significant figures, note that the following values are exact: 32°F, 1°C/1.8°F, and 1°C/1 K.

(a) $? \text{ °C} = \dfrac{1 \text{°C}}{1.8 \text{°F}} \times (0 \text{°F} - 32 \text{°F}) = \textbf{-18°C}$

(b) $? \text{ °C} = \dfrac{1 \text{°C}}{1.8 \text{°F}} \times (98.6 \text{°F} - 32 \text{°F}) = 37.0 \text{°C}$

$? \text{ K} = \dfrac{1 \text{ K}}{1 \text{°C}} \times (37.0 \text{°C} + 273.2 \text{°C}) = \textbf{310.2 K}$ since 0°C = 273.15 K

(c) $? \text{ °C} = \dfrac{1 \text{°C}}{1 \text{ K}} \times (298 \text{ K} - 273 \text{ K}) = 25 \text{°C}$

$? \text{ °F} = \left[25 \text{°C} \times \dfrac{1.8 \text{°F}}{1 \text{°C}} \right] + 32 \text{°F} = \textbf{77°F}$

(d) $? \text{ °F} = \left[23.4 \text{°C} \times \dfrac{1.8 \text{°F}}{1 \text{°C}} \right] + 32 \text{°F} = \textbf{74.1°F}$

	Freezing Point of Water (FP)	Boiling Point of Water (BP)
Celsius Scale	0°C	100°C
Fahrenheit Scale	32°F	212°F
Rėamur Scale	0°R	80°R

(a) $\dfrac{BP_{water} - FP_{water} \text{ on Celsius Scale}}{BP_{water} - FP_{water} \text{ on Rėamur Scale}} = \dfrac{100°C - 0°C}{80°R - 0°R} = \dfrac{100°C}{80°R} = \dfrac{1.0°C}{0.8°R} = \dfrac{5°C}{4°R}$

Therefore, since both scales set the freezing point of water $= 0°$, then $\; ? \; °C = \left[x°R \times \dfrac{5°C}{4°R} \right]$

(b) $\dfrac{BP_{water} - FP_{water} \text{ on Fahrenheit Scale}}{BP_{water} - FP_{water} \text{ on Rėamur Scale}} = \dfrac{212°F - 32°F}{80°R - 0°R} = \dfrac{180°F}{80°R} = \dfrac{9°F}{4°R}$

Therefore, $? \; °F = \left[x°R \times \dfrac{9°F}{4°R} + 32°F \right]$

Note that we must add 32°F to account for the fact that 0°R is equivalent to 32°F.

(c) From (a), $? \; °C = \left[x°R \times \dfrac{5°C}{4°R} \right]$.   Rearranging, we have $\; ? \; °R = \left[x°C \times \dfrac{4°R}{5°C} \right]$

$BP_{mercury} \; (°R) = 356.6°C \times \dfrac{4°R}{5°C} = \mathbf{285.3°R}$

For Al: $? \; °C = \dfrac{1°C}{1 \, K} \times (933.6 \, K - 273.2 \, K) = \mathbf{660.4°C}$

$? \; °F = \left[660.4°C \times \dfrac{1.8°F}{1°C} \right] + 32°F = \mathbf{1221°F}$

For Ag: $? \; °C = \dfrac{1°C}{1 \, K} \times (1235.1 \, K - 273.2 \, K) = \mathbf{961.9°C}$

$? \; °F = \left[961.9°C \times \dfrac{1.8°F}{1°C} \right] + 32°F = \mathbf{1763°F}$

amount of heat *gained* (J) $=$ (mass of substance)(specific heat)(temp. change)

$= 22.1 \, g \times 0.895 \, J/g \cdot °C \times (44.3°C - 27.0°C)$

$= \mathbf{342 \; J}$

(a) amount of heat *gained* (J) = (mass of substance)(specific heat)(temp. change)

$$= (78,700 \text{ g})(0.818 \text{ J/g} \cdot °C)(43.0°C - 25.0°C)$$

$$= 1.16 \times 10^6 \text{ J}$$

(b) Note that we will follow the convention of representing temperature (°C) as t and temperature (K) as T.

In any insulated system, the Law of Conservation of Energy states:

the amount of heat lost by Substance 1 = amount of heat gained by Substance 2

As will be discussed in later chapters, "heat lost" is a negative quantity and "heat gained" is a positive quantity. However, the "*amount* of heat lost" and the "*amount* of heat gained" quoted here call for *absolute* quantities without a sign associated with them:

| the amount of heat lost by Substance 1 | = | amount of heat gained by Substance 2 |

| (mass)(Sp. ht.)(temp. change) | $_1$ = | (mass)(Sp. ht.)(temp. change) | $_2$

In this exercise,

| (mass)(Sp. ht.)(temp. change) | $_{limestone}$ = | (mass)(Sp. ht.)(temp. change) | $_{air}$

Since any "change" is always defined as the final value minus the initial value, we have

(temp. change)$_{limestone}$ = (30.0°C - 43.0°C) and (temp. change)$_{air}$ = (t_{final} - 10.0°C)

for the limestone, | 30.0°C - 45.0°C | = | negative value | = (43.0°C - 30.0°C) = 13.0°C
for the interior air, | t_{final} - 10.0°C | = | positive value | = (t_{final} - 10.0°C)

Before we start, we must first calculate the mass of air inside the house:

$$? \text{ g air} = 2.83 \times 10^5 \text{ liters} \times \frac{1000 \text{ mL}}{1 \text{ L}} \times \frac{1.20 \times 10^{-3} \text{ g}}{1 \text{ mL}} = 3.40 \times 10^5 \text{ g}$$

$$78,700 \text{ g limestone} \times 0.818 \text{ J/g} \cdot °C \times (43.0°C - 30.0°C) = 3.40 \times 10^5 \text{ g air} \times 1.004 \text{ J/g} \cdot °C \times (t_{final} - 10.0°C)$$

$$8.37 \times 10^5 = (3.40 \times 10^5 \times t_{final}) - 3.40 \times 10^6$$

$$4.24 \times 10^6 = 3.40 \times 10^5 \times t_{final}$$

$$t_{final} = 12.5°C$$

All of the properties (color, luster, odor, ability to sublime and form a purple vapor, density, melting point and solubility) are physical properties except for (i) noncombustibility, (j) forming ions in aqueous solution and (k) poisonous, which are chemical properties. A comment on (k): many substances are poisonous to living tissue because they partake in biochemical reactions which interfere with normal functions.

(a) $? \text{ g Al} = 2.11 \text{ in} \times 6.25 \text{ in} \times 12.00 \text{ in} \times \frac{(2.54 \text{ cm})^3}{(1 \text{ in})^3} \times \frac{2.70 \text{ g}}{1 \text{ cm}^3} = 7.00 \times 10^3 \text{ g}$

(b) $? \text{ cm}^3 \text{ Hg} = 2.00 \text{ lb Hg} \times \frac{453.6 \text{ g}}{1 \text{ lb}} \times \frac{1 \text{ cm}^3}{13.59 \text{ g}} = 66.8 \text{ cm}^3$

1-70. *Refer to Section 1-9 and Table 1-7.*

$$? \text{ lethal dose} = 165 \text{ lb body wt} \times \frac{453.6 \text{ g body wt}}{1 \text{ lb body wt}} \times \frac{1 \text{ kg body wt}}{1000 \text{ g body wt}} \times \frac{1.6 \text{ mg KCN}}{1 \text{ kg body wt}} = \mathbf{120 \text{ mg KCN}}$$

1-72. *Refer to Section 1-9, Table 1-7 and Appendix D.*

$$? \text{ km/light year} = \frac{2.9979 \times 10^8 \text{ m}}{1 \text{ s}} \times \frac{3600 \text{ s}}{1 \text{ hr}} \times \frac{24 \text{ hr}}{1 \text{ day}} \times \frac{365 \text{ day}}{1 \text{ yr}} \times \frac{1 \text{ km}}{10^3 \text{ m}} = \mathbf{9.4542 \times 10^{12} \text{ km/yr}}$$

$$? \text{ mile/light year} = \frac{9.4542 \times 10^{12} \text{ km}}{1 \text{ yr}} \times \frac{1000 \text{ m}}{1 \text{ km}} \times \frac{100 \text{ cm}}{1 \text{ m}} \times \frac{1 \text{ in}}{2.54 \text{ cm}} \times \frac{1 \text{ ft}}{12 \text{ in}} \times \frac{1 \text{ mile}}{5280 \text{ ft}} = \mathbf{5.8746 \times 10^{12} \text{ miles/yr}}$$

1-74. *Refer to Sections 1-9 and 1-11.*

$$\text{Length of Al foil roll} = 66 \,^2/_3 \text{ yards} \times \frac{3 \text{ ft}}{1 \text{ yard}} \times \frac{12 \text{ in}}{1 \text{ ft}} = 2.400 \times 10^3 \text{ in}$$

$$\text{Density of Al (g/cm}^3) = (\text{specific gravity})(\text{density of H}_2\text{O}) = 2.70 \times \frac{0.988 \text{ g}}{1 \text{ cm}^3} = 2.69 \text{ g/cm}^3$$

$$? \text{ g Al} = 2.400 \times 10^3 \text{ in} \times 12 \text{ in} \times 6.5 \times 10^{-4} \text{ in} \times \frac{(2.54 \text{ cm})^3}{(1 \text{ in})^3} \times \frac{2.69 \text{ g}}{1 \text{ cm}^3} = \mathbf{8.3 \times 10^2 \text{ g}}$$

1-76. *Refer to Section 1-4, Exercise 1-8, and your common sense.*

As a student writes out an End-of-Chapter Exercise, the direct chemical changes that occur include (1) reactions (including irreversible adsorption) of the ink in the pen with the paper, (2) the body's biochemical reactions, (3) the creation of new neural pathways in the student's brain due to the new information she/he is learning. More indirect chemical changes include the burning of coal or natural gas to provide the power for electricity, heat and light. If the student is doing a problem outside on a beautiful day, chemical changes might involve photosynthesis occurring in the plants around her/him providing oxygen for the student to breathe and the fusion reactions in the sun which provide heat and light, etc. The complete answer is limited only by the student's imagination and understanding of the meaning of chemical changes. So, definitely yes, the answer involves knowledge not covered in Chapter 1.

2 Chemical Formulas and Composition Stoichiometry

2-2. *Refer to Section 2-1.*

Dalton developed the atomic theory, the thesis that each element is composed of tiny, indivisible particles called atoms that are all alike in properties and have the same atomic weight. Nuclear chemistry now tells us that (1) atoms are divisible, (2) atoms of a given element can have slightly different mass and properties due to having different numbers of neutrons, and (3) atoms of one element can be changed to those of other elements. However, his understanding that elements combine in whole number ratios to form compounds and for a given compound, these ratios are constant, has been proven to be correct. Even with the many shortcomings, Dalton's postulates provided a framework that could be modified and expanded by later scientists.

2-4. *Refer to Section 2-1.*

Examples of molecules that exist as diatomic:

(1) N_2, O_2, H_2, F_2, Cl_2, Br_2, I_2 in which the atoms are identical (homonuclear)
(2) HF, HCl, NO in which the atoms are different (heteronuclear)

Examples of more complex molecules:

(1) P_4, S_8 (homonuclear)
(2) H_2SO_4, H_3PO_4 (heteronuclear)

2-6. *Refer to Section 2-2 and Table 2-2.*

(a) HNO_3 nitric acid   (b) C_2H_6 ethane   (c) NH_3 ammonia

(d) CH_3COCH_3 acetone   (e) CH_3CH_2OH ethanol

2-8. *Refer to Section 2-3 and the Key Terms for Chapter 2.*

(a) There are no "molecules" in ionic compounds. Ionic compounds are extended arrays of cations (positively charged ions) and anions (negatively charged ions). Recall that a molecule is the smallest particle of a compound that can have a stable, independent existence. Therefore, the term "molecule" does not apply to ionic substances. Instead, we use "formula unit" to represent the smallest repeating unit of an ionic compound.

(b) A polyatomic molecule is a molecule containing more than 1 atom. A formula unit of an ionic compound is the smallest repeating unit of an ionic compound. For example, CH_3Cl is a polyatomic molecule, while sodium chloride, NaCl, represents a formula unit of an ionic compound consisting of one Na^+ and one Cl^- ion.

2-10. *Refer to Section 2-3 and Table 2-3.*

(a) K^+ monatomic cation   (b) SO_4^{2-} polyatomic anion   (c) Fe^{3+} monatomic cation

(d) NH_4^+ polyatomic cation   (e) CO_3^{2-} polyatomic anion

2-12.   *Refer to Section 2-4, Table 2-2 and Example 2-1.*

(a) KCH_3COO (b) $(NH_4)_2CO_3$ (c) $Zn_3(PO_4)_2$ (d) CaO (e) $Al(NO_3)_3$

2-14.   *Refer to Section 2-4, Table 2-3, and Examples 2-1 and 2-2.*

(a) $CuCO_3$ copper(II) carbonate or cupric carbonate   (b) $Mg(OH)_2$ magnesium hydroxide

(c) $(NH_4)_2CO_3$ ammonium carbonate   (d) $ZnCl_2$ zinc chloride

(e) $Fe(CH_3COO)_2$ iron(II) acetate or ferrous acetate

2-16.   *Refer to Section 2-5.*

The mass ratio of a silver atom (107.8682 amu) to a chlorine atom (35.4527 amu) is 107.8682/35.4527 = 3.04259 (to 6 significant figures) or **3.043** (to 4 significant figures).

2-18.   *Refer to Section 2-6.*

(1) $\dfrac{?\ g}{1\ dozen} = \dfrac{4.11\ g}{1\ screw} \times \dfrac{12\ screws}{1\ dozen} = \mathbf{49.3\ g}$

(2) $\dfrac{?\ g}{1\ gross} = \dfrac{4.11\ g}{1\ screw} \times \dfrac{144\ screws}{1\ gross} = \mathbf{592\ g}$

(3) $\dfrac{?\ g}{1\ mole} = \dfrac{4.11\ g}{1\ screw} \times \dfrac{6.022 \times 10^{23}\ screws}{1\ mole} = \mathbf{2.48 \times 10^{24}\ g}$

   Therefore, 1 mole of screws would be about 0.04 % of the earth's mass, 5.98×10^{24} kg.

2-20.   *Refer to Section 2-6 and the inside front cover of this textbook.*

	Element	Atomic Weight (amu)	Mass of 1 mole of atoms (g)
(a)	Ca	<u>**40.078**</u>	<u>**40.078**</u>
(b)	<u>**Br**</u>	79.904	<u>**79.904**</u>
(c)	P	<u>**30.9738**</u>	<u>**30.9738**</u>
(d)	<u>**Cr**</u>	<u>**51.9961**</u>	51.9961

2-22.   *Refer to Section 2-6 and Example 2-3.*

Method 1:   Recall, moles of element = $\dfrac{\text{mass of element (g)}}{\text{AW of element}}$

   $?\ mol\ Ni = \dfrac{245.2\ g\ Ni}{58.69\ g\ Ni/mol\ Ni} = \mathbf{4.178\ mol\ Ni}$

Method 2:   Dimensional Analysis

   $?\ mol\ Ni = 245.2\ g\ Ni \times \dfrac{1\ mol\ Ni}{58.69\ g\ Ni} = \mathbf{4.178\ mol\ Ni}$

12

(a) iodine, I_2 $2 \times I = 2 \times 126.9045$ amu $= \textbf{253.809}$ **amu**

(b) water, H_2O $2 \times H = 2 \times 1.0079$ amu $=$ 2.0158 amu
                                                $1 \times O = 1 \times 15.9994$ amu $=$ 15.9994 amu
                                                                formula weight $=$ **18.0152 amu**

(c) saccharin, $C_7H_5NSO_3$ $7 \times C = 7 \times 12.011$ amu $=$ 84.077   amu
                                                $5 \times H = 5 \times 1.0079$ amu $=$ 5.0395 amu
                                                $1 \times N = 1 \times 14.0067$ amu $=$ 14.0067 amu
                                                $1 \times S = 1 \times 32.066$ amu $=$ 32.066   amu
                                                $3 \times O = 3 \times 15.9994$ amu $=$ 47.9982 amu
                                                                formula weight $= \textbf{183.187}$ **amu**

(d) sodium dichromate, $Na_2Cr_2O_7$ $2 \times Na = 2 \times 22.9898$ amu $=$ 45.9796 amu
                                                $2 \times Cr = 2 \times 51.9961$ amu $=$ 103.9922 amu
                                                $7 \times O = 7 \times 15.9994$ amu $=$ 111.9958 amu
                                                                formula weight $=$ **261.9676 amu**

Method 1:   Use the units of formula weight to derive a formula relating grams, moles and formula weight:

$$\text{formula weight, FW} \left[\frac{g}{mol}\right] = \frac{\text{grams of substance}}{\text{moles of substance}}$$

Therefore,

$$\text{moles of substance} = \frac{\text{grams of substance}}{\text{formula weight (g/mol)}}$$

(a) ? mol $NH_3 = \dfrac{16.8 \text{ g}}{17.0 \text{ g/mol}} = \textbf{0.988 mol } NH_3$

(b) ? mol $NH_4Br = \dfrac{3250 \text{ g}}{97.9 \text{ g/mol}} = \textbf{33.2 mol } NH_4Br$ Note: 1 kg $=$ 1000 g

(c) ? mol $PCl_5 = \dfrac{5.6 \text{ g}}{208 \text{ g/mol}} = \textbf{0.027 mol } PCl_5$

(d) ? mol $Sn = \dfrac{126.5 \text{ g}}{118.710 \text{ g/mol}} = \textbf{1.066 mol } Sn$

Method 2: Dimensional Analysis

(a) ? mol $NH_3 = 16.8 \text{ g } NH_3 \times \dfrac{1 \text{ mol } NH_3}{17.0 \text{ g } NH_3} = \textbf{0.988 mol } NH_3$

(b) ? mol $NH_4Br = 3.25 \text{ kg } NH_4Br \times \dfrac{1000 \text{ g}}{1 \text{ kg}} \times \dfrac{1 \text{ mol } NH_4Br}{97.9 \text{ g } NH_4Br} = \textbf{33.2 mol } NH_4Br$

(c) ? mol $PCl_5 = 5.6 \text{ g } PCl_5 \times \dfrac{1 \text{ mol } PCl_5}{208 \text{ g } PCl_5} = \textbf{0.027 mol } PCl_5$

(d) ? mol $Sn = 126.5 \text{ g } Sn \times \dfrac{1 \text{ mol } Sn}{118.710 \text{ g } Sn} = \textbf{1.066 mol } Sn$

2-28. *Refer to Section 2-6 and Example 2-4.*

Method 1: Recall: Avogadro's Number, $N = 6.022 \times 10^{23}$ atoms/mol.

? Ni atoms $= 4.178$ mol Ni $\times$ $(6.022 \times 10^{23}$ atoms/mol$) = 2.516 \times 10^{24}$ Ni atoms

Method 2: Dimensional Analysis

? Ni atoms $= 4.178$ mol Ni $\times \dfrac{6.022 \times 10^{23} \text{ atoms}}{1 \text{ mol Ni}} = \mathbf{2.516 \times 10^{24}}$ **Ni atoms**

2-30. *Refer to Section 2-6 and Example 2-5.*

Recall: (1) 1 mol of Ni contains 6.022×10^{23} Ni atoms

(2) 1 mol of Ni has a mass of 58.69 g

? g Ni $= 1$ Ni atom $\times \dfrac{1 \text{ mol Ni atoms}}{6.022 \times 10^{23} \text{ Ni atoms}} \times \dfrac{58.69 \text{ g Ni}}{1 \text{ mol Ni atoms}} = \mathbf{9.746 \times 10^{-23}}$ **g Ni**

2-32. *Refer to Section 2-6 and Exercise 2-30.*

Plan: g Ni $\overset{(1)}{\Longrightarrow}$ moles Ni $\overset{(2)}{\Longrightarrow}$ atoms of Ni

? Ni atoms $= 1.0 \times 10^{-6}$ g Ni $\times \dfrac{1 \text{ mol Ni atoms}}{58.69 \text{ g Ni}} \times \dfrac{6.02 \times 10^{23} \text{ Ni atoms}}{1 \text{ mol Ni atoms}} = \mathbf{1.0 \times 10^{16}}$ **Ni atoms**

2-34. *Refer to Section 2-7, Exercise 2-30, and Example 2-8.*

Plan: molecules CH_4 $\overset{(1)}{\Longrightarrow}$ moles CH_4 $\overset{(2)}{\Longrightarrow}$ g CH_4

The molecular mass of CH_4 is 16.0 g/mol.

? g $CH_4 = 10.0 \times 10^6$ molecules $CH_4 \times \dfrac{1 \text{ mol } CH_4}{6.02 \times 10^{23} \text{ molecules } CH_4} \times \dfrac{16.0 \text{ g } CH_4}{1 \text{ mol } CH_4} = \mathbf{2.66 \times 10^{-16}}$ **g CH_4**

2-36. *Refer to Section 2-7 and Example 2-9.*

Plan: g substance $\overset{(1)}{\Longrightarrow}$ moles substance $\overset{(2)}{\Longrightarrow}$ molecules substance

Method 1: Recall: mol substance $= \dfrac{\text{g substance}}{\text{formula weight}}$ and Avogadro's Number, $N = 6.02 \times 10^{23}$ molecules/mol

As an example:

(a) (1) ? mol CO $= \dfrac{\text{g CO}}{\text{FW CO}} = \dfrac{14.0 \text{ g}}{28.0 \text{ g/mol}} = \mathbf{0.500}$ **mol CO**

(2) ? molecules CO $= 0.500$ mol CO $\times (6.02 \times 10^{23}$ molecules/mol$) = \mathbf{3.01 \times 10^{23}}$ **molecules CO**

Method 2: Dimensional Analysis. Each unit factor corresponds to a step in the Plan.

Step 1   Step 2

(a) ? molecules CO $= 14.0 \text{ g CO} \times \dfrac{1 \text{ mol CO}}{28.0 \text{ g CO}} \times \dfrac{6.02 \times 10^{23} \text{ molecules CO}}{1 \text{ mol CO}} = 3.01 \times 10^{23} \text{ molecules CO}$

(b) ? molecules $N_2 = 14.0 \text{ g } N_2 \times \dfrac{1 \text{ mol } N_2}{28.0 \text{ g } N_2} \times \dfrac{6.02 \times 10^{23} \text{ molecules } N_2}{1 \text{ mol } N_2} = 3.01 \times 10^{23} \text{ molecules } N_2$

(c) ? molecules $P_4 = 14.0 \text{ g } P_4 \times \dfrac{1 \text{ mol } P_4}{124 \text{ g } P_4} \times \dfrac{6.02 \times 10^{23} \text{ molecules } P_4}{1 \text{ mol } P_4} = 6.80 \times 10^{22} \text{ molecules } P_4$

(d) ? molecules $P_2 = 14.0 \text{ g } P_2 \times \dfrac{1 \text{ mol } P_2}{62.0 \text{ g } P_2} \times \dfrac{6.02 \times 10^{23} \text{ molecules } P_2}{1 \text{ mol } P_2} = 1.36 \times 10^{23} \text{ molecules } P_2$

(e) ? atoms P in (c) $= 6.80 \times 10^{22} \text{ molecules } P_4 \times \dfrac{4 \text{ atoms P}}{1 \text{ } P_4 \text{ molecule}} = 2.72 \times 10^{23} \text{ atoms P in (c)}$

? atoms P in (d) $= 1.36 \times 10^{23} \text{ molecules } P_2 \times \dfrac{2 \text{ atoms P}}{1 \text{ } P_2 \text{ molecule}} = 2.72 \times 10^{23} \text{ atoms P in (d)}$

Yes, there are the same number of P atoms in 14.0 g of pure phosphorus, regardless of whether the phosphorus is in the form of P_4 or P_2.

2-38. *Refer to Section 2-7, and Examples 2-9 and 2-10.*

The molecular mass of C_3H_8 is 44.1 g/mol. Each C_3H_8 molecule contains 8 hydrogen atoms.

Plan: $g \ C_3H_8 \Longrightarrow mol \ C_3H_8 \Longrightarrow molecules \ C_3H_8 \Longrightarrow atoms \ H$

? H atoms $= 75.0 \text{ g } C_3H_8 \times \dfrac{1 \text{ mol } C_3H_8}{44.1 \text{ g } C_3H_8} \times \dfrac{6.02 \times 10^{23} \text{ } C_3H_8 \text{ molecules}}{1 \text{ mol } C_3H_8} \times \dfrac{8 \text{ H atoms}}{1 \text{ } C_3H_8 \text{ molecule}}$

$= 8.19 \times 10^{24} \text{ H atoms}$

2-40. *Refer to Section 2-7 and Example 2-11.*

The molecular mass of H_2SO_4 is 98.1 g/mol.

Plan: $g \ H_2SO_4 \Longrightarrow mol \ H_2SO_4 \Longrightarrow mmol \ H_2SO_4$

? millimoles $H_2SO_4 = 2.34 \text{ g } H_2SO_4 \times \dfrac{1 \text{ mol } H_2SO_4}{98.1 \text{ g } H_2SO_4} \times \dfrac{1 \times 10^3 \text{ mmol } H_2SO_4}{1 \text{ mol } H_2SO_4} = 23.9 \text{ mmol } H_2SO_4$

2-42. *Refer to Section 2-7 and Table 2-6.*

Moles of compound	Moles of cations	Moles of anions
1 mol NaCl	1 mol Na^+	1 mol Cl^-
2 mol Na_2SO_4	4 mol Na^+	2 mol SO_4^{2-}
0.2 mol $Ca(NO_3)_2$	0.2 mol Ca^{2+}	0.4 mol NO_3^-
0.25 mol $(NH_4)_2SO_4$	0.50 mol NH_4^+	0.25 mol SO_4^{2-}

(a) mass of 1 mol $C_{10}H_{14}N_2$

$10 \times C = 10 \times 12.01$ g	= 120.1 g	
$14 \times H = 14 \times 1.008$ g	= 14.11 g	
$2 \times N = 1 \times 14.01$ g	= 28.02 g	
mass of 1 mol	= 162.2 g	

? % C = (120.1 g/162.2 g) × 100 = **74.04 % C**
? % H = (14.11 g/162.2 g) × 100 = **8.699 % H**
? % N = (28.02 g/162.2 g) × 100 = **17.27 % N**

(b) mass of 1 mole $C_{18}H_{21}NO_3$

$18 \times C = 18 \times 12.01$ g	= 216.2 g
$21 \times H = 21 \times 1.008$ g	= 21.17 g
$1 \times N = 1 \times 14.01$ g	= 14.01 g
$3 \times O = 3 \times 16.00$ g	= 48.00 g
mass of 1 mol	= 299.4 g

? % C = (216.2 g/299.4 g) × 100 = **72.21 % C**
? % H = (21.17 g/299.4 g) × 100 = **7.071 % H**
? % N = (14.01 g/299.4 g) × 100 = **4.679 % N**
? % O = (48.00 g/299.4 g) × 100 = **16.03 % O**

(c) mass of 1 mol $C_8H_8O_3$

$8 \times C = 8 \times 12.01$ g	= 96.08 g
$8 \times H = 8 \times 1.008$ g	= 8.064 g
$3 \times O = 3 \times 16.00$ g	= 48.00 g
mass of 1 mol	= 152.14 g

? % C = (96.08 g/152.14 g) × 100 = **63.15 % C**
? % H = (8.064 g/152.14 g) × 100 = **5.300 % H**
? % O = (48.00 g/152.14 g) × 100 = **31.55 % O**

mass of 1 mol $Cu_3(CO_3)_2(OH)_2$

$3 \times Cu = 3 \times 63.55$ g = 190.6 g	
$2 \times C = 2 \times 12.01$ g = 24.02 g	
$8 \times O = 8 \times 16.00$ g = 128.0 g	
$2 \times H = 2 \times 1.01$ g = 2.02 g	
mass of 1 mol = 344.6 g	

percent Cu by mass

%Cu = (190.6/344.6) × 100 = 55.31%

mass of 1 mol Cu_2S

$2 \times Cu = 2 \times 63.55$ g = 127.1 g
$1 \times S = 1 \times 32.07$ g = 32.07 g
mass of 1 mol = 159.2 g

percent Cu by mass

%Cu = (127.1/159.2) × 100 = 79.84%

mass of 1 mol $CuFeS_2$

$1 \times Cu = 1 \times 63.55$ g = 63.55 g
$1 \times Fe = 1 \times 55.85$ g = 55.85 g
$2 \times S = 2 \times 32.07$ g = 64.14 g
mass of 1 mol = 183.54 g

percent Cu by mass

% Cu = (63.55/183.5) × 100 = 34.63%

mass of 1 mol CuS

$1 \times Cu = 1 \times 63.55$ g = 63.55 g
$1 \times S = 1 \times 32.07$ g = 32.07 g
mass of 1 mol = 95.62 g

percent Cu by mass

% Cu = (63.55/95.62) × 100 = 66.46%

mass of 1 mol Cu$_2$O

2 × Cu = 2 × 63.55 g = 127.1 g
1 × 0 = 1 × 16.00 g = 16.00 g
_____
mass of 1 mol   = 143.1 g

percent Cu by mass

% Cu = (127.1/143.1) × 100 = 88.82%

mass of 1 mol Cu$_2$CO$_3$(OH)$_2$

2 × Cu = 2 × 63.55 g = 127.1 g
1 × C = 1 × 12.01 g = 12.01 g
5 × O = 5 × 16.00 g = 80.00 g
2 × H = 2 × 1.01 g =   2.02 g
_____
mass of 1 mol   = 221.1 g

percent Cu by mass

% Cu = (127.1/221.1) × 100 = 57.49%

Therefore, **cuprite, Cu$_2$O**, has the highest copper content on a percent by mass basis.

2-48.   *Refer to Section 2-9 and Example 2-13.*

Plan: (1) If percentage composition instead of sample mass is given, assume a 100 g sample.
(2) Calculate the moles of each element in the 100 g sample.
(3) Divide each of the mole values by the smallest number obtained as a mole value for the 100 g sample. General Rule: do not round off further than 0.1 from a whole number.
(4) Determine a whole number ratio.

Let us assume we have a 100.0 g sample of epinephrine containing 56.8 g C, 6.56 g H, 28.4 g O and 8.28 g N.

$? \text{ mol C} = \dfrac{\text{g C}}{\text{AW C}} = \dfrac{56.8 \text{ g}}{12.0 \text{ g/mol}} = 4.73 \text{ mol}$ $\text{Ratio} = \dfrac{4.73}{0.591} = 8$

$? \text{ mol H} = \dfrac{\text{g}}{\text{AW H}} = \dfrac{6.56 \text{ g}}{1.008 \text{ g/mol}} = 6.51 \text{ mol}$ $\text{Ratio} = \dfrac{6.51}{0.591} = 11$

$? \text{ mol O} = \dfrac{\text{g O}}{\text{AW O}} = \dfrac{28.4 \text{ g}}{16.00 \text{ g/mol}} = 1.78 \text{ mol}$ $\text{Ratio} = \dfrac{1.78}{0.591} = 3.01 = 3$

$? \text{ mol N} = \dfrac{\text{g N}}{\text{AW N}} = \dfrac{8.28 \text{ g}}{14.01 \text{ g/mol}} = 0.591 \text{ mol}$ $\text{Ratio} = \dfrac{0.591}{0.591} = 1$

Therefore, the simplest formula is **C$_8$H$_{11}$O$_3$N**.

2-50.   *Refer to Sections 2-9 and 2-10 and Examples 2-16 and 2-17.*

Plan: (1) Use the mass of CO$_2$ to calculate the mass of C in the original sample.
(2) Calculate the mass of H in the sample by difference: mg H = mg compound - mg C for a hydrocarbon containing only carbon and hydrogen.
(3) Determine the simplest formula.

(1) $? \text{ mg C} = 1.790 \text{ mg CO}_2 \times \dfrac{12.01 \text{ mg C}}{44.01 \text{ mg CO}_2} = 0.4885 \text{ mg C}$

(2) $? \text{ mg H} = 0.5707 \text{ mg compound} - 0.4885 \text{ mg C} = 0.0822 \text{ mg H}$

(3) $? \text{ mmol C} = \dfrac{0.4885 \text{ mg C}}{12.01 \text{ mg/mmol}} = 0.04067 \text{ mmol C}$ $\text{Ratio} = \dfrac{0.04067}{0.04067} = 1$

$$? \text{ mmol H} = \frac{0.0822 \text{ mg H}}{1.0079 \text{ mg/mmol}} = 0.0816 \text{ mmol H} \qquad \text{Ratio} = \frac{0.0816}{0.04067} = 2.01$$

Therefore the simplest formula is CH_2.

2-52. *Refer to Sections 2-9 and 2-10 and Examples 2-16 and 2-17.*

Plan: (1) Use the masses of CO_2 and H_2O to calculate the masses of C and H respectively.
(2) Calculate the percentages of C and H in the sample.

$$(1) \ ? \text{ g C} = 0.5694 \text{ g CO}_2 \times \frac{12.01 \text{ g C}}{44.01 \text{ g CO}_2} = 0.1554 \text{ g C}$$

$$? \text{ g H} = 0.0826 \text{ g H}_2\text{O} \times \frac{2.016 \text{ g H}}{18.02 \text{ g H}_2\text{O}} = 0.00924 \text{ g H}$$

(2) ? g sample = mass of C + mass of H = 0.1554 g C + 0.00924 g H = 0.1646 g sample

$$? \text{ \% C} = \frac{0.1554 \text{ g C}}{0.1647 \text{ g sample}} \times 100 = \mathbf{94.35 \ \% \ C}$$

$$? \text{ \% H} = \frac{0.00924 \text{ g H}}{0.1647 \text{ g sample}} \times 100 = \mathbf{5.61 \ \% \ H}$$

2-54. *Refer to Sections 2-9 and 2-10 and Examples 2-16 and 2-17.*

Plan: (1) Use the masses of CO_2 and H_2O to calculate the masses of C and H respectively.
(2) Calculate the mass of O in the sample by difference: g O = g sample - g C - g H since the compound contains only C, H and O.
(3) Determine the simplest formula.

$$(1) \ ? \text{ g C} = 1.913 \text{ g CO}_2 \times \frac{12.01 \text{ g C}}{44.01 \text{ g CO}_2} = 0.5220 \text{ g C}$$

$$? \text{ g H} = 1.174 \text{ g H}_2\text{O} \times \frac{2.016 \text{ g H}}{18.02 \text{ g H}_2\text{O}} = 0.1313 \text{ g H}$$

(2) ? g O = 1.000 g compound - 0.5220 g C - 0.1313 g H = 0.347 g O

$$(3) \ ? \text{ mol C} = \frac{0.5220 \text{ g C}}{12.01 \text{ g/mol}} = 0.04346 \text{ mol C} \qquad \text{Ratio} = \frac{0.04346}{0.0217} = 2$$

$$? \text{ mol H} = \frac{0.1313 \text{ g H}}{1.008 \text{ g/mol}} = 0.1303 \text{ mol H} \qquad \text{Ratio} = \frac{0.1303}{0.0217} = 6$$

$$? \text{ mol O} = \frac{0.347 \text{ g O}}{16.00 \text{ g/mol}} = 0.0217 \text{ mol O} \qquad \text{Ratio} = \frac{0.0217}{0.0217} = 1$$

The simplest formula for this alcohol is $\mathbf{C_2H_6O}$.

2-56. *Refer to Section 2-10 and Examples 2-16 and 2-17.*

(a) First, we must calculate the % by mass of N in skatole.

? % N = 100.00 - (% C + % H) = 100.00 - (82.40 + 6.92) = 10.68% N

To find the simplest formula, assume 100 g of skatole.

$? \text{ mol C} = \dfrac{\text{g C}}{\text{AW C}} = \dfrac{82.40 \text{ g}}{12.01 \text{ g/mol}} = 6.861 \text{ mol C}$ $\text{Ratio} = \dfrac{6.861}{0.7623} = 9$

$? \text{ mol H} = \dfrac{\text{g H}}{\text{AW H}} = \dfrac{6.92 \text{ g}}{1.008 \text{ g/mol}} = 6.87 \text{ mol H}$ $\text{Ratio} = \dfrac{6.87}{0.7623} = 9$

$? \text{ mol N} = \dfrac{\text{g N}}{\text{AW N}} = \dfrac{10.68 \text{ g}}{14.01 \text{ g/mol}} = 0.7623 \text{ mol N}$ $\text{Ratio} = \dfrac{0.7623}{0.7623} = 1$

The simplest formula is the true formula, C_9H_9N.

(b) The molecular weight of skatole:

$$
\begin{aligned}
9 \times C &= 9 \times 12.01 \text{ g} &= 108.09 \text{ g} \\
9 \times H &= 9 \times 1.01 \text{ g} &= 9.09 \text{ g} \\
1 \times N &= 1 \times 14.01 \text{ g} &= 14.01 \text{ g} \\
\hline
\text{mass of 1 mol } C_9H_9N &&= \mathbf{131.19 \text{ g}}
\end{aligned}
$$

2-58.   *Refer to Section 2-10, Example 2-13 and Exercise 2-56 Solution.*

(a) Assume 100 g of timolol.

$? \text{ mol C} = \dfrac{47.2 \text{ g C}}{12.0 \text{ g/mol}} = 3.93 \text{ mol C}$ $\text{Ratio} = \dfrac{3.93}{0.232} = 17$

$? \text{ mol H} = \dfrac{6.55 \text{ g H}}{1.008 \text{ g/mol}} = 6.50 \text{ mol H}$ $\text{Ratio} = \dfrac{6.50}{0.232} = 28$

$? \text{ mol N} = \dfrac{13.0 \text{ g N}}{14.0 \text{ g/mol}} = 0.929 \text{ mol N}$ $\text{Ratio} = \dfrac{0.929}{0.232} = 4$

$? \text{ mol O} = \dfrac{25.9 \text{ g O}}{16.0 \text{ g/mol}} = 1.62 \text{ mol O}$ $\text{Ratio} = \dfrac{1.62}{0.232} = 7$

$? \text{ mol S} = \dfrac{7.43 \text{ g S}}{32.0 \text{ g/mol}} = 0.232 \text{ mol S}$ $\text{Ratio} = \dfrac{0.232}{0.232} = 1$

The simplest formula for timolol is $C_{17}H_{28}N_4O_7S$ (FW = 432 g/mol)

(b) $\text{MW (g/mol)} = \dfrac{\text{g timolol}}{\text{mol timolol}} = \dfrac{4.32 \text{ g}}{0.0100 \text{ mol}} = 432 \text{ g/mol}$ $n = \dfrac{\text{molecular weight}}{\text{simplest formula weight}} = \dfrac{432}{432} = 1$

The simplest formula is therefore the true molecular formula, $C_{17}H_{28}N_4O_7S$.

2-60.   *Refer to Section 2-10 and Example 2-18.*

(a) in NO: $? \text{ g O} = 3.00 \text{ g N} \times \dfrac{16.0 \text{ g O}}{14.0 \text{ g N}} = \mathbf{3.43 \text{ g O}}$

(b) in NO_2: $? \text{ g O} = 3.00 \text{ g N} \times \dfrac{32.0 \text{ g O}}{14.0 \text{ g N}} = \mathbf{6.86 \text{ g O}}$

One can easily see that the ratio: $\dfrac{\text{g O in NO}}{\text{g O in NO}_2} = \dfrac{3.43}{6.86} = \dfrac{1}{2}$

This result illustrates the **Law of Multiple Proportions** which states that when elements form more than one compound, the ratio of the masses of one element that combine with a given mass of another element in each of the compounds can be expressed by small whole numbers.

(a) in SO_2: $? \text{ g O} = 1.50 \text{ g S} \times \dfrac{32.0 \text{ g O}}{32.066 \text{ g S}} = \mathbf{1.50 \text{ g O}}$

(b) in SO_3: $? \text{ g O} = 1.50 \text{ g S} \times \dfrac{48.0 \text{ g O}}{32.066 \text{ g S}} = \mathbf{2.25 \text{ g O}}$

2-64. *Refer to Section 2-11 and Example 2-19.*

Plan: $\underset{(1)}{\text{g HgS}} \Longrightarrow \underset{(2)}{\text{mol HgS}} \Longrightarrow \underset{(3)}{\text{mol Hg}} \Longrightarrow \text{g Hg}$

 Step 1 Step 2 Step 3

$? \text{ g Hg} = 225.0 \text{ g HgS} \times \dfrac{1 \text{ mol HgS}}{232.7 \text{ g HgS}} \times \dfrac{1 \text{ mol Hg}}{1 \text{ mol HgS}} \times \dfrac{200.6 \text{ g Hg}}{1 \text{ mol Hg}} = \mathbf{194.0 \text{ g Hg}}$

2-66. *Refer to Section 2-11 and Example 2-20.*

Plan: $\underset{(1)}{\text{g Mn}} \Longrightarrow \underset{(2)}{\text{mol Mn}} \Longrightarrow \underset{(3)}{\text{mol KMnO}_4} \Longrightarrow \text{g KMnO}_4$

 Step 1 Step 2 Step 3

$? \text{ g KMnO}_4 = 10.5 \text{ g Mn} \times \dfrac{1 \text{ mol Mn}}{54.9 \text{ g Mn}} \times \dfrac{1 \text{ mol KMnO}_4}{1 \text{ mol Mn}} \times \dfrac{158 \text{ g KMnO}_4}{1 \text{ mol KMnO}_4} = \mathbf{30.2 \text{ g KMnO}_4}$

2-68. *Refer to Section 2-11 and Example 2-21.*

Plan: $\text{tons CuFeS}_2 \underset{(1)}{\Longrightarrow} \text{tons Cu in CuFeS}_2 \underset{(2)}{=} \text{tons Cu in Cu}_2\text{S} \underset{(3)}{\Longrightarrow} \text{tons Cu}_2\text{S}$

Since the formula weights are: $CuFeS_2$ (183.5 g/mol), Cu_2S (159.2 g/mol) and Cu (63.55 g/mol), we have

 Step 1 Step 2 Step 3

$? \text{ tons Cu}_2\text{S} = 175 \text{ tons CuFeS}_2 \times \dfrac{63.55 \text{ tons Cu in CuFeS}_2}{183.5 \text{ tons CuFeS}_2} \times \dfrac{1 \text{ ton Cu in Cu}_2\text{S}}{1 \text{ ton Cu in CuFeS}_2} \times \dfrac{159.2 \text{ tons Cu}_2\text{S}}{2 \times 63.55 \text{ tons Cu in Cu}_2\text{S}}$

$\phantom{? \text{ tons Cu}_2\text{S} } = \mathbf{75.9 \text{ tons Cu}_2\text{S}}$

2-70. *Refer to Section 2-11.*

Plan: g $MgCl_2 \Longrightarrow$ mol $MgCl_2 \Longrightarrow$ mol ions $\Longrightarrow$ mol NaCl $\Longrightarrow$ g NaCl (FW of $MgCl_2$ is 95.2 g/mol)

$? \text{ g NaCl} = 245 \text{ g MgCl}_2 \times \dfrac{1 \text{ mol MgCl}_2}{95.2 \text{ g MgCl}_2} \times \dfrac{3 \text{ mol ions}}{1 \text{ mol MgCl}_2} \times \dfrac{1 \text{ mol NaCl}}{2 \text{ mol ions}} \times \dfrac{58.4 \text{ g NaCl}}{1 \text{ mol NaCl}} = \mathbf{225 \text{ g NaCl}}$

2-72. *Refer to Section 2-11 and Example 2-22.*

(a) Plan: g $CuSO_4 \cdot 5H_2O \Longrightarrow$ mol $CuSO_4 \cdot 5H_2O \Longrightarrow$ mol $CuSO_4 \cdot H_2O \Longrightarrow$ g $CuSO_4 \cdot H_2O$

$? \text{ g CuSO}_4 \cdot \text{H}_2\text{O} = 556 \text{ g CuSO}_4 \cdot 5\text{H}_2\text{O} \times \dfrac{1 \text{ mol CuSO}_4 \cdot 5\text{H}_2\text{O}}{249.7 \text{ g CuSO}_4 \cdot 5\text{H}_2\text{O}} \times \dfrac{1 \text{ mol CuSO}_4 \cdot \text{H}_2\text{O}}{1 \text{ mol CuSO}_4 \cdot 5\text{H}_2\text{O}} \times \dfrac{177.6 \text{ g CuSO}_4 \cdot \text{H}_2\text{O}}{1 \text{ mol CuSO}_4 \cdot \text{H}_2\text{O}}$

$\phantom{? \text{ g CuSO}_4 \cdot \text{H}_2\text{O} } = \mathbf{395 \text{ g CuSO}_4 \cdot \text{H}_2\text{O}}$

(b) Plan: g $CuSO_4 \cdot 5H_2O \Longrightarrow$ mol $CuSO_4 \cdot 5H_2O \Longrightarrow$ mol $CuSO_4 \Longrightarrow$ g $CuSO_4$

$$? \text{ g } CuSO_4 = 556 \text{ g } CuSO_4 \cdot 5H_2O \times \frac{1 \text{ mol } CuSO_4 \cdot 5H_2O}{249.7 \text{ g } CuSO_4 \cdot 5H_2O} \times \frac{1 \text{ mol } CuSO_4}{1 \text{ mol } CuSO_4 \cdot 5H_2O} \times \frac{159.6 \text{ g } CuSO_4}{1 \text{ mol } CuSO_4}$$

$$= \textbf{355 g } CuSO_4$$

2-74. *Refer to Section 2-12 and Example 2-23.*

(a) Let us assume that we have 1 mole of $(COOH)_2 \cdot 2H_2O$

$$\% \ (COOH)_2 \text{ by mass} = \frac{FW \ (COOH)_2}{FW \ (COOH)_2 \cdot 2H_2O} \times 100 = \frac{90.03 \text{ g } (COOH)_2}{126.1 \text{ g } (COOH)_2 \cdot 2H_2O} \times 100 = \textbf{71.40\%}$$

(b) $\% \ (COOH)_2$ by mass $= \dfrac{72.4 \text{ g } (COOH)_2 \cdot 2H_2O}{100.0 \text{ g sample}} \times \dfrac{71.41 \text{ g } (COOH)_2}{100.0 \text{ g } (COOH)_2 \cdot 2H_2O} = \textbf{51.7\%}$

2-76. *Refer to Section 2-12 and Example 2-23.*

(a) ? lb $MgCO_3 = 671$ lb ore $\times \dfrac{27.7 \text{ lb } MgCO_3}{100 \text{ lb ore}} = \textbf{186 lb } MgCO_3$

(b) ? lb impurities $= 671$ lb ore - 186 lb $MgCO_3$ = **485 lb impurity**

(c) ? lb Mg $= 671$ lb ore $\times \dfrac{27.7 \text{ lb } MgCO_3}{100 \text{ lb ore}} \times \dfrac{24.3 \text{ lb Mg}}{84.3 \text{ lb } MgCO_3} = \textbf{53.6 lb Mg}$

2-78. *Refer to Section 2-8.*

Plan: mol compound $\Longrightarrow$ mol C = mol CO_2

(a) ? mol $CO_2 = 2.00$ mol $Fe_2(CO_3)_3 \times \dfrac{3 \text{ mol C}}{1 \text{ mol } Fe_2(CO_3)_3} \times \dfrac{1 \text{ mol } CO_2}{1 \text{ mol C}} = \textbf{6.00 mol } CO_2$

(b) ? mol $CO_2 = 2.00$ mol $CaCO_3 \times \dfrac{1 \text{ mol C}}{1 \text{ mol } CaCO_3} \times \dfrac{1 \text{ mol } CO_2}{1 \text{ mol C}} = \textbf{2.00 mol } CO_2$

(c) ? mol $CO_2 = 2.00$ mol $Ni(CO)_4 \times \dfrac{4 \text{ mol C}}{1 \text{ mol } Ni(CO)_4} \times \dfrac{1 \text{ mol } CO_2}{1 \text{ mol C}} = \textbf{8.00 mol } CO_2$

2-80. *Refer to Sections 2-7.*

(a) Formula Weight, FW $\left[\dfrac{g}{mol} \right] = \dfrac{g \text{ substance}}{mol \text{ substance}}$

? mol $O_3 = \dfrac{g \ O_3}{FW} = \dfrac{94.0 \text{ g } O_3}{48.0 \text{ g/mol}} = \textbf{1.96 mol } O_3$

(b) Plan: g $O_3 \Longrightarrow$ mol $O_3 \Longrightarrow$ mol O

? mol O $= 94.0$ g $O_3 \times \dfrac{1 \text{ mol } O_3}{48.0 \text{ g } O_3} \times \dfrac{3 \text{ mol O}}{1 \text{ mol } O_3} = \textbf{5.88 mol O}$

(c) Plan: $g\ O_3 \Longrightarrow mol\ O_3 \Longrightarrow mol\ O \Longrightarrow mol\ O_2 \Longrightarrow g\ O_2$

$$? \ mol\ O_2 = 94.0\ g\ O_3 \times \frac{1\ mol\ O_3}{48.0\ g\ O_3} \times \frac{3\ mol\ O}{1\ mol\ O_3} \times \frac{1\ mol\ O_2}{2\ mol\ O} \times \frac{32.0\ g\ O_2}{1\ mol\ O_2} = \textbf{94.0 g } O_2$$

(d) Plan: $g\ O_3 \Longrightarrow mol\ O_3 \Longrightarrow molecules\ O_3 = molecules\ O_2 \Longrightarrow mol\ O_2 \Longrightarrow g\ O_2$

$$? \ g\ O_2 = 94.0\ g\ O_3 \times \frac{1\ mol\ O_3}{48.0\ g\ O_3} \times \frac{6.02 \times 10^{23}\ molecules\ O_3}{1\ mol\ O_3} \times \frac{1\ molecule\ O_2}{1\ molecule\ O_3} \times \frac{1\ mole\ O_2}{6.02 \times 10^{23}\ molecules\ O_2}$$
$$\times \frac{32.0\ g\ O_2}{1\ mol\ O_2} = \textbf{62.7 g } O_2$$

2-82. *Refer to Sections 2-7 and 2-11.*

(a) $? \ mol\ Ag = 0.555\ mol\ Ag_2S \times \dfrac{2\ mol\ Ag}{1\ mol\ Ag_2S} = \textbf{1.11 mol Ag}$

(b) $? \ mol\ Ag = 0.555\ mol\ Ag_2O \times \dfrac{2\ mol\ Ag}{1\ mol\ Ag_2O} = \textbf{1.11 mol Ag}$

(c) $? \ mol\ Ag = 0.555\ g\ Ag_2S \times \dfrac{1\ mol\ Ag_2S}{248\ g\ Ag_2S} \times \dfrac{2\ mol\ Ag}{1\ mol\ Ag_2S} = \textbf{4.48} \times \textbf{10}^{-3} \textbf{ mol Ag}$

(d) $? \ mol\ Ag = 5.55 \times 10^{20}\ formula\ units\ Ag_2S \times \dfrac{1\ mol\ Ag_2S}{6.02 \times 10^{23}\ formula\ units\ Ag_2S} \times \dfrac{2\ mol\ Ag}{1\ mol\ Ag_2S}$
$$= \textbf{1.84} \times \textbf{10}^{-3} \textbf{ mol Ag}$$

2-84. *Refer to Section 2-9.*

Plan: (1) Use the H_2O, NH_3 and CO_2 data to calculate the masses of H, N and C, respectively, in the sample.
      (2) From the AgCl and AgBr information, calculate the mass of Cl and the mass of Br.
      (3) Determine the empirical (simplest) formula for the organic compound.

(1) $? \ mg\ H = 1.99\ mg\ H_2O \times \dfrac{2.02\ mg\ H}{18.0\ mg\ H_2O} = 0.223\ mg\ H$

    $? \ mg\ N = 1.25\ mg\ NH_3 \times \dfrac{14.0\ mg\ N}{17.0\ mg\ NH_3} = 1.03\ mg\ N$

    $? \ mg\ C = 6.47\ mg\ CO_2 \times \dfrac{12.0\ mg\ C}{44.0\ mg\ CO_2} = 1.76\ mg\ C$

(2) We know: (mg AgCl + mg AgBr) - (mg AgCl + mg $AgCl_{from\ AgBr}$) = 48.8 mg - 42.3 mg = 6.5 mg

    Therefore, we can say,   mg AgBr - mg $AgCl_{from\ AgBr}$ = 6.5 mg

    But, ? mg $AgCl_{from\ AgBr}$ = mg AgBr $\times \dfrac{FW\ AgCl}{FW\ AgBr}$ = mg AgBr $\times \dfrac{143.3\ mg\ AgCl}{187.8\ mg\ AgBr}$ = mg AgBr $\times$ 0.763

    Substituting, we have   mg AgBr - 0.763 $\times$ mg AgBr = 6.5 mg
$$0.237 \times mg\ AgBr = 6.5\ mg$$
$$? \ mg\ AgBr = 27\ mg\ (to\ 2\ significant\ figures)$$
$$? \ mg\ AgCl = 48.8\ mg\ \text{-}\ 27\ mg = 22\ mg$$

    $? \ mg\ Br = 27\ mg\ AgBr \times \dfrac{79.9\ mg\ Br}{187.8\ mg\ AgBr} = 11\ mg$

    $? \ mg\ Cl = 22\ mg\ AgCl \times \dfrac{35.45\ mg\ Cl}{143.3\ mg\ AgCl} = 5.4\ mg$

Check: total mass of sample (mg) = 1.03 mg N + 1.76 mg C + 0.223 mg H + 11 mg Br + 5.2 mg Cl = 19 mg
(This is close considering the loss of significant figures that occurred during subtraction.)

(3) $? \text{ mmol C} = \dfrac{1.76 \text{ mg C}}{12.0 \text{ mg/mmol}} = 0.147 \text{ mmol C}$ $\qquad$ $\text{Ratio} = \dfrac{0.147}{0.0736} = 2$

$\quad ? \text{ mmol H} = \dfrac{0.223 \text{ mg H}}{1.01 \text{ mg/mmol}} = 0.221 \text{ mmol H}$ $\qquad$ $\text{Ratio} = \dfrac{0.221}{0.0736} = 3$

$\quad ? \text{ mmol N} = \dfrac{1.03 \text{ mg N}}{14.0 \text{ mg/mmol}} = 0.0736 \text{ mmol N}$ $\qquad$ $\text{Ratio} = \dfrac{0.0736}{0.0736} = 1$

$\quad ? \text{ mmol Br} = \dfrac{11 \text{ mg Br}}{79.9 \text{ mg/mmol}} = 0.14 \text{ mmol Br}$ $\qquad$ $\text{Ratio} = \dfrac{0.14}{0.0736} = 2$

$\quad ? \text{ mmol Cl} = \dfrac{5.4 \text{ mg Cl}}{35.5 \text{ mg/mmol}} = 0.15 \text{ mmol Cl}$ $\qquad$ $\text{Ratio} = \dfrac{0.15}{0.0736} = 2$

Therefore, the empirical (simplest) formula for the organic compound is $C_2H_3NBr_2Cl_2$.

2-86. *Refer to Sections 2-2 and 1-5.*

Sample 1: $\dfrac{1.60 \text{ g O}}{2.43 \text{ g Mg}} = 0.658$ $\qquad$ Sample 2: $\dfrac{0.658 \text{ g O}}{1.00 \text{ g Mg}} = 0.658$ $\qquad$ Sample 3: $\dfrac{2.29 \text{ g O}}{3.48 \text{ g Mg}} = 0.658$

All three samples of magnesium oxide had the same O/Mg mass ratio. This is an example of the **Law of Constant Composition.**

2-88. *Refer to Section 2-11 and Exercise 2-72.*

$? \text{ g CuSO}_4 \cdot 5\text{H}_2\text{O} = 10.0 \text{ g CuSO}_4 \times \dfrac{1 \text{ mol CuSO}_4}{159.6 \text{ g CuSO}_4} \times \dfrac{1 \text{ mol CuSO}_4 \cdot 5\text{H}_2\text{O}}{1 \text{ mol CuSO}_4} \times \dfrac{249.6 \text{ g CuSO}_4 \cdot 5\text{H}_2\text{O}}{1 \text{ mol CuSO}_4 \cdot 5\text{H}_2\text{O}}$
$\qquad\qquad\qquad = \textbf{15.6 g CuSO}_4 \cdot \textbf{5H}_2\textbf{O}$

2-90. *Refer to Section 2-11.*

Plan: $\text{g BaCrO}_4 \Longrightarrow \text{mol BaCrO}_4 \Longrightarrow \text{mol Cr} \Longrightarrow \text{mol Cr}_2\text{O}_3 \Longrightarrow \text{g Cr}_2\text{O}_3$

$? \text{ g Cr}_2\text{O}_3 = 1.00 \text{ g BaCrO}_4 \times \dfrac{1 \text{ mol BaCrO}_4}{253 \text{ g BaCrO}_4} \times \dfrac{1 \text{ mol Cr}}{1 \text{ mol BaCrO}_4} \times \dfrac{1 \text{ mol Cr}_2\text{O}_3}{2 \text{ mol Cr}} \times \dfrac{152 \text{ g Cr}_2\text{O}_3}{1 \text{ mol Cr}_2\text{O}_3} = \textbf{0.300 g Cr}_2\textbf{O}_3$

2-92. *Refer to Section 2-10, Examples 2-16 and 2-17, and Exercise 2-54 Solution.*

(a) (1) $? \text{ g C} = 2.960 \text{ g CO}_2 \times \dfrac{12.01 \text{ g C}}{44.01 \text{ g CO}_2} = 0.8078 \text{ g C}$

$\quad ? \text{ g H} = 1.010 \text{ g H}_2\text{O} \times \dfrac{2.016 \text{ g H}}{18.02 \text{ g H}_2\text{O}} = 0.1130 \text{ g H}$

(2) $? \text{ g O} = 1.6380 \text{ g adipic acid} - 0.8078 \text{ g C} - 0.1130 \text{ g H} = 0.7172 \text{ g O}$

(3) $? \text{ mol C} = \dfrac{0.8078 \text{ g C}}{12.01 \text{ g/mol}} = 0.06726 \text{ mol C}$ $\quad\quad \text{Ratio} = \dfrac{0.06726}{0.04483} = 1.5$

$? \text{ mol H} = \dfrac{0.1130 \text{ g H}}{1.008 \text{ g/mol}} = 0.1121 \text{ mol H}$ $\quad\quad \text{Ratio} = \dfrac{0.1121}{0.04483} = 2.5$

$? \text{ mol O} = \dfrac{0.7172 \text{ g O}}{16.00 \text{ g/mol}} = 0.04483 \text{ mol O}$ $\quad\quad \text{Ratio} = \dfrac{0.04483}{0.04483} = 1$

A 1.5:2.5:1 ratio converts to 3:5:2 by multiplying by 2. Therefore, the simplest formula for adipic acid is $C_3H_5O_2$ (FW = 73.07 g/mol).

(b) $n = \dfrac{\text{molecular weight}}{\text{simplest formula weight}} = \dfrac{146.1 \text{ g/mol}}{73.07 \text{ g/mol}} = 2$

The true molecular formula for adipic acid is $(C_3H_5O_2)_2 = C_6H_{10}O_4$.

2-94.   *Refer to Sections 2-9 and 1-11, and Table 1-8.*

Plan: (1) Determine the simplest formula for the unknown compound.
(2) Calculate the density of the unknown and compare it with the value for ethanol obtained from Table 1-8.
(3) Determine whether the compound is likely to be ethanol.

(1) Let us assume we have a 100 g sample containing 53 g C, 11 g H and 36 g O.

$? \text{ mol C} = \dfrac{53 \text{ g C}}{12.0 \text{ g/mol}} = 4.42 \text{ mol C}$ $\quad\quad \text{Ratio} = \dfrac{4.42}{2.25} = 1.96$

$? \text{ mol H} = \dfrac{11 \text{ g H}}{1.01 \text{ g/mol}} = 10.9 \text{ mol H}$ $\quad\quad \text{Ratio} = \dfrac{10.9}{2.25} = 4.84$

$? \text{ mol O} = \dfrac{36 \text{ g O}}{16.0 \text{ g/mol}} = 2.25 \text{ mol O}$ $\quad\quad \text{Ratio} = \dfrac{2.25}{2.25} = 1.00$

The simplest formula is likely to be C_2H_5O, not C_2H_6O.

(2) Density of unknown compound $= \dfrac{\text{mass (g)}}{\text{volume (mL)}} = \dfrac{19 \text{ g}}{22 \text{ mL}} = 0.86 \text{ g/mL}$

The density for ethanol obtained from Table 1-8 is 0.789 g/mL.

(3) Both the simplest formula and the density of the unknown are different from those of ethanol. It is **NOT** likely to be ethanol.

2-96.   *Refer to Sections 2-6, 2-7 and 1-11.*

Plan: (1) Determine the number of molecules in 125 mL of H_2O.
(2) Determine the volume of ethanol that contains the same number of molecules.

(1) $? H_2O \text{ molecules} = 125 \text{ mL } H_2O \dfrac{1.00 \text{ g } H_2O}{1.00 \text{ mL } H_2O} \times \dfrac{1 \text{ mol } H_2O}{18.0 \text{ g } H_2O} \times \dfrac{6.02 \times 10^{23} \text{ } H_2O \text{ molecules}}{1 \text{ mol } H_2O}$
$= 4.18 \times 10^{24} \text{ molecules}$

(2) $? \text{ mL ethanol} = 4.18 \times 10^{24} \text{ molecules} \times \dfrac{1 \text{ mol}}{6.02 \times 10^{23} \text{ molecules}} \times \dfrac{46.1 \text{ g}}{1 \text{ mol}} \times \dfrac{1.00 \text{ mL}}{0.789 \text{ g}}$
$= \textbf{406 mL ethanol}$

3 Chemical Equations and Reaction Stoichiometry

3-2.	*Refer to Section 3-1.*

The Law of Conservation of Matter provides the basis for balancing a chemical equation. It states that matter is neither created nor destroyed during an ordinary chemical reaction. Therefore, a balanced chemical equation must always contain the same number of each kind of atom on both sides of the equation.

3-4.	*Refer to Section 3-1.*

Hints for balancing equations:

(1) Use smallest whole number coefficients. However, it may be useful to temporarily use a fractional coefficient, then for the last step, multiply all the terms by a factor to change the fractions to whole numbers.

(2) Look for special groups of elements that appear unchanged on both sides of the equation, e.g., NO_3, PO_4, SO_4. Treat them as units when balancing.

(3) Begin by balancing both the special groups and the elements that appear only once on both sides of the equation.

(4) Any element that appears more than once on one side of the equation is normally the last element to be balanced.

(5) If free, uncombined elements appear on either side, balance them last.

(6) When an element has an "odd" number of atoms on one side of the equation and an "even" number on the other side, it is often advisable to multiply the "odd" side by 2.

(a) unbalanced: $Al + O_2 \rightarrow Al_2O_3$

      Step 1: $Al + \boxed{3}O_2 \rightarrow \boxed{2}Al_2O_3$ balance O

      Step 2: $\boxed{4}Al + 3O_2 \rightarrow 2Al_2O_3$ balance Al

(b) unbalanced: $N_2 + O_2 \rightarrow N_2O$

      Step 1: $N_2 + \boxed{1/2}O_2 \rightarrow N_2O$ balance O

      Step 2: $\boxed{2}N_2 + \boxed{1}O_2 \rightarrow \boxed{2}N_2O$ multiply by 2
                                                           whole number coefficients

(c) unbalanced: $K + KNO_3 \rightarrow K_2O + N_2$

      Step 1: $K + \boxed{2}KNO_3 \rightarrow K_2O + N_2$ balance N

      Step 2: $K + 2KNO_3 \rightarrow \boxed{6}K_2O + N_2$ balance O

      Step 3: $\boxed{10}K + 2KNO_3 \rightarrow 6K_2O + N_2$ balance K

(d)   unbalanced: $H_2O + KO_2 \rightarrow KOH + O_2$

       Step 1: $H_2O + KO_2 \rightarrow \boxed{2}KOH + O_2$ balance H

       Step 2: $H_2O + \boxed{2}KO_2 \rightarrow 2KOH + O_2$ balance K

       Step 3: $H_2O + 2KO_2 \rightarrow 2KOH + \boxed{3/2}O_2$ balance O

       Step 4: $\boxed{2}H_2O + \boxed{4}KO_2 \rightarrow \boxed{4}KOH + \boxed{3}O_2$ multiply by 2
                                                         whole number coefficients

(e)   unbalanced: $H_2SO_4 + NH_3 \rightarrow (NH_4)_2SO_4$

       Step 1: $H_2SO_4 + \boxed{2}NH_3 \rightarrow (NH_4)_2SO_4$ balance N, H

3-6.   Refer to Section 3-1 and Exercise 3-4 Solution.

(a)   unbalanced: $NaCl + H_2O \rightarrow NaOH + Cl_2 + H_2$

       Step 1: $\boxed{2}NaCl + H_2O \rightarrow NaOH + Cl_2 + H_2$ balance Cl

       Step 2: $2NaCl + H_2O \rightarrow \boxed{2}NaOH + Cl_2 + H_2$ balance Na

       Step 3: $2NaCl + \boxed{2}H_2O \rightarrow 2NaOH + Cl_2 + H_2$ balance H, O

(b)   unbalanced: $RbOH + SO_2 \rightarrow Rb_2SO_3 + H_2O$

       Step 1: $\boxed{2}RbOH + SO_2 \rightarrow Rb_2SO_3 + H_2O$ balance Rb, H, O

(c)   unbalanced: $Ba(OH)_2 + P_4O_{10} \rightarrow Ba_3(PO_4)_2 + H_2O$

       Step 1: $Ba(OH)_2 + P_4O_{10} \rightarrow \boxed{2}Ba_3(PO_4)_2 + H_2O$ balance P

       Step 2: $\boxed{6}Ba(OH)_2 + P_4O_{10} \rightarrow 2Ba_3(PO_4)_2 + H_2O$ balance Ba

       Step 3: $6Ba(OH)_2 + P_4O_{10} \rightarrow 2Ba_3(PO_4)_2 + \boxed{6}H_2O$ balance H, O

(d)   unbalanced: $(NH_4)_2Cr_2O_7 \rightarrow N_2 + H_2O + Cr_2O_3$

       Step 1: $(NH_4)_2Cr_2O_7 \rightarrow N_2 + \boxed{4}H_2O + Cr_2O_3$ balance H, O

(e)   unbalanced: $Al + Cr_2O_3 \rightarrow Al_2O_3 + Cr$

       Step 1: $\boxed{2}Al + Cr_2O_3 \rightarrow Al_2O_3 + Cr$ balance Al

       Step 2: $2Al + Cr_2O_3 \rightarrow Al_2O_3 + \boxed{2}Cr$ balance Cr

---

3-8.   Refer to Section 3-2 and Example 3-1.

(a) $N_2 + 3H_2 \rightarrow 2NH_3$

(b)   ? molecules $H_2 = 200$ molecules $N_2 \times \dfrac{3 \text{ molecules } H_2}{1 \text{ molecule } N_2} = $ **600 molecules H_2**

(c)   ? molecules $NH_3 = 200$ molecules $N_2 \times \dfrac{2 \text{ molecules } NH_3}{1 \text{ molecule } N_2} = $ **400 molecules N_2**

3-10.   *Refer to Section 3-2 and Example 3-2.*

(a) $CaO + 2HCl \rightarrow CaCl_2 + H_2O$

(b) $? \text{ mol HCl} = 8.8 \text{ mol CaO} \times \dfrac{2 \text{ mol HCl}}{1 \text{ mol CaO}} = \textbf{18 mol HCl}$ (2 significant figures)

(c) $? \text{ mol } H_2O = 8.8 \text{ mol CaO} \times \dfrac{1 \text{ mol } H_2O}{1 \text{ mol CaO}} = \textbf{8.8 mol } \textbf{H}_2\textbf{O}$

3-12.   *Refer to Section 3-2 and Example 3-2.*

(a)   balanced equation: $2KClO_3 \rightarrow 2 KCl + 3O_2$

$? \text{ mol } O_2 = 10.0 \text{ mol KClO}_3 \times \dfrac{3 \text{ mol } O_2}{2 \text{ mol KClO}_3} = \textbf{15.0 mol } \textbf{O}_2$

(b)   balanced equation: $2H_2O_2 \rightarrow 2H_2O + O_2$

$? \text{ mol } O_2 = 10.0 \text{ mol } H_2O_2 \times \dfrac{1 \text{ mol } O_2}{2 \text{ mol } H_2O_2} = \textbf{5.00 mol } \textbf{O}_2$

(c)   balanced equation: $2HgO \rightarrow 2Hg + O_2$

$? \text{ mol } O_2 = 10.0 \text{ mol HgO} \times \dfrac{1 \text{ mol } O_2}{2 \text{ mol HgO}} = \textbf{5.00 mol } \textbf{O}_2$

(d)   balanced equation: $2NaNO_3 \rightarrow 2NaNO_2 + O_2$

$? \text{ mol } O_2 = 10.0 \text{ mol NaNO}_3 \times \dfrac{1 \text{ mol } O_2}{2 \text{ mol NaNO}_3} = \textbf{5.00 mol } \textbf{O}_2$

(e)   balanced equation: $KClO_4 \rightarrow KCl + 2O_2$

$? \text{ mol } O_2 = 10.0 \text{ mol KClO}_4 \times \dfrac{2 \text{ mol } O_2}{1 \text{ mol KClO}_4} = \textbf{20.0 mol } \textbf{O}_2$

3-14.   *Refer to Section 3-2 and Example 3-2.*

unbalanced:	$NH_3 + O_2 \rightarrow NO + H_2O$	
Step 1:	$\boxed{2}NH_3 + O_2 \rightarrow NO + \boxed{3}H_2O$	balance H
Step 2:	$2NH_3 + O_2 \rightarrow \boxed{2}NO + 3H_2O$	balance N
Step 3:	$2NH_3 + \boxed{5/2}O_2 \rightarrow 2NO + 3H_2O$	balance O
Step 4:	$\boxed{4}NH_3 + \boxed{5}O_2 \rightarrow \boxed{4}NO + \boxed{6}H_2O$	whole number coefficients

(a) $? \text{ mol } O_2 = 10.0 \text{ mol NH}_3 \times \dfrac{5 \text{ mol } O_2}{4 \text{ mol NH}_3} = \textbf{12.5 mol } \textbf{O}_2$

(b) $? \text{ mol NO} = 10.0 \text{ mol NH}_3 \times \dfrac{4 \text{ mol NO}}{4 \text{ mol NH}_3} = \textbf{10.0 mol NO}$

(c) $? \text{ mol } H_2O = 10.0 \text{ mol NH}_3 \times \dfrac{6 \text{ mol } H_2O}{4 \text{ mol NH}_3} = \textbf{15.0 mol } \textbf{H}_2\textbf{O}$

Balanced equation: $H_2 + Cl_2 \rightarrow 2HCl$

Method 1: Use the units of formula weight as an equation.

$$FW \left(\frac{g}{mol} \right) = \frac{g \text{ substance}}{mol \text{ substance}}$$

Plan: $\underset{(1)}{g \ H_2 \Longrightarrow} \underset{(2)}{mol \ H_2 \Longrightarrow} \underset{(3)}{mol \ Cl_2 \Longrightarrow} g \ Cl_2$

(1) $? \ mol \ H_2 = \dfrac{g \ H_2}{FW \ H_2} = \dfrac{4.77 \ g}{2.02 \ g/mol} = 2.36 \ mol \ H_2$

(2) $? \ mol \ Cl_2 = mol \ H_2 = 2.36 \ mol \ Cl_2$

(3) $? \ g \ Cl_2 = mol \ Cl_2 \times FW \ Cl_2 = 2.36 \ mol \times 70.9 \ g/mol = \textbf{167 g } \textbf{Cl}_2$

Method 2: Dimensional Analysis (Each unit factor corresponds to a step in Method 1.)

$$? \ g \ Cl_2 = 4.77 \ g \ H_2 \times \overset{\text{Step 1}}{\frac{1 \ mol \ H_2}{2.02 \ g \ H_2}} \times \overset{\text{Step 2}}{\frac{1 \ mol \ Cl_2}{1 \ mol \ H_2}} \times \overset{\text{Step 3}}{\frac{70.9 \ g \ Cl_2}{1 \ mol \ Cl_2}} = \textbf{167 g } \textbf{Cl}_2$$

Method 3: Proportion or Ratio Method

$$\frac{? \ g \ Cl_2}{g \ H_2} = \frac{1 \times FW \ Cl_2}{1 \times FW \ H_2} \qquad \text{Solving, } ? \ g \ Cl_2 = g \ H_2 \times \frac{1 \times FW \ Cl_2}{1 \times FW \ H_2} = 4.77 \ g \times \frac{70.9 \ g}{2.02 \ g} = \textbf{167 g } \textbf{Cl}_2$$

Balanced equation: $Fe_3O_4 + 4H_2 \rightarrow 3Fe + 4H_2O$

Method 1: Plan: $\underset{(1)}{g \ H_2O \Longrightarrow} \underset{(2)}{mol \ H_2O \Longrightarrow} \underset{(3)}{mol \ Fe_3O_4 \Longrightarrow} g \ Fe_3O_4$

(1) $? \ mol \ H_2O = \dfrac{g \ H_2O}{FW \ H_2O} = \dfrac{11.25 \ g}{18.02 \ g/mol} = 0.6243 \ mol \ H_2O$

(2) $? \ mol \ Fe_3O_4 = mol \ H_2O \times 1/4 = 0.6243 \ mol \times 1/4 = 0.1561 \ mol \ Fe_3O_4$

(3) $? \ g \ Fe_3O_4 = mol \ Fe_3O_4 \times FW \ Fe_3O_4 = 0.1561 \ mol \times 231.55 \ g/mol = \textbf{36.14 g } \textbf{Fe}_3\textbf{O}_4$

Method 2: Dimensional Analysis (Each unit factor corresponds to a step in Method 1.)

$$? \ g \ Fe_3O_4 = 11.25 \ g \ H_2O \times \overset{\text{Step 1}}{\frac{1 \ mol \ H_2O}{18.02 \ g \ H_2O}} \times \overset{\text{Step 2}}{\frac{1 \ mol \ Fe_3O_4}{4 \ mol \ H_2O}} \times \overset{\text{Step 3}}{\frac{231.55 \ g \ Fe_3O_4}{1 \ mol \ Fe_3O_4}} = \textbf{36.14 g } \textbf{Fe}_3\textbf{O}_4$$

Method 3: Proportion or Ratio Method

$$\frac{? \ g \ Fe_3O_4}{g \ H_2O} = \frac{1 \times FW \ Fe_3O_4}{4 \times FW \ H_2O} \qquad \text{Solving, } ? \ g \ Fe_3O_4 = g \ H_2O \times \frac{FW \ Fe_3O_4}{4 \times FW \ H_2O} = 11.25 \ g \times \frac{1 \times 231.55 \ g}{4 \times 18.02 \ g}$$

$$= \textbf{36.14 g } \textbf{Fe}_3\textbf{O}_4$$

3-20. *Refer to Section 3-2 and Example 3-5.*

Balanced equation: $CH_4 + 2O_2 \rightarrow CO_2 + 2H_2O$

Method 1: Plan: $g\ CH_4 \overset{(1)}{\Longrightarrow} mol\ CH_4 \overset{(2)}{\Longrightarrow} mol\ O_2 \overset{(3)}{\Longrightarrow} g\ O_2$

(1) $?\ mol\ CH_4 = \dfrac{g\ CH_4}{FW\ CH_4} = \dfrac{32.0\ g}{16.0\ g/mol} = 2.00\ mol\ CH_4$

(2) $?\ mol\ O_2 = mol\ CH_4 \times 2 = 2.00\ mol \times 2 = 4.00\ mol\ O_2$

(3) $?\ g\ O_2 = mol\ O_2 \times FW\ O_2 = 4.00\ mol \times 32.0\ g/mol = \mathbf{128\ g\ O_2}$

Method 2: Dimensional Analysis (Each unit factor corresponds to a step in Method 1.)

$$? g\ O_2 = 32.0\ g\ CH_4 \times \overset{\textbf{Step 1}}{\frac{1\ mol\ CH_4}{16.0\ g\ CH_4}} \times \overset{\textbf{Step 2}}{\frac{2\ mol\ O_2}{1\ mol\ CH_4}} \times \overset{\textbf{Step 3}}{\frac{32.0\ g\ O_2}{1\ mol\ O_2}} = \mathbf{128\ g\ O_2}$$

Method 3: Proportion or Ratio Method

$$\frac{?\ g\ O_2}{g\ CH_4} = \frac{2 \times FW\ O_2}{1 \times FW\ CH_4}$$

Solving, $?\ g\ O_2 = g\ CH_4 \times \dfrac{2 \times FW\ O_2}{1 \times FW\ CH_4} = 32.0\ g \times \dfrac{2 \times 32.0\ g}{1 \times 16.0\ g} = \mathbf{128\ g\ O_2}$

3-22. *Refer to Section 3-2 and Example 3-6.*

Balanced equation: $2KBr + Cl_2 \rightarrow 2KCl + Br_2$

Method 1: Plan: $g\ Cl_2 \overset{(1)}{\Longrightarrow} mol\ Cl_2 \overset{(2)}{\Longrightarrow} mol\ Br_2 \overset{(3)}{\Longrightarrow} g\ Br_2$

(1) $?\ mol\ Cl_2 = \dfrac{g\ Cl_2}{FW\ Cl_2} = \dfrac{0.361\ g}{70.9\ g/mol} = 5.09 \times 10^{-3}\ mol\ Cl_2$

(2) $?\ mol\ Br_2 = mol\ Cl_2 = 5.09 \times 10^{-3}\ mol\ Br_2$

(3) $?\ g\ Br_2 = mol\ Br_2 \times FW\ Br_2 = (5.09 \times 10^{-3}\ mol) \times 159.8\ g/mol = \mathbf{0.814\ g\ Br_2}$

Method 2: Dimensional Analysis (The unit factors correspond to the steps in Method 1.)

$$? g\ Br_2 = 0.361\ g\ Cl_2 \times \overset{\textbf{Step 1}}{\frac{1\ mol\ Cl_2}{70.9\ g\ Cl_2}} \times \overset{\textbf{Step 2}}{\frac{1\ mol\ Br_2}{1\ mol\ Cl_2}} \times \overset{\textbf{Step 3}}{\frac{159.8\ g\ Br_2}{1\ mol\ Br_2}} = \mathbf{0.814\ g\ Br_2}$$

Method 3: Proportion or Ratio Method

$$\frac{?\ g\ Br_2}{g\ Cl_2} = \frac{1 \times FW\ Br_2}{1 \times FW\ Cl_2}$$

Solving, $?\ g\ Br_2 = g\ Cl_2 \times \dfrac{1 \times FW\ Br_2}{1 \times FW\ Cl_2} = 0.361\ g \times \dfrac{159.8\ g}{70.9\ g} = \mathbf{0.814\ g\ Br_2}$

3-24. *Refer to Section 3-2 and Example 3-6.*

Balanced equation: $2C_4H_{10} + 13O_2 \rightarrow 8CO_2 + 10H_2O$

Method 1: Plan: $g\ C_4H_{10} \overset{(1)}{\Longrightarrow} mol\ C_4H_{10} \overset{(2)}{\Longrightarrow} mol\ H_2O \overset{(3)}{\Longrightarrow} g\ H_2O$

(1) $?\ mol\ C_4H_{10} = \dfrac{g\ C_4H_{10}}{FW\ C_4H_{10}} = \dfrac{26.0\ g}{58.1\ g/mol} = 0.448\ mol\ C_4H_{10}$

(2) $?\ mol\ H_2O = mol\ C_4H_{10} \times 10/2 = 0.448\ mol \times 10/2 = 2.24\ mol\ H_2O$

(3) $?\ g\ H_2O = mol\ H_2O \times FW\ H_2O = 2.24\ mol \times 18.0\ g/mol =$ **40.3 g H₂O**

Method 2: Dimensional Analysis (Each unit factor corresponds to a step in Method 1.)

$$?\ g\ H_2O = 26.0\ g\ C_4H_{10} \times \overset{\textbf{Step 1}}{\frac{1\ mol\ C_4H_{10}}{58.1\ g\ C_4H_{10}}} \times \overset{\textbf{Step 2}}{\frac{10\ mol\ H_2O}{2\ mol\ C_4H_{10}}} \times \overset{\textbf{Step 3}}{\frac{18.0\ g\ H_2O}{1\ mol\ H_2O}} = \textbf{40.3 g H}_2\textbf{O}$$

Method 3: Proportion or Ratio Method

$$\frac{?\ g\ H_2O}{g\ C_4H_{10}} = \frac{10 \times FW\ H_2O}{2 \times FW\ C_4H_{10}}$$

Solving, $?\ g\ H_2O = g\ C_4H_{10} \times \dfrac{10 \times FW\ H_2O}{2 \times FW\ C_4H_{10}} = 26.0\ g \times \dfrac{10 \times 18.0\ g}{2 \times 58.1\ g} =$ **40.3 g H₂O**

3-26. *Refer to Sections 3-2 and 2-7, and Example 3-7.*

Balanced equation: $2C_3H_8 + 10O_2 \rightarrow 6CO_2 + 8H_2O$

Method 1: Plan: $g\ H_2O \overset{(1)}{\Longrightarrow} mol\ H_2O \overset{(2)}{\Longrightarrow} mol\ C_3H_8 \overset{(3)}{\Longrightarrow} molecules\ C_3H_8$

(1) $?\ mol\ H_2O = \dfrac{g\ H_2O}{FW\ H_2O} = \dfrac{3.20\ g}{18.0\ g/mol} = 0.178\ mol\ H_2O$

(2) $?\ mol\ C_3H_8 = mol\ H_2O \times 2/8 = 0.178\ mol \times 2/8 = 0.0445\ mol\ C_3H_8$

(3) $?\ molecules\ C_3H_8 = mol\ C_3H_8 \times Avogadro's\ Number = 0.0445\ mol \times (6.02 \times 10^{23}\ molecules/mol)$

$= $ **2.68 × 10²² molecules C₃H₈**

Method 2: Dimensional Analysis (Each unit factor corresponds to a step in Method 1.)

$$?\ molecules\ C_3H_8 = 3.20\ g\ H_2O \times \overset{\textbf{Step 1}}{\frac{1\ mol\ H_2O}{18.0\ g\ H_2O}} \times \overset{\textbf{Step 2}}{\frac{2\ mol\ C_3H_8}{8\ mol\ H_2O}} \times \overset{\textbf{Step 3}}{\frac{6.02 \times 10^{23}\ molecules\ C_3H_8}{1\ mol\ C_3H_8}}$$

$= $ **2.68 × 10²² molecules C₃H₈**

Balanced equation: $N_2 + 3H_2 \rightarrow 2NH_3$

This is a limiting reactant problem.

Plan: (1) Find the limiting reactant.

(2) Do the stoichiometric problem based on amount of limiting reactant.

(1) Convert the mass of reactants to moles and compare the required ratio to the available ratio.

$$? \text{ mol } N_2 = \frac{\text{g } N_2}{\text{FW } N_2} = \frac{85.5 \text{ g}}{28.0 \text{ g/mol}} = 3.05 \text{ mol } N_2$$

$$? \text{ mol } H_2 = \frac{\text{g } H_2}{\text{FW } H_2} = \frac{17.3 \text{ g}}{2.02 \text{ g/mol}} = 8.56 \text{ mol } H_2$$

$$\text{Required ratio} = \frac{3 \text{ mol } H_2}{1 \text{ mol } N_2} = 3 \qquad\qquad \text{Available ratio} = \frac{8.56 \text{ mol } H_2}{3.05 \text{ mol } N_2} = 2.81$$

The available ratio < required ratio. Therefore, we do not have enough H_2 to react with all the N_2, and so H_2 is the limiting reactant.

(2) The amount of NH_3 is determined by the amount of limiting reactant, 8.56 moles of H_2.

$$? \text{ g } NH_3 = 8.56 \text{ mol } H_2 \times \frac{2 \text{ mol } NH_3}{3 \text{ mol } H_2} \times \frac{17.0 \text{ g } NH_3}{1 \text{ mol } NH_3} = \textbf{97.0 g } \textbf{NH}_3$$

Balanced equation: $Ca_3(PO_4)_2 + 2H_2SO_4 \rightarrow Ca(H_2PO_4)_2 + 2CaSO_4$

This is a limiting reactant problem.

(1) Convert to moles and compare required ratio to available ratio to find the limiting reactant.

$$? \text{ mol } Ca_3(PO_4)_2 = \frac{\text{g } Ca_3(PO_4)_2}{\text{FW } Ca_3(PO_4)_2} = \frac{450 \text{ g}}{310 \text{ g/mol}} = 1.45 \text{ mol } Ca_3(PO_4)_2$$

$$? \text{ mol } H_2SO_4 = \frac{\text{g } H_2SO_4}{\text{FW } H_2SO_4} = \frac{300 \text{ g}}{98.1 \text{ g/mol}} = 3.06 \text{ mol } H_2SO_4$$

$$\text{Required ratio} = \frac{2 \text{ mol } H_2SO_4}{1 \text{ mol } Ca_3(PO_4)_2} = 2 \qquad\qquad \text{Available ratio} = \frac{3.06 \text{ mol } H_2SO_4}{1.45 \text{ mol } Ca_3(PO_4)_2} = 2.11$$

Available ratio > required ratio; $Ca_3(PO_4)_2$ is the limiting reactant.

(2) First, find the mass of H_2SO_4 that reacted, then determine the mass of superphosphate.

The Law of Conservation of Mass states that the mass of reactants that react equal the mass of products formed. Therefore, we can calculate the mass of superphosphate.

$$? \text{ g } H_2SO_4 = 450 \text{ g } Ca_3(PO_4)_2 \times \frac{1 \text{ mol } Ca_3(PO_4)_2}{310 \text{ g } Ca_3(PO_4)_2} \times \frac{2 \text{ mol } H_2SO_4}{1 \text{ mol } Ca_3(PO_4)_2} \times \frac{98.1 \text{ g } H_2SO_4}{1 \text{ mol } H_2SO_4} = 285 \text{ g } H_2SO_4$$

$$? \text{ g } [Ca(H_2PO_4)_2 + 2CaSO_4]_{\text{formed}} = \text{g } [H_2SO_4 + Ca_3(PO_4)_2]_{\text{reacted}} \qquad = 285 \text{ g } H_2SO_4 + 450 \text{ g } Ca_3(PO_4)_2$$

$$= \textbf{735 g superphosphate}$$

3-32. *Refer to Section 3-3, Examples 3-8 and 3-9, and Exercise 3-28 Solution.*

Balanced equation: $Na + KCl \rightarrow NaCl + K$

This is a limiting reactant problem.

(1) Convert to moles and compare the required ratio to the available ratio to find the limiting reactant.

$$? \text{ mol Na} = \frac{g\ Na}{AW\ Na} = \frac{150.0\ g}{22.99\ g/mol} = 6.525 \text{ mol Na}$$

$$? \text{ mol KCl} = \frac{g\ KCl}{FW\ KCl} = \frac{150.0\ g}{74.55\ g/mol} = 2.012 \text{ mol KCl}$$

Required ratio $= \dfrac{1\ mol\ Na}{1\ mol\ KCl} = 1$ 　　　　　 Available ratio $= \dfrac{6.525\ mol\ Na}{2.012\ mol\ KCl} = 3.243$

Available ratio > required ratio; KCl is the limiting reactant.

(2) The mass of K produced is determined by the mass of KCl.

$$? \text{ g K} = 150.0 \text{ g KCl} \times \frac{1\ mol\ KCl}{74.55\ g\ KCl} \times \frac{1\ mol\ K}{1\ mol\ KCl} \times \frac{39.10\ g\ K}{1\ mol\ K} = \textbf{78.67 g K}$$

3-34. *Refer to Section 3-4 and Example 3-10.*

Balanced equation: $PCl_3 + Cl_2 \rightarrow PCl_5$

Step 1. Calculate the theoretical yield of PCl_5.

Plan: g $PCl_3 \overset{(1)}{\Longrightarrow}$ mol $PCl_3 \overset{(2)}{\Longrightarrow}$ mol $PCl_5 \overset{(3)}{\Longrightarrow}$ g PCl_5 (theoretical)

(1) $? \text{ mol } PCl_3 = \dfrac{g\ PCl_3}{FW\ PCl_3} = \dfrac{56.7\ g}{137\ g/mol} = 0.414 \text{ mol } PCl_3$

(2) $? \text{ mol } PCl_5 = \text{mol } PCl_3 = 0.414 \text{ mol } PCl_5$

(3) $? \text{ g } PCl_5 = \text{mol } PCl_5 \times FW\ PCl_5 = 0.414 \text{ mol} \times 208 \text{ g/mol} = 86.1 \text{ g } PCl_5$

Step 2. Solve for the actual yield of PCl_5.

$? \text{ \% yield} = \dfrac{\text{actual yield}}{\text{theoretical yield}} \times 100$ 　　　 Substituting, $83.2\% = \dfrac{?\ \text{actual yield}}{86.1\ g} \times 100$

Therefore, $? \text{ actual yield} = \dfrac{83.2\% \times 86.1\ g}{100} = \textbf{71.6 g } PCl_5$

3-36. *Refer to Section 3-4 and Example 3-10.*

Balanced equation: $2AgNO_3 \rightarrow 2Ag + 2NO_2 + O_2$

Step 1. Calculate the theoretical yield of Ag.

Plan: g $AgNO_3 \overset{(1)}{\Longrightarrow}$ mol $AgNO_3 \overset{(2)}{\Longrightarrow}$ mol Ag $\overset{(3)}{\Longrightarrow}$ g Ag (theoretical)

(1) $? \text{ mol } AgNO_3 = \dfrac{g\ AgNO_3}{FW\ AgNO_3} = \dfrac{0.722\ g}{169.9\ g/mol} = 4.25 \times 10^{-3} \text{ mol } AgNO_3$

(2) $?$ mol Ag $=$ mol AgNO$_3$ $= 4.25 \times 10^{-3}$ mol Ag

(3) $?$ g Ag $=$ mol Ag $\times$ AW Ag $= (4.25 \times 10^{-3}$ mol$) \times 107.9$ g/mol $= 0.459$ g Ag (theoretical yield)

Step 2.   Calculate the percent yield of Ag.

$$\% \text{ yield} = \frac{\text{actual yield}}{\text{theoretical yield}} \times 100 = \frac{0.443 \text{ g}}{0.459 \text{ g}} = \textbf{96.5\%}$$

3-38.   *Refer to Section 3-4.*

Balanced equation: $C_2H_5OBr + NaOH \rightarrow C_2H_4O + NaBr + H_2O$

Step 1.   Solve for the theoretical yield of C_2H_4O.

$$\% \text{ yield} = \frac{\text{actual yield}}{\text{theoretical yield}} \times 100 \qquad \text{Substituting, } 88.1\% = \frac{383 \text{ g}}{? \text{ theoretical yield}}$$

$$\text{Therefore, } ? \text{ theoretical yield} = \frac{383 \text{ g}}{88.1\%} \times 100 = 435 \text{ g } C_2H_4O$$

This means that if we start with just enough C_2H_5OBr to theoretically prepare 435 g C_2H_4O, we would actually obtain 383 g C_2H_4O because the percentage yield is only 88.1%.

Step 2.   Solve for the mass of C_2H_5OBr required to prepare 435 g C_2H_4O.

Plan: g $C_2H_4O \overset{(1)}{\Longrightarrow}$ mol $C_2H_4O \overset{(2)}{\Longrightarrow}$ mol $C_2H_5OBr \overset{(3)}{\Longrightarrow}$ g C_2H_5OBr

(1) $?$ mol $C_2H_4O = \dfrac{\text{g } C_2H_4O}{\text{FW } C_2H_4O} = \dfrac{435 \text{ g}}{44.1 \text{ g/mol}} = 9.86$ mol C_2H_4O

(2) $?$ mol $C_2H_5OBr =$ mol $C_2H_4O = 9.86$ mol C_2H_5OBr

(3) $?$ g $C_2H_5OBr =$ mol $C_2H_5OBr \times$ FW $C_2H_5OBr = 9.86$ mol $\times 125$ g/mol $= \textbf{1.23} \times \textbf{10}^{\textbf{3}}$ **g** C_2H_5OBr

---

3-40.   *Refer to Section 3-4.*

Balanced equation: $CaO + 3C \rightarrow CaC_2 + CO$

(a) Plan: (1) Calculate the masses of CaC_2 and CaO in the crude product.
(2) Use stoichiometry and mass balance to determine the initial mass of CaO.

(1) $?$ g CaC_2 in product $= 0.85 \times$ g product $= 0.85 \times (4.50 \times 10^5 \text{ g}) = 3.8 \times 10^5$ g CaC_2

$?$ g CaO in product $=$ g product - g $CaC_2 = 4.50 \times 10^5$ g - 3.8×10^5 g $= 7 \times 10^4$ g

(2) $?$ g $CaO_{\text{reacted}} = 3.8 \times 10^5$ g $CaC_2 \times \dfrac{1 \text{ mol } CaC_2}{64 \text{ g } CaC_2} \times \dfrac{1 \text{ mol } CaO}{1 \text{ mol } CaC_2} \times \dfrac{56 \text{ g } CaO}{1 \text{ mol } CaO} = 3.3 \times 10^5$ g CaO

$?$ g $CaO_{\text{initial}} =$ g $CaO_{\text{reacted}} +$ g $CaO_{\text{unreacted}} = 3.3 \times 10^5$ g + 7×10^4 g $= \textbf{4.0} \times \textbf{10}^{\textbf{5}}$ **g or 400 kg** CaO

(b) From (a), the mass of CaC_2 in the crude product is $\textbf{3.8} \times \textbf{10}^{\textbf{5}}$ **or 380 kg.**

3-42. *Refer to Section 3-5 and Example 3-11.*

Balanced equations:
$$TeO_2 + 2OH^- \rightarrow TeO_3^{2-} + H_2O$$
$$TeO_3^{2-} + 2H^+ \rightarrow H_2TeO_3$$

Plan: $\underset{(1)}{g\ TeO_2} \Longrightarrow \underset{(2)}{mol\ TeO_2} \Longrightarrow \underset{(3)}{mol\ TeO_3^{2-}} \Longrightarrow \underset{(4)}{mol\ H_2TeO_3} \Longrightarrow g\ H_2TeO_3$

$$? \text{ g } H_2TeO_3 = 72.1 \text{ g } TeO_2 \times \underset{\text{Step 1}}{\frac{1 \text{ mol } TeO_2}{159.6 \text{ g } TeO_2}} \times \underset{\text{Step 2}}{\frac{1 \text{ mol } TeO_3^{2-}}{1 \text{ mol } TeO_2}} \times \underset{\text{Step 3}}{\frac{1 \text{ mol } H_2TeO_3}{1 \text{ mol } TeO_3^{2-}}} \times \underset{\text{Step 4}}{\frac{177.6 \text{ g } H_2TeO_3}{1 \text{ mol } H_2TeO_3}} = \textbf{80.2 g } H_2TeO_3$$

3-44. *Refer to Section 3-5 and Example 3-11.*

Balanced equations:
$$2KClO_3 \rightarrow 2KCl + 3O_2$$
$$CH_4 + 2O_2 \rightarrow CO_2 + 2H_2O$$

Plan: $\underset{(1)}{g\ CH_4} \Longrightarrow \underset{(2)}{mol\ CH_4} \Longrightarrow \underset{(3)}{mol\ O_2} \Longrightarrow \underset{(4)}{mol\ KClO_3} \Longrightarrow g\ KClO_3$

$$? \text{ g } KClO_3 = 33.2 \text{ g } CH_4 \times \underset{\text{Step 1}}{\frac{1 \text{ mol } CH_4}{16.0 \text{ g } CH_4}} \times \underset{\text{Step 2}}{\frac{2 \text{ mol } O_2}{1 \text{ mol } CH_4}} \times \underset{\text{Step 3}}{\frac{2 \text{ mol } KClO_3}{3 \text{ mol } O_2}} \times \underset{\text{Step 4}}{\frac{122.6 \text{ g } KClO_3}{1 \text{ mol } KClO_3}} = \textbf{339 g } KClO_3$$

3-46. *Refer to Section 3-5 and Example 3-12.*

Balanced equations:
$$CH_3CH_2Cl + Mg \rightarrow CH_3CH_2MgCl \qquad\qquad\quad 79.5\% \text{ efficient}$$
$$CH_3CH_2MgCl + H_2O \rightarrow CH_3CH_3 + Mg(OH)Cl \qquad 78.8\% \text{ efficient}$$

Plan: g $CH_3CH_2Cl \Longrightarrow$ mol $CH_3CH_2Cl \Longrightarrow$ mol CH_3CH_2MgCl (theoretical) $\Longrightarrow$ mol CH_3CH_2MgCl (actual) $\Longrightarrow$ mol CH_3CH_3 (theoretical) $\Longrightarrow$ mol CH_3CH_3 (actual) $\Longrightarrow$ g CH_3CH_3

$$? \text{ g } CH_3CH_3 = 27.2 \text{ g } CH_3CH_2Cl \times \frac{1 \text{ mol } CH_3CH_2Cl}{64.4 \text{ g } CH_3CH_2Cl} \times \frac{1 \text{ mol } CH_3CH_2MgCl(\text{theoretical})}{1 \text{ mol } CH_3CH_2Cl}$$

$$\times \frac{79.5 \text{ mol } CH_3CH_2MgCl(\text{actual})}{100. \text{ mol } CH_3CH_2MgCl(\text{theoretical})} \times \frac{1 \text{ mol } CH_3CH_3}{1 \text{ mol } CH_3CH_2MgCl} \times \frac{78.8 \text{ mol } CH_3CH_3(\text{actual})}{100. \text{ mol } CH_3CH_3(\text{theoretical})}$$

$$\times \frac{30.0 \text{ g } CH_3CH_3}{1 \text{ mol } CH_3CH_3} = \textbf{7.94 g } CH_3CH_3$$

3-48. *Refer to Section 3-5 and Example 3-12.*

Balanced equations:

ZnS in ore → ZnS	flotation	90.6% efficient
$2ZnS + 3O_2 \rightarrow 2ZnO + 2SO_2$	heat in air	100% efficient
$ZnO + H_2SO_4 \rightarrow ZnSO_4 + H_2O$	acid treatment	100% efficient
$2ZnSO_4 + 2H_2O \rightarrow 2Zn + 2H_2SO_4 + O_2$	electrolysis	98.2% efficient

Plan: kg ZnS in ore $\Longrightarrow$ kg ZnS $\Longrightarrow$ g ZnS $\Longrightarrow$ mol ZnS $\Longrightarrow$ mol ZnO $\Longrightarrow$ mol $ZnSO_4$ $\Longrightarrow$ mol Zn (theoretical) $\Longrightarrow$ mol Zn (actual) $\Longrightarrow$ g Zn $\Longrightarrow$ kg Zn

$$? \text{ kg Zn} = 225 \text{ kg ZnS in ore} \times \frac{90.6 \text{ kg ZnS}}{100 \text{ kg ZnS in ore}} \times \frac{1000 \text{ g ZnS}}{1 \text{ kg ZnS}} \times \frac{1 \text{ mol ZnS}}{97.5 \text{ g ZnS}} \times \frac{2 \text{ mol ZnO}}{2 \text{ mol ZnS}}$$

$$\times \frac{1 \text{ mol } ZnSO_4}{1 \text{ mol ZnO}} \times \frac{2 \text{ mol Zn}}{2 \text{ mol } ZnSO_4} \times \frac{0.982 \text{ mol Zn (actual)}}{1 \text{ mol Zn (theoretical)}} \times \frac{65.39 \text{ g Zn}}{1 \text{ mol Zn}} \times \frac{1 \text{ kg Zn}}{1000 \text{ g Zn}} = \textbf{134 kg Zn}$$

3-50. *Refer to Section 3-6 and Examples 3-13, 3-14 and 3-15.*

Recall: $\% \text{ by mass} = \dfrac{\text{g Na}_2\text{S}}{\text{g soln}} \times 100 = \dfrac{\text{g Na}_2\text{S}}{100 \text{ g soln}}$

(a) $? \text{ mol Na}_2\text{S} = 100 \text{ g soln} \times \dfrac{0.250 \text{ g Na}_2\text{S}}{100 \text{ g soln}} \times \dfrac{1 \text{ mol Na}_2\text{S}}{78.0 \text{ g Na}_2\text{S}} = \mathbf{3.21 \times 10^{-3} \text{ mol Na}_2\text{S}}$

(b) $? \text{ g Na}_2\text{S} = 100 \text{ g soln} \times \dfrac{0.250 \text{ g Na}_2\text{S}}{100 \text{ g soln}} = \mathbf{0.250 \text{ g Na}_2\text{S}}$

(c) $? \text{ g H}_2\text{O} = 100.000 \text{ g soln} - 0.250 \text{ g Na}_2\text{S} = \mathbf{99.750 \text{ g H}_2\text{O}}$

3-52. *Refer to Section 3-6 and Example 3-15.*

We know $D\left(\dfrac{\text{g}}{\text{mL}}\right) = \dfrac{\text{g soln}}{\text{mL soln}}$ and $\% \text{ by mass} = \dfrac{\text{g (NH}_4)_2\text{SO}_4}{\text{g soln}} \times 100$

$? \text{ g (NH}_4)_2\text{SO}_4 = 425 \text{ mL soln} \times \dfrac{1.10 \text{ g soln}}{1.00 \text{ mL soln}} \times \dfrac{18.0 \text{ g (NH}_4)_2\text{SO}_4}{100 \text{ g soln}} = \mathbf{84.2 \text{ g (NH}_4)_2\text{SO}_4}$

3-54. *Refer to Section 3-6, Exercise 3-52, and Example 3-16.*

$? \text{ mL soln} = 90.0 \text{ g (NH}_4)_2\text{SO}_4 \times \dfrac{100 \text{ g soln}}{18.0 \text{ g (NH}_4)_2\text{SO}_4} \times \dfrac{1.00 \text{ mL soln}}{1.10 \text{ g soln}} = \mathbf{455 \text{ mL soln}}$

3-56. *Refer to Section 3-6 and Example 3-17.*

Method 1: Use the units of molarity as an equation: $M\left(\dfrac{\text{mol}}{\text{L}}\right) = \dfrac{\text{mol substance}}{\text{L soln}}$

Plan: $\text{g H}_3\text{PO}_4 \xrightarrow{(1)} \text{mol H}_3\text{PO}_4 \xrightarrow{(2)} M \text{ H}_3\text{PO}_4$

(1) $? \text{ mol H}_3\text{PO}_4 = \dfrac{\text{g H}_3\text{PO}_4}{\text{FW H}_3\text{PO}_4} = \dfrac{650 \text{ g}}{98.0 \text{ g/mol}} = 6.63 \text{ mol H}_3\text{PO}_4$

(2) $? M \text{ H}_3\text{PO}_4 = \dfrac{\text{mol H}_3\text{PO}_4}{\text{L soln}} = \dfrac{6.63 \text{ mol}}{3.00 \text{ L}} = \mathbf{2.21 \; \mathit{M} \text{ H}_3\text{PO}_4}$

Method 2: Dimensional Analysis

$? M \text{ H}_3\text{PO}_4 = \dfrac{650 \text{ g H}_3\text{PO}_4}{3.00 \text{ L soln}} \times \dfrac{1 \text{ mol H}_3\text{PO}_4}{98.0 \text{ g H}_3\text{PO}_4} = \mathbf{2.21 \; \mathit{M} \text{ H}_3\text{PO}_4}$

3-58. *Refer to Section 3-6 and Example 3-17.*

Use the units of molarity as an equation: $M\left(\dfrac{\text{mol}}{\text{L}}\right) = \dfrac{\text{mol substance}}{\text{L soln}}$

$? M \text{ H}_3\text{PO}_4 = \dfrac{\text{mol H}_3\text{PO}_4}{\text{L soln}} = \dfrac{0.335 \text{ mol}}{0.250 \text{ L soln}} = \mathbf{1.34 \; \mathit{M} \text{ H}_3\text{PO}_4}$

3-60. *Refer to Section 3-6, Example 3-17, and Exercise 3-56 Solution.*

(a) *Method 1:* Plan: g $(CH_3)_2CHOH \overset{(1)}{\Longrightarrow}$ mol $(CH_3)_2CHOH \overset{(2)}{\Longrightarrow} M \ (CH_3)_2CHOH$

(1) ? mol $(CH_3)_2CHOH = \dfrac{g \ (CH_3)_2CHOH}{FW \ (CH_3)_2CHOH} = \dfrac{2.25 \ g}{60.1 \ g/mol} = 0.0374$ mol $(CH_3)_2CHOH$

(2) ? $M \ (CH_3)_2CHOH = \dfrac{mol \ (CH_3)_2CHOH}{L \ soln} = \dfrac{0.0374 \ mol}{0.150 \ L} =$ **0.250 M $(CH_3)_2CHOH$**

Method 2: Dimensional Analysis

? $M \ (CH_3)_2CHOH = \dfrac{2.25 \ g \ (CH_3)_2CHOH}{150 \ mL \ soln} \times \dfrac{1000 \ mL}{1 \ L} \times \dfrac{1 \ mol \ (CH_3)_2CHOH}{60.1 \ g \ (CH_3)_2CHOH} =$ **0.250 M $(CH_3)_2CHOH$**

(b) ? mol $(CH_3)_2CHOH = M \times L$ soln $= 0.250 \ M \ (CH_3)_2CHOH \times 0.00200 \ L =$ **5.00×10^{-4} mol $(CH_3)_2CHOH$**

3-62. *Refer to Section 3-6 and Example 3-18.*

(a) *Method 1:* Plan: M, L Na_3PO_4 soln $\overset{(1)}{\Longrightarrow}$ mol $Na_3PO_4 \overset{(2)}{\Longrightarrow}$ g Na_3PO_4

We know $M \ Na_3PO_4 = \dfrac{mol \ Na_3PO_4}{L \ soln}$

(1) ? mol $Na_3PO_4 = M \ Na_3PO_4 \times L$ soln $= 0.40 \ M \times 0.250 \ L = 0.10$ mol Na_3PO_4

(2) ? g $Na_3PO_4 = $ mol $Na_3PO_4 \times FW \ Na_3PO_4 = 0.10$ mol $\times 164 \ g/mol =$ **16 g Na_3PO_4**

Method 2: Dimensional Analysis

? g $Na_3PO_4 = 0.250 \ L$ soln $\times \dfrac{0.40 \ mol \ Na_3PO_4}{1 \ L} \times \dfrac{164 \ g \ Na_3PO_4}{1 \ mol \ Na_3PO_4} =$ **16 g Na_3PO_4**

(b) same as (a)

3-64. *Refer to Section 3-6 and Example 3-19.*

(a) % by mass $= \dfrac{g \ CaCl_2}{g \ soln} \times 100 = \dfrac{16.0 \ g \ CaCl_2}{64.0 \ g \ H_2O + 16.0 \ g \ CaCl_2} \times 100 =$ **20.0% $CaCl_2$**

(b) Assume we have 1 liter of solution

Plan: 1 L soln $\overset{(1)}{\Longrightarrow}$ g soln in 1 L $\overset{(2)}{\Longrightarrow}$ g $CaCl_2$ in 1 L $\overset{(3)}{\Longrightarrow}$ mol $CaCl_2$ in 1 L $= M \ CaCl_2$

(1) ? g soln in 1 L $= 1000 \ mL \times \dfrac{1.180 \ g}{mL} = 1180$ g soln

(2) ? g $CaCl_2$ in 1 L $= 1180$ g soln $\times 20.0\% \ CaCl_2 = 236$ g $CaCl_2$

(3) ? mol $CaCl_2$ in 1 L $= \dfrac{236 \ g \ CaCl_2 \ in \ 1 \ L \ soln}{111 \ g/mol} = 2.13$ mol $CaCl_2$ in 1 L $=$ **2.13 $M \ CaCl_2$**

3-66. *Refer to Section 3-6, Example 3-19, and Exercise 3-64 Solution.*

$$? \, M \text{ HF} = \frac{1000 \text{ mL soln}}{1 \text{ L}} \times \frac{1.17 \text{ g soln}}{1 \text{ mL soln}} \times \frac{49.0 \text{ g HF}}{100 \text{ g soln}} \times \frac{1 \text{ mol HF}}{20.0 \text{ g HF}} = \textbf{28.7 } \textbf{\textit{M}} \textbf{ HF}$$

3-68. *Refer to Section 3-6, Example 3-17, and Exercise 3-56 Solution.*

Method 1: Plan: $\text{g BaCl}_2 \cdot 2\text{H}_2\text{O} \overset{(1)}{\Longrightarrow} \text{mol BaCl}_2 \cdot 2\text{H}_2\text{O} \overset{(2)}{=} \text{mol BaCl}_2 \overset{(3)}{\Longrightarrow} M \text{ BaCl}_2$

(1) $? \text{ mol BaCl}_2 \cdot 2\text{H}_2\text{O} = \dfrac{\text{g BaCl}_2 \cdot 2\text{H}_2\text{O}}{\text{FW BaCl}_2 \cdot 2\text{H}_2\text{O}} = \dfrac{5.50 \text{ g}}{244 \text{ g/mol}} = 0.0225 \text{ mol BaCl}_2 \cdot 2\text{H}_2\text{O}$

(2) $? \text{ mol BaCl}_2 = \text{mol BaCl}_2 \cdot 2\text{H}_2\text{O} = 0.0225 \text{ mol BaCl}_2$

(3) $? \, M \text{ BaCl}_2 = \dfrac{\text{mol BaCl}_2}{\text{L soln}} = \dfrac{0.0225 \text{ mol}}{0.600 \text{ L}} = \textbf{0.0376 } \textbf{\textit{M}} \textbf{ BaCl}_2$

Method 2: Dimensional Analysis

$$? \, M \text{ BaCl}_2 = \frac{5.50 \text{ g BaCl}_2 \cdot 2\text{H}_2\text{O}}{600 \text{ mL soln}} \times \frac{1000 \text{ mL}}{1 \text{ L}} \times \frac{1 \text{ mol BaCl}_2 \cdot 2\text{H}_2\text{O}}{244 \text{ g BaCl}_2 \cdot 2\text{H}_2\text{O}} \times \frac{1 \text{ mol BaCl}_2}{1 \text{ mol BaCl}_2 \cdot 2\text{H}_2\text{O}} = \textbf{0.0376 } \textbf{\textit{M}} \textbf{ BaCl}_2$$

3-70. *Refer to Section 3-7 and Example 3-20.*

For a dilution problem, $M_1 \times V_1 = M_2 \times V_2$

Therefore, $V_1 = \dfrac{M_2 \times V_2}{M_1} = \dfrac{1.80 \, M \times 4.50 \text{ L}}{12.0 \, M} = \textbf{0.675 L conc. HCl soln}$

3-72. *Refer to Section 3-7 and Example 3-20.*

We know that $M = \dfrac{\text{mol Ba(OH)}_2}{\text{L soln}}$ and $\text{mol Ba(OH)}_2 = M \times \text{L soln}$

Therefore, moles Ba(OH)_2 in Soln 1 = moles Ba(OH)_2 in Soln 2

$$M_1 \times V_1 = M_2 \times V_2$$

$$V_1 = \frac{M_2 \times V_2}{M_1} = \frac{0.0900 \, M \times 180 \text{ mL}}{0.0600 \, M} = \textbf{270 mL of 0.0600 } \textbf{\textit{M}} \textbf{ Ba(OH)}_2 \textbf{ soln}$$

3-74. *Refer to Section 3-7.*

Plan: Find the total number of moles of H_2SO_4 and divide by the total volume in liters to calculate molarity. Let V = volume of solution in liters.

$? \text{ mol H}_2\text{SO}_4 \text{ in Solution 1} = M_1 \times V_1 = 6.00 \, M \times 0.145 \text{ L} = 0.870 \text{ mol H}_2\text{SO}_4$

$? \text{ mol H}_2\text{SO}_4 \text{ in Solution 2} = M_2 \times V_2 = 3.00 \, M \times 0.245 \text{ L} = 0.735 \text{ mol H}_2\text{SO}_4$

$? \text{ total L H}_2\text{SO}_4 \text{ soln} = (145 \text{ mL} + 245 \text{ mL}) \times \dfrac{1 \text{ L}}{1000 \text{ mL}} = 0.390 \text{ L}$

$? \, M \text{ H}_2\text{SO}_4 = \dfrac{\text{mol H}_2\text{SO}_4}{\text{L H}_2\text{SO}_4 \text{ soln}} = \dfrac{(0.870 + 0.735) \text{ mol}}{0.390 \text{ L}} = \textbf{4.12 } \textbf{\textit{M}} \textbf{ H}_2\text{SO}_4$

3-76. *Refer to Section 3-8, and Examples 3-22 and 3-23.*

Balanced equation: $KOH + CH_3COOH \rightarrow KCH_3COO + H_2O$

Plan: g $CH_3COOH \overset{(1)}{\Longrightarrow}$ mol $CH_3COOH \overset{(2)}{\Longrightarrow}$ mol $KOH \overset{(3)}{\Longrightarrow}$ L KOH soln

(1) ? mol $CH_3COOH = \dfrac{g\ CH_3COOH}{FW\ CH_3COOH} = \dfrac{0.155\ g}{60.0\ g/mol} = 2.58 \times 10^{-3}$ mol CH_3COOH

(2) ? mol $KOH = 2.58 \times 10^{-3}$ mol $CH_3COOH \times 1/1 = 2.58 \times 10^{-3}$ mol KOH

(3) ? L KOH soln $= \dfrac{mol\ KOH}{M\ KOH} = \dfrac{2.58 \times 10^{-3}\ mol\ KOH}{0.225\ M} = \mathbf{0.0115\ L\ KOH\ soln}$

3-78. *Refer to Section 3-8, and Examples 3-22 and 3-23.*

Balanced equation: $Ba(OH)_2 + 2HNO_3 \rightarrow Ba(NO_3)_2 + 2H_2O$

Plan: M, L $Ba(OH)_2$ soln $\overset{(1)}{\Longrightarrow}$ mol $Ba(OH)_2 \overset{(2)}{\Longrightarrow}$ mol $HNO_3 \overset{(3)}{\Longrightarrow}$ L HNO_3 soln

(1) ? mol $Ba(OH)_2 = 0.0515\ M \times 0.0320\ L = 1.65 \times 10^{-3}$ mol $Ba(OH)_2$

(2) ? mol $HNO_3 = 1.65 \times 10^{-3}$ mol $Ba(OH)_2 \times 2/1 = 3.30 \times 10^{-3}$ mol HNO_3

(3) ? L $HNO_3 = \dfrac{mol\ HNO_3}{M\ HNO_3} = \dfrac{3.30 \times 10^{-3}\ mol}{0.246\ M} = \mathbf{0.0134\ L\ HNO_3\ soln}$

3-80. *Refer to Section 3-8.*

Balanced equation: $AlCl_3 + 3AgNO_3 \rightarrow 3AgCl + Al(NO_3)_3$

Plan: g $AgCl \overset{(1)}{\Longrightarrow}$ mol $AgCl \overset{(2)}{\Longrightarrow}$ mol $AlCl_3 \overset{(3)}{\Longrightarrow}$ M $AlCl_3$

Recall: $M = \dfrac{mol\ substance}{L\ soln}$

Method 1: (1) ? mol $AgCl = \dfrac{g\ AgCl}{FW\ AgCl} = \dfrac{0.325\ g}{143.3\ g/mol} = 2.27 \times 10^{-3}$ mol $AgCl$

(2) ? mol $AlCl_3 =$ mol $AgCl \times 1/3 = 7.56 \times 10^{-4}$ mol $AlCl_3$

(3) ? M $AlCl_3 = \dfrac{mol\ AlCl_3}{L\ soln} = \dfrac{7.56 \times 10^{-4}\ mol\ AlCl_3}{0.1000\ L\ soln} = \mathbf{0.00756\ \textit{M}\ AlCl_3\ soln}$

Method 2: ? mol $AlCl_3 = 0.325$ g $AgCl \times \dfrac{1\ mol\ AgCl}{143.3\ g\ AgCl} \times \dfrac{1\ mol\ AlCl_3}{3\ mol\ AgCl} = 7.56 \times 10^{-4}$ mol $AlCl_3$

? M $AlCl_3 = \dfrac{mol\ AlCl_3}{L\ soln} = \dfrac{7.56 \times 10^{-4}\ mol\ AlCl_3}{0.1000\ L\ soln} = \mathbf{0.00756\ \textit{M}\ AlCl_3\ soln}$

3-82. *Refer to Section 3-8.*

Balanced equation: $ZnCl_2 + 2AgNO_3 \rightarrow Zn(NO_3)_2 + 2AgCl$

Plan: $M, V\ AgNO_3 \overset{(1)}{\Longrightarrow} mol\ AgNO_3 \overset{(2)}{\Longrightarrow} mol\ AgCl \overset{(3)}{\Longrightarrow} g\ AgCl$

Method 1: (1) ? mol $AgNO_3 = M\ AgNO_3 \times L\ soln = 0.425\ M \times 0.0450\ L = 0.0191\ mol\ AgNO_3$

(2) ? mol $AgCl = mol\ AgNO_3 = 0.0191\ mol\ AgCl$

(3) ? g $AgCl = mol\ AgCl \times FW\ AgCl = 0.0191\ mol \times 143.4\ g/mol = $ **2.74 g AgCl**

Method 2: ? g $AgCl = 0.0450\ L\ AgNO_3\ soln \times \dfrac{0.425\ mol\ AgNO_3}{1\ L\ AgNO_3\ soln} \times \dfrac{2\ mol\ AgCl}{2\ mol\ AgNO_3} \times \dfrac{143.4\ g\ AgCl}{1\ mol\ AgCl}$

$= $ **2.74 g AgCl**

3-84. *Refer to Section 3-9.*

The *equivalence point* in an acid-base titration is the point at which stoichiometrically equivalent amounts of acid and base have reacted, whereas the *end point* is the point at which the indicator changes color and the titration is stopped. The specific indicator is chosen so that its color change occurs at or near the equivalence point.

3-86. *Refer to Section 3-8, and Examples 3-22 and 3-23.*

Balanced equation: $KOH + HClO_4 \rightarrow KClO_4 + H_2O$

Plan: $M, L\ KOH\ soln \overset{(1)}{\Longrightarrow} mol\ KOH \overset{(2)}{\Longrightarrow} mol\ HClO_4 \overset{(3)}{\Longrightarrow} L\ HClO_4\ soln$

(1) ? mol $KOH = 0.505\ M\ KOH \times 0.0250\ L = 0.0126\ mol\ KOH$

(2) ? mol $HClO_4 = 0.0126\ mol\ KOH \times 1 = 0.0126\ mol\ HClO_4$

(3) ? L $HClO_4 = \dfrac{mol\ HClO_4}{M\ HClO_4} = \dfrac{0.0126\ mol}{0.0496\ M} = $ **0.255 L HClO₄ soln**

3-88. *Refer to Section 3-8, and Examples 3-24 and 3-25.*

Balanced equation: $HCl + NaOH \rightarrow NaCl + H_2O$

Plan: $M, L\ HCl\ soln \overset{(1)}{\Longrightarrow} mol\ HCl \overset{(2)}{\Longrightarrow} mol\ NaOH \overset{(3)}{\Longrightarrow} M\ NaOH\ soln$

(1) ? mol $HCl = 0.0513\ M \times 0.0250\ L = 1.28 \times 10^{-3}\ mol\ HCl$

(2) ? mol $NaOH = 1.28 \times 10^{-3}\ mol\ HCl \times 1 = 1.28 \times 10^{-3}\ mol\ NaOH$

(3) ? M $NaOH = \dfrac{mol\ NaOH}{L\ NaOH} = \dfrac{1.28 \times 10^{-3}\ mol}{0.0362\ L} = $ **0.0354 M NaOH soln**

3-90. *Refer to Section 3-8.*

Balanced equation: $2HCl + CaCO_3 \rightarrow CaCl_2 + CO_2 + H_2O$

$$\text{Plan: } M, \text{ L HCl} \overset{(1)}{\Longrightarrow} \text{mol HCl} \overset{(2)}{\Longrightarrow} \text{mol } CaCO_3 \overset{(3)}{\Longrightarrow} \text{g } CaCO_3$$

(1) ? mol HCl = 0.0887 M HCl $\times$ 0.0267 L = 2.37×10^{-3} mol HCl

(2) ? mol $CaCO_3$ = 1/2 $\times$ mol HCl = 1.18×10^{-3} mol $CaCO_3$

(3) ? g $CaCO_3$ = 1.18×10^{-3} mol $CaCO_3$ $\times$ 100. g/mol = **0.118 g $CaCO_3$**

3-92. *Refer to Section 3-2.*

Balanced equation: $CS_2 + 3O_2 \rightarrow CO_2 + 2SO_2$

(1) The Law of Conservation of Mass states that the mass of reactants that react equals the mass of products formed. If we determine the mass of O_2, then

$$\text{mass } CS_2 + \text{mass } O_2 = \text{mass of products}$$

$$? \text{ g } O_2 = 33.8 \text{ g } CS_2 \times \frac{1 \text{ mol } CS_2}{76.1 \text{ g } CS_2} \times \frac{3 \text{ mol } O_2}{1 \text{ mol } CS_2} \times \frac{32.0 \text{ g } O_2}{1 \text{ mol } O_2} = 42.6 \text{ g } O_2$$

Therefore, mass of products = g CS_2 + g O_2 = 33.8 g + 42.6 g = **76.4 g products**

(2) Use a mass ratio.

$$\frac{? \text{ g } CS_2}{\text{g products}} = \frac{\text{g } CS_2 \text{ (from Step 1)}}{\text{g products (from Step 1)}}$$

Solving,

$$? \text{ g } CS_2 = \text{g products} \times \frac{\text{g } CS_2 \text{ (from Step 1)}}{\text{g products (from Step 1)}} = 54.2 \text{ g} \times \frac{33.8 \text{ g}}{76.4 \text{ g}} = \textbf{24.0 g } CS_2$$

3-94. *Refer to Sections 3-2 and 3-4.*

Balanced equation: $Fe_3O_4 + 2C \rightarrow 3Fe + 2CO_2$

Plan: g Fe $\Longrightarrow$ mol Fe $\Longrightarrow$ mol Fe_3O_4 $\Longrightarrow$ g Fe_3O_4 $\Longrightarrow$ % Fe_3O_4 in ore

$$? \text{ g } Fe_3O_4 = 2.09 \text{ g Fe in ore} \times \frac{1 \text{ mol Fe}}{55.85 \text{ g Fe}} \times \frac{1 \text{ mol } Fe_3O_4}{3 \text{ mol Fe}} \times \frac{231.55 \text{ g } Fe_3O_4}{1 \text{ mol } Fe_3O_4} = 2.89 \text{ g } Fe_3O_4$$

$$? \text{ \% } Fe_3O_4 \text{ in ore} = \frac{2.89 \text{ g } Fe_3O_4}{50.0 \text{ g ore}} \times 100 = \textbf{5.78\% } Fe_3O_4 \textbf{ in ore}$$

3-96. *Refer to Sections 3-3, 3-4 and 1-11, and Tables 1-1 and 1-8.*

Balanced equation:

$$CH_3COOH + CH_3CH_2OH \rightarrow CH_3COOCH_2CH_3 + H_2O$$

	acetic acid	ethanol	ethyl acetate
FW (g/mol)	60.0	46.0	88.0
Density (g/mL)	1.05	0.789	0.902
	(from Table 1-1)	(from Table 1-8)	

(a) Plan: (1) Convert the volume of reactants to the mass of reactants.

           (2) Evaluate the limiting reactant.

(1) ? g CH_3COOH = 20.2 mL CH_3COOH × 1.05 g/mL = 21.2 g CH_3COOH

   ? g CH_3CH_2OH = 20.1 mL CH_3CH_2OH × 0.789 g/mL = 15.9 g CH_3CH_2OH

(2) Convert the mass of reactants to moles and compare the required ratio to the available ratio.

$$? \text{ mol } CH_3COOH = \frac{g\ CH_3COOH}{FW\ CH_3COOH} = \frac{21.2\ g}{60.0\ g/mol} = 0.353 \text{ mol } CH_3COOH$$

$$? \text{ mol } CH_3CH_2OH = \frac{g\ CH_3CH_2OH}{FW\ CH_3CH_2OH} = \frac{15.9\ g}{46.0\ g/mol} = 0.346 \text{ mol } CH_3CH_2OH$$

$$\text{Required ratio} = \frac{1\ mol\ CH_3CH_2OH}{1\ mol\ CH_3COOH} = 1 \qquad \text{Available ratio} = \frac{0.346\ mol\ CH_3CH_2OH}{0.353\ mol\ CH_3COOH} = 0.980$$

Available ratio < required ratio; CH_3CH_2OH is the limiting reactant.

(b) Plan: (1) Calculate the actual yield of ethyl acetate, $CH_3COOCH_2CH_3$, in grams.

           (2) Calculate the theoretical yield of ethyl acetate from 0.346 mol CH_3CH_2OH.

           (3) Calculate the percent yield of ethyl acetate.

(1) ? g ethyl acetate = 27.5 mL ethyl acetate × 0.902 g/mL = 24.8 g ethyl acetate (actual yield)

(2) ? mol ethyl acetate = mol CH_3CH_2OH × 1 = 0.346 mol ethyl acetate

   ? g ethyl acetate = mol ethyl acetate × FW ethyl acetate = 0.346 mol × 88.0 g/mol

               = 30.4 g ethyl acetate (theoretical yield)

$$(3)\ ? \text{ \% yield ethyl acetate} = \frac{\text{actual yield}}{\text{theoretical yield}} \times 100 = \frac{24.8\ g}{30.4\ g} \times 100 = \mathbf{81.6\%}$$

3-98. *Refer to Section 3-2.*

Balanced equations: $Zn + 2HCl \rightarrow ZnCl_2 + H_2$

                        $2Al + 6HCl \rightarrow 2AlCl_3 + 3H_2$

Assume: (1) 1 mol of H_2 is produced.

           (2) Zn costs \$1.00/g Zn; Al costs \$2.00/g Al

Plan: mol H_2 $\Longrightarrow$ mol metal $\Longrightarrow$ g metal $\Longrightarrow$ \$ required/mol H_2

$$(1)\ \text{For Zn:} \quad 1 \text{ mol } H_2 \times \frac{1\ mol\ Zn}{1\ mol\ H_2} \times \frac{65.4\ g\ Zn}{1\ mol\ Zn} \times \frac{\$1.00}{1\ g\ Zn} = \mathbf{\$65.4/mol\ H_2}$$

$$(2)\ \text{For Al:} \quad 1 \text{ mol } H_2 \times \frac{2\ mol\ Al}{3\ mol\ H_2} \times \frac{27.0\ g\ Al}{1\ mol\ Al} \times \frac{\$2.00}{1\ g\ Al} = \mathbf{\$36.0/mol\ H_2}$$

Therefore, Al is less expensive for the production of equal amounts of hydrogen gas.

3-100. *Refer to Section 3-5 and Example 3-12.*

(a) Balanced equations: $P_4 + 5O_2 \rightarrow P_4O_{10}$ 89.5% efficient

  $P_4O_{10} + 6H_2O \rightarrow 4H_3PO_4$ 97.8% efficient

(b) Plan: $g\ P_4 \Longrightarrow mol\ P_4 \Longrightarrow mol\ P_4O_{10}$ (theoretical) $\Longrightarrow mol\ P_4O_{10}$ (actual) $\Longrightarrow mol\ H_3PO_4$ (theoretical)

  $\Longrightarrow mol\ H_3PO_4$ (actual) $\Longrightarrow g\ H_3PO_4$

$$? \text{ g } H_3PO_4 = 272 \text{ g } P_4 \times \frac{1 \text{ mol } P_4}{124 \text{ g } P_4} \times \frac{1 \text{ mol } P_4O_{10} \text{ (theoretical)}}{1 \text{ mol } P_4} \times \frac{89.5 \text{ mol } P_4O_{10} \text{ (actual)}}{100. \text{ mol } P_4O_{10} \text{ (theoretical)}}$$

$$\times \frac{4 \text{ mol } H_3PO_4 \text{ (theoretical)}}{1 \text{ mol } P_4O_{10}} \times \frac{97.8 \text{ mol } H_3PO_4 \text{ (actual)}}{100. \text{ mol } H_3PO_4 \text{ (theoretical)}} \times \frac{98.0 \text{ g } H_3PO_4 \text{ (actual)}}{1 \text{ mol } H_3PO_4}$$

$$= \textbf{753 g } \mathbf{H_3PO_4}$$

4 Some Types of Chemical Reactions

4-2. *Refer to Section 4-1 and Figure 4-1.*

Mendeleev arranged the known elements in order of increasing atomic weight in sequence so that elements with similar chemical and physical properties fell in the same column or group. To achieve this chemical periodicity, it was necessary for Mendeleev to leave blank spaces for elements undiscovered at that time and to make assumptions concerning atomic weights not known with certainty. The modern periodic table has elements arranged in order of increasing atomic number so that elements with similar chemical properties fall in the same column.

4-4. *Refer to Section 4-1.*

The atomic weight of an element is a weighted average of the mass of the naturally occurring isotopes of that element. Therefore, the atoms in a naturally occurring sample of argon must be heavier than the atoms in a naturally occurring sample of potassium. Atoms of argon have 18 protons, whereas atoms of potassium have 19 protons. In order for the atomic weight of argon to be greater than that of potassium, argon atoms must have more neutrons.

Consider the isotopes for these elements:

	Isotope	Percent Composition	Number of Protons	Number of Neutrons
argon (AW: 39.948 amu)	$^{40}_{18}Ar$	99.60%	18	22
potassium (AW: 39.0983 amu)	$^{39}_{19}K$	93.1%	19	20
	$^{41}_{19}K$	6.88%	19	22

From these data, we can see that argon would have a higher atomic weight than potassium.

4-6. *Refer to Section 4-1 and the Periodic Table.*

The periodic trends of the element properties also apply to compound containing the elements. Therefore, the melting points of CF_4, CCl_4, CBr_4 and CI_4 should follow a trend.

If we graph the melting points of these compounds versus molecular weight, we can estimate the melting point of CBr_4.

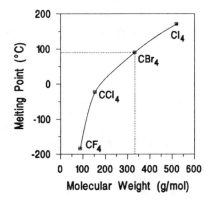

Compound	MW (g/mol)	MP (°C)
CF_4	88.0	-184
CCl_4	153.8	-23
CBr_4	331.6	?
CI_4	519.6	171

The estimated melting point of CBr_4 is about **90°C**. The actual value is 90.1°C.

Hydride formulas are related to the group number of the central element, e.g.,

IIA	IIIA	IVA	VA	VIA	VIIA
BeH_2	BH_3	CH_4	NH_3	H_2O	HF

Arsine, the hydride of arsenic, has the formula AsH_x. Arsenic, the central element, is a VA element. Therefore, its structure should be similar to NH_3 and is predicted to be AsH_3.

4-10.   *Refer to Section 4-1.*

(a) alkaline earth metals: beryllium (Be), magnesium (Mg), calcium (Ca), strontium (Sr), barium (Ba) and radium (Ra)

(b) Group IVA elements: carbon (C), silicon (Si), germanium (Ge), tin (Sn) and lead (Pb)

(c) Group VIB elements: chromium (Cr), molybdenum (Mo) and tungsten (W)

4-12.   *Refer to Section 4-1, the Key Terms for Chapter 4 and Tables 4-2 and 4-3.*

(a) Metals are the elements below and to the left of the stepwise division (metalloids) in the upper right corner of the periodic table. They possess metallic bonding. Approximately 80% of the known elements are metals, including potassium (K), calcium (Ca), scandium (Sc) and vanadium (V).

(b) Nonmetals are the elements above and to the right of the metalloids in the periodic table, including carbon (C), nitrogen (N), sulfur (S) and chlorine (Cl).

(c) The halogens, meaning "salt-formers," are the elements of Group VIIA. They include fluorine (F), chlorine (Cl), bromine (Br), iodine (I) and astatine (At).

4-14.   *Refer to Section 4-2.*

Three major classes of compounds are electrolytes:

	Strong Electrolytes	Weak Electrolytes
(1) acids	HCl, $HClO_4$	CH_3COOH, HF
(2) soluble bases	NaOH, $Ba(OH)_2$	NH_3, $(CH_3)_3N$
(3) soluble salts	NaCl, KNO_3	$Pb(CH_3COO)_2$*

* This is one of the very few soluble salts that is a weak electrolyte.

Therefore, the three classes of compounds which are *strong* electrolytes are strong acids, strong soluble bases and soluble salts.

4-16.   *Refer to Sections 4-2 and 4-5.*

A salt is a compound that contains a cation other than H^+ and an anion other than the hydroxide ion, OH^-, or the oxide ion, O^{2-}. A salt is a product of the reaction between a particular acid and base and consists of the cation of the base and the anion of the acid. For example,

$$NaOH \; + \; HCl \; \rightarrow \; NaCl \; + \; H_2O$$
$$\textbf{base} \qquad \textbf{acid} \qquad \textbf{salt}$$

(a) $HCl(aq) \rightarrow H^+(aq) + Cl^-(aq)$

(b) $HNO_3(aq) \rightarrow H^+(aq) + NO_3^-(aq)$

(c) $HClO_3(aq) \rightarrow H^+(aq) + ClO_3^-(aq)$

4-20.   *Refer to Section 4-2 and Table 4-7.*

Common strong soluble bases include:

lithium hydroxide	LiOH	calcium hydroxide	$Ca(OH)_2$
sodium hydroxide	NaOH	strontium hydroxide	$Sr(OH)_2$
potassium hydroxide	KOH	barium hydroxide	$Ba(OH)_2$
cesium hydroxide	CsOH		

4-22.   *Refer to Section 4-2.*

Household ammonia is the most common weak base. It ionizes as follows:

$$NH_3(aq) + H_2O(\ell) \rightleftarrows NH_4^+(aq) + OH^-(aq)$$

4-24.   *Refer to Table 4-8 and the Solubility Rules in Section 4-2.*

Ionic Substance	Soluble	Insoluble
chloride	NaCl, KCl	$AgCl$, Hg_2Cl_2
sulfate	Na_2SO_4, K_2SO_4	$BaSO_4$, $PbSO_4$
hydroxide	NaOH, KOH	$Cu(OH)_2$, $Mg(OH)_2$

4-26.   *Refer to Section 4-2 and Tables 4-5, 4-6 and 4-7.*

(a) The acids are: HBr, H_2SeO_4, H_3SbO_4, H_2S, H_3BO_3 and HCN.

(b) The bases are: NaOH and NH_3.

4-28.   *Refer to Section 4-2.*

NaCl	strong electrolyte	RbOH	strong electrolyte	NH_3	weak electrolyte
$MgSO_4$	strong electrolyte	HNO_3	strong electrolyte	KOH	strong electrolyte
HCl	strong electrolyte	HI	strong electrolyte	$Mg(CH_3COO)_2$	strong electrolyte
$H_2C_2O_4$	weak electrolyte	$Ba(OH)_2$	strong electrolyte	HCN	weak electrolyte
$Ba(NO_3)_2$	strong electrolyte	LiOH	strong electrolyte	$HClO_4$	strong electrolyte
H_3PO_4	weak electrolyte	C_2H_5COOH	weak electrolyte		

(a) strong acids: HCl, HNO_3, HI, $HClO_4$ (c) weak acids: $H_2C_2O_4$, H_3PO_4, C_2H_5COOH, HCN

(b) strong bases: RbOH, $Ba(OH)_2$, LiOH, KOH   (d) weak bases: NH_3

4-30.   *Refer to Section 4-3 and Table 4-8.*

	Formula Unit Equation	Net Ionic Equation
(a)	$(NH_4)_2SO_4(aq)$	$2NH_4^+(aq) + SO_4^{2-}(aq)$
(b)	$NaBr(aq)$	$Na^+(aq) + Br^-(aq)$
(c)	$Sr(OH)_2(aq)$	$Sr^{2+}(aq) + 2OH^-(aq)$
(d)	$Mg(OH)_2(s)$	$Mg(OH)_2(s)$
(e)	$K_2CO_3(aq)$	$2K^+(aq) + CO_3^{2-}(aq)$

4-32.   *Refer to Section 4-2 and the Solubility Rules summerized in Table 4-8.*

(a) $BaSO_4$ insoluble   (d) Na_2S soluble

(b) $Al(NO_3)_3$ soluble   (e) $Ca(CH_3COO)_2$ soluble

(c) CuS   insoluble

4-34.   *Refer to Section 4-2 and the Solubility Rules summerized in Table 4-8.*

(a) $KClO_3$ soluble   (d) HNO_2 soluble

(b) NH_4Cl soluble   (e) PbS   insoluble

(c) NH_3 soluble

4-36.   *Refer to Sections 4-2, 4-3 and 4-4, and Examples 4-3 and 4-4.*

(a) formula unit: $3CaCl_2(aq) + 2K_3PO_4(aq) \rightarrow Ca_3(PO_4)_2(s) + 6KCl(aq)$

    total ionic: $3Ca^{2+}(aq) + 6Cl^-(aq) + 6K^+(aq) + 2PO_4^{3-}(aq) \rightarrow Ca_3(PO_4)_2(s) + 6K^+(aq) + 6Cl^-(aq)$

    net ionic: $3Ca^{2+}(aq) + 2PO_4^{3-}(aq) \rightarrow Ca_3(PO_4)_2(s)$

(b) formula unit: $Hg(NO_3)_2(aq) + Na_2S(aq) \rightarrow HgS(s) + 2NaNO_3(aq)$

    total ionic: $Hg^{2+}(aq) + 2NO_3^-(aq) + 2Na^+(aq) + S^{2-}(aq) \rightarrow HgS(s) + 2Na^+(aq) + 2NO_3^-(aq)$

    net ionic: $Hg^{2+}(aq) + S^{2-}(aq) \rightarrow HgS(s)$

(c) formula unit: $2CrCl_3(aq) + 3Ca(OH)_2(aq) \rightarrow 2Cr(OH)_3(s) + 3CaCl_2(aq)$

    total ionic: $2Cr^{3+}(aq) + 6Cl^-(aq) + 3Ca^{2+}(aq) + 6OH^-(aq) \rightarrow 2Cr(OH)_3(s) + 3Ca^{2+}(aq) + 6Cl^-(aq)$

    net ionic: $2Cr^{3+}(aq) + 6OH^-(aq) \rightarrow 2Cr(OH)_3(s)$

      therefore, $Cr^{3+}(aq) + 3OH^-(aq) \rightarrow Cr(OH)_3(s)$

4-38. *Refer to Sections 4-3 and 4-4.*

(a) formula unit:
$$Cu(NO_3)_2(aq) + Na_2S(aq) \rightarrow CuS(s) + 2NaNO_3(aq)$$

total ionic:
$$Cu^{2+}(aq) + 2NO_3^-(aq) + 2Na^+(aq) + S^{2-}(aq) \rightarrow CuS(s) + 2Na^+(aq) + 2NO_3^-(aq)$$

net ionic:
$$Cu^{2+}(aq) + S^{2-}(aq) \rightarrow CuS(s)$$

(b) formula unit:
$$CdSO_4(aq) + H_2S(aq) \rightarrow CdS(s) + H_2SO_4(aq)$$

total ionic:
$$Cd^{2+}(aq) + SO_4^{2-}(aq) + H_2S(aq) \rightarrow CdS(s) + 2H^+(aq) + SO_4^{2-}(aq)$$

net ionic:
$$Cd^{2+}(aq) + H_2S(aq) \rightarrow CdS(s) + 2H^+(aq)$$

(c) formula unit:
$$Bi_2(SO_4)_3(aq) + 3(NH_4)_2S(aq) \rightarrow Bi_2S_3(s) + 3(NH_4)_2SO_4(aq)$$

total ionic:
$$2Bi^{3+}(aq) + 3SO_4^{2-}(aq) + 6NH_4^+(aq) + 3S^{2-}(aq) \rightarrow Bi_2S_3(s) + 6NH_4^+(aq) + 3SO_4^{2-}(aq)$$

net ionic:
$$2Bi^{3+}(aq) + 3S^{2-}(aq) \rightarrow Bi_2S_3(s)$$

4-40. *Refer to Sections 4-2, 4-3, 4-4, the Solubility Rules, and Example 4-3.*

Plan: We know that at the time of mixing, the solution contains only ions because the compounds are water soluble. A precipitate will form only if a combination of these ions (in reasonable concentration) produces an insoluble substance.

(a) ions in solution: $NH_4^+(aq)$, $Br^-(aq)$, $Hg_2^{2+}(aq)$ and $NO_3^-(aq)$
new possible combinations of ions: Hg_2Br_2 and NH_4NO_3
Solubility Rule 4 says that Hg_2Br_2 is insoluble, while Solubility Rules 2 and 3 say that NH_4NO_3 is soluble. Therefore, a **precipitate will form** and it is **Hg_2Br_2**.

(b) ions in solution: $K^+(aq)$, $OH^-(aq)$, $Na^+(aq)$ and $S^{2-}(aq)$
new possible combinations of ions: K_2S and $NaOH$
Solubility Rules 1, 6 and 8 say that both K_2S and $NaOH$ are soluble. Therefore a **precipitate will not form**.

(c) ions in solution: $Cs^+(aq)$, $SO_4^{2-}(aq)$, $Mg^{2+}(aq)$ and $Cl^-(aq)$
new possible combinations of ions: $CsCl$ and $MgSO_4$
Solubility Rules 2, 4 and 5 say that $CsCl$ and $MgSO_4$ are soluble. Therefore, a **precipitate will not form**.

4-42. *Refer to Sections 4-2 and 4-5, and Examples 4-5 and 4-6.*

(a) formula unit:
$$CH_3COOH(aq) + NaOH(aq) \rightarrow NaCH_3COO(aq) + H_2O(\ell)$$

total ionic:
$$CH_3COOH(aq) + Na^+(aq) + OH^-(aq) \rightarrow Na^+(aq) + CH_3COO^-(aq) + H_2O(\ell)$$

net ionic:
$$CH_3COOH(aq) + OH^-(aq) \rightarrow CH_3COO^-(aq) + H_2O(\ell)$$

(b) formula unit:
$$H_2SO_3(aq) + 2NaOH(aq) \rightarrow Na_2SO_3(aq) + 2H_2O(\ell)$$

total ionic:
$$H_2SO_3(aq) + 2Na^+(aq) + 2OH^-(aq) \rightarrow 2Na^+(aq) + SO_3^{2-}(aq) + 2H_2O(\ell)$$

net ionic:
$$H_2SO_3(aq) + 2OH^-(aq) \rightarrow SO_3^{2-}(aq) + 2H_2O(\ell)$$

(c) formula unit:
$$HF(aq) + LiOH(aq) \rightarrow LiF(aq) + H_2O(\ell)$$

total ionic:
$$HF(aq) + Li^+(aq) + OH^-(aq) \rightarrow Li^+(aq) + F^-(aq) + H_2O(\ell)$$

net ionic:
$$HF(aq) + OH^-(aq) \rightarrow F^-(aq) + H_2O(\ell)$$

4-44. **Refer to Sections 4-2 and 4-5, and Examples 4-5 and 4-6.**

(a) formula unit: $2NaOH(aq) + H_2SO_4(aq) \rightarrow Na_2SO_4(aq) + 2H_2O(\ell)$

total ionic: $2Na^+(aq) + 2OH^-(aq) + 2H^+(aq) + SO_4^{2-}(aq) \rightarrow 2Na^+(aq) + SO_4^{2-}(aq) + 2H_2O(\ell)$

net ionic: $2OH^-(aq) + 2H^+(aq) \rightarrow 2H_2O(\ell)$

therefore, $OH^-(aq) + H^+(aq) \rightarrow H_2O(\ell)$

(b) formula unit: $3Ca(OH)_2(aq) + 2H_3PO_4(aq) \rightarrow Ca_3(PO_4)_2(s) + 6H_2O(\ell)$

total ionic: $3Ca^{2+}(aq) + 6OH^-(aq) + 2H_3PO_4(aq) \rightarrow Ca_3(PO_4)_2(s) + 6H_2O(\ell)$

net ionic: $3Ca^{2+}(aq) + 6OH^-(aq) + 2H_3PO_4(aq) \rightarrow Ca_3(PO_4)_2(s) + 6H_2O(\ell)$

(c) formula unit: $Cu(OH)_2(s) + 2HNO_3(aq) \rightarrow Cu(NO_3)_2(aq) + 2H_2O(\ell)$

total ionic: $Cu(OH)_2(s) + 2H^+(aq) + 2NO_3^-(aq) \rightarrow Cu^{2+}(aq) + 2NO_3^-(aq) + 2H_2O(\ell)$

net ionic: $Cu(OH)_2(s) + 2H^+(aq) \rightarrow Cu^{2+}(aq) + 2H_2O(\ell)$

4-46. **Refer to Sections 4-2 and 4-5, and Example 4-7.**

(a) formula unit: $2HClO_4(aq) + Ca(OH)_2(aq) \rightarrow Ca(ClO_4)_2(aq) + 2H_2O(\ell)$

total ionic: $2H^+(aq) + 2ClO_4^-(aq) + Ca^{2+}(aq) + 2OH^-(aq) \rightarrow Ca^{2+}(aq) + 2ClO_4^-(aq) + 2H_2O(\ell)$

net ionic: $2H^+(aq) + 2OH^-(aq) \rightarrow 2H_2O(\ell)$

therefore, $H^+(aq) + OH^-(aq) \rightarrow H_2O(\ell)$

(b) formula unit: $H_2SO_4(aq) + 2NH_3(aq) \rightarrow (NH_4)_2SO_4(aq)$

total ionic: $2H^+(aq) + SO_4^{2-}(aq) + 2NH_3(aq) \rightarrow 2NH_4^+(aq) + SO_4^{2-}(aq)$

net ionic: $2H^+(aq) + 2NH_3(aq) \rightarrow 2NH_4^+(aq)$

therefore, $H^+(aq) + NH_3(aq) \rightarrow NH_4^+(aq)$

(c) formula unit: $2CH_3COOH(aq) + Cu(OH)_2(s) \rightarrow Cu(CH_3COO)_2(aq) + 2H_2O(\ell)$

total ionic: $2CH_3COOH(aq) + Cu(OH)_2(s) \rightarrow Cu^{2+}(aq) + 2CH_3COO^-(aq) + 2H_2O(\ell)$

net ionic: $2CH_3COOH(aq) + Cu(OH)_2(s) \rightarrow Cu^{2+}(aq) + 2CH_3COO^-(aq) + 2H_2O(\ell)$

4-48. **Refer to Sections 4-2 and 4-5, and Example 4-7.**

(a) formula unit: $H_2S(aq) + 2NaOH(aq) \rightarrow Na_2S(aq) + 2H_2O(\ell)$

total ionic: $H_2S(aq) + 2Na^+(aq) + 2OH^-(aq) \rightarrow 2Na^+(aq) + S^{2-}(aq) + 2H_2O(\ell)$

net ionic: $H_2S(aq) + 2OH^-(aq) \rightarrow S^{2-}(aq) + 2H_2O(\ell)$

(b) formula unit: $H_3PO_4(aq) + Al(OH)_3(s) \rightarrow AlPO_4(s) + 3H_2O(\ell)$

total ionic: $H_3PO_4(aq) + Al(OH)_3(s) \rightarrow AlPO_4(s) + 3H_2O(\ell)$

net ionic: $H_3PO_4(aq) + Al(OH)_3(s) \rightarrow AlPO_4(s) + 3H_2O(\ell)$

(c) formula unit: $2H_3AsO_4(aq) + 3Pb(OH)_2(s) \rightarrow Pb_3(AsO_4)_2(s) + 6H_2O(\ell)$

total ionic: $2H_3AsO_4(aq) + 3Pb(OH)_2(s) \rightarrow Pb_3(AsO_4)_2(s) + 6H_2O(\ell)$

net ionic: $2H_3AsO_4(aq) + 3Pb(OH)_2(s) \rightarrow Pb_3(AsO_4)_2(s) + 6H_2O(\ell)$

4-50. ***Refer to Section 4-5.***

(a) $NiCO_3$ is the result of H_2CO_3 (acid) reacting with $Ni(OH)_2$ (base).

(b) Ag_2CrO_4 is the result of H_2CrO_4 (acid) reacting with AgOH (base).

(c) $Hg_3(PO_4)_2$ is the result of H_3PO_4 (acid) reacting with $Hg(OH)_2$ (base).

4-52. ***Refer to Section 4-6 and Example 4-8.***

For a compound, the sum of the oxidation numbers of the component elements must be equal to zero.

(a) Let x = oxidation number of P

PCl_3 $0 = x + 3(\text{ox. no. Cl}) = x + 3(-1) = x - 3$
$x = +3$

P_4O_6 $0 = 4x + 6(\text{ox. no. O}) = 4x + 6(-2) = 4x - 12$
$x = +3$

P_4O_{10} $0 = 4x + 10(\text{ox. no. O}) = 4x + 10(-2) = 4x - 20$
$x = +5$

HPO_3 $0 = 1(\text{ox. no. H}) + x + 3(\text{ox. no. O}) = 1(+1) + x + 3(-2) = x - 5$
$x = +5$

H_3PO_3 $0 = 3(\text{ox. no. H}) + x + 3(\text{ox. no. O}) = 3(+1) + x + 3(-2) = x - 3$
$x = +3$

$POCl_3$ $0 = x + 1(\text{ox. no. O}) + 3(\text{ox. no. Cl}) = x + 1(-2) + 3(-1) = x - 5$
$x = +5$

$H_4P_2O_7$ $0 = 4(\text{ox. no. H}) + 2x + 7(\text{ox. no. O}) = 4(+1) + 2x + 7(-2) = 2x - 10$
$x = +5$

$Mg_3(PO_4)_2$ $0 = 3(\text{ox. no. Mg}) + 2x + 8(\text{ox. no. O}) = 3(+2) + 2x + 8(-2) = 2x - 10$
$x = +5$

(b) Let x = oxidation number of Cl

Cl_2 $0 = 2x$
$x = 0$

HCl $0 = 1(\text{ox. no. H}) + x = 1(+1) + x = x + 1$
$x = -1$

HClO $0 = 1(\text{ox. no. H}) + x + 1(\text{ox. no. O}) = 1(+1) + x + 1(-2) = x - 1$
$x = +1$

$HClO_2$ $0 = 1(\text{ox. no. H}) + x + 2(\text{ox. no. O}) = 1(+1) + x + 2(-2) = x - 3$
$x = +3$

$KClO_3$ $0 = 1(\text{ox. no. K}) + x + 3(\text{ox. no. O}) = 1(+1) + x + 3(-2) = x - 5$
$x = +5$

Cl_2O_7 $0 = 2x + 7(\text{ox. no. O}) = 2x + 7(-2) = 2x - 14$
$x = +7$

$Ca(ClO_4)_2$ $0 = 1(\text{ox. no. Ca}) + 2x + 8(\text{ox. no. O}) = 1(+2) + 2x + 8(-2) = 2x - 14$
$x = +7$

(c) Let x = oxidation number of Mn

    MnO $0 = x + 1(\text{ox. no. O}) = x + 1(-2) = x - 2$
                        $x = +2$

    MnO_2 $0 = x + 2(\text{ox. no. O}) = x + 2(-2) = x - 4$
                        $x = +4$

    $Mn(OH)_2$ $0 = x + 2(\text{ox. no. O}) + 2(\text{ox. no. H}) = x + 2(-2) + 2(+1) = x - 2$
                        $x = +2$

    K_2MnO_4 $0 = 2(\text{ox. no. K}) + x + 4(\text{ox. no. O}) = 2(+1) + x + 4(-2) = x - 6$
                        $x = +6$

    $KMnO_4$ $0 = 1(\text{ox. no. K}) + x + 4(\text{ox. no. O}) = 1(+1) + x + 4(-2) = x - 7$
                        $x = +7$

    Mn_2O_7 $0 = 2x + 7(\text{ox. no. O}) = 2x + 7(-2) = 2x - 14$
                        $x = +7$

(d) Let x = oxidation of O

    OF_2 $0 = x + 2(\text{ox. no. F}) = x + 2(-1) = x - 2$
                        $x = +2$

    Na_2O $0 = 2(\text{ox. no. Na}) + x = 2(+1) + x = x + 2$
                        $x = -2$

    Na_2O_2 $0 = 2(\text{ox. no. Na}) + 2x = 2(+1) + 2x = 2x + 2$
                        $x = -1$

    KO_2 $0 = 1(\text{ox. no. K}) + 2x = 1(+1) + 2x = 2x + 1$
                        $x = -1/2$

4-54.   *Refer to Section 4-6 and Example 4-8.*

For an ion, the sum of the oxidation numbers of the component elements must equal the charge on the ion.

(a) Let x = oxidation number of N

    N^{3-} $x = -3$

    NO_2^- $-1 = x + 2(\text{ox. no. O}) = x + 2(-2) = x - 4$
                        $x = +3$

    NO_3^- $-1 = x + 3(\text{ox. no. O}) = x + 3(-2) = x - 6$
                        $x = +5$

    N_3^- $-1 = 3x$
                        $x = -1/3$

    NH_4^+ $+1 = x + 4(\text{ox. no. H}) = x + 4(+1) = x + 4$
                        $x = -3$

(b) Let x = oxidation number of Br

    Br^- $x = -1$

    BrO^- $-1 = x + 1(\text{ox. no. O}) = x + 1(-2) = x - 2$
                        $x = +1$

BrO_2^- $-1 = x + 2(ox.\ no.\ O) = x + 2(-2) = x - 4$
　　　　　　$x = +3$

BrO_3^- $-1 = x + 3(ox.\ no.\ O) = x + 3(-2) = x - 6$
　　　　　　$x = +5$

BrO_4^- $-1 = x + 4(ox.\ no.\ O) = x + 4(-2) = x - 8$
　　　　　　$x = +7$

4-56.　*Refer to Section 4-7.*

Due to the Law of Conservation of Matter, electrons cannot be created or destroyed in chemical reactions.　The electrons that cause the reduction of one substance must be produced from the oxidation of another substance. Therefore, oxidation and reduction always occur simultaneously in ordinary chemical reactions.

4-58.　*Refer to Sections 4-7 and Example 4-9.*

Reaction (b) is the only oxidation-reduction reaction.　In reactions (a), (c) and (d), there are no elements that are changing oxidation number.

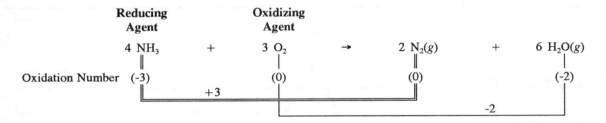

4-60.　*Refer to Section 4-8, Table 4-13 and Example 4-10.*

Zn and Fe are more active metals than Cu and will displace Cu from an aqueous solution of $CuSO_4$.

$Hg(\ell) + CuSO_4(aq) \rightarrow$ no reaction　　　　　$Fe(s) + Cu^{2+}(aq) \rightarrow Fe^{2+}(aq) + Cu(s)$

$Zn(s) + Cu^{2+}(aq) \rightarrow Zn^{2+}(aq) + Cu(s)$　　　$Pt(s) + CuSO_4(aq) \rightarrow$ no reaction

---

4-62.　*Refer to Section 4-8, Table 4-13, Exercise 4-60 and Example 4-11.*

In order of increasing activity:

　　　Pt < Hg < Cu < Fe < Zn

4-64.　*Refer to Section 4-8, Table 4-13, Exercise 4-63 and Example 4-12.*

In order of increasing activity:

　　　Cr < Zn < Na < Ca

Each halogen will displace less electronegative (heavier) halogens from their binary salts. Hence, reactions (b), (c) and (d) will occur and reaction (a) will not occur.

4-68. *Refer to Section 4-8 and Table 4-13.*

(a) $3Cd^{2+}(aq) + 2Al(s) \rightarrow 2Al^{3+}(aq) + 3Cd(s)$ (Note: masses and charges must both be balanced)

(b) $2K(s) + 2H_2O(\ell) \rightarrow 2K^+(aq) + 2OH^-(aq) + H_2(g)$

(c) $Ni(s) + H_2O(\ell) \rightarrow$ no reaction

(d) $Hg(\ell) + HCl(aq) \rightarrow$ no reaction

(e) $Ni(s) + 2H^+(aq) \rightarrow Ni^{2+}(aq) + H_2(g)$

(f) $Fe(s) + 2H^+(aq) \rightarrow Fe^{2+}(aq) + H_2(g)$

4-70. *Refer to Section 4-8 and Table 4-13.*

(a) no   (b) no   (c) no   (d) no

4-72. *Refer to Section 4-5.*

The acid-base reactions are (a) and (k) only, in which an acid reacts with a base to give a salt and water. In all acid-base reactions, no oxidation or reduction is involved.

(a) $H_2SO_4(aq) + 2KOH(aq) \rightarrow K_2SO_4(aq) + 2H_2O(\ell)$

(k) $RbOH(aq) + HNO_3(aq) \rightarrow RbNO_3(aq) + H_2O(\ell)$

4-74. *Refer to Section 4-7, Example 4-9 and Exercise 4-73.*

The oxidation-reduction reactions are:

Net Ionic Equation	Oxidizing Agent	Reducing Agent
(b) $2Rb(s) + Br_2(\ell) \xrightarrow{\Delta} 2RbBr(s)$	$Br_2(\ell)$	$Rb(s)$
(c) $2I^-(aq) + F_2(g) \rightarrow 2F^-(aq) + I_2(s)$	$F_2(g)$	$I^-(aq)$
(e) $S(s) + O_2(g) \xrightarrow{\Delta} SO_2(g)$	$O_2(g)$	$S(s)$
(g) $HgS(s) + O_2(g) \xrightarrow{\Delta} Hg(\ell) + SO_2(g)$	$\underline{Hg}S(s),\ O_2(g)$	$Hg\underline{S}(s)$
(i) $Pb(s) + 2H^+(aq) + 2Br^-(aq) \rightarrow PbBr_2(s) + H_2(g)$	$H^+(aq)$	$Pb(s)$
(j) $2H^+(aq) + 2I^-(aq) + H_2O_2(aq) \rightarrow I_2(s) + 2H_2O(\ell)$	$H_2\underline{O}_2(aq)$	$I^-(aq)$
(m) $H_2O(g) + CO(g) \xrightarrow{\Delta} H_2(g) + CO_2(g)$	$\underline{H}_2O(g)$	$\underline{C}O(g)$
(o) $PbSO_4(s) + PbS(s) \xrightarrow{\Delta} 2Pb(s) + 2SO_2(g)$	$\underline{Pb}\ \underline{S}O_4(s),\ \underline{Pb}\underline{S}(s)$	$Pb\underline{S}(s)$

Some reactions can fit into more than one class of reaction because the classification criteria generally only refer to one aspect of the reaction, while the actual reaction can be viewed from more than one aspect. Although oxidation-reduction reactions are never acid-base or precipitation reactions, they do frequently overlap with other types, such as displacement reactions. A displacement reaction is classified according to the format or appearance of its overall chemical equation, while an oxidation-reduction reaction is characterized by the change in oxidation number of certain elements in the reaction process. As it turns out, displacement by free elements always involve changes in oxidation numbers.

4-78.   *Refer to Section 4-9 and Table 4-14.*

(a) Li^+ lithium ion

(b) Au^{3+} gold(III) ion

(c) Ba^{2+} barium ion

(d) Zn^{2+} zinc ion

(e) Ag^+ silver ion

4-80.   *Refer to Section 4-9 and Table 4-14.*

In simple binary ionic compounds, IA, IIA and IIIA elements exhibit $+1$, $+2$ and $+3$ oxidation states, respectively. The VA, VIA and VIIA elements exhibit -3, -2 and -1 oxidation states, respectively.

(a) chloride ion Cl^-

(b) sulfide ion S^{2-}

(c) telluride ion Te^{2-}

(d) iodide ion I^-

(e) oxide ion O^{2-}

4-82.   *Refer to Section 4-9 and Table 4-14.*

(a) NaI   sodium iodide

(b) Hg_2S mercury(I) sulfide

(c) Li_3N lithium nitride

(d) $MnCl_2$ manganese(II) chloride

(e) $CuCO_3$ copper(II) carbonate

(f) FeO   iron(II) oxide

4-84.   *Refer to Sections 4-9 and 4-10.*

(a) copper(II) chlorate $Cu(ClO_3)_2$

(b) potassium nitrite KNO_2

(c) barium phosphate $Ba_3(PO_4)_2$

(d) copper(I) sulfate Cu_2SO_4

(e) sodium carbonate Na_2CO_3

4-86.   *Refer to Section 4-10.*

H_3PO_3 phosphorous acid

HPO_3^{2-} hydrogen phosphite ion

4-88.   *Refer to Section 4-10.*

(a) AsF_3 arsenic trifluoride

(b) Br_2O dibromine oxide

(c) BrF_5 bromine pentafluoride

(d) CSe_2 carbon diselenide

(e) Cl_2O_7 dichlorine heptoxide

(a) diboron trioxide B_2O_3 (d) sulfur tetrachloride SCl_4 (g) tetraphosphorus hexoxide P_4O_6

(b) dinitrogen pentasulfide N_2S_5 (e) silicon sulfide SiS

(c) phosphorus triiodide PI_3 (f) hydrogen sulfide H_2S

(a) NH_4F ammonium fluoride CuF_2 copper(II) fluoride or cupric fluoride
    KF potassium fluoride FeF_3 iron(III) fluoride or ferric fluoride
    MgF_2 magnesium fluoride AgF silver fluoride

(b) - $Cu(OH)_2$ copper(II) hydroxide or cupric hydroxide
    KOH potassium hydroxide $Fe(OH)_3$ iron(III) hydroxide or ferric hydroxide
    $Mg(OH)_2$ magnesium hydroxide $AgOH$ silver hydroxide

(c) $(NH_4)_2SO_3$ ammonium sulfite $CuSO_3$ copper(II) sulfite or cupric sulfite
    K_2SO_3 potassium sulfite $Fe_2(SO_3)_3$ iron(III) sulfite or ferric sulfite
    $MgSO_3$ magnesium sulfite Ag_2SO_3 silver sulfite

(d) $(NH_4)_3PO_4$ ammonium phosphate $Cu_3(PO_4)_2$ copper(II) phosphate or cupric phosphate
    K_3PO_4 potassium phosphate $FePO_4$ iron(III) phosphate or ferric phosphate
    $Mg_3(PO_4)_2$ magnesium phosphate Ag_3PO_4 silver phosphate

(e) NH_4NO_3 ammonium nitrate $Cu(NO_3)_2$ copper(II) nitrate or cupric nitrate
    KNO_3 potassium nitrate $Fe(NO_3)_3$ iron(III) nitrate or ferric nitrate
    $Mg(NO_3)_2$ magnesium nitrate $AgNO_3$ silver nitrate

(a) $N_2(g) + O_2(g) \overset{\Delta}{\rightarrow} 2NO(g)$

(b) $PbS(s) + PbSO_4(s) \overset{\Delta}{\rightarrow} 2Pb(s) + 2SO_2(g)$

(a) Balanced equation: $2KClO_3(s) \rightarrow 2KCl(s) + 3O_2(g)$

$$? \text{ mol } O_2 = 10.0 \text{ g } KClO_3 \times \frac{1 \text{ mol } KClO_3}{122.6 \text{ g } KClO_3} \times \frac{3 \text{ mol } O_2}{2 \text{ mol } KClO_3} = \textbf{0.122 mol } O_2$$

(b) Balanced equation: $2H_2O_2(aq) \rightarrow 2H_2O(\ell) + O_2(g)$

$$? \text{ mol } O_2 = 10.0 \text{ g } H_2O_2 \times \frac{1 \text{ mol } H_2O_2}{34.02 \text{ g } H_2O_2} \times \frac{1 \text{ mol } O_2}{2 \text{ mol } H_2O_2} = \textbf{0.147 mol } O_2$$

(c) Balanced equation: $2HgO(s) \rightarrow 2 \text{ Hg}(\ell) + O_2(g)$

$$? \text{ mol } O_2 = 10.0 \text{ g } HgO \times \frac{1 \text{ mol } HgO}{216.6 \text{ g } HgO} \times \frac{1 \text{ mol } O_2}{2 \text{ mol } HgO} = \textbf{0.0231 mol } O_2$$

Plan: (1) Write the equation of the displacement reaction. Note: In aqueous solution, the attached water molecules detach themselves.
(2) Balance the equation.
(3) Calculate the mass of Zn needed.

(1) Balanced equation: $Zn + CuSO_4 \rightarrow Cu + ZnSO_4$

(2) The equation is already balanced.

(3) Plan: $\underset{(1)}{g\ Cu \Longrightarrow}\ \underset{(2)}{mol\ Cu \Longrightarrow}\ \underset{(3)}{mol\ Zn \Longrightarrow}\ g\ Zn$

$$? \text{ g Zn} = 14.5 \text{ g Cu} \times \frac{1 \text{ mol Cu}}{63.55 \text{ g Cu}} \times \frac{1 \text{ mol Zn}}{1 \text{ mol Cu}} \times \frac{65.39 \text{ g Zn}}{1 \text{ mol Zn}} = \textbf{14.9 g Zn}$$

5 The Structure of Atoms

5-2.   *Refer to Section 5-2 and Figure 5-2.*

If any oil droplets in Millikan's oil drop experiment had possessed a deficiency of electrons, the droplets would have been positively charged and would have been attracted to, not repelled by, the negatively charged plate. There would have been no voltage setting possible where the electrical and gravitational forces on the drop would have balanced.

5-4.   *Refer to Section 5-3 and Figure 5-3.*

(a) Canal rays, also produced in the cathode ray tube, move toward the cathode (the negative electrode). Therefore, they must be positively charged. Canal rays are positively charged ions created when cathode rays knock electrons from the gaseous atoms in the tube.

(b) Cathode rays are electrons and are independent of source. Canal rays are the positive ions from the specific gas used after a loss of electrons; they are therefore dependent upon the gas used.

5-6.   *Refer to Sections 5-2 and 5-3.*

(a) We must modify the Millikan oil drop experiment in order to determine the charge-to-mass ratio of the positively charged whizatron by

    (1) in some way producing an excess of whizatrons on the oil droplets, and
    (2) switching the leads to the plates to make the bottom plate positively charged.

The positively charged whizatrons on the oil droplets will be repulsed by the plate.

(b) Since all of the charges on the droplets will be integral multiples of the charge on the whizatron, we will identify the droplet with the smallest charge and test to see if the other droplets have charges that are multiples of its charge.

From the results shown in the table below, we can deduce that the charge on the whizatron is 1/2 of the smallest observed charge:

$$1/2 \times (3.26 \times 10^{-19}) = 1.63 \times 10^{-19} \text{ coulombs.}$$

All the droplets have charges that are integral multiples of **1.63×10^{-19} coulombs**.

Charge on Droplets (coulombs)	Ratio
6.52×10^{-19}	$\dfrac{6.52 \times 10^{-19}}{3.26 \times 10^{-19}} = 2.0$
8.16×10^{-19}	$\dfrac{8.16 \times 10^{-19}}{3.26 \times 10^{-19}} = 2.5$
3.26×10^{-19}	$\dfrac{3.26 \times 10^{-19}}{3.26 \times 10^{-19}} = 1.0$
11.40×10^{-19}	$\dfrac{11.40 \times 10^{-19}}{3.26 \times 10^{-19}} = 3.5$
9.78×10^{-19}	$\dfrac{9.78 \times 10^{-19}}{3.26 \times 10^{-19}} = 3.0$

volume of a hydrogen atom $= (4/3)\pi r^3 = (4/3)\pi(5.29 \times 10^{-11} \text{ m})^3 = 6.20 \times 10^{-31} \text{ m}^3$ $(1 \text{ nm} = 1 \times 10^{-9} \text{ m})$

volume of a hydrogen nucleus = volume of a proton $= (4/3)\pi r^3 = (4/3)\pi(1.5 \times 10^{-15} \text{ m})^3 = 1.4 \times 10^{-44} \text{ m}^3$

Therefore, the fraction of space in a hydrogen atom occupied by the nucleus is:

$$\frac{V_{\text{hydrogen nucleus}}}{V_{\text{hydrogen atom}}} = \frac{1.4 \times 10^{-44} \text{ m}^3}{6.20 \times 10^{-31} \text{ m}^3} = \mathbf{2.3 \times 10^{-14}}$$

As can be seen, an atom is mostly empty space.

Calculation of the charge-to-mass ratio:

Species	Charge	Mass Number	Charge-to-Mass Ratio
$^{12}C^+$	+1	12	$1/12 = 0.0833$
$^{12}C^{2+}$	+2	12	$2/12 = 0.167$
$^{13}C^+$	+1	13	$1/13 = 0.0769$
$^{13}C^{2+}$	+2	13	$2/13 = 0.154$

The order of increasing charge-to-mass ratios is $^{13}C^+ < {}^{12}C^+ < {}^{13}C^{2+} < {}^{12}C^{2+}$.

A neutral atom of $^{197}_{79}$Au contains 79 electrons, 79 protons and $(197-79) = 118$ neutrons.

If we assume that the mass of the atom is simply the sum of the masses of its subatomic particles, then

$$\begin{aligned}
\text{mass of } ^{197}\text{Au} &= (79 \; e^- \times \text{mass } e^-) + (79 \; p \times \text{mass } p) + (118 \; n \times \text{mass } n) \\
&= (79 \times 0.00054858 \text{ amu}) + (79 \times 1.0073 \text{ amu}) + (118 \times 1.0087 \text{ amu}) \\
&= 198.65 \text{ amu/atom}
\end{aligned}$$

(a) % by mass $e^- = \dfrac{\text{mass } e^-}{\text{mass Au}} \times 100 = \dfrac{79 \; e^- \times 0.00054858 \text{ amu}/e^-}{198.65 \text{ amu}} \times 100 = \mathbf{0.021816\%}$

(b) % by mass $p = \dfrac{\text{mass } p}{\text{mass Au}} \times 100 = \dfrac{79 \; p \times 1.0073 \text{ amu}/p}{198.65 \text{ amu}} \times 100 = \mathbf{40.059\%}$

(c) % by mass $n = \dfrac{\text{mass } n}{\text{mass Au}} \times 100 = \dfrac{118 \; n \times 1.0087 \text{ amu}/n}{198.65 \text{ amu}} \times 100 = \mathbf{59.918\%}$

(a) The atomic number of an element is the integral number of protons in the nucleus. It defines the identity of that element. For example, oxygen has an atomic number of 8 and therefore has 8 protons. All oxygen atoms have exactly 8 protons and there is no other element that has 8 protons in its nucleus (**Section 5-5**).

(b) Isotopes are two or more forms of atoms of the same element with different masses. In other words, they are atoms containing the same number of protons but different numbers of neutrons. ^{16}O and ^{17}O are isotopes; both have 8 protons but ^{16}O has $(16 - 8) = 8$ neutrons while ^{17}O has $(17 - 8) = 9$ neutrons (**Section 5-7**).

(c) The mass number of an element is the integral sum of the numbers of protons and neutrons in that atom. The mass number of ^{17}O is 17, the sum of protons and neutrons in the nucleus (**Section 5-7**).

(d) Nuclear charge refers to the number of protons or positive charges in the nucleus. The nuclear charge of an oxygen atom is $+8$.

5-16. *Refer to Section 5-7, Table 5-2 and the Periodic Table on the inside front cover of the textbook.*

From the Periodic Table, we see that the atomic number of strontium is 38. Therefore, each strontium atom has 38 protons. If it is a neutral atom then it also has 38 electrons. If we assume that these isotopes are neutral, then

Isotope	Number of Protons	Number of Electrons	Number of Neutrons (Mass Number - Atomic Number)
$^{84}_{38}Sr$	38	38	46 (= 84 - 38)
$^{86}_{38}Sr$	38	38	48 (= 86 - 38)
$^{87}_{38}Sr$	38	38	49 (= 87 - 38)
$^{88}_{38}Sr$	38	38	50 (= 88 - 38)

5-18. *Refer to Section 5-7.*

Remember: atomic number = number of protons = number of electrons in a neutral atom
mass number = number of protons + number of neutrons

Name of Element	Atomic Number	Mass Number	Isotope	Number of Protons	Number of Electrons	Number of Neutrons
cobalt	**27**	**59**	$^{59}_{27}Co$	**27**	**27**	32
iridium	**77**	**193**	$^{193}_{77}Ir$	**77**	**77**	**116**
manganese	**25**	**55**	$^{55}_{25}Mn$	**25**	25	30
platinum	**78**	182	$^{182}_{78}Pt$	**78**	78	**104**

5-20. *Refer to Section 5-7.*

	Symbol of Species	Number of Protons	Number of Neutrons	Number of Electrons
(a)	$^{24}_{12}Mg$	12	12	12
(b)	$^{45}_{21}Sc$	21	24	21
(c)	$^{91}_{40}Zr$	40	51	40
(d)	$^{27}_{13}Al^{3+}$	13	14	10
(e)	$^{65}_{30}Zn^{2+}$	30	35	28
(f)	$^{108}_{47}Ag^{+}$	47	61	46

5-22. *Refer to Section 5-7.*

	Number of Protons	Number of Neutrons	Number of Electrons	Z	A	Charge	Symbol
(a)	25	30	25	25	55	0	$^{55}_{25}Mn$
(b)	20	20	18	20	40	+2	$^{40}_{20}Ca^{2+}$
(c)	33	42	33	33	75	0	$^{75}_{33}As$
(d)	53	74	54	53	127	-1	$^{127}_{53}I^{-}$

5-24. *Refer to Section 5-9 and Example 5-2.*

We know: AW Li = (mass ^{6}Li × fraction of ^{6}Li) + (mass ^{7}Li × fraction of ^{7}Li)

let   x = fraction of ^{6}Li
then (1-x) = fraction of ^{7}Li

Substituting,

$$6.941 \text{ amu} = (6.01512 \text{ amu})x + (7.01600 \text{ amu})(1 - x)$$
$$= 6.01512x + 7.01600 - 7.01600x$$
$$0.075 = 1.00088x$$

fraction of ^{6}Li = x = 0.075   % abundance of ^{6}Li = **7.5%**

5-26. *Refer to Section 5-8.*

The net charge on a species is the result of unequal numbers of protons and electrons.

charge on species = (charge on proton)(no. of protons) + (charge on electron)(no. of electrons)
= (+1)(no. of protons) + (-1)(no. of electrons)
= (no. of protons) - (no. of electrons)

(a) $_{20}$Ca   charge = 20 - 18 = **+2**   (d) $_{78}$Pt   charge = 78 - 74 = **+4**   (g) $_6$C   charge = 6 - 7 = **-1**

(b) $_{29}$Cu   charge = 29 - 28 = **+1**   (e) $_9$F   charge = 9 - 10 = **-1**   (h) $_6$C   charge = 6 - 5 = **+1**

(c) $_{29}$Cu   charge = 29 - 26 = **+3**   (f) $_6$C   charge = 6 - 6 = **0**

5-28. *Refer to Section 5-9 and Example 5-1.*

If the mass spectrum were complete for germanium, the calculated atomic weight would be the weighted average of the isotopes:

$$AW \text{ (amu)} = \sum \frac{\text{(relative abundance)}}{\text{(total relative abundance)}} \times \text{isotope mass (amu)}$$

$$= (5.49/15.90)71.9217 + (1.55/15.90)72.9234 + (7.31/15.90)73.9219 + (1.55/15.90)75.9219$$

$$= \textbf{73.3 amu}$$

However, the true atomic weight of germanium is 72.61 amu. The observed data gives a value that is too high. Therefore, the spectrum given is incomplete and data must have been lost at the low end of the plot when the recorder jammed.

5-30. *Refer to Section 5-9 and Example 5-1.*

Plan: (1) For each isotope, convert the % abundance to a fraction.
      (2) Multiply the fraction of each isotope by its mass and add the terms to obtain the atomic weight of Cr.

? AW Cr = (mass ^{50}Cr × fraction of ^{50}Cr) + (mass ^{52}Cr × fraction of ^{52}Cr) + (mass ^{53}Cr × fraction of ^{53}Cr)
        + (mass ^{54}Cr × fraction of ^{54}Cr)

? AW Cr = (49.9461 amu × $\frac{4.35}{100}$) + (51.9405 amu × $\frac{83.79}{100}$) + (52.9407 amu × $\frac{9.50}{100}$)

      + (53.9389 amu × $\frac{2.36}{100}$)

    = **52.0 amu** (to 3 significant figures)

5-32. *Refer to Section 5-8, Example 5-1 and the* **Handbook of Chemistry and Physics, 75th Edition.**

(a) There are 5 known naturally occurring isotopes of germanium (see below).

(b) 72.61 amu

(c)

Isotope	% Natural Abundance	Atomic Mass (amu)
$^{70}_{32}$Ge	20.52	69.9243
$^{72}_{32}$Ge	27.43	71.9217
$^{73}_{32}$Ge	7.76	72.9234
$^{74}_{32}$Ge	36.54	73.9219
$^{75}_{32}$Ge	7.76	75.9214

(d) ? AW Ge = (mass of ^{70}Ge × fraction of ^{70}Ge) + (mass of ^{72}Ge × fraction of ^{72}Ge)
        + (mass of ^{73}Ge × fraction of ^{73}Ge) + (mass of ^{74}Ge × fraction of ^{74}Ge)
        + (mass of ^{75}Ge × fraction of ^{75}Ge)

    = (69.9243 amu × 0.2052) + (71.9217 amu × 0.2743) + (72.9234 amu × 0.0776)
        + (73.9219 amu × 0.3654) + (75.9214 amu × 0.0776)

    = 14.35 amu + 19.73 amu + 5.66 amu + 27.01 amu + 5.89 amu

    = **72.64 amu**

5-34. *Refer to Section 5-10.*

For electromagnetic radiation: frequency × wavelength = speed of light

$$\nu \ (s^{-1}) \times \lambda \ (m) = c \ (m/s)$$

$$\nu \ (s^{-1}) = \frac{c \ (m/s)}{\lambda \ (m)}$$

(a) $\lambda \ (m) = 9774 \ \text{Å} \times \frac{10^{-10} \ m}{1 \ \text{Å}} = 9.774 \times 10^{-7} \ m$ $\nu \ (s^{-1}) = \frac{3.00 \times 10^8 \ m/s}{9.774 \times 10^{-7} \ m} = \mathbf{3.07 \times 10^{14} \ s^{-1}}$

(b) $\lambda \ (m) = 492 \ nm \times \frac{10^{-9} \ m}{1 \ nm} = 4.92 \times 10^{-7} \ m$ $\nu \ (s^{-1}) = \frac{3.00 \times 10^8 \ m/s}{4.92 \times 10^{-7} \ m} = \mathbf{6.10 \times 10^{14} \ s^{-1}}$

(c) λ (m) = 4.92 cm $\times$ $\dfrac{1 \text{ m}}{100 \text{ cm}}$ = 0.0492 m ν (s^{-1}) = $\dfrac{3.00 \times 10^8 \text{ m/s}}{0.0492 \text{ m}}$ = **6.10 $\times$ 10^9 s^{-1}**

(d) λ (m) = 4.92 $\times$ 10^{-9} cm $\times$ $\dfrac{1 \text{ m}}{100 \text{ cm}}$ = 4.92 $\times$ 10^{-11} m ν (s^{-1}) = $\dfrac{3.00 \times 10^8 \text{ m/s}}{4.92 \times 10^{-11} \text{ m}}$ = **6.10 $\times$ 10^{18} s^{-1}**

5-36.   *Refer to Sections 5-10 and 5-12, Examples 5-4 and 5-5, and Figure 5-12b.*

(a) ν (s^{-1}) = $\dfrac{c \text{ (m/s)}}{\lambda \text{ (m)}}$ = $\dfrac{3.00 \times 10^8 \text{ m/s}}{670.8 \text{ nm} \times 10^{-9} \text{ m/1 nm}}$ = **4.47 $\times$ 10^{14} s^{-1}**

(b) $E = h\nu$ = (6.63 $\times$ 10^{-34} J·s)(4.47 $\times$ 10^{14} s^{-1}) = **2.96 $\times$ 10^{-19} J/photon**

(c) From Figure 5-12b, the color corresponding to λ = 670.8 nm or 6708Å is **red**.

5-38.   *Refer to Sections 5-10 and 5-12, and Example 5-5.*

E (J/photon) = $\dfrac{hc}{\lambda}$ = $\dfrac{(6.63 \times 10^{-34} \text{ J·s})(3.00 \times 10^8 \text{ m/s})}{3400\text{Å} \times (10^{-10} \text{ m/1 Å})}$ = **5.85 $\times$ 10^{-19} J/photon**

E (J/mol) = 5.85 $\times$ 10^{-19} J/photon $\times$ 6.02 $\times$ 10^{23} photons/mol = **3.52 $\times$ 10^5 J/mol or 352 kJ/mol**

5-40.   *Refer to Sections 5-10 and 5-12, and Examples 5-4 and 5-5.*

Plan: (1) Calculate the energy of 1 photon of green light.
    (2) Determine the number of photons of green light required for the eye to "see."

(1) E (J/photon) = $\dfrac{hc}{\lambda}$ = $\dfrac{(6.63 \times 10^{-34} \text{ J·s})(3.00 \times 10^8 \text{ m/s})}{495 \text{ nm} \times (10^{-9} \text{ m/1 nm})}$ = **4.02 $\times$ 10^{-19} J/photon**

(2) E (J) required to "see" = E (J/photon) $\times$ number of photons

Therefore, ? photons = $\dfrac{E \text{ (J)}}{E \text{ (J/photon)}}$ = $\dfrac{1.00 \times 10^{-17} \text{ J}}{4.02 \times 10^{-19} \text{ J/photon}}$ = **25 photons**

5-42.   *Refer to Section 5-12, Example 5-5, and Exercise 5-40 Solution.*

Plan: (1) Calculate the energy, in joules, required to heat the water.
    (2) Calculate the energy, in joules, of one photon of the microwave radiation.
    (3) Determine the number of photons required to heat the water.

(1) energy absorbed by the water (J) = specific heat of water (J/g·°C) $\times$ mass (g) $\times$ ΔT (°C)
    = 4.184 J/g·°C $\times$ 250 g $\times$ (75°C - 25°C)
    = 5.2 $\times$ 10^4 J

(2) E (J/photon) = $\dfrac{hc}{\lambda}$ = $\dfrac{(6.63 \times 10^{-34} \text{ J·s})(3.00 \times 10^8 \text{ m/s})}{3 \text{ mm} \times (10^{-3} \text{ m/1 mm})}$ = 6.63 $\times$ 10^{-23} J

(3) ? photons = $\dfrac{E \text{ (J)}}{E \text{ (J/photon)}}$ = $\dfrac{5.2 \times 10^4 \text{ J}}{6.63 \times 10^{-23} \text{ J/photon}}$ = **7.9 $\times$ 10^{26} photons**

5-44. *Refer to Section 5-11.*

The photoelectric effect is the emission of an electron from a metal surface caused by impinging electromagnetic radiation. This radiation must have a certain minimum energy, i.e., its frequency must be greater than the threshold frequency, which is characteristic of a particular metal, for current to flow. If the frequency is below the threshold frequency, no current flows. As long as this criterion is met, the current increases with increasing intensity (brightness) of the light.

5-46. *Refer to Sections 5-10, 5-11 and 5-12, and Exercise 45.*

We know: $E = h\nu = \dfrac{hc}{\lambda}$ Therefore, $\lambda \text{ (m)} = \dfrac{hc}{E} = \dfrac{(6.63 \times 10^{-34} \text{ J·s})(3.00 \times 10^8 \text{ m/s})}{(3.89 \text{ eV})(1.60 \times 10^{-19} \text{ J/eV})} = 3.20 \times 10^{-7} \text{ m}$

$$\lambda \text{ (nm)} = 3.20 \times 10^{-7} \text{ m} \times \frac{1 \text{ nm}}{10^{-9} \text{ m}} = \textbf{320 nm} \text{ (violet/near ultraviolet)}$$

5-48. *Refer to Section 5-12 and Figure 5.16b.*

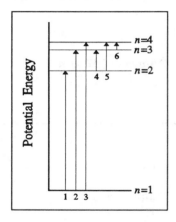

Transitions for Absorption Spectrum

1	$n = 1 \rightarrow n = 2$
2	$n = 1 \rightarrow n = 3$
3	$n = 1 \rightarrow n = 4$
4	$n = 2 \rightarrow n = 3$
5	$n = 2 \rightarrow n = 4$
6	$n = 3 \rightarrow n = 4$

5-50. *Refer to Sections 5-10 and 5-12.*

Using the equations, $E = h\nu$ and $c = \lambda\nu$, we can qualitatively deduce the relationships between the energies of photons, E, their wavelengths, λ, and frequency, ν. Consider the diagram at the right, drawn not entirely to scale.

(a) The photon with the smallest energy is produced by an electron in the $n = 7$ major energy level falling to the $n = 6$ major energy level. This photon therefore also has the smallest frequency since $E \propto \nu$ and the longest wavelength since $\nu \propto 1/\lambda$.

(b) The photon with the highest frequency is produced by the transition, $n = 7 \rightarrow n = 1$, since this transition involves the release of the most energy ($\nu \propto E$)

(c) The photon with the shortest wavelength also has the highest frequency ($\lambda \propto 1/\nu$) and releases the most energy. It is produced by an electron in the $n = 7$ major energy level falling to the $n = 1$ major energy level.

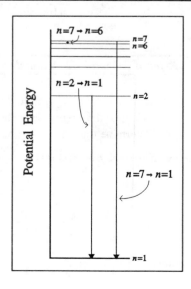

(d) Plan: (1) Use the Rydberg equation, $\frac{1}{\lambda} = R\left[\frac{1}{n_1^2} - \frac{1}{n_2^2}\right]$ where $R = 1.097 \times 10^7$ m^{-1} to evaluate $\frac{1}{\lambda}$.

(2) Solve for λ and ν.

(1) $\frac{1}{\lambda} = 1.097 \times 10^7$ m^{-1} $\left[\frac{1}{1^2} - \frac{1}{6^2}\right] = (1.097 \times 10^7$ m$^{-1})(0.9722) = 1.066 \times 10^7$ m^{-1}

(2) $\lambda = 9.376 \times 10^{-8}$ m

$$\nu = \frac{c}{\lambda} = \frac{3.00 \times 10^8 \text{ m/s}}{9.376 \times 10^{-8} \text{ m}} = \textbf{3.20} \times \textbf{10}^{\textbf{15}} \textbf{ s}^{-1}$$

5-52. *Refer to Section 5-10 and Figures 5-12b and 5-15.*

(a) lithium	$\lambda = 4603$ Å	**blue**
(b) neon	$\lambda = 540.0$ nm or 5400 Å	**yellow**
(c) calcium	$\lambda = 6573$ Å	**red**
(d) cesium	$\nu = 3.45 \times 10^{14}$ s^{-1}	**red**
(e) potassium	$\nu = 3.90 \times 10^{14}$ s^{-1}	**red**

5-54. *Refer to Section 5-12 and Example 5-5.*

The energy loss due to 1 atom emitting a photon is $E = hc/\lambda$.
The energy loss due to 1 mole of atoms each emitting a photon is $E = (hc/\lambda)N$, where N is Avogadro's Number.

Substituting, $E = \dfrac{(6.63 \times 10^{-34} \text{ J·s})(3.00 \times 10^8 \text{ m/s})}{(6.64 \times 10^3 \text{ Å})(1 \times 10^{-10} \text{ m/Å})} \times 6.02 \times 10^{23}$ mol$^{-1} = 1.80 \times 10^5$ J/mol $= \textbf{180 kJ/mol}$

5-56. *Refer to Section 5-12.*

The energy emitted by 1 photon is $E = hc/\lambda$. The energy emitted by n photons is $E = (hc/\lambda)$n.
The energy emitted by this laser in 2 seconds is

$$E = \text{power} \times \text{time} = 515 \text{ milliwatts} \times \frac{1 \text{ watt}}{1000 \text{ milliwatts}} \times \frac{1 \text{ J/s}}{1 \text{ watt}} \times 2.00 \text{ s} = 1.03 \text{ J}$$

Substituting,

$$1.03 \text{ J} = \frac{(6.63 \times 10^{-34} \text{ J·s})(3.00 \times 10^8 \text{ m/s})}{(488.0 \text{ nm})(10^{-9} \text{ m/nm})} \times \text{n}$$

$$\text{n} = \textbf{2.53} \times \textbf{10}^{\textbf{18}} \textbf{ photons}$$

5-58. *Refer to Section 5-13, Example 5-6 and Figure 6-1.*

(a) The de Broglie wavelength is given by λ (m) $= \dfrac{h(\text{J·s})}{m(\text{kg})\nu(\text{m/s})}$ where h is Planck's constant, ν is velocity

The units are as stated because 1 J $= 1$ kg·m^2/s^2.

The mass of a proton is 1.67×10^{-24} g $\times 1$ kg/1000 g $= 1.67 \times 10^{-27}$ kg.
The velocity of the proton, $\nu = 2.50 \times 10^7$ m/s (1/12 of the speed of light)

Substituting,

$$\lambda = \frac{h}{mv} = \frac{6.63 \times 10^{-34} \text{ J·s}}{(1.67 \times 10^{-27} \text{ kg})(2.50 \times 10^7 \text{ m/s})} = \mathbf{1.59 \times 10^{-14} \text{ m}}$$

(b) For the stone,   mass (kg) = 30.0 g × 1 kg/1000 g = 0.0300 kg

$$v \text{ (m/s)} = \frac{2.00 \times 10^3 \text{ m}}{1 \text{ h}} \times \frac{1 \text{ h}}{60 \text{ min}} \times \frac{1 \text{ min}}{60 \text{ s}} = 0.556 \text{ m/s}$$

$$\lambda = \frac{h}{mv} = \frac{6.63 \times 10^{-34} \text{ J·s}}{(0.0300 \text{ kg})(0.556 \text{ m/s})} = \mathbf{3.97 \times 10^{-32} \text{ m}}$$

(c) Radii of atoms range from 0.4 Å (4×10^{-11} m) to 3 Å (3×10^{-10} m). The wavelength of a proton, calculated in (a), is 4 orders of magnitude smaller than the radius of a typical atom. The wavelength of a 30 g stone is much smaller, by 22 orders of magnitude, than a typical atom's radius.

5-60.   _Refer to Section 5-13._

Plan: (1) Calculate the mass (M) of an alpha particle (He nucleus, He^{2+}) in kilograms.

(2) Calculate the velocity (v) from $\lambda = \dfrac{h}{mv}$ (Note: 1 J = 1 kg·m²/s²)

(1) M = 4.002 g/mol He^{2+} × $\dfrac{1 \text{ mol He}^{2+}}{6.02 \times 10^{23} \text{ He}^{2+} \text{ ions}}$ × $\dfrac{1 \text{ kg}}{10^3 \text{ g}}$ = 6.65 × 10⁻²⁷ kg/He²⁺ ion

$$(2) \quad v = \frac{h}{m\lambda} \quad = \frac{6.63 \times 10^{-34} \text{ J·s}}{(6.65 \times 10^{-27} \text{ kg})(0.529 \text{ Å})(1 \times 10^{-10} \text{ m/Å})}$$

$$= \frac{6.63 \times 10^{-34} \text{ kg·m}^2/\text{s}^2 \text{·s}}{(6.65 \times 10^{-27} \text{ kg})(5.29 \times 10^{-11} \text{ m})} = \mathbf{1.88 \times 10^3 \text{ m/s}}$$

5-62.   _Refer to Section 5-15._

The subsidiary (or azimuthal) quantum number, ℓ, for a particular energy level as defined by the principle quantum number, n, depends on the value of n. ℓ can take integral values from 0 up to and including $(n - 1)$. For example, when $n = 3$, $\ell = 0$, 1, or 2.

5-64.   _Refer to Sections 5-15 and 5-16, and Table 5-4._

(a) For a particular electron, $m\ell = -\ell, -\ell + 1, -\ell + 2, \ldots 0, \ldots \ell - 2, \ell - 1, \ell$

(b) The values of the principle quantum number, n, have letter designations (used in the past) for the corresponding electron shells:

value of n:	1	2	3	4
letter designation:	K	L	M	N

(c) The letters corresponding to the values of the subsidiary quantum number, ℓ, are

value of ℓ:	0	1	2	3
letter designation:	s	p	d	f

5-66. *Refer to Sections 5-14 and 5-15, and Table 5-4.*

There are 9 individual orbitals in the third major energy level, $n = 3$

n	ℓ	m_ℓ	orbital
3	0	0	3s
3	1	-1	3p
3	1	0	3p
3	1	1	3p

n	ℓ	m_ℓ	orbital
3	2	-2	3d
3	2	-1	3d
3	2	0	3d
3	2	+1	3d
3	2	+2	3d

5-68. *Refer to Sections 5-15 and 5-16, and Table 5-4.*

(a) $n = 3$, $\ell = 0$ 3s subshell   (c) $n = 7$, $\ell = 0$ 7s subshell

(b) $n = 3$, $\ell = 1$ 3p subshell   (d) $n = 4$, $\ell = 3$ 4f subshell

5-70. *Refer to Section 5-16.*

	Designation	Number of Orbitals		Designation	Number of Orbitals
(a)	4p	3	(e)	6d	5
(b)	3p	3	(f)	5d	5
(c)	$3p_x$	1	(g)	5f	7
(d)	$n = 5$	25	(h)	7s	1

5-72. *Refer to Section 5-16 and Figures 5-21, 5-22 and 5-23.*

(a) A 1s and a 2s orbital, like all s orbitals, can be described as *spherically symmetrical*, i.e., round like a ball. A 2s orbital is larger than a 1s orbital.

(b) All p orbitals resemble equal-arm dumbbells. A $2p_x$ and a $2p_y$ orbital are identical in size, shape and energy. They differ only in their orientation: a $2p_x$ orbital lies along the x axis whereas a $2p_y$ orbital lies along the y axis.

5-74. *Refer to Sections 5-16 and 5-17.*

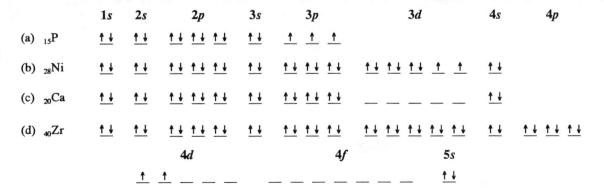

65

(a) $_{15}$P $\quad 1s^22s^22p^63s^23p^3$ $\qquad\qquad$ or $\quad$ [Ne]$3s^23p^3$

(b) $_{28}$Ni $\quad 1s^22s^22p^63s^23p^63d^84s^2$ $\qquad$ or $\quad$ [Ar]$3d^84s^2$

(c) $_{20}$Ca $\quad 1s^22s^22p^63s^23p^63d^04s^2$ $\qquad$ or $\quad$ [Ar]$4s^2$

(d) $_{40}$Zr $\quad 1s^22s^22p^63s^23p^63d^{10}4s^24p^64d^24d^05s^2$ $\qquad$ or $\quad$ [Kr]$4d^25s^2$

Hund's Rule states that electrons must occupy all the orbitals of a given sublevel singly before pairing begins. These unpaired electrons have parallel spins. The electron configuration that violates Hund's Rule is (b) $1s^22s^22p_x^2$. The second electron in $2p_x$ should have occupied other $2p$ orbitals before pairing.

(a) $_{35}$Br $\qquad$ (b) $_{108}$Uno $\qquad$ (c) $_{85}$At $\qquad$ (d) $_{43}$Tc $\qquad$ (e) $_{23}$V

		Number of Electrons		
		s	p	d
(a) $_{14}$Si	$1s^22s^22p^63s^23p^2$	6	8	0
(b) $_{18}$Ar	$1s^22s^22p^63s^23p^6$	6	12	0
(c) $_{28}$Ni	$1s^22s^22p^63s^23p^63d^84s^2$	8	12	8
(d) $_{30}$Zn	$1s^22s^22p^63s^23p^63d^{10}4s^2$	8	12	10
(e) $_{37}$Rb	$1s^22s^22p^63s^23p^63d^{10}4s^24p^64d^05s^1$	9	18	10

Given below are the shorthand notations and number of unpaired electrons present for the elements in their ground states.

	Shorthand	Unpaired Electrons
$_{11}$Na	[Ne]$3s^1$	1
$_{10}$Ne	$1s^22s^22p^6$	0
$_5$B	$1s^22s^22p^1$	1
$_4$Be	$1s^22s^2$	0
$_{34}$Se	[Ar]$3d^{10}4s^24p^4$	2
$_{22}$Ti	[Ar]$3d^24s^2$	2

Given below are the shorthand notations and number of unpaired electrons present for the elements and ions in their ground states.

		Shorthand	Unpaired Electrons
(a)	$_9F$	$1s^2 2s^2 2p^5$	1
(b)	$_{10}Ne$	$1s^2 2s^2 2p^6$	0
(c)	$_{10}Ne^+$	$1s^2 2s^2 2p^5$	1
(d)	$_{30}Zn$	$[Ar]3d^{10}4s^2$	0
(e)	$_{16}S^{2-}$	$[Ne]3s^2 3p^6$ or $[Ar]$	0

Therefore, only $_9F$ and $_{10}Ne^+$ have at least 1 unpaired electron and exhibit paramagnetic properties.

(a) $_7N$ $1s^2 2s^2 2p^3$

Electron	n	ℓ	m_ℓ	m_s
1	1	0	0	+1/2
2	1	0	0	−1/2
3	2	0	0	+1/2
4	2	0	0	−1/2
5	2	1	−1	+1/2
6	2	1	0	+1/2
7	2	1	+1	+1/2

(b) $_{16}S$ $1s^2 2s^2 2p^6 3s^2 3p^4$

Electron	n	ℓ	m_ℓ	m_s
1	1	0	0	+1/2
2	1	0	0	−1/2
3	2	0	0	+1/2
4	2	0	0	−1/2
5	2	1	−1	+1/2
6	2	1	0	+1/2
7	2	1	+1	+1/2
8	2	1	−1	−1/2
9	2	1	0	−1/2
10	2	1	+1	−1/2
11	3	0	0	+1/2
12	3	0	0	−1/2
13	3	1	−1	+1/2
14	3	1	0	+1/2
15	3	1	+1	+1/2
15	3	1	−1	−1/2

(c) $_{30}Zn$ $1s^2 2s^2 2p^6 3s^2 3p^6 4s^2 3d^{10}$

Electron	n	ℓ	m_ℓ	m_s
1	1	0	0	+1/2
2	1	0	0	−1/2
3	2	0	0	+1/2
4	2	0	0	−1/2
5	2	1	−1	+1/2
6	2	1	0	+1/2
7	2	1	+1	+1/2
8	2	1	−1	−1/2
9	2	1	0	−1/2
10	2	1	+1	−1/2
11	3	0	0	+1/2
12	3	0	0	−1/2
13	3	1	−1	+1/2
14	3	1	0	+1/2
15	3	1	+1	+1/2
16	3	1	−1	−1/2
17	3	1	0	−1/2
18	3	1	+1	−1/2
19	4	0	0	+1/2
20	4	0	0	−1/2
21	3	2	−2	+1/2
22	3	2	−1	+1/2
23	3	2	0	+1/2
24	3	2	+1	+1/2
25	3	2	+2	+1/2
26	3	2	−2	−1/2
27	3	2	−1	−1/2
28	3	2	0	−1/2
29	3	2	+1	−1/2
30	3	2	+2	−1/2

5-90. *Refer to Sections 5-17 and 5-18, and Table 5-5.*

	ns	np			ns	np
IA	↑	__ __ __		VA	↑↓	↑ ↑ ↑
IIA	↑↓	__ __ __		VIA	↑↓	↑↓ ↑ ↑
IIIA	↑↓	↑ __ __		VIIA	↑↓	↑↓ ↑↓ ↑
IVA	↑↓	↑ ↑ __		O	↑↓	↑↓ ↑↓ ↑↓

5-92. *Refer to Section 5-17, Examples 5-7 and 5-8, Table 5-5 and Appendix B.*

(a) $_{15}P$ $[Ne]3s^23p^3$ The "last" electron entered a $3p$ orbital.

                                    Therefore, $n = 3$, $\ell = 1$, and $m_\ell = -1, 0$ or $+1$.

(b) $_{41}Nb$ $[Kr]4d^35s^2$ The "last" electron went into a $4d$ orbital.

                                    Therefore, $n = 4$, $\ell = 2$, and $m_\ell = -2, -1, 0, +1,$ or $+2$.

(c) $_{17}Cl$ $[Ne]3s^23p^5$ The "last" electron entered a $3p$ orbital.

                                    Therefore, $n = 3$, $\ell = 1$, and $m_\ell = -1, 0$ or $+1$.

(d) $_{59}Pr$ $[Kr]4d^{10}4f^35s^25p^66s^2$ The "last" electron entered a $4f$ orbital.

                                    Therefore, $n = 4$, $\ell = 3$, and $m_\ell = -3, -2, -1, 0, +1, +2$ or $+3$.

5-94. *Refer to Sections 5-17 and 5-18, Table 5-5, and Appendix B.*

$A \equiv {}_{29}Cu \;\; 1s^22s^22p^63s^23p^63d^{10}4s^1$

$B \equiv {}_{40}Zr \;\; 1s^22s^22p^63s^23p^63d^{10}4s^24p^64d^25s^2$

$C \equiv {}_{54}Xe \;\; 1s^22s^22p^63s^23p^63d^{10}4s^24p^64d^{10}5s^25p^6$

$D \equiv {}_{83}Bi \;\; 1s^22s^22p^63s^23p^63d^{10}4s^24p^64d^{10}4f^{14}5s^25p^65d^{10}6s^26p^3$

$E \equiv {}_4Be \;\; 1s^22s^2$

5-96. *Refer to Section 5-8.*

(a),(b) There are four different HCl molecules which can be formed from the naturally-occurring hydrogen and chlorine isotopes:

Molecules	$^1H^{35}Cl$	$^1H^{37}Cl$	$^2H^{35}Cl$	$^2H^{37}Cl$
Approximate Masses	36 amu	38 amu	37 amu	39 amu

(c) From the relative abundance of the isotopes, the expected abundances of the molecules are in the following decreasing order:

$$^1H^{35}Cl \;>\; ^1H^{37}Cl \;>\; ^2H^{35}Cl \;>\; ^2H^{37}Cl$$

5-98. Refer to Section 5-10.

We know: $c \text{ (m/s)} = \lambda \text{ (m)} \times \nu \text{ (s}^{-1})$

Since $\nu = 89.5 \text{ MHz} = 8.95 \times 10^7 \text{ s}^{-1}$ and $c = 3.00 \times 10^8 \text{ m/s}$

$$\lambda \text{ (m)} = \frac{c \text{ (m/s)}}{\nu \text{ (s}^{-1})} = \frac{3.00 \times 10^8 \text{ m/s}}{8.95 \times 10^7 \text{ s}^{-1}} = \mathbf{3.35 \text{ m}}$$

5-100. Refer to Section 5-13.

Plan: (1) Use $E = hc/\lambda$ to calculate the energy loss in J/atom
        (2) Convert J/atom into kJ/mol

(1) $\lambda = 554 \text{ nm} = 554 \times 10^{-9} \text{ m} = 5.54 \times 10^{-7} \text{ m}$

$$E \text{ lost per atom} = \frac{(6.63 \times 10^{-34} \text{ J·s})(3.00 \times 10^8 \text{ m/s})}{(5.54 \times 10^{-7} \text{ m})} = 3.59 \times 10^{-19} \text{ J/atom}$$

(2) ? kJ/mol Ba atoms $= 3.59 \times 10^{-19} \text{ J/atom} \times \dfrac{6.02 \times 10^{23} \text{ atoms}}{1 \text{ mol}} \times \dfrac{1 \text{ kJ}}{1000 \text{ J}} = \mathbf{216 \text{ kJ/mol}}$

6 Chemical Periodicity

6-2. *Refer to Section 6-1.*

Period 1 contains two elements because electrons are being added to the $n = 1$ energy level, which can only hold $2n^2 = 2(1)^2 = 2$ electrons.

Period 2 contains eight elements, because electrons are being added to the $n = 2$ energy level. This energy level can hold $2n^2 = 2(2)^2 = 8$ electrons in its $2s$ and $2p$ orbitals.

6-4. *Refer to Section 6-1*

The general order in which the energy levels are filled starting with the $n = 3$ major energy level is:

$$3s \quad 3p \quad 4s \quad 3d \quad 4p \quad \text{etc.}$$

Period 3 includes the elements whose outer electrons are in $3s$ or $3p$ energy sublevels. The maximum number of electrons in these sublevels is a total of 8. Hence Period 3 contains only 8 elements. Since the $3d$ sublevel is higher in energy than the $4s$ sublevel, it is not going to be filled until Period 4.

6-6. *Refer to Section 6-1 and Table 5-5.*

The atomic number of the yet-to-be discovered alkaline earth element in period 8 is 120. The last portion of its electron configuration after [Rn] (atomic number = 86) should be:

$$7s^2 \, 5f^{14} \, 6d^{10} \, 7p^6 \, 8s^2$$

6-8. *Refer to Section 6-1.*

(a) ns^2np^5 — Group VIIA (halogens)

(b) ns^2 — Group IIA (alkaline earth metals)

(c) $ns^2(n-1)d^{1-10}$ — d-transition elements

(d) ns^2np^1 — Group IIIA

6-10. *Refer to Section 6-1.*

(a) alkali metals — B

(b) outer configuration of d^8s^2 — H

(c) lanthanides — A

(d) p-block representative elements — C, F, G, I

(e) incompletely filled f-subshells — A, D

(f) halogens — I

(g) s-block representative elements — B, J

(h) actinides — D

(i) d-transition elements — E, H, K

(j) noble gases — G

Electrons that are in filled sets of orbitals between the nucleus and outer shell electrons shield the outer shell electrons partially from the effect of the protons in the nucleus; this effect is called nuclear shielding. As we move from left to right along a period, the outer shell electrons do experience a progressively stronger force of attraction to the nucleus due to the combination of an increase in the number of protons and a constant nuclear shielding by inner electrons. As a result the atomic radii decrease. As we move down a group, the outer electrons are partially shielded from the attractive force of the nucleus by an increasing number of inner electrons. This effect is *partially* responsible for the observed increase in atomic radii going down a group.

As we move down a group, atomic radii increase *primarily* because electrons are added to orbitals in higher energy levels which are further and further away from the nucleus.

Atomic radii increase from top to bottom within a group and from right to left within a period. Therefore, in order of increasing size, we have:

(a) Be < Mg < Ca < Sr < Ba < Ra

(b) He < Ne < Ar < Kr < Xe < Rn

(c) Simple prediction: Ne < F < O < N < C < B < Be < Li

Actual Size: F < O < Ne $\approx$ N < C < B < Be < Li

(d) N < B < Te < Sb < Sr

(a) The first ionization energy, IE_1, also called the first ionization potential, is the minimum amount of energy required to remove the most loosely bound electron from an isolated gaseous atom to form an ion with a 1+ charge.

$$X(g) + IE_1 \rightarrow X^+(g) + e^-$$

(b) The second ionization energy, IE_2, is the amount of energy required to remove a second electron from an isolated gaseous atom; i.e. to remove an electron from an ion with a 1+ charge to give an ion with a 2+ charge.

$$X^+(g) + IE_2 \rightarrow X^{2+}(g) + e^-$$

As we move down a given group, the valence electrons are further and further away from the nucleus. The first ionization energies of the elements, which is the energy required to remove an electron from an isolated gaseous atom, decrease while the atomic radii increase.

Likewise, from left to right across a period, the forces of attraction between the outermost electron and the nucleus increase. Therefore the ionization energies increase while the atomic radii decrease. Refer to Figure 6-2 for the exceptions to the general trends for ionization energy.

6-22. *Refer to Section 6-3 and Table 6-1.*

First ionization energies increase from left to right and bottom to top in the periodic table. However, there are exceptions: elements of Group IIIA generally have lower first ionization energies than elements of Group IIA, and elements of Group VIA generally have lower first ionization energies than elements of Group VA. Therefore, we obtain the following orders of increasing first ionization energies:

(a) Fr < Cs < Rb < K < Na < Li

(b) At < I < Br < Cl < F

(c) Li < B < Be < C < O < N < F < Ne

(d) Cs < Ga < B < Br < H < F

6-24. *Refer to Section 6-3.*

As atomic radii increase moving down a group, first ionization energies decrease because the valence electrons are further from the attractive force of the nucleus. The force of attraction of the positively charged nucleus for the valence electrons is inversely proportional to the square of the distance between them. This effect is enhanced by the decrease in effective nuclear charge down the group.

6-26. *Refer to Section 6-3 and Exercise 6-22 Solution.*

As we move from left to right across Period 2 of the periodic table, there is an increase in effective nuclear charge and a decrease in atomic radii. Outer valence electrons are held more tightly and first ionization energies *generally* increase. Therefore, as the atomic radii decrease, the first ionization energies increase. Refer to Figure 6-2 for the exceptions to the general trend for ionization energy.

6-28. *Refer to Section 6-3.*

It is difficult to prepare compounds containing Ca^{3+} due to the immense amount of energy that is required to remove a third electron from an atom of calcium, i.e., there is a very large amount of energy (the third ionization energy) required for this reaction:

$$Ca^{2+}(g) + 4912.4 \text{ kJ/mol} \rightarrow Ca^{3+}(g) + e^-$$

This energy is not likely to be repaid during compound formation. The reason for such a high third ionization energy for Ca^{3+} is because the electron configuration of Ca^{2+} is $1s^2 2s^2 2p^6 3s^2 3p^6$ which has a filled set of outer p orbitals. It is the special stability of the filled p orbitals which prevents the formation of Ca^{3+} ions.

On the other hand, Al^{2+} has an electron configuration of $1s^2 2s^2 2p^6 3s^1$ where the single e^- in $3s$ could be easily removed to give a stable Al^{3+} ion and consequently an Al^{3+} compound.

6-30. *Refer to Section 6-4, Figure 6-3, Table 6-2, and Example 6-3.*

The electron affinity of an element is defined as the amount of energy absorbed when an electron is added to an isolated gaseous atom to form an ion with a $1-$ charge.

In general, electron affinities become more negative from bottom to top and from left to right in the periodic table, but there are many exceptions. According to Table 6-2, the order of increasing negative values of electron affinity is:

(least negative EA)   O < S < Br < Cl   (most negative EA)

Elements that gain electrons easily to form negative ions have very negative electron affinities. The halogens, with electronic configurations of $ns^2 np^5$, easily gain one electron to form stable ions with a filled set of p orbitals. These ions are isoelectronic with the noble gases and have noble gas electronic configurations, $ns^2 np^6$. Therefore, the halogens have the most negative electron affinities. This does not occur when a Group VIA element gains an electron.

6-34. *Refer to Section 6-4 and Table 6-2.*

			Electronic Configuration		
(a)	$O(g) + e^- \rightarrow O^-(g) + 142$ kJ/mol	O	$1s^2 2s^2 2p^4$	O^-	$1s^2 2s^2 2p^5$
(b)	$Cl(g) + e^- \rightarrow Cl^-(g) + 348$ kJ/mol	Cl	$[Ne]\, 3s^2 3p^5$	Cl^-	$[Ar]$
(c)	$Ca(g) + e^- + 156$ kJ/mol $\rightarrow Ca^-(g)$	Ca	$[Ar]\, 4s^2$	Ca^-	$[Ar]\, 4s^2 3d^1$

6-36. *Refer to Section 6-5 and Figure 6-1.*

(a) Within an isoelectronic series, ionic radii increase with decreasing atomic number. Therefore, in order of increasing ionic radii, we have
$$Ga^{3+} < Ca^{2+} < K^+$$

(b) Ionic radii increase down a group. So, $Be^{2+} < Mg^{2+} < Ca^{2+} < Ba^{2+}$

(c) $Al^{3+} < Sr^{2+} < K^+ < Rb^+$ (See Figure 6-1)

6-38. *Refer to Section 6-5 and Figure 6-1.*

(a) In an isoelectronic series, ionic radii increase with decreasing atomic number. Therefore, in order of increasing ionic radii, we have
$$Cl^- < S^{2-} < P^{3-}$$

(b) Ionic radii increase down a group. So, $O^{2-} < S^{2-} < Se^{2-}$

(c) $N^{3-} < S^{2-} < Br^- < P^{3-}$ (See Figure 6-1)

6-40. *Refer to Section 6-5.*

The Fe^{2+} ion has 26 protons pulling on 24 electrons, whereas the Fe^{3+} ion has 26 protons pulling on 23 electrons. The electrons in the Fe^{3+} ion are more tightly held and therefore, Fe^{3+} is the smaller ion.

Likewise, the Cu^+ ion has 29 protons pulling on 28 electrons, whereas the Cu^{2+} ion has 29 protons attracting 27 electrons. The electrons in the Cu^{2+} ion are more tightly held and therefore, Cu^{2+} is the smaller ion.

6-42. *Refer to Section 6-6 and Table 6-3.*

Electronegativities usually increase from left to right across periods and from bottom to top within groups. Exceptions are explained in Section 6-6.

(a) $Al < In < Ga < B$ (b) $Na < Mg < S < Cl$ (c) $Bi < Sb < P < N$ (d) $Ba < Sc < Si < Se < F$

6-44. *Refer to Sections 6-5 and 6-6.*

Both statements, "Chlorine has a high electronegativity," and "It forms chloride ions, Cl^-, readily" are correct. However the quoted sentence, "Chlorine has a high electronegativity *because* it forms chloride ions, Cl^-, readily," is a poor statement. The high electronegativity is part of the fundamental nature of chlorine due to the electron configuration of the element, while the ease of forming chlorides is a resultant property of chlorine due to its high electron pulling force. Therefore, the correct statement should be, "Chlorine forms chloride ions, Cl^-, readily *because* it has a high electronegativity."

6-46. *Refer to Sections 5-2 and 6-5, and Figure 6-1.*

Ion	Electronic Charge (coulombs)	Ionic Radii (Å)	$V = (4/3)\pi r^3$ (Å^3)	Charge Density = Charge/V (coulombs/Å^3)
Li^+	$+1(1.60 \times 10^{-19})$ $= 1.60 \times 10^{-19}$	0.60	0.90	1.8×10^{-19}
Mg^{2+}	$+2(1.60 \times 10^{-19})$ $= 3.20 \times 10^{-19}$	0.65	1.2	2.7×10^{-19}
Be^{2+}	$+2(1.60 \times 10^{-19})$ $= 3.20 \times 10^{-19}$	0.31	0.12	2.7×10^{-18}
Al^{3+}	$+3(1.60 \times 10^{-19})$ $= 4.80 \times 10^{-19}$	0.50	0.52	9.2×10^{-19}

The charge densities of Li^+ and Mg^{2+} agree within a factor of 1.5, and the charge densities of Be^{2+} and Al^{3+} agree within a factor of 3.

6-48. *Refer to Section 6-2.*

atomic radius of F = 1/2 × bond length of F_2 = 1/2 × 1.42 Å = **0.71 Å**

atomic radius of Cl = 1/2 × bond length of Cl_2 = 1/2 × 1.98 Å = **0.99 Å**

predicted Cl-F bond length = (atomic radius of F) + (atomic radius of Cl) = 0.71 Å + 0.99 Å = **1.70 Å**
        (actual Cl-F bond length = 1.64 Å)

6-50. *Refer to Sections 6-3 and 6-4, and Tables 6-1 and 6-2.*

If we compare the values of the first ionization energy and electron affinity for the Period 3 elements, we have

	Na	Mg	Al	Si	P	S	Cl	Ar
First Ionization Energy (kJ/mol)	497	738	578	786	1012	1000	1251	1521
Electron Affinity (kJ/mol)	-53	(231)	-44	-119	-74	-200	-348	(35)

The magnitude of the electron affinity values is less than that of the first ionization energies. It is much more difficult and hence more energy is required to remove an electron from a neutral gaseous atom, quantified by the first ionization energy, than to add an electron to a neutral gaseous atom, quantified by the electron affinity. In fact, many atoms actually release energy when an extra electron is added as denoted by the negative sign attached to the electron affinity value.

The electron affinities (in kJ/mol) of the halogens are:

$$F (-322) > Cl (-348) < Br (-323) < I (-295).$$

The actual electron affinity of fluorine is less negative than expected from the trend set by the other halogens. This implies that it is more difficult to add an electron to fluorine than to chlorine. Although the variations in electron affinity down a group (family) are not so easily explained, one plausible reason is that because fluorine is small and the electron is added to the small $2p$ sublevel which is already crowded with five other electrons, the repulsion between the incoming electron and the others will cause a slightly lower value in electron affinity.

Elemental hydrogen exists as a colorless, odorless, tasteless, diatomic gas with the lowest atomic weight and density of any known substance. It melts at -259.14°C and boils at -252.8°C.

(a) Hydrogen gas reacts with the alkali metals and heavier alkaline earth metals to form ionic hydrides:

$$2Li(molten) + H_2(g) \rightarrow 2LiH(s)$$

(b) Hydrogen gas reacts with other nonmetals to form molecular (covalent) hydrides:

$$H_2(g) + Cl_2(g) \rightarrow 2HCl(g)$$

NaH, sodium hydride, is the product of hydrogen gas reacting with an active metal, sodium. A compound consisting of a metal and a nonmetal is ionic, hence this is an ionic hydride.

H_2S, hydrogen sulfide, is the product of hydrogen gas reacting with a nonmetal, sulfur. A compound consisting of two nonmetals is covalent, and therefore this is a molecular (covalent) hydride.

(a) H_2S hydrogen sulfide   (d) NH_3 ammonia   (g) CaH_2 calcium hydride

(b) HF   hydrogen fluoride   (e) H_2Se hydrogen selenide

(c) KH   potassium hydride   (f) MgH_2 magnesium hydride

H_2, hydrogen, is a colorless, odorless, tasteless, nonpolar, diamagnetic, diatomic gas with the lowest atomic weight and density of any known substance. It has low solubility in water and is very flammable. Hydrogen is prepared by reactions of metals with water, steam or various acids, electrolysis of water, the water gas reaction and thermal cracking of hydrocarbons. It combines with metals and nonmetals to form hydrides.

O_2, oxygen, is nearly colorless, odorless, tasteless, nonpolar, paramagnetic, diatomic gas. It is nonflammable but participates in all combustion reactions. It is prepared by fractional distillation of liquid air, electrolysis of water and thermal decomposition of certain oxygen-containing salts. Oxygen combines with almost all other elements to form oxides and can be converted to an allotropic form, ozone, O_3.

The elements that react with oxygen to form primarily normal oxides include (a) Li, (d) Mg, (e) Zn and (f) Al.

6-66.   *Refer to Section 6-8 and Exercise 6-76 Solution.*

(a) $2C(s) + O_2(g) \rightarrow 2CO(g)$ (O_2 is limited)

(b) $As_4(s) + 3O_2(g) \rightarrow As_4O_6(s)$ (O_2 is limited)

(c) $2Ge(s) + O_2(g) \rightarrow 2GeO(s)$ (O_2 is limited)

6-68.   *Refer to Section 6-8 and Table 6-4.*

A normal oxide is a binary (two element) compound containing oxygen in the -2 oxidation state. BaO is an example of an ionic oxide and SO_2 is an example of a molecular (covalent) oxide.

A peroxide can be a binary ionic compound containing the O_2^{2-} ion, such as Na_2O_2, or a covalent compound, such as H_2O_2, with oxygen in the -1 oxidation state.

A superoxide is a binary ionic compound containing the O_2^- ion with oxygen in the -1/2 oxidation state, such as KO_2.

6-70.   *Refer to Section 6-8 and Example 6-10.*

(a) $SO_2(g) + H_2O(\ell) \rightarrow H_2SO_3(aq)$ sulfurous acid

(b) $SO_3(\ell) + H_2O(\ell) \rightarrow H_2SO_4(aq)$ sulfuric acid

(c) $SeO_3(s) + H_2O(\ell) \rightarrow H_2SeO_4(aq)$ selenic acid

(d) $N_2O_5(s) + H_2O(\ell) \rightarrow 2HNO_3(aq)$ nitric acid

(e) $Cl_2O_7(\ell) + H_2O(\ell) \rightarrow 2HClO_4(aq)$ perchloric acid

6-72.   *Refer to Section 6-8.*

The acid anhydrides are:
(a) SO_3 (b) CO_2 (c) SO_2 (d) As_2O_5 (e) N_2O_3

6-74.   *Refer to Section 6-8.*

Combustion is an oxidation-reduction reaction in which oxygen gas combines rapidly with oxidizable materials in highly exothermic reactions usually with a visible flame. A combustion reaction is a redox reaction. The oxygen atoms are being reduced since the oxidation number of oxygen is changed from 0 in O_2 to -2 in the products, usually CO_2 and H_2O, while the other reactants have elements being oxidized.

6-76.   *Refer to Section 6-8.*

(a) $2C_2H_6(g) + 5O_2(g) \rightarrow 4CO(g) + 6H_2O(g)$ (O_2 is limited)

(b) $2C_3H_8(g) + 7O_2(g) \rightarrow 6CO(g) + 8H_2O(g)$ (O_2 is limited)

6-78. *Refer to Section 6-8.*

(a)
$$\overset{-2.5\ +1}{2C_4H_{10}(g)} + \overset{0}{13O_2(g)} \rightarrow \overset{+4\ -2}{8CO_2(g)} + \overset{+1\ -2}{10H_2O(g)} \qquad \text{(O}_2 \text{ is in excess)}$$

(b)
$$\overset{-2.5\ +1}{2C_4H_{10}(g)} + \overset{0}{9O_2(g)} \rightarrow \overset{+2\ -2}{8CO(g)} + \overset{+1\ -2}{10H_2O(g)} \qquad \text{(O}_2 \text{ is limited)}$$

(c)
$$\overset{-2.5\ +1}{C_4H_{10}(g)} + \overset{0}{O_2(g)} \rightarrow \overset{-1\ +1}{C_2H_2(g)} + \overset{+2\ -2}{2CO(g)} + \overset{0}{4H_2(g)} \qquad \text{(O}_2 \text{ is very limited)}$$

6-80. *Refer to Section 6-8.*

(a) $4C_6H_5N(\ell) + 31O_2(g) \rightarrow 24CO_2(g) + 10\ H_2O(\ell) + 4NO(g)$

(b) $2C_2H_5SH(\ell) + 9O_2(g) \rightarrow 4CO_2(g) + 6H_2O(g) + 2SO_2(g)$

(c) $C_7H_{10}NO_2S(\ell) + 10O_2(g) \rightarrow 7CO_2(g) + 5H_2O(g) + NO(g) + SO_2(g)$

6-82. *Refer to Section 6-8.*

When fossil fuels are burned, oxides of sulfur and nitrogen (acid anhydrides) are released into the atmosphere and dissolve in atmospheric moisture producing "acid rain." The reactions converting sulfur to two common oxides, SO_2 and SO_3, with their final conversion to acid rain are shown below:

$$S(s) + O_2(g) \rightarrow SO_2(g) \qquad \text{(burning of sulfur compounds in fossil fuels)}$$
$$2SO_2(g) + O_2(g) \rightarrow 2SO_3(\ell) \qquad \text{(occurs slowly in air)}$$
$$SO_2(g) + H_2O(\ell, \text{ moisture in air}) \rightarrow H_2SO_3(\ell)$$
$$SO_3(\ell) + H_2O(\ell, \text{ moisture in air}) \rightarrow H_2SO_4(\ell)$$

6-84. *Refer to Sections 6-3 and 5-9.*

Recall: For 1 atom, $E \text{ (J/atom)} = h \text{ (J·s)} \times \nu \text{ (s}^{-1})$
For 1 mole of atoms, $E \text{ (J/mol)} = h\nu N$ where N is Avogadro's Number

Solving for ν, we have

$$\nu \text{ (s}^{-1}) = \frac{E}{hN} = \frac{419 \text{ kJ/mol} \times 1000 \text{ J/kJ}}{(6.63 \times 10^{-34} \text{ J·s})(6.02 \times 10^{23} \text{ mol}^{-1})} = 1.05 \times 10^{15} \text{ s}^{-1}$$

6-86. *Refer to Section 6-3 and Table 6-1.*

First Ionization Energy for Mg (kJ/mol) = 738 kJ/mol   So, $Mg(g) + 738 \text{ kJ/mol} \rightarrow Mg^+(g) + e^-$

Second Ionization Energy for Mg (kJ/mol) = 1451 kJ/mol   So, $\underline{Mg^+(g) + 1451 \text{ kJ/mol} \rightarrow Mg^{2+}(g) + e^-}$

$$Mg(g) + 2189 \text{ kJ/mol} \rightarrow Mg^{2+}(g) + 2e^-$$

And so, 2189 kJ/mol of energy is required to produce 1 mole of gaseous Mg^{2+} ions from gaseous Mg atoms. To convert this energy into units of kJ/g, which is the energy required per 1 gram of gaseous Mg atoms, we apply dimensional analysis:

$$? \text{ energy (kJ/g)} = 2189 \text{ kJ/mol Mg} \times 1 \text{ mol Mg}/24.305 \text{ g} = \textbf{90.06 kJ/g}$$

7   Chemical Bonding

7-2.   *Refer to the Introduction to Chapter 7.*

The two major types of bonding are ionic bonding and covalent bonding. Examples of each can be found in NaCl and CO_2, respectively. Ionic bonding results from electrostatic interactions between ions, which can be formed by the *transfer* of one or more electrons from one atom or group of atoms to another. Covalent bonding, on the other hand, results from *sharing* one or more electron pairs between two atoms. Some general properties of ionic and covalent bonding are:

ionic: high solubility in polar solvents, such as water, and low solubility in nonpolar solvents, such as hexane
high melting points
molten compounds and aqueous solutions conduct electricity well due to the presence of mobile ions, solids do not

covalent: low solubility in polar solvents, high solubility in nonpolar solvents
low melting points
liquids or molten compounds do not conduct electricity, aqueous solutions usually conduct electricity poorly

---

7-4.   *Refer to Section 7-1.*

An alkali metal atom (Group IA), represented by M with electronic configuration ns^1, can attain noble gas configuration by losing the ns^1 electron and becoming M^+ ion.

7-6.   *Refer to Section 7-1 and Table 7-1.*

(a)   Lewis dot representations for the representative elements show only the valence electrons in the outermost occupied *s* and *p* orbitals. Paired and unpaired electrons are also indicated.

(b)   He: Ṡi·   ·P̈·   :N̈e: Mg: ·C̈l:

7-8.   *Refer to Sections 7-2 and 7-3.*

In NaOCl, there is ionic bonding occurring between the Na^+ ion and the OCl^- ion, and covalent bonding between the O and Cl atoms in the OCl^- ion.

7-10.   *Refer to Sections 7-2 and 7-3.*

In general, the bond between a metal and a nonmetal is ionic, whereas the bond between two nonmetals is covalent. In other words, the further apart across the periodic table the two elements are, the more likely they are to form an ionic bond.

(a)   Ba (metal) and Cl (nonmetal)   ionic bond

(b)   P (nonmetal) and O (nonmetal)   covalent bond

(c)   Br (nonmetal) and I (nonmetal)   covalent bond

(d)   Li (metal) and I (nonmetal)   ionic bond

(e)   Si (metalloid) and Br (nonmetal)   covalent bond

(f)   Ca (metal) and F (nonmetal)   ionic bond

In general, whenever a metal and a nonmetal are together in a compound, it is ionic. If the compound consists only of nonmetals, it is covalent. In other words, the further apart two elements are on the periodic table, the more likely they are to form an ionic compound.

(a) $CaSO_4$ metal + nonmetals   ionic (within the SO_4^{2-} ion, there are covalent bonds)

(b) SO_2 nonmetals   covalent

(c) KNO_3 metal + nonmetals   ionic (within the NO_3^- ion, there are covalent bonds)

(d) $NiCl_2$ metal + nonmetal   ionic

(e) H_2CO_3 nonmetals   covalent (H is not a metal)

(f) PCl_3 nonmetals   covalent

(g) Li_2O metal + nonmetal   ionic

(h) N_2H_4 nonmetals   covalent

(i) $SOCl_2$ nonmetals   covalent

7-14. *Refer to Section 7-2 and Chapter 13.*

An ionic crystal is a solid characterized by a regular, ordered arrangement of ions in three-dimensional space. The specific geometrical arrangement of the ions is controlled by

(1) the compound formula, i.e., the ratio of cations to anions,
(2) the size of the ions and
(3) the conditions (temperature and pressure) under which the solid exists.

7-16. *Refer to Section 7-2 and Table 7-2.*

(a) $Cs + 1/2Br_2 \rightarrow CsBr$ (b) $Ba + S \rightarrow BaS$ (c) $K + 1/2Cl_2 \rightarrow KCl$

7-18. *Refer to Section 7-2 and Appendix B.*

(a) Cr^{3+} [Ar] $3d^3$ (d) Fe^{3+} [Ar] $3d^5$ (g) Cu^+ [Ar]$3d^{10}$

(b) Mn^{2+} [Ar] $3d^5$ (e) Cu^{2+} [Ar] $3d^9$

(c) Ag^+ [Kr] $4d^{10}$ (f) Sc^{3+} [Ar] $3d^0$

7-20. *Refer to Section 7-2.*

Stable binary ionic compounds are formed from ions that have noble gas configurations. None of the compounds, except CO_2, meet this requirement. CO_2 is not an ionic compound at all because it is a covalent compound, made from 2 nonmetals. Consider the following:

MgI ($Mg^+ + I^-$) $Al(OH)_2$ ($Al^{2+} + 2OH^-$) InF_2 ($In^{2+} + 2F^-$)

$RbCl_2$ ($Rb^{2+} + 2Cl^-$) $CsSe$ ($Cs^{2+} + Se^{2-}$) Be_2O ($2Be^+ + O^{2-}$).

Neither Mg^+, Al^{2+}, In^{2+}, Rb^{2+}, Cs^{2+} nor Be^+ have noble gas configurations.

7-22. *Refer to Section 7-2.*

(a) Species are isoelectronic if they have the same total number of electrons, and have the same electronic configuration. Three examples of isoelectronic species are (1) Na^+ and Ne, (2) Cl^- and Ar, and (3) N^{3-} and O^{2-}.

(b) All of the following chemical species, O^{2-}, Ne, Na^+, Mg^{2+} and Al^{3+}, have 10 electrons and are isoelectronic. The fluorine atom, F, has 9 electrons.

7-24. *Refer to Section 7-2.*

The following species have 18 electrons and are therefore isoelectronic with argon:

        cations: K^+, Ca^{2+}, Sc^{3+} anions: Cl^-, S^{2-}, P^{3-}

7-26. *Refer to Section 7-2.*

(a) Cations with ....$3s^2 3p^6$ electronic configurations are isoelectronic with argon. Examples: K^+, Ca^{2+}

(b) Cations with ....$4s^2 4p^6$ electronic configurations are isoelectronic with krypton. Examples: Rb^+, Sr^{2+}

7-28. *Refer to Section 7-3 and Figure 7-4.*

Figure 7-4 is a plot of potential energy versus the distance between 2 hydrogen atoms. The resulting function is the sum of two opposing forces: (1) the attractive force between the negatively charged electron of one H atom and the positively charged nucleus of the other H atom, and (2) the repulsive force between the two positively charged nuclei.

When the two atoms are relatively far apart, there is essentially no interaction at all between them; both the attractive and repulsive forces are about zero. As the two atoms get closer, the attractive forces dominate, and the potential energy decreases to a minimum at a distance of 0.74 Å, which is the H−H bond length. At distances less than 0.74 Å, the repulsive forces become more important and the energy increases sharply.

7-30. *Refer to Section 7-3.*

(a) A single covalent bond contains 2 shared electrons.

(b) A double covalent bond contains 4 shared electrons.

(c) A triple covalent bond contains 6 shared electrons.

7-32. *Refer to Section 7-4.*

Lewis formulas are representations of molecules or ions which show the element symbols, the order in which the atoms are connected, the number of valence electrons linking the atoms together, the number of lone pairs of valence electrons not used for bonding, and the number and kind of bonds. They do not show the shape of a chemical species.

7-34. *Refer to Sections 7-4 and 7-5, and Examples 7-1, 7-2 and 7-3.*

H_2O

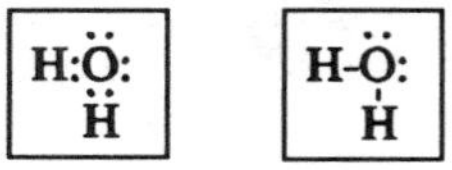

H:Ö:
  H

H-Ö:
  H

$S = N - A$
$= [2 \times 2(\text{for H}) + 1 \times 8(\text{for O})] - [2 \times 1(\text{for H}) + 1 \times 6(\text{for O})]$
$= 12 - 8$
$= 4$ and there are 4 electrons shared in the molecule

NH_3

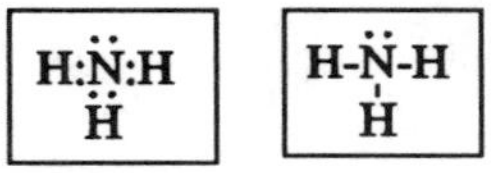

H:N:H
  H

H-N-H
  H

$S = N - A$
$= [3 \times 2(\text{for H}) + 1 \times 8(\text{for N})] - [3 \times 1(\text{for H}) + 1 \times 5(\text{for N})]$
$= 14 - 8$
$= 6$ and there are 6 electrons shared in the molecule

OH^-

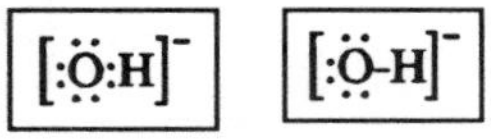

[:Ö:H]⁻

[:Ö-H]⁻

$S = N - A$
$= [1 \times 2(\text{for H}) + 1 \times 8(\text{for O})] - [1 \times 1(\text{for H}) + 1 \times 6(\text{for O})$
$\qquad\qquad + 1e^-]$
$= 10 - 8$
$= 2$ and there are 2 electrons shared in the diatomic ion

F^-

:F:⁻

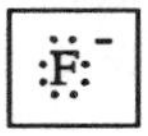

7-36. *Refer to Sections 7-4 and 7-5, and Example 7-1 and 7-2.*

(a) SCl_2

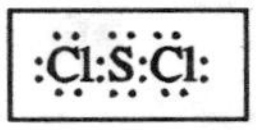

:Cl:S:Cl:

$S = N - A$
$= [1 \times 8(\text{for S}) + 2 \times 8(\text{for Cl})] - [1 \times 6(\text{for S}) + 2 \times 7(\text{for Cl})]$
$= 4$ and there are 4 electrons shared

(b) AsF_3

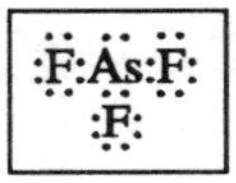

:F:As:F:
  :F:

$S = N - A$
$= [1 \times 8(\text{for As}) + 3 \times 8(\text{for F})] - [1 \times 5(\text{for As}) + 3 \times 7(\text{for F})]$
$= 6$ and there are 6 electrons shared

(c) ICl

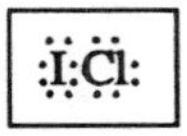

:I:Cl:

$S = N - A$
$= [1 \times 8(\text{for I}) + 1 \times 8(\text{for Cl})] - [1 \times 7(\text{for I}) + 1 \times 7(\text{for Cl})]$
$= 2$ and there are 2 electrons shared

(d) NCl_3

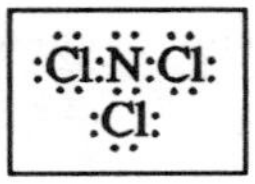

:Cl:N:Cl:
  :Cl:

$S = N - A$
$= [1 \times 8(\text{for N}) + 3 \times 8(\text{for Cl})] - [1 \times 5(\text{for N}) + 3 \times 7(\text{for Cl})]$
$= 6$ and there are 6 electrons shared

7-38. *Refer to Sections 7-4, 7-5 and 7-7, and Examples 7-1, 7-2, 7-6 and 7-7.*

(a) H_2S

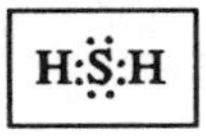

H:S:H

$S = N - A$
$= [2 \times 2(\text{for H}) + 1 \times 8(\text{for S})] - [2 \times 1(\text{for H}) + 1 \times 6(\text{for S})]$
$= 4$ shared electrons

(b) PCl_3

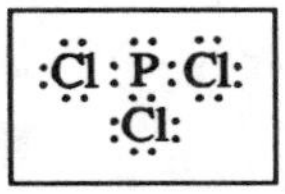

:Cl:P:Cl:
  :Cl:

$S = N - A$
$= [1 \times 8(\text{for P}) + 3 \times 8(\text{for Cl})] - [1 \times 5(\text{for P}) + 3 \times 7(\text{for Cl})]$
$= 6$ shared electrons

(c) BCl_3

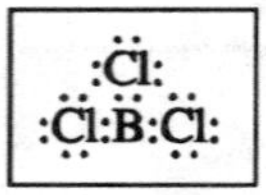

:$\ddot{C}l$:
:$\ddot{C}l$:B:$\ddot{C}l$:

The octet rule is not valid without modification (Section 7-7, Limitation 2).

$A = 1 \times 3(\text{for B}) + 3 \times 7(\text{for Cl}) = 24$ (total number of dots)

(d) SiH_4

H
H:$\ddot{S}i$:H
H

$S = N - A$
$= [1 \times 8(\text{for Si}) + 4 \times 2(\text{for H})] - [1 \times 4(\text{for Si}) + 4 \times 1(\text{for H})]$
$= 8$ shared electrons

(e) $ClNO$

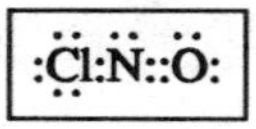

:$\ddot{C}l$:N::$\ddot{O}$:

$S = N - A$
$= [1 \times 8(\text{for Cl}) + 1 \times 8(\text{for N}) + 1 \times 8(\text{for O})]$
$\qquad - [1 \times 7(\text{for Cl}) + 1 \times 5(\text{for N}) + 1 \times 6(\text{for O})]$
$= 6$ shared electrons

7-40.   Refer to Sections 7-5, 7-6 and 7-7.

(a) H_2O_2

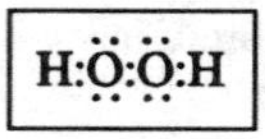

H:$\ddot{O}$:$\ddot{O}$:H

$S = N - A$
$= [2 \times 2(\text{for H}) + 2 \times 8(\text{for O})] - [2 \times 1(\text{for H}) + 2 \times 6(\text{for O})]$
$= 6$ shared electrons

(b) IO_4^-

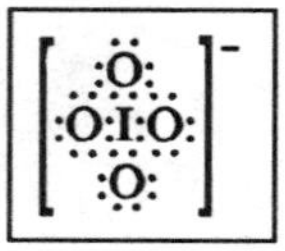

$\left[\begin{array}{c} :\ddot{O}: \\ :\ddot{O}:I:\ddot{O}: \\ :\ddot{O}: \end{array} \right]^-$

$S = N - A$
$= [1 \times 8(\text{for I}) + 4 \times 8(\text{for O})] - [1 \times 7(\text{for I}) + 4 \times 6(\text{for O}) + 1e^-]$
$= 8$ shared electrons

(c) BeH_2

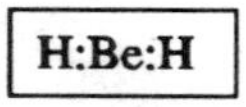

H:Be:H

The octet rule is not valid without modification (Section 7-10, Limitation 1).

$A = 1 \times 2(\text{for Be}) + 2 \times 1(\text{for H}) = 4$ (total number of dots)

(d) NCl_3

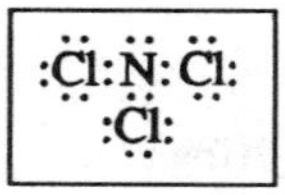

:$\ddot{C}l$:$\ddot{N}$:$\ddot{C}l$:
:$\ddot{C}l$:

$S = N - A$
$= [1 \times 8(\text{for N}) + 3 \times 8(\text{for Cl})] - [1 \times 5(\text{for N}) + 3 \times 7(\text{for Cl})]$
$= 6$ shared electrons

(e) $HClO$

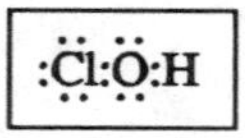

:$\ddot{C}l$:$\ddot{O}$:H

$S = N - A$
$= [1 \times 2(\text{for H}) + 1 \times 8(\text{for Cl}) + 1 \times 8(\text{for O})]$
$\qquad - [1 \times 1(\text{for H}) + 1 \times 7(\text{for Cl}) + 1 \times 6(\text{for O})]$
$= 4$ shared electrons

(f) XeF_4

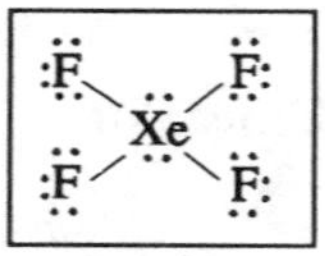

:$\ddot{F}$:   :$\ddot{F}$:
        Xe
:$\ddot{F}$:   :$\ddot{F}$:

The octet rule is not valid without modification (Section 7-7, Limitation 4).

$A = 1 \times 8(\text{for Xe}) + 4 \times 7(\text{for F}) = 36$ (total number of dots)

(g) C_3H_4

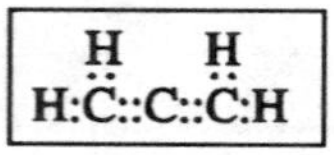

H   H
H:$\ddot{C}$::C::$\ddot{C}$:H

or

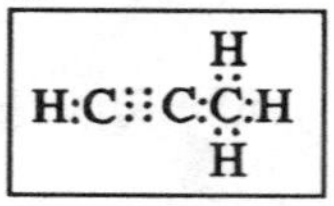

$\qquad$H
H:C::C:$\ddot{C}$:H
$\qquad$H

or

H   H
:C:
:C::C:
H   H
$=$
H   H
C
C=C
H   H

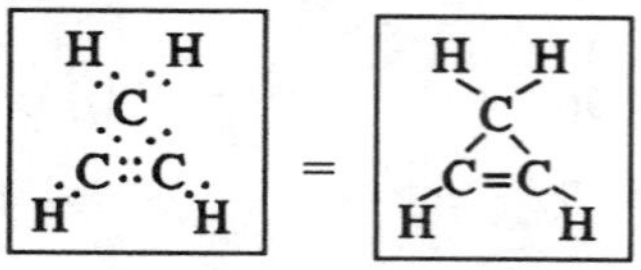

$S = N - A$
$= [3 \times 8(\text{for C}) + 4 \times 2(\text{for H})] - [3 \times 4(\text{for C}) + 4 \times 1(\text{for H})]$
$= 16$ shared electrons

82

ClO_2 :Ö:Cl:Ö: The octet rule is not valid without modification (Section 7-7, Limitation 3).

$A = 1 \times 7$(for Cl) $+ 2 \times 6$(for O) $= 19$ (total number of dots)

As the number of electrons in a bond increases, the energy of the bond increases, and the length of the bond decreases. Therefore,

$$C-C > C=C > C\equiv C \quad \text{in bond length}$$

From the discussion of resonance, the carbon-carbon bond length in the six-membered ring of toluene is intermediate in length between a single bond and a double bond. Therefore, this bond would be shorter than a regular single bond found between the CH_3 group and the carbon atom on the ring.

Resonance structures for NO_3^- ion: $S = N - A$

$= [1 \times 8$(for N) $+ 3 \times 8$(for O)$] - [1 \times 5$(for N) $+ 3 \times 6$(for O) $+ 1e^-]$

$= 8$ shared electrons

$$\left[\ddot{O}::N:\ddot{O}: \atop :\ddot{O}:\right]^- \leftrightarrow \left[:\ddot{O}:N:\ddot{O}: \atop :\ddot{O}:\right]^- \leftrightarrow \left[:\ddot{O}:N::\ddot{O} \atop :\ddot{O}:\right]^- \quad \text{or} \quad \left[\ddot{O}=N-\ddot{O}: \atop :\ddot{O}:\right]^- \leftrightarrow \left[:\ddot{O}-N-\ddot{O}: \atop :\ddot{O}:\right]^- \leftrightarrow \left[:\ddot{O}-N=\ddot{O} \atop :\ddot{O}:\right]^-$$

(a) $BeBr_2$:Br:Be:Br: :Br-Be-Br: The octet rule is not valid without modification (Section 7-7, Limitation 1).

$A = 1 \times 2$(for Be) $+ 2 \times 7$(for Br) $= 16$ (total number of dots)

(b) BBr_3 :Br: / :Br:B:Br: :Br: / :Br-B-Br: The octet rule is not valid without modification (Section 7-7, Limitation 2).

$A = 1 \times 3$(for B) $+ 3 \times 7$(for Br) $= 24$ (total number of dots)

(c) BCl_3 :Cl: / :Cl:B:Cl: :Cl: / :Cl-B-Cl: The octet rule is not valid without modification (Section 7-7, Limitation 2).

$A = 1 \times 3$(for B) $+ 3 \times 7$(for Cl) $= 24$ (total number of dots)

(d) $AlCl_3$:Cl: / :Cl:Al:Cl: :Cl: / :Cl-Al-Cl: The octet rule is not valid without modification (Section 7-7, Limitation 2).

$A = 1 \times 3$(for Al) $+ 3 \times 7$(for Cl) $= 24$ (total number of dots)

All compounds have a central atom that disobeys the octet rule with a share in less than an octet of valence electrons.

7-50. *Refer to Section 7-7, and Examples 7-5 and 7-6.*

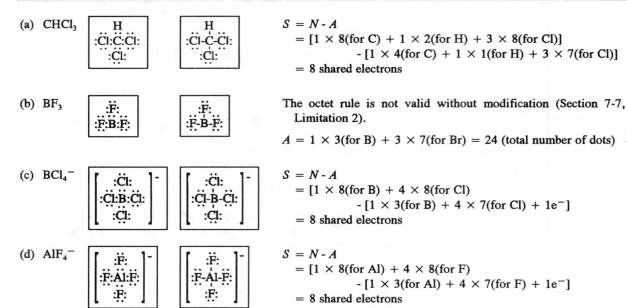

(a) $CHCl_3$

$$S = N - A$$
$$= [1 \times 8(\text{for C}) + 1 \times 2(\text{for H}) + 3 \times 8(\text{for Cl})]$$
$$- [1 \times 4(\text{for C}) + 1 \times 1(\text{for H}) + 3 \times 7(\text{for Cl})]$$
$$= 8 \text{ shared electrons}$$

(b) BF_3

The octet rule is not valid without modification (Section 7-7, Limitation 2).

$$A = 1 \times 3(\text{for B}) + 3 \times 7(\text{for Br}) = 24 \text{ (total number of dots)}$$

(c) BCl_4^-

$$S = N - A$$
$$= [1 \times 8(\text{for B}) + 4 \times 8(\text{for Cl})$$
$$- [1 \times 3(\text{for B}) + 4 \times 7(\text{for Cl}) + 1e^-]$$
$$= 8 \text{ shared electrons}$$

(d) AlF_4^-

$$S = N - A$$
$$= [1 \times 8(\text{for Al}) + 4 \times 8(\text{for F})$$
$$- [1 \times 3(\text{for Al}) + 4 \times 7(\text{for F}) + 1e^-]$$
$$= 8 \text{ shared electrons}$$

Only **Compound (b)** has a central atom that disobeys the octet rule with a share in less than an octet of valence electrons.

7-52. *Refer to Sections 7-4, 7-5, 7-7 and Example 7-7.*

(a) $AlCl_4^-$

$$S = N - A$$
$$= [1 \times 8(\text{for Al}) + 4 \times 8(\text{for Cl})$$
$$- [1 \times 3(\text{for Al}) + 4 \times 7(\text{for Cl}) + 1e^-]$$
$$= 8 \text{ shared electrons}$$

(b) $AsCl_5$

The octet rule is not valid without modification (Section 7-7, Limitation 4).

$$A = 1 \times 5(\text{for As}) + 5 \times 7(\text{for Cl}) = 40 \text{ (total number of dots)}$$

(c) SF_4

The octet rule is not valid without modification (Section 7-7, Limitation 4).

$$A = 1 \times 6(\text{for S}) + 4 \times 7(\text{for F}) = 34 \text{ (total number of dots)}$$

(d) C_2H_6

$$S = N - A$$
$$= [2 \times 8(\text{for C}) + 6 \times 2(\text{for H})]$$
$$- [2 \times 4(\text{for C}) + 6 \times 1(\text{for H})]$$
$$= 14 \text{ shared electrons}$$

Compounds (b) and (c) have central atoms that disobey the octet rule with a share in more than an octet of valence electrons.

7-54. *Refer to Sections 7-4, 7-5 and 7-7, and Examples 7-7 and 7-8.*

(a) KCl_3

$K^+ \left[:\ddot{\underset{..}{C}l}:\ddot{\underset{..}{C}l}:\ddot{\underset{..}{C}l}: \right]^-$ $K^+ \left[:\ddot{\underset{..}{C}l}-\ddot{\underset{..}{C}l}-\ddot{\underset{..}{C}l}: \right]^-$

This is an ionic compound. For Cl_3^-, the octet rule is not valid without modification (Section 7-7, Limitation 4).

$A = 3 \times 7(\text{for Cl}) + 1\ e^- = 22$ (total number of dots)

(b) KrF_2

$:\ddot{F}:\ddot{Kr}:\ddot{F}:$ $:\ddot{F}-Kr-\ddot{F}:$

The octet rule is not valid without modification (Section 7-7, Limitation 4).

$A = 1 \times 8(\text{for Kr}) + 2 \times 7(\text{for F}) = 22$ (total number of dots)

(c) BrF_5

The octet rule is not valid without modification (Section 7-7, Limitation 4).

$A = 1 \times 7(\text{for Br}) + 5 \times 7(\text{for F}) = 42$ (total number of dots)

(d) PF_6^-

$\left[\begin{array}{c} :\ddot{F}: \\ :\ddot{F}:\overset{..}{P}:\ddot{F}: \\ :\ddot{F}:\ \ \ddot{F}: \\ :\ddot{F}: \end{array} \right]^-$ $\left[\begin{array}{c} :\ddot{F}: \\ :\ddot{F}\diagdown P\diagup\ddot{F}: \\ :\ddot{F}\diagup\ |\ \diagdown\ddot{F}: \\ :\ddot{F}: \end{array} \right]^-$

The octet rule is not valid without modification (Section 7-7, Limitation 4).

$A = 1 \times 5(\text{for P}) + 6 \times 7(\text{for F}) + 1\ e^- = 48$

(total number of dots)

Compounds (a), (b), (c) and (d) have central atoms that disobey the octet rule with a share in more than an octet of valence electrons.

7-56. *Refer to Sections 7-4, 7-5, 7-6 and 7-7, and Examples 7-3 and 7-4.*

(a) SO_2 exhibits resonance and obeys the octet rule:

$\ddot{\underset{..}{O}}::\ddot{S}:\ddot{\underset{..}{O}}: \quad \leftrightarrow \quad :\ddot{\underset{..}{O}}:\ddot{S}::\ddot{\underset{..}{O}}$

$S = N - A = [1 \times 8(\text{for S}) + 2 \times 8(\text{for O})] - [1 \times 6(\text{for S}) + 2 \times 6(\text{for O})] = 6$ shared electrons

(b) NO_2 exhibits resonance, but it violates the octet rule because the compound contains an odd number of valence electrons, 17 (Section 7-10, Limitation 3).

$:\ddot{\underset{..}{O}}::N:\ddot{\underset{..}{O}}: \quad \leftrightarrow \quad :\ddot{\underset{..}{O}}:N::\ddot{\underset{..}{O}}: \quad \leftrightarrow \quad \cdot\ddot{\underset{..}{O}}:N::\ddot{\underset{..}{O}}: \quad \leftrightarrow \quad :\ddot{\underset{..}{O}}::N:\ddot{\underset{..}{O}}\cdot$

$A = 1 \times 5(\text{for N}) + 2 \times 6(\text{for O}) = 17$ (total number of dots)

(c) CO exhibits resonance. It is known from experiments that the C-O bond in CO is intermediate between a typical double and triple bond length. Only one resonance structure obeys the octet rule.

$:C:::O: \quad \leftrightarrow \quad :C::\ddot{O}:$

$S = N - A$
$= [1 \times 8(\text{for C}) + 1 \times 8(\text{for O})] - [1 \times 4(\text{for C}) + 1 \times 6(\text{for O})]$
$= 16 - 10$
$= 6$ shared electrons

(d) O_3 exhibits resonance and obeys the octet rule.

$:\ddot{\underset{..}{O}}::\ddot{O}:\ddot{\underset{..}{O}}: \quad \leftrightarrow \quad :\ddot{\underset{..}{O}}:\ddot{O}::\ddot{\underset{..}{O}}:$

$S = N - A = [3 \times 8(\text{for O})] - [3 \times 6(\text{for O})] = 24 - 18 = 6$ shared electrons

(e) SO₃ exhibits resonance and obeys the octet rule.

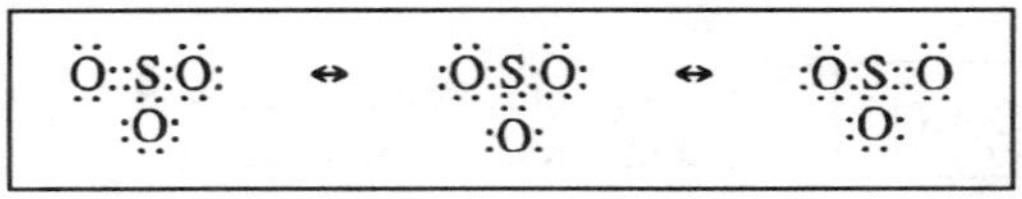

$$\ddot{O}::S:\ddot{O}: \quad \leftrightarrow \quad :\ddot{O}:S:\ddot{O}: \quad \leftrightarrow \quad :\ddot{O}:S::\ddot{O}$$
$$:\ddot{O}: \qquad\qquad :\ddot{O}: \qquad\qquad :\ddot{O}:$$

$S = N - A.$
$= [1 \times 8(\text{for S}) + 3 \times 8(\text{for O})] - [1 \times 6(\text{for S}) + 3 \times 6(\text{for O})]$
$= 8$ shared electrons

(f) $(NH_4)_2SO_4$ is an ionic solid composed of covalently bonded polyatomic ions:

NH_4^+

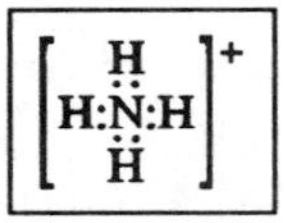

$$\left[\begin{array}{c} H \\ H:\ddot{N}:H \\ H \end{array} \right]^+$$

$S = N - A$
$= [4 \times 2(\text{for H}) + 1 \times 8(\text{for N})] - [4 \times 1(\text{for H}) + 1 \times 5(\text{for N}) - 1\ e^-]$
$= 16 - 8$
$= 8$ shared electrons

SO_4^{2-}

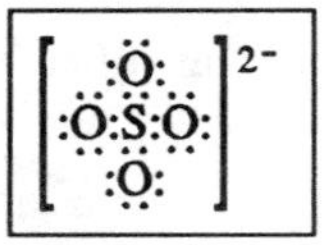

$$\left[\begin{array}{c} :\ddot{O}: \\ :\ddot{O}:S:\ddot{O}: \\ :\ddot{O}: \end{array} \right]^{2-}$$

$S = N - A$
$= [4 \times 8(\text{for O}) + 1 \times 8(\text{for S})] - [4 \times 6(\text{for O}) + 1 \times 6(\text{for S}) + 2\ e^-]$
$= 40 - 32$
$= 8$ shared electrons

7-58.   Refer to Section 7-6.

The formal charge, FC = (Group No.) - [(No. of bonds) + (No. of unshared e^-)]

(a)

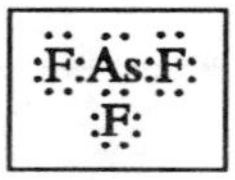

$:\ddot{F}:\ddot{A}s:\ddot{F}:$
$:\ddot{F}:$

for As,   FC = 5 - (3 + 2) = 0
for F,   FC = 7 - (1 + 6) = 0

(b)

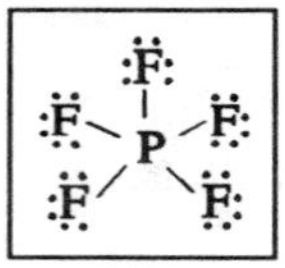

$$\begin{array}{c} :\ddot{F}: \\ | \\ :\ddot{F}\diagdown \!\! \underset{\diagup}{\overset{|}{P}} \!\! \diagup\ddot{F}: \\ :\ddot{F} \qquad \ddot{F}: \end{array}$$

for P,   FC = 5 - (5 + 0) = 0
for F,   FC = 7 - (1 + 6) = 0

(c)

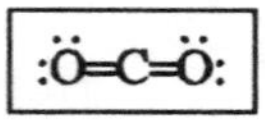

$:\ddot{O}=C=\ddot{O}:$

for C,   FC = 4 - (4 + 0) = 0
for O,   FC = 6 - (2 + 4) = 0

(d)

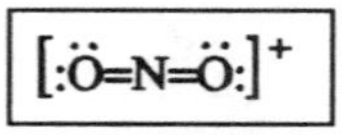

$[:\ddot{O}=N=\ddot{O}:]^+$

for N,   FC = 5 - (4 + 0) = +1
for O,   FC = 6 - (2 + 4) = 0

(e)

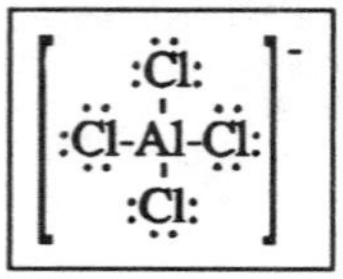

$$\left[\begin{array}{c} :\ddot{C}l: \\ | \\ :\ddot{C}l\text{-}Al\text{-}\ddot{C}l: \\ | \\ :\ddot{C}l: \end{array} \right]^-$$

for Al,   FC = 3 - (4 + 0) = -1
for Cl,   FC = 7 - (1 + 6) = 0

7-60.   Refer to Section 7-6.

(1)

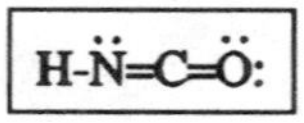

$H\text{-}\ddot{N}=C=\ddot{O}:$

for H,   FC = 1 - (1 + 0) = 0
for N,   FC = 5 - (3 + 2) = 0
for C,   FC = 4 - (4 + 0) = 0
for O,   FC = 6 - (2 + 4) = 0

(2)

$\boxed{:N\equiv C\text{-}\ddot{O}\text{-}H}$

for N,   FC = 5 - (3 + 2) = 0
for C,   FC = 4 - (4 + 0) = 0
for O,   FC = 6 - (2 + 4) = 0
for H,   FC = 1 - (1 + 0) = 0

(3)

$\boxed{:C\equiv N\text{-}\ddot{O}\text{-}H}$

for C,   FC = 4 - (3 + 2) = -1
for N,   FC = 5 - (4 + 0) = +1
for O,   FC = 6 - (2 + 4) = 0
for H,   FC = 1 - (1 + 0) = 0

This structure is predicted to be the least stable because it contains elements having formal charges that are not equal to zero, and because the more electronegative N has a positive FC while the less electronegative C has a negative FC.

7-62.   *Refer to Section 7-8.*

An HBr molecule is a heteronuclear diatomic molecule composed of H (EN = 2.1) and Br (EN = 2.8). Because the electronegativities of the elements are different, the pull on the electrons in the covalent bond between them is unequal. Hence HBr is a polar molecule. A homonuclear diatomic molecule contains a nonpolar bond, since the electron pair between the two atoms is shared equally. Br_2 is an example of a homonuclear diatomic molecule.

7-64.   *Refer to Sections 7-8 and 7-9.*

(a) The two pairs of elements most likely to form ionic bonds are (1) Ba (metal) and F (nonmetal) and (2) K (metal) and O (nonmetal).

(b) We know that bond polarity increases with increasing Δ(EN), the difference in electronegativity between 2 atoms that are bonded together.

bond	Te – H		C – F		N - F	
EN	2.1	2.1	2.5	4.0	3.0	4.0
Δ(EN)	0.0		1.5		1.0	

Therefore, the least polar bond is Te-H and the most polar bond is C-F.

7-66.   *Refer to Section 7-8.*

The use of ΔEN alone to distinguish between ionic and polar covalent bonds will lead to the mis-labeling of some bonds, especially when the elements, H and F are involved. Position on the periodic table can also be used as an indicator: metal + nonmetal → ionic bond, and nonmetal + nonmetal → covalent bond. However, there are also many exceptions, especially when Be is involved.

	ΔEN	Bonding Type
(a) Li (metal, EN = 1.0) and O (nonmetal, EN = 3.5)	2.5	ionic
(b) Br (nonmetal, EN = 2.8) and I (nonmetal, EN = 2.5)	0.3	polar covalent
(c) Na (metal, EN = 1.0) and H (nonmetal, EN = 2.1)	1.1	ionic (NaH is a solid)
(d) O (nonmetal, EN = 3.5) and O (nonmetal, EN = 3.5)	0.0	nonpolar covalent
(e) H (nonmetal, EN = 2.1) and O (nonmetal, EN = 3.5)	1.4	polar covalent (H_2O is a liquid)

In this handbook, you will find the following tabulated data:

Compound	Form	Density	MP (°C)	BP (°C)	Solubility	
					Water	Ether
NaCl	colorless cubic crystal	2.165 g/mL (at 25°C)	801	1413	soluble	insoluble
PCl_3	colorless fuming liquid	1.574 g/mL (at 21°C)	-112	75.5	decomposes	soluble

NaCl is a crystalline solid with relatively high melting and boiling points. It is also soluble in water, a polar solvent and is relatively insoluble in alcohol and ether, solvents that are less polar. Hence it is consistent with the properties of an ionic compound.

PCl_3 is a colorless fuming liquid with low melting and boiling points. Unlike NaCl, it is soluble in ether and decomposes in water. Therefore, it is consistent with the properties of a covalent compound.

7-70. *Refer to Sections 7-8 and 7-9.*

(a) Ca_3N_2 — calcium nitride (ionic)

(b) Al_2O_3 — aluminum oxide (ionic)

(c) K_2Se — potassium selenide (ionic)

(d) $SrCl_2$ — strontium chloride (ionic)

7-72. *Refer to Sections 4-4, 4-5, 7-4, 7-5 and 7-6.*

(a) formula unit:
$$HCN(aq) + NaOH(aq) \rightarrow NaCN(aq) + H_2O(\ell)$$

total ionic:
$$HCN(aq) + Na^+(aq) + OH^-(aq) \rightarrow Na^+(aq) + CN^-(aq) + H_2O(\ell)$$

net ionic:
$$HCN(aq) + OH^-(aq) \rightarrow CN^-(aq) + H_2O(\ell)$$

(b) formula unit:
$$HCl(aq) + NaOH(aq) \rightarrow NaCl(aq) + H_2O(\ell)$$

total ionic:
$$H^+(aq) + Cl^-(aq) + Na^+(aq) + OH^-(aq) \rightarrow Na^+(aq) + Cl^-(aq) + H_2O(\ell)$$

net ionic:
$$H^+(aq) + OH^-(aq) \rightarrow H_2O(\ell)$$

(c) formula unit:
$$CaCl_2(aq) + Na_2CO_3(aq) \rightarrow 2NaCl(aq) + CaCO_3(s)$$

total ionic:
$$Ca^{2+}(aq) + 2Cl^-(aq) + 2Na^+(aq) + CO_3^{2-}(aq) \rightarrow 2Na^+(aq) + 2Cl^-(aq) + CaCO_3(s)$$

net ionic:
$$Ca^{2+}(aq) + CO_3^{2-}(aq) \rightarrow CaCO_3(s)$$

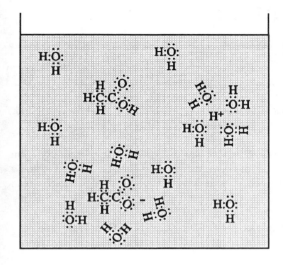

There are three solute species present due to the partial dissociation of acetic acid, CH_3COOH:

$$CH_3COOH(aq) \rightleftarrows CH_3COO^-(aq) + H^+(aq)$$

The solvent species, H_2O, is a very polar molecule. The water molecules arrange themselves around the ions so that the slightly positive ends of the water molecules point toward the negative ions, and the slightly negative ends of the water molecules point toward the positive ions.

8 Molecular Structure and Covalent Bonding Theories

8-2.   *Refer to Sections 7-8 and 8-8.*

(a) "Bonding pair" is a term that refers to a pair of electrons that is shared between two nuclei in a covalent bond, while the term "lone pair" refers to an unshared pair of electrons that is associated with a single nucleus.

(b) Lone pairs of electrons occupy more space than bonding pairs. This fact was determined experimentally from measurements of bond angles of many molecules and polyatomic ions. An explanation for this is the fact that a lone pair has only one atom exerting strong attractive forces on it, and it exists closer to the nucleus than bonding pairs.

(c) The relative magnitudes of the repulsive forces between pairs of electrons on an atom are as follows:

$$bp/bp < lp/bp << lp/lp$$

where lp refers to lone pairs and bp refers to bonding pairs of valence shell electrons.

8-4.   *Refer to Sections 8-5, 8-6 and 8-7.*

The molecular (or ionic) geometry is identical to the electronic geometry when there are no lone pairs on the central atom.

8-6.   *Refer to Section 8-2.*

When VSEPR theory is used to predict molecular geometries, double and triple bonds are treated identically to single bonds: as a single region of high electron density. A nonbonding electron is also counted as one region of high electron density.

8-8.   *Refer to Sections 8-2, 8-11 and 8-15, and Tables 8-3 and 8-4.*

Three possible arrangements of AB_2U_3 with 2 atoms of B and 3 lone pairs around the central A atom are:

(1) 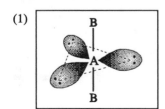   (2) 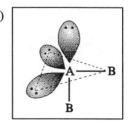   (3)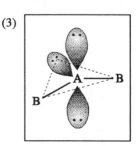

According to VSEPR theory, the most stable arrangement of the three lone pairs of electrons would be in the equatorial position, as shown in (1), where they would be less crowded. Therefore, a linear structure is the correct molecular geometry of the molecule.

(a) *s-s* overlap   (b) *s-p* overlap   (c) *p-p* overlap along bond axis   (d) *p-p* overlap perpendicular to bond axis

 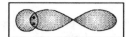  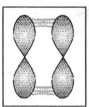

8-12. *Refer to Table 8-4.*

(a) *sp*

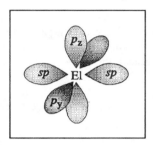

(b) *sp²*

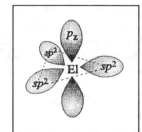

(c) *sp³*

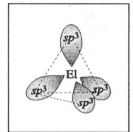

(d) *sp³d*

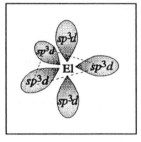

(e) *sp³d²*

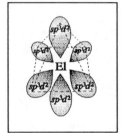

8-14. *Refer to Table 8-4 and Exercise 8-12 Solution.*

(a) *sp* 180°   (b) *sp²* 120°   (c) *sp³* 109.5°   (d) *sp³d* 90°, 120°, 180°   (e) *sp³d²* 90°, 180°

8-16. *Refer to Table 8-4.*

(a) ABU₅   *sp³d²*   (b) AB₂U₄   *sp³d²*   (c) AB₂   **B-A-B**   *sp*

(d) AB₃U₂   *sp³d*   (e) AB₅   *sp³d*

(a) NCl_3

$S = N - A = [3 \times 8(\text{for Cl}) + 1 \times 8(\text{for N})] - [3 \times 7(\text{for Cl}) + 1 \times 5(\text{for N})]$
$= 6$ shared electrons

The Lewis formula for the molecule (type AB_3U) predicts 4 regions of high electron density around the central N atom and a tetrahedral electronic geometry. There is 1 lone pair of electrons on the N atom, so the molecular geometry is pyramidal (Section 8-8).

(b) $AlCl_3$

This molecule (type AB_3) does not obey the octet rule, since Al is a IIIA element.

$A = 3 \times 7(\text{for Cl}) + 1 \times 3(\text{for Al}) = 24$ (total number of dots)

The Lewis formula predicts 3 regions of high electron density around the central Al atom and a trigonal planar electronic geometry. Since there are no lone pairs on Al, the molecular geometry is also trigonal planar (Section 8-6).

(c) SiH_4

$S = N - A = [4 \times 2(\text{for H}) + 1 \times 8(\text{for Si})] - [4 \times 1(\text{for H}) + 1 \times 4(\text{for Si})]$
$= 8$ shared electrons

The Lewis formula for the molecule (type AB_4) predicts 4 regions of high electron density around the central Si atom and a tetrahedral electronic geometry. The molecular geometry is the same as the electronic geometry because there are no lone pairs on Si (Section 8-7).

(d) SF_6

This molecule (type AB_6) does not obey the octet rule since 12 electrons must be shared to form 6 S-F bonds.

Available electrons, $A = 6 \times 7(\text{for F}) + 1 \times 6(\text{for S}) = 48$ (total number of dots)

The Lewis formula predicts 6 regions of high electron density around the central S atom and an octahedral electronic geometry. There are no lone pairs on the S atom, so the molecular geometry is the same as the electronic geometry (Section 8-12).

(e) IO_4^-

$S = N - A = [4 \times 8(\text{for O}) + 1 \times 8(\text{for I})] - [4 \times 6(\text{for O}) + 1 \times 7(\text{for I}) + 1\ e^-]$
$= 8$ shared electrons

The Lewis formula for the ion (type AB_4) predicts 4 regions of high electron density around the central I atom with no lone pairs of electrons. The ionic geometry is the same as the electronic geometry: tetrahedral (Section 8-7).

(a) SeF_6

This molecule (type AB_6) does not obey the octet rule since 12 electrons must be shared to form 6 Se-F bonds.

Available electrons, $A = 6 \times 7(\text{for F}) + 1 \times 6(\text{for Se}) = 48$ (total number of dots)

The Lewis formula predicts 6 regions of high electron density around the central Se atom and an octahedral electronic geometry. There are no lone pairs on the Se atom, so the molecular geometry is the same as the electronic geometry (Section 8-12).

(b) $ONCl$

$S = N - A = [1 \times 8(\text{for O}) + 1 \times 8(\text{for N}) + 1 \times 8(\text{for Cl})]$
$\qquad - [3 \times 6(\text{for O}) + 1 \times 5(\text{for N}) + 1 \times 7(\text{for Cl})]$
$= 6$ shared electrons

The Lewis formula for the molecule (type AB_2U) predicts 3 regions of high electron density around the central N atom and a trigonal planar electronic geometry. There is 1 lone pair of electrons around N, therefore the molecular geometry is angular or bent (Section 8-13).

(c) Cl_2CO

$S = N - A = [2 \times 8(\text{for Cl}) + 1 \times 8(\text{for C}) + 1 \times 8(\text{for O})]$
$\qquad\qquad - [2 \times 7(\text{for Cl}) + 1 \times 4(\text{for C}) + 1 \times 6(\text{for O})]$
$\qquad\quad = 8 \text{ shared electrons}$

The Lewis formula for the molecule (type AB_3) predicts 3 regions of high electron density around the central C atom and a trigonal planar electronic geometry. The molecular geometry is also trigonal planar since there are no lone pairs of electrons on the C atom (Section 8-13).

(d) $AsCl_3$

$S = N - A = [3 \times 8(\text{for Cl}) + 1 \times 8(\text{for As})] - [3 \times 7(\text{for Cl}) + 1 \times 5(\text{for As})]$
$\qquad\quad = 6 \text{ shared electrons}$

The Lewis formula for the molecule (type AB_3U) predicts 4 regions of high electron density around the central As atom and a tetrahedral electronic geometry. There is 1 lone pair of electrons on the As atom, so the molecular geometry is pyramidal (Section 8-8).

(e) BCl_3

This compound (type AB_3) does not obey the octet rule since only 6 electrons are shared to form 3 B-Cl bonds.

$A = 3 \times 7(\text{for Cl}) + 1 \times 3(\text{for B}) = 24 \text{ (total number of dots)}$

The Lewis formula predicts 3 regions of high electron density around the central B atom with no lone pairs of electrons. The electronic geometry is therefore the same as its molecular geometry: trigonal planar (Section 8-6).

(f) ClO_4^-

$S = N - A = [4 \times 8(\text{for O}) + 1 \times 8(\text{for Cl})] - [4 \times 6(\text{for O}) + 1 \times 7(\text{for Cl}) + 1e^-]$
$\qquad\quad = 8 \text{ shared electrons}$

The Lewis formula for this polyatomic ion (type AB_4) predicts 4 regions of high electron density around the central Cl atom. Because there are no lone pairs of electrons on Cl, the electronic and ionic geometry are the same: tetrahedral (Section 8-7).

8-22. *Refer to Table 8-4 and the Sections as stated.*

(a) H_2O

$S = N - A = [2 \times 2(\text{for H}) + 1 \times 8(\text{for O})] - [2 \times 1(\text{for H}) + 1 \times 6(\text{for O})]$
$\qquad\quad = 4 \text{ shared electrons}$

The Lewis formula for the molecule (type AB_2U_2) predicts 4 regions of high electron density around the central O atom and a tetrahedral electronic geometry. There are two lone pairs on the O atom, so the molecular geometry is bent or angular (Section 8-9).

(b) $SnCl_4$

Sn is a IVA element and has 4 valence electrons.

$S = N - A = [4 \times 8(\text{for Cl}) + 1 \times 8(\text{for Sn})] - [4 \times 7(\text{for Cl}) + 1 \times 4(\text{for Sn})]$
$\qquad\quad = 8 \text{ shared electrons}$

The Lewis formula for the molecule (type AB_4) predicts 4 regions of high electron density around the central Sn atom and a tetrahedral electronic geometry. Since there are no lone pairs on Sn, the molecular geometry is also tetrahedral (Section 8-7).

(c) BrF_3

This molecule (type AB_3U_2) does not obey the octet rule without modification.

Available electrons, $A = 3 \times 7(\text{for F}) + 1 \times 7(\text{for Br}) = 28 \text{ (total number of dots)}$

The Lewis formula predicts 5 regions of high electron density around the central Br atom and a trigonal bipyramidal electronic geometry. There are two lone pairs on the Br atom, so the molecular geometry is T-shaped (Section 8-11).

(d) SbF₆⁻

This polyatomic ion (type AB₆), like (c), does not obey the octet rule without modification since 12 electrons must be shared to form 6 Sb-F bonds. Sb is a VA element, but the charge on the ion gives an extra electron which participates in bonding. The Lewis formula predicts 6 regions of high electron density around the central Sb atom and an octahedral electronic geometry. There are no lone pairs on the Sb atom, so the ionic geometry is the same as the electronic geometry (Section 8-12).

8-24.   *Refer to Exercise 8-22 Solution, Table 8-1 and Section 8-9.*

(a) H₂O   The ideal bond angles would be those for a perfect tetrahedral structure, 109.5°.

     SnCl₄   The ideal bond angles would be 109.5° since the structure is tetrahedral.

     BrF₃   The ideal bond angles would be those for a trigonal bipyramidal geometry (type AB₃U₂): 90° and 180°. One F atom and the 2 lone pairs on the Br atom are separated by 120°.

     SbF₆⁻   The ideal bond angles would be those for an octahedron, 90° and 180°.

(b) These bond angles differ from the actual bond angles for H₂O and BrF₃, since these species have lone pairs of electrons on the central atom. Lone pairs of electrons require more space than bonding pairs of electrons: the H-O-H bond angle in H₂O is reduced from 109.5° to 104.5°. The F-Br-F bond angles in BrF₃ are also slightly reduced.

8-26.   *Refer to Table 8-4.*

C₃O₂ $S = N - A = [2 \times 8(\text{for O}) + 3 \times 8(\text{for C})] - [2 \times 6(\text{for O}) + 3 \times 4(\text{for C})] = 16$ shared electrons

$$:\ddot{O}::C::C::C::\ddot{O}: \quad \text{or} \quad :\ddot{O}=C=C=C=\ddot{O}:$$

The Lewis formula predicts 2 regions of high electron density around each C atom, resulting in the linear structure of C₃O₂.

8-28.   *Refer to Tables 8-3 and 8-4, and the Sections as stated.*

(a) AsCl₄⁻

This polyatomic ion (type AB₄U) does not obey the octet rule.

Available electrons, $A = 4 \times 7(\text{for Cl}) + 1 \times 5(\text{for As}) + 1\ e^- = 34$ (total no. of dots)

The Lewis formula shows 5 regions of high electron density around the central As atom, including 1 lone pair of electrons. The electronic geometry is trigonal bipyramidal and the ionic geometry is a see-saw (Section 8-11).

(b) PCl₆⁻

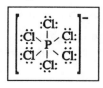

The ion (type AB₆) does not obey the octet rule.

$A = 6 \times 7(\text{for Cl}) + 1 \times 5(\text{for P}) + 1\ e^- = 48$ (total number of dots)

The Lewis formula shows 6 regions of high electron density around the central P atom. The electronic geometry and the ionic geometry are both octahedral because there are no lone pairs of electrons on P (Section 8-12).

(c) PCl_4^-

The ion (type AB_4U) does not obey the octet rule.

$A = 4 \times 7(\text{for Cl}) + 1 \times 5(\text{for P}) + 1\,e^- = 34$ (total number of dots)

The Lewis formula shows 5 regions of high electron density around the central P atom and its electronic geometry is trigonal bipyramidal. The ionic geometry is a seesaw due to the presence of 1 lone pair of electrons on the central P atom (Section 8-11).

(d) $SbCl_4^+$

$S = N - A = [4 \times 8(\text{for Cl}) + 1 \times 8(\text{for Sb})] - [4 \times 7(\text{for Cl}) + 1 \times 5(\text{for Sb}) - 1e^-]$
$= 8$ shared electrons

The Lewis formula for the ion (type AB_4) predicts 4 regions of high electron density around the central Sb atom and a tetrahedral electronic geometry. Since there are no lone pairs on Sb, the ionic geometry is also tetrahedral (Section 8-7).

(a)

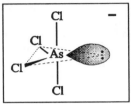

(b)

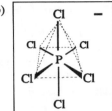

(c)

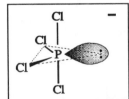

(d)

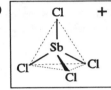

8-30. *Refer to Sections 8-11 and 8-12, and Table 8-3.*

None of the following interhalogen polyatomic ions obey the octet rule.

(a) IF_4^+

Available electrons, $A = 4 \times 7(\text{for F}) + 1 \times 7(\text{for I}) - 1\,e^- = 34$ (total number of dots)

The Lewis formula for the ion (type AB_4U) predicts 5 regions of high electron density around I and a trigonal bipyramidal electronic geometry. Due to the presence of a lone pair, the ionic geometry is a seesaw.

(b) ICl_4^-

$A = 4 \times 7(\text{for Cl}) + 1 \times 7(\text{for I}) + 1\,e^- = 36$ (total number of dots)

The Lewis formula for the ion (type AB_4U_2) predicts 6 regions of high electron density and an octahedral electronic geometry. The ionic geometry is square planar, due to the presence of 2 lone pairs.

(c) BrF_4^-

$A = 4 \times 7(\text{for F}) + 1 \times 7(\text{for Br}) + 1\,e^- = 36$ (total number of dots)

The Lewis formula for the ion (type AB_4U_2) predicts 6 regions of high electron density and an octahedral electronic geometry. The ionic geometry is square planar, due to the presence of 2 lone pairs.

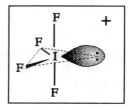

(a) F, F, I, F, F, +

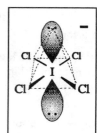

(b) Cl, Cl, I, Cl, Cl, −

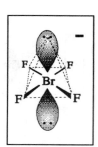

(c) F, F, Br, F, F, −

8-32. Refer to Sections 8-7, 8-11 and 8-12.

(a) GeF_4

$S = N - A = [4 \times 8(\text{for F}) + 1 \times 8(\text{for Ge})] - [4 \times 7(\text{for F}) + 1 \times 4(\text{for Ge})]$
$= 8$ shared electrons

The Lewis formula for the molecule (type AB_4) predicts 4 regions of high electron density around the central Ge atom with no lone pairs of electrons. The molecular geometry is the same as the electronic geometry: tetrahedral.

SF_4

The molecule (type AB_4U) does not obey the octet rule.

$A = 4 \times 7(\text{for F}) + 1 \times 6(\text{for S}) = 34$ (total number of dots)

The Lewis formula shows 5 regions of high electron density around the central S atom and its electronic geometry is trigonal bipyramidal. The molecular geometry is a seesaw due to the presence of 1 lone pair of electrons on the central S atom.

XeF_4

This molecule (type AB_4U_2) does not obey the octet rule.

$A = 4 \times 7(\text{for F}) + 1 \times 8(\text{for Xe}) = 36$ (total number of dots)

The Lewis formula predicts 6 regions of high electron density around the central Xe atom and its electronic geometry is octahedral. The molecular geometry is square planar due to the presence of 2 lone pairs of electrons on the central Xe atom.

(b) It is obvious that the molecular geometries of GeF_4, SF_4 and XeF_4 are not the same even though their molecular formulas are similar. The differences are due to the different number of valence electrons that must be accommodated by the central atom as lone pairs of electrons.

8-34. Refer to Table 8-3 and the Sections as stated.

(a) ICl_2^-

The ion (type AB_2U_3) does not obey the octet rule.

$A = 2 \times 7(\text{for Cl}) + 1 \times 7(\text{for I}) + 1\ e^- = 22$ (total number of dots)

The Lewis formula predicts 5 regions of high electron density around the central I atom and a trigonal bipyramidal electronic geometry. This ionic geometry is probably linear.

(b) $TeCl_4$

The molecule (type AB_4U) does not obey the octet rule.

$A = 4 \times 7(\text{for Cl}) + 1 \times 6(\text{for Te}) = 34$ (total number of dots)

The Lewis formula predicts 5 regions of high electron density around Te with 1 lone pair of electrons. The electronic geometry is trigonal bipyramidal and the molecular geometry is a seesaw (Section 8-11).

(c) XeO₃

$S = N - A = [3 \times 8(\text{for O}) + 1 \times 8(\text{for Xe})] - [3 \times 6(\text{for O}) + 1 \times 8(\text{for Xe})]$
$= 6 \text{ shared electrons}$

The Lewis formula for the molecule (type AB₃U) predicts 4 regions of high electron density around Xe including 1 lone pair of electrons. The electronic geometry is tetrahedral and the molecular geometry is pyramidal (Section 8-8).

(d) BrNO

$S = N - A = [1 \times 8(\text{for Br}) + 1 \times 8(\text{for N}) + 1 \times 8(\text{for O})]$
$\quad - [1 \times 7(\text{for Br}) + 1 \times 5(\text{for N}) + 1 \times 6(\text{for O})]$
$= 6 \text{ shared electrons}$

The Lewis formula for the molecule (type AB₂U) predicts 3 regions of high electron density around the central N atom including 1 lone pair of electrons. The electronic geometry is trigonal planar and the molecular geometry is angular or bent (Table 8-3).

(e) ClNO₂

$S = N - A = [1 \times 8(\text{for Cl}) + 1 \times 8(\text{for N}) + 2 \times 8(\text{for O})]$
$\quad - [1 \times 7(\text{for Cl}) + 1 \times 5(\text{for N}) + 2 \times 6(\text{for O})]$
$= 8 \text{ shared electrons}$

The Lewis formula for the molecule (type AB₃) predicts 3 regions of high electron density around the central N atom. Only 1 of the two resonance structures is shown. The electronic and molecular geometries are the same: trigonal planar because there are no lone pairs of electrons on the N atom (Section 8-6).

(f) Cl₂SO

$S = N - A = [2 \times 8(\text{for Cl}) + 1 \times 8(\text{for S}) + 1 \times 8(\text{for O})]$
$\quad - [2 \times 7(\text{for Cl}) + 1 \times 6(\text{for S}) + 1 \times 6(\text{for O})]$
$= 6 \text{ shared electrons}$

The Lewis formula for the molecule (type AB₃U) predicts 4 regions of high electron density around the central S atom including 1 lone pair of electrons. The electronic geometry is tetrahedral and the molecular geometry is pyramidal (Section 8-8).

8-36.   Refer to Table 8-4.

(a) H₃O⁺

$S = N - A = [3 \times 2(\text{for H}) + 1 \times 8(\text{for O})] - [3 \times 1(\text{for H}) + 1 \times 6(\text{for O}) - 1e^-]$
$= 6 \text{ shared electrons}$

The Lewis formula for the ion (type AB₃U) predicts 4 regions of high electron density around O including 1 lone pair of electrons. The electronic geometry is tetrahedral and the ionic geometry is pyramidal.

(b) GeF₃⁻

$S = N - A = [3 \times 8(\text{for F}) + 1 \times 8(\text{for Ge})] - [3 \times 7(\text{for F}) + 1 \times 4(\text{for Ge}) + 1e^-]$
$= 6 \text{ shared electrons}$

The Lewis formula for the ion (type AB₃U) predicts 4 regions of high electron density around Ge including 1 lone pair of electrons. The electronic geometry is tetrahedral and the ionic geometry is pyramidal.

(c) ClF₃²⁻

The ion (type AB₃U₃) does not obey the octet rule.

$A = 3 \times 7(\text{for F}) + 1 \times 7(\text{for Cl}) + 2\ e^- = 30 \text{ (total number of dots)}$

The Lewis formula predicts 6 regions of high electron density around Cl including 3 lone pairs of electrons. The electronic geometry is octahedral and the ionic geometry is T-shaped.

(d) $IO_2F_2^-$

The ion (type AB_4U) does not obey the octet rule.

$A = 2 \times 6(\text{for O}) + 2 \times 7(\text{for F}) + 1 \times 7(\text{for I}) + 1\ e^- = 34$ (total number of dots)

The Lewis formula predicts 5 regions of high electron density around the I atom including 1 lone pair of electrons. The electronic geometry is trigonal bipyramidal and the ionic geometry is a seesaw.

8-38. *Refer to Section 8-3 and the Sections as stated.*

(a) CdI_2

This molecule (type AB_2) has a linear electronic and molecular geometry. The Cd-I bond are polar. Since the molecule is symmetric, the bond dipoles cancel to give a nonpolar molecule (Section 8-5).

(b) BCl_3

This molecule (type AB_3) has a trigonal planar electronic geometry and trigonal planar molecular geometry. The B-Cl bonds are polar, but since the molecule is symmetrical, the bond dipoles cancel to give a nonpolar molecule (Section 8-6).

(c) NF_3

This molecule (type AB_3U) has a tetrahedral electronic geometry and a pyramidal molecular geometry. F (EN = 4.0) is more electronegative than N (EN = 3.0). The polar N-F bond dipoles oppose the effect of the lone pair. The molecule is only slightly polar (Section 8-8).

(d) H_2O

This molecule (type AB_2U_2) has a tetrahedral electronic geometry and an angular molecular geometry. Oxygen (EN = 3.5) is more electronegative than H (EN = 2.1). The O-H bond dipole reinforces the effect of the two lone pairs of electrons and so, H_2O is very polar (Section 8-9).

(e) SF_6

This molecule (type AB_6) has an octahedral electronic and molecular geometry. The S-F bonds are polar, but the molecule is symmetrical. The S-F bond dipoles cancel to give a nonpolar molecule (Section 8-12).

8-40. *Refer to Sections 8-3 and 8-11.*

PF_2Cl_3

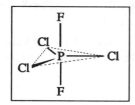

The Lewis formula for this molecule (type AB_5) predicts 5 regions of high electron density with no lone pairs of electrons on the central P atom. The molecular geometry is the same as the electronic geometry: trigonal bipyramidal. For the molecule to be nonpolar (dipole moment = 0), the P-F and P-Cl bonds must be symmetrically arranged, i.e., the three Cl atoms must be at the equatorial positions and the two F atoms must be at the axial positions.

8-42. *Refer to Table 8-2, Section 8-4 and Exercise 8-18 Solution.*

(a) NCl_3 sp^3 **(b)** $AlCl_3$ sp^2 **(c)** SiH_4 sp^3

(d) SF_6 sp^3d^2 **(e)** IO_4^- sp^3

8-44. *Refer to Table 8-2, Section 8-4 and Exercise 8-20 Solution.*

(a) SeF_6	sp^3d^2	(b) ONCl	sp^2	(c) Cl_2CO	sp^2
(d) $AsCl_3$	sp^3	(e) BCl_3	sp^2	(f) ClO_4^-	sp^3

8-46. *Refer to Table 8-2.*

	Lewis Formula	Hybridization	Electronic Geometry	Molecular Geometry
(1) $CHCl_3$	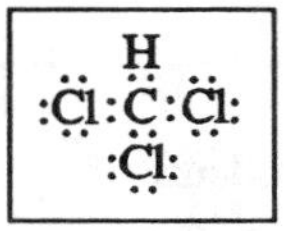	sp^3	tetrahedral	tetrahedral (distorted)
(2) CH_2Cl_2	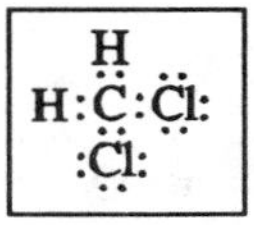	sp^3	tetrahedral	tetrahedral (distorted)
(3) NF_3	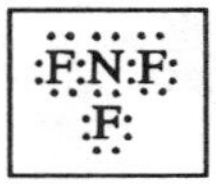	sp^3	tetrahedral	pyramidal (1 lone pair of e^-)
(4) ClO_4^-	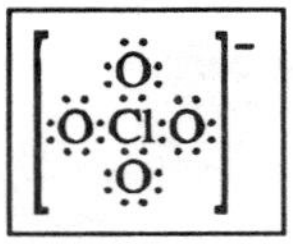	sp^3	tetradedral	tetrahedral
(5) IF_6^+	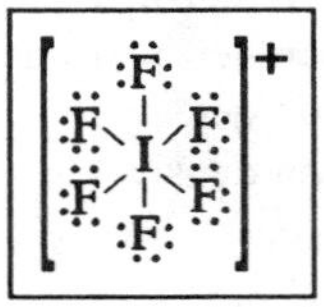	sp^3d^2	octahedral	octahedral
(6) SiF_6^{2-}	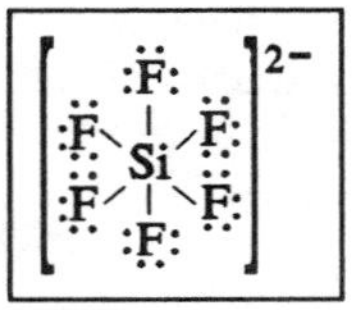	sp^3d^2	octahedral	octahedral

8-48 *Refer to Table 8-2.*

(a) NO_2^+

[:Ö::N::Ö:]$^+$

$S = N - A = [1 \times 8(\text{for N}) + 2 \times 8(\text{for O})]$
$\quad - [1 \times 5(\text{for N}) + 2 \times 6(\text{for O}) - 1e^-]$
$\quad = 8$ shared electrons

The Lewis formula of NO_2^+ (type AB_2) shows 2 regions of high electron density around the central N atom. The hybridization of N is *sp*.

NO_2^-

$$\left[:\ddot{\text{O}}:\text{N}::\ddot{\text{O}}:\right]^- \leftrightarrow \left[:\ddot{\text{O}}::\text{N}:\ddot{\text{O}}:\right]^-$$

$$S = N - A = [1 \times 8(\text{for N}) + 2 \times 8(\text{for O})]$$
$$- [1 \times 5(\text{for N}) + 2 \times 6(\text{for O}) + 1e^-]$$
$$= 6 \text{ shared electrons}$$

The Lewis resonance structures of NO_2^- (type AB_2U) show 3 regions of high electron density around the central N atom. The hybridization of N is sp^2.

(b) The bond angle for the linear ion, NO_2^+, is 180°. The ideal bond angle for the angular ion, NO_2^-, is 120°. The actual O-N-O bond angle is slightly less than 120°, because a lone pair of electrons requires more room than a bonding pair.

8-50. *Refer to Table 8-2 and Sections 8-13 and 8-14.*

(a)

$$\begin{array}{c} \text{H H} \\ | \ | \\ \text{H-C}_1\text{-C}_2\text{-}\ddot{\text{O}}\text{-H} \\ | \ | \\ \text{H H} \end{array}$$

C_1 sp^3 hybridization (4 regions of high electron density)
C_2 sp^3 hybridization (4 regions of high electron density)

(b)

$$\begin{array}{c} \text{H} \\ | \\ :\text{N-H} \ \ \ddot{\text{O}}. \\ | \ \ \ \ // \\ \text{H-C}_1\text{-C}_2 \\ \ \ \ \ \ \ \ \ \ddot{\text{O}}\text{-H} \end{array}$$

C_1 sp^3 hybridization (4 regions of high electron density)
C_2 sp^2 hybridization (3 regions of high electron density)

(c)

$$\begin{array}{c} :\text{N}\equiv\text{C} \ \ \ \ \ \ \ \ \text{C}\equiv\text{N}: \\ \ \ \ \ \ \ \ \ \backslash \ \ \ \ / \\ \ \ \ \ \ \ \ \ \text{C}=\text{C} \\ \ \ \ \ \ \ \ / \ \ \ \ \backslash \\ :\text{N}\equiv\text{C} \ \ \ \ \ \ \ \ \text{C}\equiv\text{N}: \end{array}$$

C_1 sp hybridization (2 regions of high electron density)
C_2 sp^2 hybridization (3 regions of high electron density)

(d)

$$\begin{array}{c} \text{H} \ \ \ \ \ \ \ \ \text{H} \\ \ \backslash \ \ \ \ \ \ / \\ \ \ \text{C}=\text{C}_2 \ \ \ \ \text{H} \\ / \ \ \ \ \ \ \ \ \ / \\ \text{H} \ \ \ \ \text{C}_3=\text{C}_4 \\ \ \ \ :\ddot{\text{Cl}} \ \ \ \ \text{H} \end{array}$$

C_1 sp^2 hybridization (3 regions of high electron density)
C_2 sp^2 hybridization (3 regions of high electron density)
C_3 sp^2 hybridization (3 regions of high electron density)
C_4 sp^2 hybridization (3 regions of high electron density)

(e)

$$\begin{array}{c} \text{H H H} \\ | \ | \ | \\ \text{H-C-C}=\text{C-C}\equiv\text{C-H} \\ | \\ \text{H} \end{array}$$

C_1 sp^3 hybridization (4 regions of high electron density)
C_2 sp^2 hybridization (3 regions of high electron density)
C_3 sp^2 hybridization (3 regions of high electron density)
C_4 sp hybridization (2 regions of high electron density)
C_5 sp hybridization (2 regions of high electron density)

8-52. *Refer to Sections 8-13 and 8-14, and Figures 8-6 and 8-8.*

(a) The molecule contains 7 single bonds (7 sigma bonds) and 1 double bond (1 sigma bond, 1 pi bond) for a total of 8 sigma bonds and 1 pi bond.

(b) The molecule contains 4 single bonds (4 sigma bonds) and 2 double bonds (2 sigma bonds, 2 pi bonds) for a total of 6 sigma bonds and 2 pi bonds.

(c) The molecule contains 9 single bonds (9 sigma bonds) and 1 double bond (1 sigma bond, 1 pi bond) for a total of 10 sigma bonds and 1 pi bond.

(d) The molecule contains 3 single bonds (3 sigma bonds) and 2 triple bonds (2 sigma bonds, 4 pi bonds) for a total of 5 sigma bonds and 4 pi bonds.

8-54. *Refer to Section 8-14 and Figure 8-8b.*

N_2 $\boxed{\text{:N::N:}}$ $S = N - A = [2 \times 8(\text{for N})] - [2 \times 5(\text{for N})] = 6$ shared electrons

There are 2 regions of high electron density around each N, so each N atom is *sp* hybridized and the bond angle is 180°.

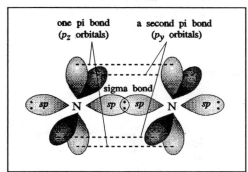

The sigma bond results from the head-on overlap of the *sp* hybrid orbitals located on each N atom.

One pi bond results from the side-on overlap of the $2p_y$ orbitals located on each N atom. The other pi bond results from the overlap of the $2p_z$ orbitals on each N atom.

8-56. *Refer to Sections 8-5, 8-6 and 8-7.*

Note: Cd has two valence electrons.

	Lewis Formula	Hybridization	Molecular (or Ionic) Geometry
(1) $CdBr^+$	$\left[\text{Cd:}\ddot{\text{B}}\ddot{\text{r}}\text{:}\right]^+$	-	linear
(2) $CdBr_2$	$\text{:}\ddot{\text{B}}\ddot{\text{r}}\text{:Cd:}\ddot{\text{B}}\ddot{\text{r}}\text{:}$	*sp*	linear
(3) $CdBr_3^-$	$\left[\begin{array}{c}\text{:}\ddot{\text{B}}\ddot{\text{r}}\text{:}\\ \text{:}\ddot{\text{B}}\ddot{\text{r}}\text{:Cd:}\ddot{\text{B}}\ddot{\text{r}}\text{:}\end{array}\right]^-$	sp^2	trigonal planar
(4) $CdBr_4^{2-}$	$\left[\begin{array}{c}\text{:}\ddot{\text{B}}\ddot{\text{r}}\text{:}\\ \text{:}\ddot{\text{B}}\ddot{\text{r}}\text{:Cd:}\ddot{\text{B}}\ddot{\text{r}}\text{:}\\ \text{:}\ddot{\text{B}}\ddot{\text{r}}\text{:}\end{array}\right]^{2-}$	sp^3	tetrahedral

8-58. *Refer to Sections 8-13 and 8-14.*

(a) butane

C_4H_{10}

$$\begin{array}{c}\text{H H H H}\\ \text{H-C-C-C-C-H}\\ \text{H H H H}\end{array}$$

C_1, C_2, C_3, C_4 sp^3 hybridized with bond angles of 109.5°

(b) propene

$H_2C{=}CHCH_3$

$$\begin{array}{c}\text{H} \quad \text{H H}\\ \text{C=C-C-H}\\ \text{H} \quad \text{H}\end{array}$$

C_1, C_2 sp^2 hybridized with bond angles of 120°
C_3 sp^3 hybridized with bond angles of 109.5°

(c) 1-butyne

$HC \equiv CCH_2CH_3$

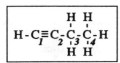

C_1, C_2 *sp* hybridized with bond angles of 180°.
C_3, C_4 *sp³* hybridized with bond angles of 109.5°.

(d) acetaldehyde

CH_3CHO

C_1 *sp³* hybridized with bond angles of 109.5°.
C_2 *sp²* hybridized with bond angles of 120°.

(a) butane, $CH_3CH_2CH_2CH_3$

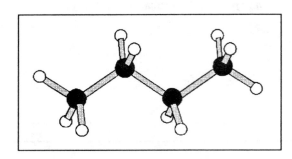

(b) propene, CH_2CHCH_3

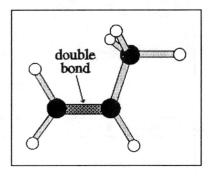

double bond

(c) 1-butyne, $CHCCH_2CH_3$

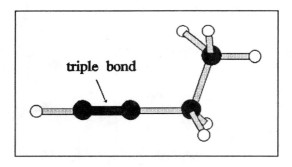

triple bond

(d) acetaldehyde, CH_3CHO

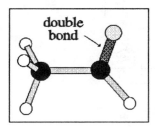

double bond

8-60.   *Refer to Sections 8-7, 8-8, 8-13 and 8-14.*

(1) NH_3

(a) The N atom is *sp³* hybridized because the molecule (type AB₃U) has 4 regions of high electron density around the central N atom.

(b) Three of the *sp³* hybrid orbitals overlap with the 1*s* orbitals of the H atoms to form sigma bonds, leaving the fourth hybrid orbital to contain 1 lone pair of electrons on the N.

(2) NH_4^+

(a) The N atom is *sp³* hybridized because the ion (type AB₄) has 4 regions of high electron density around the central N atom.

(b) Each of the four *sp³* hybrid orbitals overlap with the 1*s* orbital of an H atom to form a sigma bond.

(3) N_2H_2 (a) Each N atom is sp^2 hybridized because there are 3 regions of high electron density around each N atom (a single bond, a double bond and a lone pair of electrons).

H-N̈=N̈-H

(b) Each N-H single bond is formed from the overlap of an sp^2 hybrid orbital of N with a $1s$ orbital of H. The double bond is the result of the side-on overlap of the unhybridized $2p$ orbitals on the N atoms, yielding a pi bond, and the head-on overlap of two sp^2 hybrid orbitals, yielding a sigma bond. The third sp^2 hybrid orbital is used to accommodate the lone pair of electrons.

(4) HCN   (a) The N atom is sp hybridized because there are 2 regions of high electron density around the N atom.

H-C≡N:

(b) A lone pair of electrons occupies one sp hybrid orbital on N. The other sp hybrid orbital on N overlaps head-on with an sp hybrid orbital on C forming a sigma bond. The unhybridized $2p$ orbitals on C and N overlap side-on to form 2 pi bonds. The one sigma bond with the two pi bonds create a C-N triple bond. The H-C bond is formed from the overlap of a $1s$ H orbital with an sp hybrid orbital on C.

(5) NH_2NH_2 (a) Both N atoms are sp^3 hybridized because there are 4 regions of high electron density around each N atom.

H-N̈-N̈-H
  H H

(b) Each N-H single bond is formed from the overlap of an sp^3 hybrid orbital of N with a $1s$ orbital of H. The single N-N bond is the result of the head-on overlap of an sp^3 hybrid orbital on each N atom, yielding a sigma bond. The fourth sp^3 hybrid orbital on each N atom is used to accommodate the lone pair of electrons.

8-62.   *Refer to Sections 8-7, 8-13 and 8-14.*

(1) H_2CO

H
H-C=Ö:

(a) The carbon atom is sp^2 hybridized.

(b) Since there are no lone pairs of electrons on C, the electronic and molecular geometries are trigonal planar. The molecule is planar (flat) with H-C-H and H-C-O bond angles of 120°.

(2) HCN

H-C≡N:

(a) The carbon atom is sp hybridized.

(b) The molecular geometry is linear and the H-C-N bond angle is 180°.

(3) $CH_3CH_2CH_3$

H H H
H-C-C-C-H
H H H

(a) All three carbon atoms are sp^3 hybridized.

(b) The molecular geometry about each carbon atom is tetrahedral with H-C-H and H-C-C bond angles of 109.5°.

(4) H_2C_2O

H
H-C=C=Ö:
  1 *2*

(a) Carbon-1 is sp^2 hybridized and carbon-2 is sp hybridized.

(b) The molecular geometry at carbon-1 is trigonal planar with H-C-H and H-C-C bond angles of 120°. The molecular geometry at carbon-2 is linear with C-C-O bond angle of 180°.

	Lewis Formula	Hybridization	Molecular (or Ionic) Geometry
(1) IF	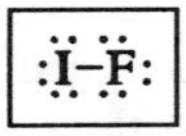 $:\ddot{I}-\ddot{F}:$	sp^3	linear
(2) IF$_3$	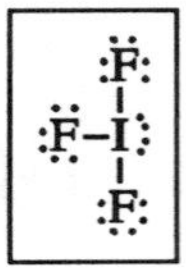 $\ddot{F}-\ddot{I}$ with F above and below	sp^3d	T-shaped
(3) IF$_4^-$	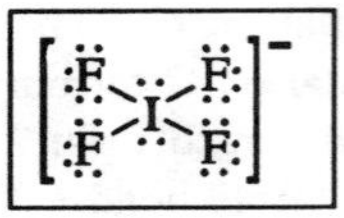 square planar IF$_4$ structure, bracketed with $-$ charge	sp^3d^2	square planar
(4) IF$_5$	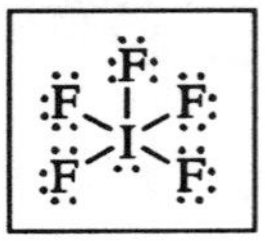 square pyramidal IF$_5$ structure	sp^3d^2	square pyramidal
(5) IF$_6^-$	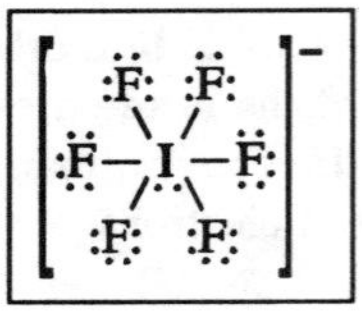 pentagonal structure, bracketed with $-$ charge	sp^3d^3	pentagonal pyramidal
(6) IF$_7$	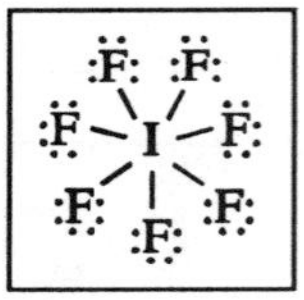 pentagonal bipyramidal IF$_7$ structure	sp^3d^3	pentagonal bipyramidal

---

8-66.   *Refer to Sections 8-8 and 8-9.*

hydroxylamine

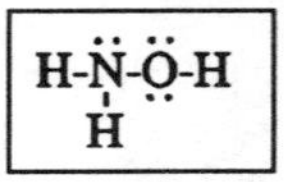

Since there are 4 regions of high electron density around both the N and O atoms, the ideal bond angles about these atoms should be 109.5°. The lone pair of electrons on the N atom would make the H-N-H and the H-N-O bond angles less than 109.5°. The two lone pairs of electrons on the O atom should make the N-O-H bond angle even smaller. In fact, the observed bond angles are 107° for H-N-H and H-N-O, and 102° for N-O-H.

8-68.   *Refer to Exercise 8-2 Solution and Section 8-8.*

Experimental evidence demonstrates that ·CH$_3$, the methyl free radical, has bond angles of about 120°, indicating a trigonal planar arrangement. This fact indicates that a single unpaired electron does not have much, if any, repulsive force. The species can be treated as if the central carbon atom had only 3 regions of high electron density, sp^2 hybridization and a resulting trigonal planar geometry.

The methyl carbanion, :CH$_3^-$, has bond angles close to that in a tetrahedral arrangement of atoms, 109°, indicating 4 regions of high electron density around the central C atom and sp^3 hybridization. The lone pair of electrons exerts significant repulsive force on the electrons in bonding orbitals and must be counted as a region of high electron density.

$CH_3CH=CH_2$

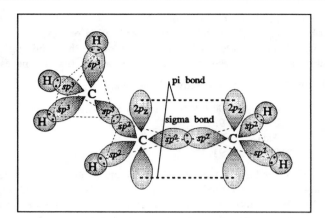

8-72. *Refer to Table 8-4.*

(a) PF_5

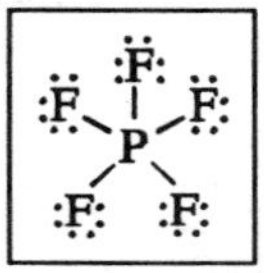

PF_6^-

The Lewis formulas predict sp^3d hybridization of P in PF_5 (5 regions of high electron density), changing to sp^3d^2 hybridization of P in PF_6^- (6 regions of high electron density).

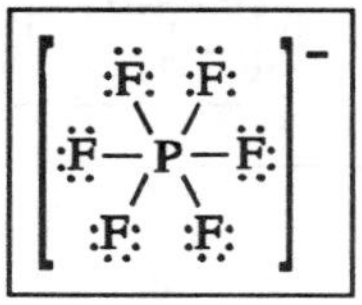

(b) CO

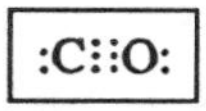

:C::O:

CO_2

:O::C::O:

The Lewis formulas predict sp hybridization of C in both molecules since both C atoms have 2 regions of high electron density around them.

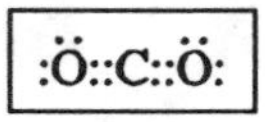

(c) AlI_3

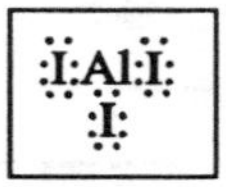

AlI_4^-

The Lewis formulas predict sp^2 hybridization of Al in AlI_3 (3 regions of high electron density) changing to sp^3 hybridization of Al in AlI_4^- (4 regions of high electron density).

(d) NH_3

BF_3

:F:B:F:
:F:

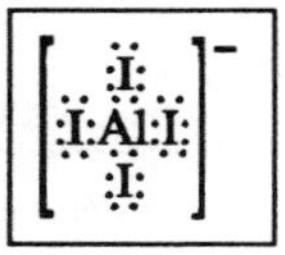

$H_3N:BF_3$

The Lewis formulas predict sp^3 hybridization of N in NH_3 (4 regions of high electron density) and sp^2 hybridization of B in BF_3 (3 regions of high electron density). The product of the reaction, $H_3N:BF_3$, contains both N and B atoms with sp^3 hybridization.

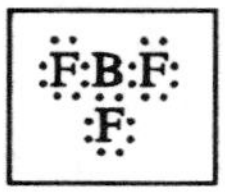

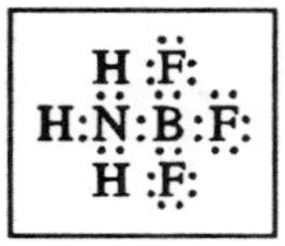

Using Valence Bond (VB) theory, the central atoms of the molecules with formulas AB_2U_2 and AB_3U should undergo sp^3 hybridized with predicted bond angles of 109.5°. If no hybridization occurs, bonds would be formed by the use of p orbitals. Since the p orbitals are oriented at 90° from each other, the bond angles would be 90°. Note that hybridization is only invoked if the actual molecular geometry data indicate that it is necessary.

The actual B-A-B bond angles for molecules of some representative elements are:

H_2O	104.45°	NH_3	106.67°
H_2S	92.2°	PH_3	93.7°
H_2Se	91.0°	AsH_3	91.8°
H_2Te	89.5°	SbH_3	91.3°

It is not necessary to invoke hybridization for the larger elements (Period 3 and greater). The above data show that since it is only the molecules containing the smaller Period 2 elements (O and N) have bond angles approaching those for sp^3 hybridization, 109.5°. The other molecules formed from the larger elements have bond angles closer to 90°, so the application of the hybridization concept is not necessary to explain their molecular geometries.

8-76.   *Refer to Sections 8-7 and 8-8.*

(1) $NH_3(g)$ + $HCl(g)$ → $NH_4Cl(s)$

(2)

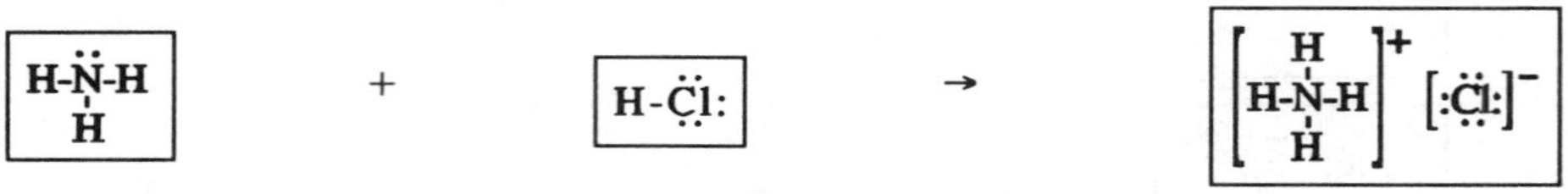

(3)

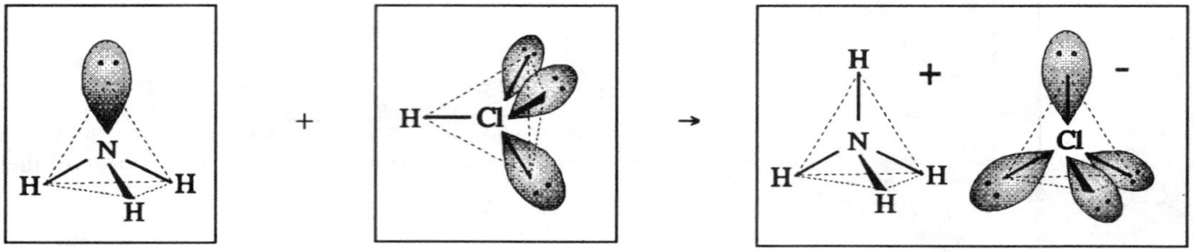

8-78.   *Refer to Sections 8-7, 8-8 and 8-9.*

(a) $H^+ + H_2O \rightarrow H_3O$ (b) $NH_3 + H^+ \rightarrow NH_4^+$

	Lewis Formula	Electronic Geometry	Molecular (Ionic) Geometry		Lewis Formula	Electronic Geometry	Molecular (Ionic) Geometry
H_2O	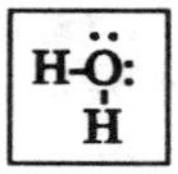	tetrahedral	angular (bent)	NH_3	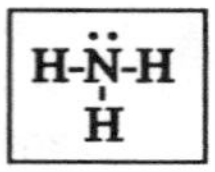	tetrahedral	pyramidal
H_3O^+	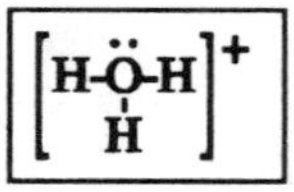	tetrahedral	pyramidal	NH_4^+	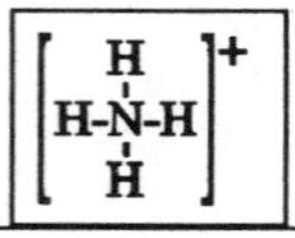	tetrahedral	tetrahedral

9 Molecular Orbitals in Chemical Bonding

9-2. *Refer to the Introduction to Chapter 9 and Section 9-1.*

A molecular orbital (MO) is an orbital resulting from the overlap and combination of atomic orbitals on different atoms. An MO and the electrons in it belong to the molecule as a whole. Molecular orbitals calculations are used to develop

(1) mathematical representations of the orbital shapes, and
(2) energy level diagrams for the molecules.

The mathematical pictures called "electron density maps" are used to determine molecular structures and the energy level diagrams are used to determine the energies of bond formation and to interpret spectroscopy data.

9-4. *Refer to Section 8-4.*

A set of hybridized atomic orbitals holds the same maximum number of electrons as the set of atomic orbitals from which the hybridized atomic orbitals were formed. A hybridized atomic orbital can hold a maximum of 2 electrons having opposite spin.

9-6. *Refer to Section 9-1.*

Bonding molecular orbitals (MOs) have energies that are lower than those of the original atomic orbitals. Antibonding molecular orbitals have energies that are higher than those of the original atomic orbitals. Therefore, bonding MOs are more stable than the original atomic orbitals, whereas, antibonding MOs are less stable than the original atomic orbitals.

9-8. *Refer to Section 9-1, and Figures 9-2 and 9-3.*

A σ (sigma) and a σ^* (sigma star) molecular orbital result from the head-on overlap and the subsequent combination of atomic orbitals on adjacent atoms. Both molecular orbitals are cylindrically symmetrical about the axis linking the two atoms. There is a high electron density in the region between the atoms for the bonding σ orbital, promoting bonding and stabilizing the system. However, the electron density between the atoms approaches zero at the nodal plane for the anti-bonding σ^* orbital, which destabilizes the system. As an example, for the overlap of 2 $1s$ orbitals (from Figure 9-2):

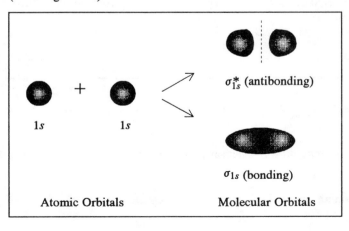

Figure 9-3 shows the bonding and antibonding sigma orbitals formed by the combining head-on of two p orbitals.

Rules for placing electrons in molecular orbitals:

(1) Choose the correct molecular orbital energy level diagram.

(2) Determine the *total* number of electrons in the molecule or ion.

(3) Put the electrons into the energy level diagram as follows:

- Each electron is placed in the lowest possible energy level.

- A maximum of 2 electrons can be placed in an orbital, but they must have opposite spins (Pauli Exclusion Principle).

- Electrons must occupy all the orbitals of a given energy level singly with the same spin before pairing begins (Hund's Rule).

9-12. *Refer to Section 9-1.*

(a) Atomic orbitals are *pure* orbitals that have not mixed with other orbitals in the same atom, molecule or ion. Molecular orbitals are orbitals resulting from the overlapping and mixing of atomic orbitals from *all* the atoms in a molecule. A molecular orbital belongs to the molecule as a whole and not simply to a particular atom in the molecule.

For example, each H atom has a $1s$ atomic orbital, but in the H_2 molecule, the σ_{1s} molecular orbital belongs to the entire H_2 molecule.

(b) A bonding molecular orbital (MO) results when two atomic orbitals overlap in phase. The energy of the bonding MO is always lower than the original atomic orbitals and is therefore more stable. An antibonding MO results when two atomic orbitals overlap out of phase; its energy is higher than the original atomic orbitals. Refer to Figure 9-2.

(c) A sigma (σ) bond results from the electron occupation of a sigma molecular orbital that is formed by the head-on overlap of two atomic orbitals. A pi (π) bond results from the electron occupation of a pi molecular orbital that is formed from the side-on overlap of atomic orbitals. See Figures 9-2, 9-3 and 9-4.

(d) A localized molecular orbital is an MO which is associated with one or two particular atoms in a molecule. The bonding orbitals are localized between the atoms that are bonded together, the nonbonding orbitals are localized on one particular atom. Delocalized molecular orbitals, on the other hand, cover the entire molecule. These orbitals cannot easily be labeled as specific bonds.

9-14. *Refer to the Key Terms for Chapter 9.*

(a) A homonuclear molecule or ion is a species containing only one element. For example, O_2, O_3 and O_2^+ are homonuclear.

(b) A heteronuclear molecule or ion is a species containing different elements. For example, HF, CN^- and H_2O are heteronuclear.

(c) A diatomic molecule or ion is a species containing two atoms. It can either be homonuclear (O_2, O_2^-) or heteronuclear (HF, CN^-).

Note: Bond Order $= \dfrac{\text{No. bonding electrons } - \text{ No. antibonding electrons}}{2}$

		No. Bonding Electrons	No. Antibonding Electrons	Bond Order
(a) He_2^+	$\sigma_{1s}^{2}\; \sigma_{1s}^{*\,1}$	2	1	0.5
(b) He_2	$\sigma_{1s}^{2}\; \sigma_{1s}^{*\,2}$	2	2	0

The ion He_2^+ has a non-zero bond order and would exist, but He_2 has a bond order of zero and would not exist.

Recall: Bond Order $= \dfrac{\text{No. bonding electrons } - \text{ No. antibonding electrons}}{2}$

		No. Bonding Electrons	No. Antibonding Electrons	Bond Order
(a) Li_2	$\sigma_{1s}^{2}\; \sigma_{1s}^{*\,2}\; \sigma_{2s}^{2}$	4	2	1
(b) Li_2^+	$\sigma_{1s}^{2}\; \sigma_{1s}^{*\,2}\; \sigma_{2s}^{1}$	3	2	0.5
(c) C_2^-	$\sigma_{1s}^{2}\; \sigma_{1s}^{*\,2}\; \sigma_{2s}^{2}\; \sigma_{2s}^{*\,2}\; \pi_{2p_y}^{2}\; \pi_{2p_z}^{2}\; \sigma_{2p}^{1}$	9	4	2.5

All these species would exist since their bond order is greater than zero.

Recall: Bond Order $= \dfrac{\text{No. bonding electrons } - \text{ No. antibonding electrons}}{2}$

		No. Bonding Electrons	No. Antibonding Electrons	Bond Order
(a) Be_2	$\sigma_{1s}^{2}\; \sigma_{1s}^{*\,2}\; \sigma_{2s}^{2}\; \sigma_{2s}^{*\,2}$	4	4	0
Be_2^+	$\sigma_{1s}^{2}\; \sigma_{1s}^{*\,2}\; \sigma_{2s}^{2}\; \sigma_{2s}^{*\,1}$	4	3	0.5
Be_2^-	$\sigma_{1s}^{2}\; \sigma_{1s}^{*\,2}\; \sigma_{2s}^{2}\; \sigma_{2s}^{*\,2}\; \pi_{2p_y}^{1}$	5	4	0.5
(b) B_2	$\sigma_{1s}^{2}\; \sigma_{1s}^{*\,2}\; \sigma_{2s}^{2}\; \sigma_{2s}^{*\,2}\; \pi_{2p_y}^{1}\; \pi_{2p_z}^{1}$	6	4	1
B_2^+	$\sigma_{1s}^{2}\; \sigma_{1s}^{*\,2}\; \sigma_{2s}^{2}\; \sigma_{2s}^{*\,2}\; \pi_{2p_y}^{1}$	5	4	0.5
B_2^-	$\sigma_{1s}^{2}\; \sigma_{1s}^{*\,2}\; \sigma_{2s}^{2}\; \sigma_{2s}^{*\,2}\; \pi_{2p_y}^{2}\; \pi_{2p_z}^{1}$	7	4	1.5
(c) CO	$\sigma_{1s}^{2}\; \sigma_{1s}^{*\,2}\; \sigma_{2s}^{2}\; \sigma_{2s}^{*\,2}\; \pi_{2p_y}^{2}\; \pi_{2p_z}^{2}\; \sigma_{2p}^{2}$	10	4	3.0
CO^+	$\sigma_{1s}^{2}\; \sigma_{1s}^{*\,2}\; \sigma_{2s}^{2}\; \sigma_{2s}^{*\,2}\; \pi_{2p_y}^{2}\; \pi_{2p_z}^{2}\; \sigma_{2p}^{1}$	9	4	2.5

9-22. *Refer to Section 9-3 and Exercise 9-20 Solution.*

In Molecular Orbital Theory, the greater the bond order, the more stable is the molecule or ion. Therefore, we predict:

Unstable: Be_2 Somewhat stable: Be_2^+, Be_2^-, B_2^+ Stable: B_2, B_2^-, CO, CO^+

This means that although Be_2 is unstable, both its $1+$ cation and $1-$ anion are somewhat stable. It also shows that the stability of the boron species is in the order: $B_2^+ < B_2 < B_2^-$. All these predictions are generally correct. The fact that many of these supposed stable species are not observed in nature is because they are chemically very reactive and are observed only at high temperatures and reduced pressures. Therefore, the chemical reactivity of a species is also instrumental when considering the survival probability of the species with stable chemical bonding.

9-24. *Refer to Section 9-3.*

Recall: Bond Order $= \dfrac{\text{No. bonding electrons} - \text{No. antibonding electrons}}{2}$

(a) X_2 bond order $= \dfrac{8 - 4}{2} = 2$

(b) X_2^+ bond order $= \dfrac{10 - 7}{2} = 1.5$

(c) X_2^- bond order $= \dfrac{10 - 5}{2} = 2.5$

9-26. *Refer to Sections 9-3 and 9-4.*

(a) N_2 $\sigma_{1s}^2\ \sigma_{1s}^{*2}\ \sigma_{2s}^2\ \sigma_{2s}^{*2}\ \pi_{2p_y}^2\ \pi_{2p_z}^2\ \sigma_{2p}^2$

N_2^- $\sigma_{1s}^2\ \sigma_{1s}^{*2}\ \sigma_{2s}^2\ \sigma_{2s}^{*2}\ \pi_{2p_y}^2\ \pi_{2p_z}^2\ \sigma_{2p}^2\ \pi_{2p_y}^{*1}$

N_2^+ $\sigma_{1s}^2\ \sigma_{1s}^{*2}\ \sigma_{2s}^2\ \sigma_{2s}^{*2}\ \pi_{2p_y}^2\ \pi_{2p_z}^2\ \sigma_{2p}^1$

(b) Recall: Bond Order $= \dfrac{\text{No. bonding electrons} - \text{No. antibonding electrons}}{2}$

N_2 bond order $= \dfrac{10 - 4}{2} = 3$

N_2^- bond order $= \dfrac{10 - 5}{2} = 2.5$

N_2^+ bond order $= \dfrac{9 - 4}{2} = 2.5$

(c) The species with the greatest bond order has the greatest bond energy and the shortest bond length. Therefore, the N_2 molecule with a bond order of 3 has the shortest bond. Both N_2^- and N_2^+ ions with bond orders of 2.5, should have slightly longer bond lengths.

in bond length: $N_2 < N_2^-$ and N_2^+

110

9-28. *Refer to Sections 9-3 and 9-4, Example 9-1, and Exercise 9-27.*

In Exercise 9-27, the following electron configurations were obtained:

F_2 $\quad \sigma_{1s}^2\ \sigma_{1s}^{*2}\ \sigma_{2s}^2\ \sigma_{2s}^{*2}\ \sigma_{2p}^2\ \pi_{2p_y}^2\ \pi_{2p_z}^2\ \pi_{2p_y}^{*2}\ \pi_{2p_z}^{*2}$

F_2^+ $\quad \sigma_{1s}^2\ \sigma_{1s}^{*2}\ \sigma_{2s}^2\ \sigma_{2s}^{*2}\ \sigma_{2p}^2\ \pi_{2p_y}^2\ \pi_{2p_z}^2\ \pi_{2p_y}^{*2}\ \pi_{2p_z}^{*1}$

F_2^- $\quad \sigma_{1s}^2\ \sigma_{1s}^{*2}\ \sigma_{2s}^2\ \sigma_{2s}^{*2}\ \sigma_{2p}^2\ \pi_{2p_y}^2\ \pi_{2p_z}^2\ \pi_{2p_y}^{*2}\ \pi_{2p_z}^{*2}\ \sigma_{2p}^{*1}$

NO $\quad \sigma_{1s}^2\ \sigma_{1s}^{*2}\ \sigma_{2s}^2\ \sigma_{2s}^{*2}\ \sigma_{2p}^2\ \pi_{2p_y}^2\ \pi_{2p_z}^2\ \pi_{2p_y}^{*1}$

NO^+ $\quad \sigma_{1s}^2\ \sigma_{1s}^{*2}\ \sigma_{2s}^2\ \sigma_{2s}^{*2}\ \sigma_{2p}^2\ \pi_{2p_y}^2\ \pi_{2p_z}^2$

(a) Recall: Bond Order $= \dfrac{\text{No. bonding electrons - No. antibonding electrons}}{2}$

	F_2	F_2^+	F_2^-	NO	NO^+
Bonding e^-	10	10	10	10	10
Antibonding e^-	8	7	9	5	4
Bond order	1	1.5	0.5	2.5	3
(b) Unpaired e^-	0	1	1	1	0
Paramagnetic or Diamagnetic	D	P	P	P	D

(c) Based on bond orders, we have the following order of stability:

Somewhat stable: F_2^-

Stable: F_2, F_2^+ , NO and NO^+

Order of stability: $F_2^- < F_2 < F_2^+ < NO < NO^+$

9-30. *Refer to Section 9-5 and the Molecular Orbital Diagram before Exercise 9-29.*

Recall: Bond Order $= \dfrac{\text{No. bonding electrons - No. antibonding electrons}}{2}$

CN $\quad \sigma_{1s}^2\ \sigma_{1s}^{*2}\ \sigma_{2s}^2\ \sigma_{2s}^{*2}\ \pi_{2p_y}^2\ \pi_{2p_z}^2\ \sigma_{2p}^1$ $\quad$ bond order $= \dfrac{9 - 4}{2} = 2.5$

CN^+ $\quad \sigma_{1s}^2\ \sigma_{1s}^{*2}\ \sigma_{2s}^2\ \sigma_{2s}^{*2}\ \pi_{2p_y}^2\ \pi_{2p_z}^2$ $\quad$ bond order $= \dfrac{8 - 4}{2} = 2$

CN^- $\quad \sigma_{1s}^2\ \sigma_{1s}^{*2}\ \sigma_{2s}^2\ \sigma_{2s}^{*2}\ \pi_{2p_y}^2\ \pi_{2p_z}^2\ \sigma_{2p}^2$ $\quad$ bond order $= \dfrac{10 - 4}{2} = 3$

The most stable species is CN^- since it has the largest bond order.

(a) NF

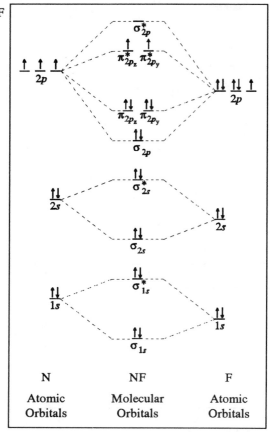

NF$^-$

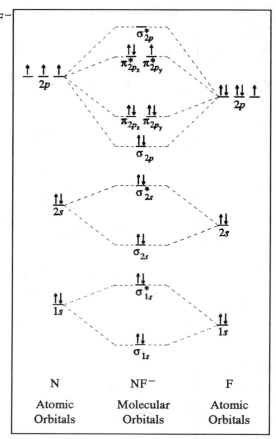

(b) NF $\quad \sigma_{1s}^{\,2}\ \sigma_{1s}^{*\,2}\ \sigma_{2s}^{\,2}\ \sigma_{2s}^{*\,2}\ \sigma_{2p}^{\,2}\ \pi_{2p_y}^{\,2}\ \pi_{2p_z}^{\,2}\ \pi_{2p_z}^{*\,1}\ \pi_{2p_y}^{*\,1}$

$\quad$ NF$^- \quad \sigma_{1s}^{\,2}\ \sigma_{1s}^{*\,2}\ \sigma_{2s}^{\,2}\ \sigma_{2s}^{*\,2}\ \sigma_{2p}^{\,2}\ \pi_{2p_y}^{\,2}\ \pi_{2p_z}^{\,2}\ \pi_{2p_z}^{*\,2}\ \pi_{2p_y}^{*\,1}$

(c) Recall: Bond Order $= \dfrac{\text{No. bonding electrons - No. antibonding electrons}}{2}$

	NF	NF$^-$
Bonding e^-	10	10
Antibonding e^-	6	7
Bond order	2	1.5
(d) Unpaired e^-	2	1
Paramagnetic or Diamagnetic	P	P

Both NF and NF$^-$ are stable, but NF is slightly more stable than NF$^-$ due to its higher bond order.

BO

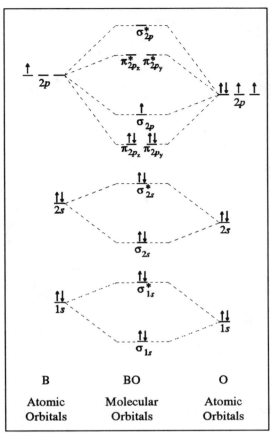

To increase the strength of the B-O bond, an electron is added to the BO molecule, forming the BO^- ion. The additional electron enters the bonding orbital, σ_{2p}.

The bond order increases from 2.5 for BO to 3.0 for BO^-. Likewise, the bond strength also increases.

9-36. *Refer to Section 9-6 and Figure 9-10.*

(a) SO_3

$$:\ddot{O}::S:\ddot{O}: \leftrightarrow :\ddot{O}:S:\ddot{O}: \leftrightarrow :\ddot{O}:S::\ddot{O}:$$
$$\quad\ :\ddot{O}: \qquad\qquad :\ddot{O}: \qquad\qquad :\ddot{O}:$$

(b) O_3

$$:\ddot{O}::\ddot{O}:\ddot{O}: \leftrightarrow :\ddot{O}:\ddot{O}::\ddot{O}:$$

(c) HCO_2^-

$$H:\overset{\cdot\cdot}{C}:\ddot{O}:^- \leftrightarrow H:\overset{\cdot\cdot}{C}::\ddot{O}:^-$$
$$\quad :\ddot{O}: \qquad\qquad\ :\ddot{O}:$$

Molecular Orbital Descriptions:

(a)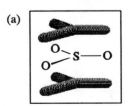

(b)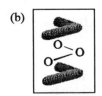

(c)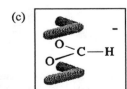

(a) O_2^{2+} $\sigma_{1s}^2 \, \sigma_{1s}^{*2} \, \sigma_{2s}^2 \, \sigma_{2s}^{*2} \, \sigma_{2p}^2 \, \pi_{2p_y}^2 \, \pi_{2p_z}^2$ bond order $= \dfrac{10-4}{2} = 3$

The ion is diamagnetic and very stable.

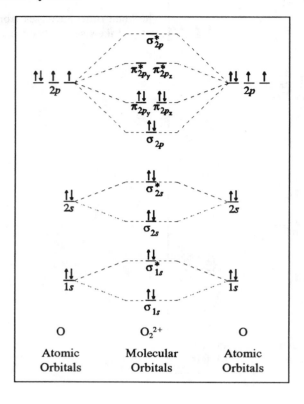

(b) HO^- $1s^2 \; 2s^2 \; \sigma_{sp}^2 \; 2p_x^2 \; 2p_y^2$ bond order $= \dfrac{2-0}{2} = 1$

The ion is diamagnetic and stable.

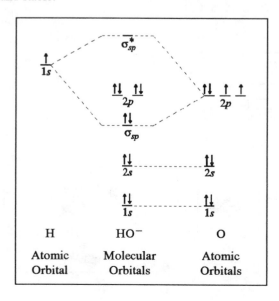

(c) HCl $1s^2$ $2s^2$ $2p_x^2$ $2p_y^2$ $2p_z^2$ $3s^2$ σ_{sp}^2 $3p_x^2$ $3p_y^2$ bond order $= \dfrac{2 - 0}{2} = 1$

The HCl molecule is diamagnetic and stable.

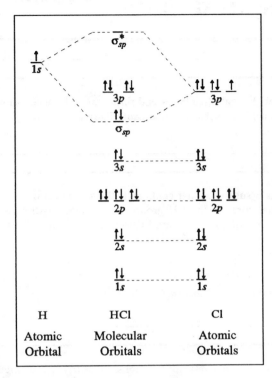

9-40.   Refer to Section 9-4.

(a) N and P are both VA elements but N_2 is much more stable than P_2 because N is a smaller atom than P and therefore can effectively participate in pi bonding, whereas P cannot. The $3p$ orbitals of a P atom do not overlap side-on in a pi bond with the corresponding $3p$ orbitals of another P atom nearly as well as the corresponding $2p$ orbitals of the much smaller nitrogen atoms. Therefore, as explained by valence bond theory, P forms 3 sigma bonds instead and acquires an octet of electrons using sp^3 hybridization to form P_4 molecules.

(b) O and S are both VIA elements. However, O_2 and O_3 are much more stable than S_2 because O is a smaller atom than S and therefore can effectively participate in pi bonding, whereas S cannot. The $3p$ orbitals of a S atom do not overlap side-on with the corresponding $3p$ orbitals of another S atom nearly as well as the corresponding $2p$ orbitals of the smaller O atoms. As a result, S forms 2 sigma bonds and obtains its octet of electrons using sp^3 hybridization to form S_8 molecules.

10 Reactions in Aqueous Solutions I: Acids, Bases, and Salts

10-2. **_Refer to Section 10-2._**

Gay-Lussac, in 1814, concluded that acids neutralize bases and these two substances should be defined in terms of their reactions with each other, i.e., an acid is a substance that neutralizes a base.

10-4. **_Refer to Section 10-2._**

(a) According to Arrhenius, an acid is a substance that contains hydrogen and produces hydrogen ions in aqueous solution. A base is a substance that contains the OH group and produces hydroxide ions in aqueous solution. Neutralization is the reaction between hydrogen ions and hydroxide ions yielding water molecules.

(b) acid: $HBr(aq) \rightarrow H^+(aq) + Br^-(aq)$

base: $Ba(OH)_2(aq) \rightarrow Ba^{2+}(aq) + 2OH^-(aq)$

neutralization: $2HBr(aq) + Ba(OH)_2(aq) \rightarrow BaBr_2(aq) + 2H_2O(\ell)$

10-6. **_Refer to Section 4-2._**

(a) A strong acid is an acid that ionizes completely into its ions or nearly so in dilute aqueous solution, e.g., $HCl(aq)$. A weak acid ionizes partially into its ions in dilute aqueous solution, e.g. $HF(aq)$ and $CH_3COOH(aq)$.

(b) A strong soluble base is a hydroxide of a Group IA or IIA metal that is soluble in water and is completely or nearly completely dissociated into its ions in dilute aqueous solutions, e.g., $NaOH(aq)$. A weak base is a molecular base that is slightly ionized in dilute aqueous solution, e.g., $NH_3(aq)$, $CH_3NH_2(aq)$.

(c) The definition of a strong soluble base is given in (b). An insoluble base is an insoluble metal hydroxide, e.g., $Cu(OH)_2(s)$, $Fe(OH)_3(s)$.

10-8. **_Refer to Section 4-2._**

To distinguish between strong electrolytes, weak electrolytes and nonelectrolytes, prepare equimolar aqueous solutions of the compounds and test their electrical conductivity. If a compound's solution conducts electricity well, it is a strong electrolyte; if its solution conducts electricity poorly, it is a weak electrolyte. A solution of a nonelectrolyte does not conduct electricity at all.

strong electrolyte: K_2CO_3, H_2SO_4

weak electrolyte: HCN, C_2H_5COOH, $HCOOH$, H_2S, NH_3

nonelectrolyte: CH_3OH

10-10. **_Refer to Section 10-3._**

A hydrated hydrogen ion containing only one water of hydration: $H^+(H_2O)$ or H_3O^+
This ion is also called the hydronium ion.

The statement, "The hydrated hydrogen ion should always be represented as H_3O^+," has two main flaws. First, the true extent of hydration of the H^+ in many solutions is unknown. Secondly, when balancing equations, it is generally much easier to use H^+ rather than H_3O^+.

10-14. *Refer to Section 10-4.*

(a) acid: a proton donor, e.g. HCl, NH_4^+, H_2O, H_3O^+

(b) conjugate base: a species that is produced when an acid donates a proton,

e.g. Cl^- is the conjugate base of HCl OH^- is the conjugate base of H_2O
         NH_3 is the conjugate base of NH_4^+ H_2O is the conjugate base of H_3O^+

(c) base: a proton acceptor, e.g., NH_3, H_2O, OH^-

(d) conjugate acid: a species that is produced when a base accepts a proton,

e.g. HCl is the conjugate acid of Cl^- H_2O is the conjugate acid of OH^-
         NH_4^+ is the conjugate acid of NH_3 H_3O^+ is the conjugate acid of H_2O

(e) conjugate acid-base pair: two species with formulas that differ only by a proton, e.g., HCl and Cl^-, NH_4^+ and NH_3, $H_2PO_4^-$ and HPO_4^{2-}. The species with the extra proton is the conjugate acid, whereas the other is the conjugate base.

10-16. *Refer to Section 10-4.*

(a) In dilute aqueous solution, ammonia is a Brønsted-Lowry base and water is a Brønsted-Lowry acid. The reaction produces an ammonium ion (the conjugate acid of ammonia) and a hydroxide ion (the conjugate base of water).

$$NH_3(aq) + H_2O(\ell) \rightleftarrows NH_4^+(aq) + OH^-(aq)$$

(b) In the gaseous state, ammonia also behaves as a Brønsted-Lowry base when reacting with gaseous hydrogen chloride. It takes a proton from the gaseous acid and produces the ionic salt, ammonium chloride.

$$NH_3(g) + HCl(g) \rightleftarrows NH_4Cl(s)$$

10-18. *Refer to Sections 10-4 and 10-8.*

Water is amphiprotic because a water molecule can either accept or donate a proton.

(a) We can describe water as amphoteric because it has the ability to react either as an acid or as a base. It can act as a Brønsted-Lowry acid by donating a proton to form a hydroxide ion or it can act as a Brønsted-Lowry base by accepting a proton to form a hydronium ion.

(b) Water as a Brønsted-Lowry acid: $H_2O(\ell) + CN^-(aq) \rightleftarrows HCN(aq) + OH^-(aq)$

     Water as a Brønsted-Lowry base: $H_2O(\ell) + HCN(aq) \rightleftarrows H_3O^+(aq) + CN^-(aq)$

10-20. *Refer to Section 10-4.*

These species are Brønsted-Lowry bases in water (none are Arrhenius bases) since they are proton acceptors and OH^- ions are produced: NH_3, HS^-, CH_3COO^- and O^{2-}.

     $NH_3 + H_2O \rightleftarrows NH_4^+ + OH^-$ $HS^- + H_2O \rightleftarrows H_2S + OH^-$

     $CH_3COO^- + H_2O \rightleftarrows CH_3COOH + OH^-$ $O^{2-} + H_2O \rightleftarrows 2OH^-$

(a) $NH_4^+ + CN^- \rightarrow NH_3 + HCN$
acid$_1$ base$_2$ base$_1$ acid$_2$

(c) $HClO_4 + [H_2NNH_3]^+ \rightarrow ClO_4^- + [H_3NNH_3]^{2+}$
acid$_1$ base$_2$ base$_1$ acid$_2$

(b) $HS^- + HSO_4^- \rightarrow H_2S + SO_4^{2-}$
base$_1$ acid$_2$ acid$_1$ base$_2$

(d) $NH_2^- + H_2O \rightarrow NH_3 + OH^-$
base$_1$ acid$_2$ acid$_1$ base$_2$

Acid	H_2O	HSe^-	HCl	PH_4^+	$HOCH_3$
Conjugate Base	OH^-	Se^{2-}	Cl^-	PH_3	CH_3O^-

	Brønsted-Lowry Acids	Brønsted-Lowry Bases
(a)	H_2O, HCN	CN^-, OH^-
(b)	H_2CO_3, H_2SO_4	HSO_4^-, HCO_3^-
(c)	$H_2C_2O_4$, HNO_2	NO_2^-, $HC_2O_4^-$

(a) $CN^- + H_2O \rightleftarrows HCN + OH^-$
base$_1$ acid$_2$ acid$_1$ base$_2$

(c) $H_2C_2O_4 + NO_2^- \rightleftarrows HNO_2 + HC_2O_4^-$
acid$_1$ base$_2$ acid$_2$ base$_1$

(b) $HCO_3^- + H_2SO_4 \rightleftarrows HSO_4^- + H_2CO_3$
base$_1$ acid$_2$ base$_2$ acid$_1$

(a) $H_2SO_4 + H_2O \rightleftarrows HSO_4^- + H_3O^+$
acid$_1$ base$_2$ base$_1$ acid$_2$

(b) $H_2SO_3 + H_2O \rightleftarrows HSO_3^- + H_3O^+$
acid$_1$ base$_2$ base$_1$ acid$_2$

$HSO_4^- + H_2O \rightleftarrows SO_4^{2-} + H_3O^+$
acid$_1$ base$_2$ base$_1$ acid$_2$

$HSO_3^- + H_2O \rightleftarrows SO_3^{2-} + H_3O^+$
acid$_1$ base$_2$ base$_1$ acid$_2$

Aqueous solutions of strong soluble bases (1) have a bitter taste, (2) have a slippery feeling, (3) change the colors of many acid-base indicators, (4) react with protonic acids to form salts and water, and (5) conduct an electrical current since they contain ions.

Aqueous ammonia reacts with protonic acids to form salts (and no water), but exhibits all the other traits to a lesser degree since it is a weak base exhibiting limited ionization and providing a lower OH^- concentration.

10-34. *Refer to Sections 10-4 and 10-6, and the Key Terms for Chapter 10.*

Solubility refers to the extent to which a solid substance will dissolve in a solvent. Substances that are dissolved in water may or may not ionize into ions. HCl, a soluble gas, ionizes almost totally into its ions, whereas glucose, a soluble molecule, does not ionize at all. Weak acids, such as HF, are soluble in water but ionize only slightly into ions.

10-36. *Refer to Section 10-4.*

Brønsted-Lowry bases: NH_3, H_2O and $:H^-$ in ionic NaH

Brønsted-Lowry acids: H_2O, HF

The hydrides, BeH_2, BH_3 and CH_4 are *generally* considered as neither Brønsted-Lowry acids nor Brønsted-Lowry bases.

10-38. *Refer to Sections 10-4 and 10-5.*

Base strength refers to the relative tendency to produce OH^- ions in aqueous solution by (1) the dissociation of soluble metal hydroxides or (2) by ionization reactions with water. A more general definition, applying Brønsted-Lowry theory, is that base strength is a measure of the relative tendency to accept a proton from any acid.

10-40. *Refer to Section 10-5.*

(a) A binary protonic acid is a covalent compound consisting of hydrogen atom(s) and one other element. The compound can act as a proton donor.

(b)

hydrofluoric acid	HF(*aq*)	hydrobromic acid	HBr(*aq*)
hydrosulfuric acid	H_2S(*aq*)	hydroselenic acid	H_2Se(*aq*)

10-42. *Refer to Section 10-5 and the Key Terms for Chapter 10.*

The leveling effect is the effect by which all bases stronger than the base formed by the autoionization of the solvent reacts with the solvent to produce that base. The same statement applies to acids. In other words, the strongest base (or acid) that can exist in a given solvent is the base (or acid) that is characteristic of that solvent.

(1) In H_2O, the strongest acid that can persistently exist is H_3O^+ and the strongest base that can persistently exist is OH^-.

(2) In liquid NH_3, the corresponding pair is NH_4^+ and NH_2^-.

(3) In liquid HF, the corresponding pair is H_2F^+ and F^-.

10-44. *Refer to Section 10-5.*

In order of decreasing acidity:

(a) $H_2Se > H_2S > H_2O$

(b) $HI > HBr > HCl > HF$

(c) $H_2S > HS^- > S^{2-}$ (not an acid)

10-46. *Refer to Section 10-5.*

Ternary acids, including nitric and sulfuric acids, can be described as hydroxides of nonmetals since they contain 1 or more -O-H groups attached to the central nonmetal atom. For example,

$$HNO_3 \equiv NO_2(OH) \qquad\qquad H_2SO_4 \equiv SO_2(OH)_2$$

10-48. *Refer to Section 10-7.*

(a) $NH_3(\ell) + NH_3(\ell) \rightarrow NH_4^+ + NH_2^-$

(b) $NH_2OH(\ell) + NH_2OH(\ell) \rightarrow NH_3OH^+ + NHOH^-$

(c) $H_2SO_4(\ell) + H_2SO_4(\ell) \rightarrow H_3SO_4^+ + HSO_4^-$

10-50. *Refer to Section 10-8.*

(a) Acid strengths of most ternary acids containing different elements in the same oxidation state from the same group in the periodic table increase with increasing electronegativity of the central element.

(b) In order of increasing acid strength:

(1) $H_3PO_4 < HNO_3$ (2) $H_3AsO_4 < H_3PO_4$ (3) $H_2SeO_4 < H_2SO_4$ (4) $HIO_3 < HBrO_3 < HClO_3$

10-52. *Refer to Section 10-6.*

Acid-base reactions are called neutralization reactions because the reaction of an acid with a base generally produces a salt with little or no acid-base character and, in many cases, water.

10-54. *Refer to Section 4-2.*

The electrolytes include NH_4Cl, HI, RaF_2, $Zn(CH_3COO)_2$, $Cu(NO_3)_2$, CH_3COOH, LiOH, $KHCO_3$ and $La_2(SO_4)_3$. The nonelectrolytes are C_6H_6, $C_{12}H_{22}O_{11}$ (table sugar), CCl_4 and I_2.

10-56. *Refer to Section 10-6 and Examples 10-1 and 10-2.*

(a) formula unit:

$$HNO_3(aq) + KOH(aq) \rightarrow KNO_3(aq) + H_2O(\ell)$$

        nitric   potassium   potassium
        acid   hydroxide   nitrate

total ionic: $H^+(aq) + NO_3^-(aq) + K^+(aq) + OH^-(aq) \rightarrow K^+(aq) + NO_3^-(aq) + H_2O(\ell)$

net ionic: $H^+(aq) + OH^-(aq) \rightarrow H_2O(\ell)$

(b) formula unit:

$$H_2SO_4(aq) + 2NaOH(aq) \rightarrow Na_2SO_4(aq) + 2H_2O(\ell)$$

        sulfuric   sodium   sodium
        acid   hydroxide   sulfate

total ionic: $2H^+(aq) + SO_4^{2-}(aq) + 2Na^+(aq) + 2OH^-(aq) \rightarrow 2Na^+(aq) + SO_4^{2-}(aq) + 2H_2O(\ell)$

net ionic: $2H^+(aq) + 2OH^-(aq) \rightarrow 2H_2O(\ell)$

     therefore, $H^+(aq) + OH^-(aq) \rightarrow H_2O(\ell)$

(c) formula unit:

$$2HCl(aq) + Ca(OH)_2(aq) \rightarrow CaCl_2(aq) + 2H_2O(\ell)$$

hydrochloric   calcium   calcium
acid   hydroxide   chloride

total ionic: $2H^+(aq) + 2Cl^-(aq) + Ca^{2+}(aq) + 2OH^-(aq) \rightarrow Ca^{2+}(aq) + 2Cl^-(aq) + 2H_2O(\ell)$

net ionic: $2H^+(aq) + 2OH^-(aq) \rightarrow 2H_2O(\ell)$

therefore, $H^+(aq) + OH^-(aq) \rightarrow H_2O(\ell)$

(d) formula unit:

$$CH_3COOH(aq) + KOH(aq) \rightarrow KCH_3COO(aq) + H_2O(\ell)$$

acetic   potassium   potassium
acid   hydroxide   acetate

total ionic: $CH_3COOH(aq) + K^+(aq) + OH^-(aq) \rightarrow K^+(aq) + CH_3COO^-(aq) + H_2O(\ell)$

net ionic: $CH_3COOH(aq) + OH^-(aq) \rightarrow CH_3COO^-(aq) + H_2O(\ell)$

(e) formula unit:

$$HI(aq) + NaOH(aq) \rightarrow NaI(aq) + H_2O(\ell)$$

hydroiodic   sodium   sodium
acid   hydroxide   iodide

total ionic: $H^+(aq) + I^-(aq) + Na^+(aq) + OH^-(aq) \rightarrow Na^+(aq) + I^-(aq) + H_2O(\ell)$

net ionic: $H^+(aq) + OH^-(aq) \rightarrow H_2O(\ell)$

10-58. *Refer to Section 10-6 and Examples 10-1 and 10-2.*

(a) formula unit:

$$2HClO_4(aq) + Ba(OH)_2(aq) \rightarrow Ba(ClO_4)_2(aq) + 2H_2O(\ell)$$

perchloric   barium   barium
acid   hydroxide   perchlorate

total ionic: $2H^+(aq) + 2ClO_4^-(aq) + Ba^{2+}(aq) + 2OH^-(aq) \rightarrow Ba^{2+}(aq) + 2ClO_4^-(aq) + 2H_2O(\ell)$

net ionic: $2H^+(aq) + 2OH^-(aq) \rightarrow 2H_2O(\ell)$

therefore, $H^+(aq) + OH^-(aq) \rightarrow H_2O(\ell)$

(b) formula unit:

$$HBr(aq) + NH_3(aq) \rightarrow NH_4Br(aq)$$

hydrobromic   ammonia   ammonium
acid   bromide

total ionic: $H^+(aq) + Br^-(aq) + NH_3(aq) \rightarrow NH_4^+(aq) + Br^-(aq)$

net ionic: $H^+(aq) + NH_3(aq) \rightarrow NH_4^+(aq)$

(c) formula unit:

$$HNO_3(aq) + NH_3(aq) \rightarrow NH_4NO_3(aq)$$

nitric   ammonia   ammonium
acid   nitrate

total ionic: $H^+(aq) + NO_3^-(aq) + NH_3(aq) \rightarrow NH_4^+(aq) + NO_3^-(aq)$

net ionic: $H^+(aq) + NH_3(aq) \rightarrow NH_4^+(aq)$

(d) formula unit:

$$3H_2SO_4(aq) + 2Fe(OH)_3(s) \rightarrow Fe_2(SO_4)_3(aq) + 6H_2O(\ell)$$

sulfuric   iron(III)   iron(III)
acid   hydroxide   sulfate

total ionic:

$$6H^+(aq) + 3SO_4^{2-}(aq) + 2Fe(OH)_3(s) \rightarrow 2Fe^{3+}(aq) + 3SO_4^{2-}(aq) + 6H_2O(\ell)$$

net ionic:

$$6H^+(aq) + 2Fe(OH)_3(s) \rightarrow 2Fe^{3+}(aq) + 6H_2O(\ell)$$

therefore,

$$3H^+(aq) + Fe(OH)_3(s) \rightarrow Fe^{3+}(aq) + 3H_2O(\ell)$$

(e) formula unit:

$$2H_3PO_4(aq) + 3Ba(OH)_2(aq) \rightarrow Ba_3(PO_4)_2(s) + 6H_2O(\ell)$$

phosphoric   barium   barium
acid   hydroxide   phosphate

total ionic:

$$2H_3PO_4(aq) + 3Ba^{2+}(aq) + 6OH^-(aq) \rightarrow Ba_3(PO_4)_2(s) + 6H_2O(\ell)$$

net ionic: same as the total ionic equation

10-60. *Refer to Section 10-6 and Examples 10-1 and 10-2.*

(a) formula unit:

$$2HNO_3(aq) + Pb(OH)_2(s) \rightarrow Pb(NO_3)_2(aq) + 2H_2O(\ell)$$

total ionic:

$$2H^+(aq) + 2NO_3^-(aq) + Pb(OH)_2(s) \rightarrow Pb^{2+}(aq) + 2NO_3^-(aq) + 2H_2O(\ell)$$

net ionic:

$$2H^+(aq) + Pb(OH)_2(s) \rightarrow Pb^{2+}(aq) + 2H_2O(\ell)$$

(b) formula unit:

$$3HCl(aq) + Al(OH)_3(s) \rightarrow AlCl_3(aq) + 3H_2O(\ell)$$

total ionic:

$$3H^+(aq) + 3Cl^-(aq) + Al(OH)_3(s) \rightarrow Al^{3+}(aq) + 3Cl^-(aq) + 3H_2O(\ell)$$

net ionic:

$$3H^+(aq) + Al(OH)_3(s) \rightarrow Al^{3+}(aq) + 3H_2O(\ell)$$

(c) formula unit:

$$H_2CO_3(aq) + 2NH_3(aq) \rightarrow (NH_4)_2CO_3(aq)$$

total ionic:

$$H_2CO_3(aq) + 2NH_3(aq) \rightarrow 2NH_4^+(aq) + CO_3^{2-}(aq)$$

net ionic: same as the total ionic equation

(d) formula unit:

$$2HClO_4(aq) + Ca(OH)_2(aq) \rightarrow Ca(ClO_4)_2(aq) + 2H_2O(\ell)$$

total ionic:

$$2H^+(aq) + 2ClO_4^-(aq) + Ca^{2+}(aq) + 2OH^-(aq) \rightarrow Ca^{2+}(aq) + 2ClO_4^-(aq) + 2H_2O(\ell)$$

net ionic:

$$2H^+(aq) + 2OH^-(aq) \rightarrow 2H_2O(\ell)$$

therefore,

$$H^+(aq) + OH^-(aq) \rightarrow H_2O(\ell)$$

(e) formula unit:

$$3H_2SO_4(aq) + 2Al(OH)_3(s) \rightarrow Al_2(SO_4)_3(aq) + 6H_2O(\ell)$$

total ionic:

$$6H^+(aq) + 3SO_4^{2-}(aq) + 2Al(OH)_3(s) \rightarrow 2Al^{3+}(aq) + 3SO_4^{2-}(aq) + 6H_2O(\ell)$$

net ionic:

$$6H^+(aq) + 2Al(OH)_3(s) \rightarrow 2Al^{3+}(aq) + 6H_2O(\ell)$$

therefore,

$$3H^+(aq) + Al(OH)_3(s) \rightarrow Al^{3+}(aq) + 3H_2O(\ell)$$

10-62. *Refer to Section 10-7.*

An acidic salt is a salt that contains an ionizable hydrogen atom. It is the product which results from reacting less than a stoichiometric amount of base with a polyprotic acid:

$$H_2SO_3(aq) + NaOH(aq) \rightarrow NaHSO_3(aq) + H_2O(\ell)$$
$$H_2CO_3(aq) + KOH(aq) \rightarrow KHCO_3(aq) + H_2O(\ell)$$
$$H_3PO_4(aq) + NaOH(aq) \rightarrow NaH_2PO_4(aq) + H_2O(\ell)$$
$$H_3PO_4(aq) + 2NaOH(aq) \rightarrow Na_2HPO_4(aq) + 2H_2O(\ell)$$

10-64. *Refer to Section 10-7.*

(a) $HNO_3 + NH_3 \rightarrow NH_4NO_3$

(b) $H_3PO_4 + NH_3 \rightarrow NH_4H_2PO_4$

(c) $H_3PO_4 + 2NH_3 \rightarrow (NH_4)_2HPO_4$

(d) $H_3PO_4 + 3NH_3 \rightarrow (NH_4)_3PO_4$

(e) $H_2SO_4 + 2NH_3 \rightarrow (NH_4)_2SO_4$

10-66. *Refer to Section 10-7.*

A basic salt is a salt containing an ionizable OH group and can therefore neutralize acids.

(a),(b) $HCl(aq)$ + $Ba(OH)_2(aq)$ $\rightarrow$ $Ba(OH)Cl(aq)$ + $H_2O(\ell)$
       1 mol   **1 mol**

       $HCl(aq)$ + $Al(OH)_3(aq)$ $\rightarrow$ $Al(OH)_2Cl(aq)$ + $H_2O(\ell)$
       1 mol   **1 mol**

       $2HCl(aq)$ + $Al(OH)_3(aq)$ $\rightarrow$ $Al(OH)Cl_2(aq)$ + $2H_2O(\ell)$
       2 mol   **1 mol**

10-68. *Refer to Section 10-8 and Table 10-2.*

(a)   for $Cr(OH)_3$

     formula unit: $Cr(OH)_3(s) + 3HNO_3(aq) \rightarrow Cr(NO_3)_3(aq) + 3H_2O(\ell)$

     total ionic: $Cr(OH)_3(s) + 3H^+(aq) + 3NO_3^-(aq) \rightarrow Cr^{3+}(aq) + 3NO_3^-(aq) + 3H_2O(\ell)$

     net ionic: $Cr(OH)_3(s) + 3H^+(aq) \rightarrow Cr^{3+}(aq) + 3H_2O(\ell)$

     for $Pb(OH)_2$

     formula unit: $Pb(OH)_2(s) + 2HNO_3(aq) \rightarrow Pb(NO_3)_2(aq) + 2H_2O(\ell)$

     total ionic: $Pb(OH)_2(s) + 2H^+(aq) + 2NO_3^-(aq) \rightarrow Pb^{2+}(aq) + 2NO_3^-(aq) + 2H_2O(\ell)$

     net ionic: $Pb(OH)_2(s) + 2H^+(aq) \rightarrow Pb^{2+}(aq) + 2H_2O(\ell)$

(b)   for $Cr(OH)_3$

     formula unit: $Cr(OH)_3(s) + KOH(aq) \rightarrow K[Cr(OH)_4](aq)$

     total ionic: $Cr(OH)_3(s) + K^+(aq) + OH^-(aq) \rightarrow K^+(aq) + [Cr(OH)_4]^-(aq)$

     net ionic: $Cr(OH)_3(s) + OH^-(aq) \rightarrow [Cr(OH)_4]^-(aq)$

for Pb(OH)$_2$

formula unit:	Pb(OH)$_2$(s) + 2KOH(aq) → K$_2$[Pb(OH)$_4$](aq)
total ionic:	Pb(OH)$_2$(s) + 2K$^+$(aq) + 2OH$^-$(aq) → 2K$^+$(aq) + [Pb(OH)$_4$]$^{2-}$(aq)
net ionic:	Pb(OH)$_2$(s) + 2OH$^-$(aq) → [Pb(OH)$_4$]$^{2-}$(aq)

10-70. *Refer to Section 10-5 and Appendix F.*

Ionization of citric acid, C$_6$H$_8$O$_7$ or C$_3$H$_5$O(COOH)$_3$:

$$C_3H_5O(COOH)_3(aq) \rightleftarrows H^+(aq) + C_3H_5O(COO)(COOH)_2^-(aq)$$

$$C_3H_5O(COO)(COOH)_2^-(aq) \rightleftarrows H^+(aq) + C_3H_5O(COO)_2(COOH)^{2-}(aq)$$

$$C_3H_5O(COO)_2(COOH)(aq) \rightleftarrows H^+(aq) + C_3H_5O(COO)_3^{3-}(aq)$$

10-72. *Refer to Section 10-9.*

(a) Hydrogen sulfide, H$_2$S(g), can be prepared by combining elemental sulfur with hydrogen gas.

$$S_8(s) + 8H_2(g) \rightarrow 8H_2S(g)$$

(b) Hydrogen chloride, HCl(g), can be prepared in small quantities by dropping concentrated nonvolatile acids such as sulfuric acid, onto an appropriate salt such as NaCl(s).

$$H_2SO_4(\ell) + NaCl(s) \rightarrow HCl(g) + NaHSO_4(s)$$

(c) An aqueous solution of the weak acid acetic acid, CH$_3$COOH(aq), can be produced by reacting an acetate salt with a strong acid:

$$NaCH_3COO(aq) + HCl(aq) \rightarrow CH_3COOH(aq) + NaCl(aq)$$

10-74. *Refer to Sections 10-9 and 6-8, and Figure 6-9.*

(a) acidic oxides: CO$_2$, SO$_2$, SO$_3$

(b) amphoteric oxides: Al$_2$O$_3$, Ga$_2$O$_3$, SnO$_2$

(c) basic oxides: Na$_2$O, K$_2$O, CaO, BaO

10-76. *Refer to Sections 10-4 and 10-10.*

The advantages of Brønsted-Lowry theory over Arrhenius theory are twofold. (1) Brønsted-Lowry theory extends the definitions of acids and bases so that they can be used in nonaqueous systems and (2) this theory also recognizes that other constituents besides the hydroxide ion can have basic properties.

A limitation of the Brønsted-Lowry theory is that it does not permit the acid-base classification to include those systems which do not involve proton transfer. This limitation is met by the Lewis theory.

(a)   H:Ö: +   H:Ö: →   H:Ö:H⁺   +   H:Ö:⁻
       |̈ |̈ |̈ |̈
       H   H   H
      base   **acid**   **acid**   **base**

(b)   H:C̈l: +   H:Ö: →   :C̈l:⁻   +   H:Ö:H⁺
                   |̈ |̈
                   H   H
      acid   **base**   **base**   **acid**

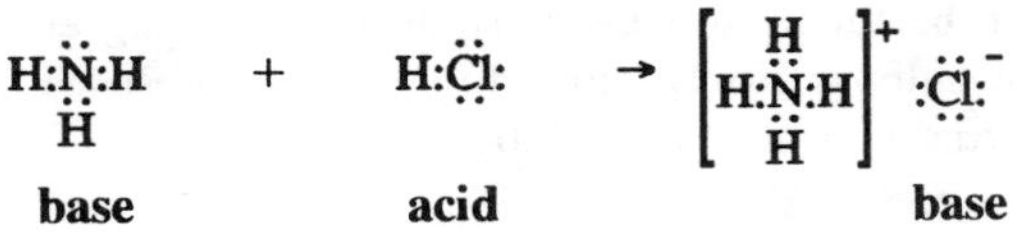

(c)   H:N̈:H   +   H:Ö: →   H:N̈:H⁺   +   H:Ö:⁻
       | |̈ | |̈
       H   H   H
      base   **acid**   **acid**   **base**

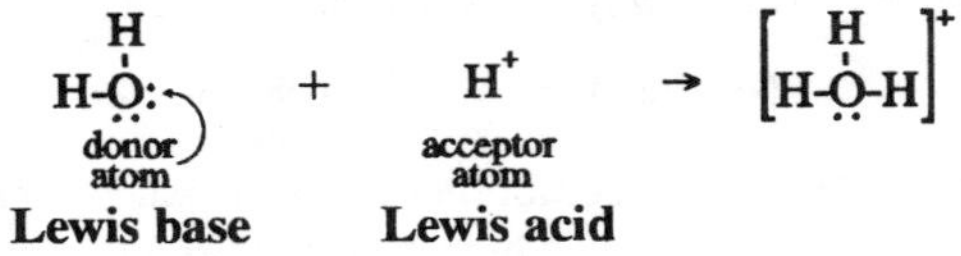

(d)   H:N̈:H   +   H:C̈l: →   [H:N̈:H]⁺ :C̈l:⁻
       | |
       H   H
      base   **acid**   **base**

(a)   H
       H-Ö:⌒   +   H⁺   →   [H-Ö-H]⁺
       donor   **acceptor**
       atom   **atom**
   Lewis base   **Lewis acid**

(b)   6[:C̈l:]⁻   +   Pt⁴⁺   →   [:C̈l: :C̈l: Pt :C̈l: :C̈l: :C̈l:]²⁻
       donor   **acceptor**
       atom   **atom**
   Lewis base   **Lewis acid**

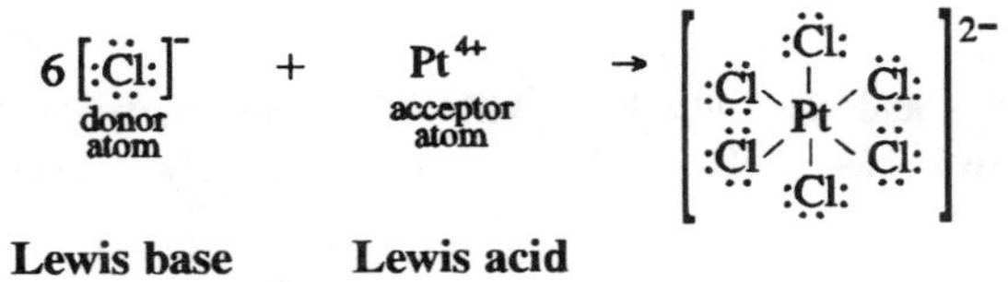

(a)   HF   +   SbF₅   →   H(SbF₆)   (b)   HF   +   BF₃   →   H(BF₄)
   Lewis   Lewis   Lewis   Lewis
   base   acid   base   acid

(b)   In H(SbF₆), H is bonded to Sb through an ionic bond. In H(BF₄), the H is bonded to B through an ionic bond.

(a) H_2S Arrhenius acid, Brønsted-Lowry acid
(b) $PO(OH)_3 \equiv H_3PO_4$ Arrhenius acid, Brønsted-Lowry acid
(c) $H_2CaO_2 \equiv Ca(OH)_2$ Arrhenius base, Brønsted-Lowry base
(d) $ClO_3(OH) \equiv HClO_4$ Arrhenius acid, Brønsted-Lowry acid
(e) $Sb(OH)_3 \equiv H_3SbO_3$ Arrhenius acid, Brønsted-Lowry acid

(a)

Acid	Conjugate base		Base	Conjugate Acid
H_3PO_4	$H_2PO_4^-$		HSO_4^-	H_2SO_4
NH_4^+	NH_3		PH_3	PH_4^+
OH^-	O^{2-}		PO_4^{3-}	HPO_4^{2-}

(b) We know that the weaker a base, the stronger is its conjugate acid. Therefore, given that NO_2^- is a stronger base than NO_3^-, then HNO_3 (the conjugate acid of NO_3^-) is a stronger acid than HNO_2 (the conjugate acid of NO_2^-).

10-88. *Refer to Section 10-6 and Figure 4-2.*

In a conductivity experiment, the indicator light bulb glows brightly when the electrodes are placed in an aqueous solution containing a high concentration of ions, such as can be found in aqueous solutions of strong electrolytes (strong acids, strong soluble bases and soluble salts). The bulb will only glow dimly in the presence of weak electrolytes because there are few ions present to conduct electricity through the solution.

(a) $NaOH(aq)$ and $HCl(aq)$ are both strong electrolytes. The light bulb glows brightly for these solutions. The neutralization reaction between NaOH and HCl that results when the two solutions are mixed can be represented as follows:

formula unit: $\qquad NaOH(aq) + HCl(aq) \rightarrow NaCl(aq) + H_2O(\ell)$

dissociation of the product, $NaCl(aq)$: $\qquad NaCl(aq) \rightarrow Na^+(aq) + Cl^-(aq)$

Even though ions are lost as the reaction proceeds, due to H^+ and OH^- ions combining to form water, there are still plenty of Na^+ and Cl^- ions remaining in the solution to cause the indicator bulb to glow brightly, but not quite as brightly as the initial solution.

(b) $NH_3(aq)$ and $CH_3COOH(aq)$ are both weak electrolytes and the light bulb will only glow dimly for these solutions. The neutralization reaction between the weak base, NH_3, and the weak acid, CH_3COOH, is as follows:

formula unit: $\qquad NH_3(aq) + CH_3COOH(aq) \rightarrow NH_4CH_3COO(aq)$

dissociation of the product, $NH_4CH_3COO(aq)$: $\qquad NH_4CH_3COO(aq) \rightarrow NH_4^+(aq) + CH_3COO^-(aq)$

As can readily be seen, the product formed is a soluble salt, the strong electrolyte, NH_4CH_3COO, which dissociates into plenty of NH_4^+ and CH_3COO^- ions. The indicator bulb glows brightly in this solution.

10-90. *Refer to Section 10-4.*

Autoionization of PCl_5:

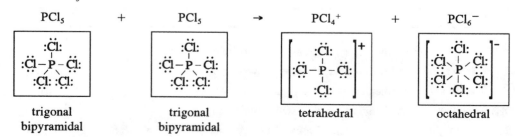

(a) $CH_3COOH(aq) + NaHCO_3(aq) \rightarrow NaCH_3COO(aq) + CO_2(g) + H_2O(\ell)$

The "fizz" is caused by the gaseous product, carbon dioxide, escaping from the solution.

(b) $CH_3CH(OH)COOH(aq) + NaHCO_3(aq) \rightarrow NaCH_3CH(OH)COO(aq) + CO_2(g) + H_2O(\ell)$

"Quick" bread "rises" during baking due to the reaction between baking soda and lactic acid found in the added milk. The resulting carbon dioxide gas bubbles are caught in the bread dough, giving the bread more volume. Yeast breads "rise" due to carbon dioxide bubbles released in the fermentation of sugar by yeast.

(a) The ionizable hydrogen atom in lactic acid, $CH_3CH(OH)COOH$, is the one in the carboxylic acid functional group, -COOH.

(b) Structural formula of lactic acid's conjugate base, the lactate ion:

(c) net ionic equation for the ionization of lactic acid:

$$CH_3CH(OH)COOH(aq) \rightleftarrows H^+(aq) + CH_3CH(OH)COO^-(aq)$$

(d) lactic acid

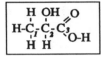

C_1 sp^3 hybridization, tetrahedral geometry
C_2 sp^3 hybridization, tetrahedral geometry
C_3 sp^2 hybridization, trigonal planar geometry

11 Reactions in Aqueous Solutions II: Calculations

11-2. *Refer to Sections 3-6 and 11-1.*

Molarity is defined as the number of moles of solute per 1 liter of solution and has units of mol/L. If we multiply molarity by unity = $10^{-3}/10^{-3}$,

$$\text{Molarity, } M \left[\frac{\text{mol}}{\text{L}}\right] = \frac{\text{mol solute}}{\text{L soln}} \times \frac{10^{-3}}{10^{-3}} = \frac{\text{mmol solute}}{\text{mL soln}}$$

11-4. *Refer to Section 3-6 and Example 3-18.*

Plan: (1) Calculate the moles of $(NH_4)_2SO_4$ present in the solution.
        (2) Calculate the mass required.

(1) ? mol $(NH_4)_2SO_4$ = 0.188 M $(NH_4)_2SO_4$ × 2.75 L = 0.517 mol $(NH_4)_2SO_4$

(2) ? g $(NH_4)_2SO_4$ = 0.517 mol $(NH_4)_2SO_4$ × 132 g/mol = **68.2 g $(NH_4)_2SO_4$**

Dimensional Analysis:

? g $(NH_4)_2SO_4$ = 2.75 L × $\dfrac{0.188 \text{ mol}}{1 \text{ L}}$ × $\dfrac{132 \text{ g}}{1 \text{ mol}}$ = **68.2 g $(NH_4)_2SO_4$**

11-6. *Refer to Section 3-6 and Example 3-19.*

Assume a 1 liter solution of 39.77% H_2SO_4 with a density of 1.305 g/mL.

? g H_2SO_4 soln in 1 L soln = $\dfrac{1.305 \text{ g soln}}{1 \text{ mL soln}}$ × 1000 mL soln = 1305 g soln

? g H_2SO_4 in 1 L soln = 1305 g soln × $\dfrac{39.77 \text{ g } H_2SO_4}{100 \text{ g soln}}$ = 519.0 g H_2SO_4

? mol H_2SO_4 in 1 L soln = $\dfrac{519.0 \text{ g } H_2SO_4}{98.08 \text{ g/mol}}$ = 5.292 mol H_2SO_4

Therefore, ? M H_2SO_4 = **5.292 M**

11-8. *Refer to Section 11-1 and Example 11-1.*

This is a possible limiting reactant problem.

Plan: (1) Calculate the number of moles of HCl and NaOH.
        (2) Determine the limiting reactant, if there is one.
        (3) Calculate the moles of NaCl formed.
        (4) Determine the molarity of NaCl in the solution.

Balanced equation: HCl(aq) + NaOH(aq) → NaCl(aq) + $H_2O(\ell)$

(1) ? mol HCl = 4.32 M HCl × 0.150 L = 0.648 mol HCl
    ? mol NaOH = 2.16 M NaOH × 0.300 L = 0.648 mol NaOH

(2) This is not a problem with a single limiting reactant since we have stoichiometric amounts of both HCl and NaOH. Our final solution is a salt solution with no excess acid or base.

(3) ? mol NaCl = mol HCl = mol NaOH = 0.648 mol NaCl

(4) ? M NaCl = $\dfrac{\text{mol NaCl}}{\text{total volume}}$ = $\dfrac{0.648 \text{ mol}}{(0.150 \text{ L} + 0.300 \text{ L})}$ = **1.44 M NaCl**

11-10. *Refer to Section 11-1, Exercise 11-8 Solution and Example 11-2.*

Balanced equation: $H_3PO_4(aq) + 3NaOH(aq) \rightarrow Na_3PO_4(aq) + 3H_2O(\ell)$

Plan: (1) Calculate the moles of H_3PO_4 and NaOH.
(2) Determine the limiting reactant, if there is one.
(3) Calculate the moles of Na_3PO_4 formed.
(4) Determine the molarity of the salt in the solution.
(5) Determine the moles and concentration of excess reactant in the solution.

(1) ? mol H_3PO_4 = 3.68 M H_3PO_4 × 0.225 L = 0.828 mol H_3PO_4
? mol NaOH = 3.68 M NaOH × 0.775 L = 2.852 mol NaOH

(2) In the balanced equation, H_3PO_4 reacts with NaOH in a 1:3 mole ratio.

mol H_3PO_4:mol NaOH = 0.828 mol:2.852 mol = 1:3.44

We do not have stoichiometric amounts of both reactants; this is a limiting reactant problem. We have less H_3PO_4 than is necessary to react with all of the NaOH, so H_3PO_4 is the limiting reactant and NaOH is in excess. The amount of salt formed is set then by the amount of H_3PO_4.

(3) ? mol Na_3PO_4 = mol H_3PO_4 = 0.828 mol Na_3PO_4

(4) ? M Na_3PO_4 = $\dfrac{0.828 \text{ mol } Na_3PO_4}{(0.225 \text{ L} + 0.775 \text{ L})}$ = **0.828 M Na_3PO_4**

(5) The moles of NaOH consumed is determined from the amount of limiting reactant, H_3PO_4.

? mol NaOH consumed = 3 × 0.828 mol H_3PO_4 = 2.48 mol NaOH

? excess mol NaOH = total mol NaOH - mol NaOH consumed by H_3PO_4 = 2.85 mol - 2.48 mol = 0.37 mol

? M NaOH = $\dfrac{0.37 \text{ mol NaOH}}{(0.225 \text{ L} + 0.775 \text{ L})}$ = **0.37 M NaOH**

11-12. *Refer to Section 11-1 and Example 11-3.*

Balanced equation: $2CH_3COOH(aq) + Ba(OH)_2(s) \rightarrow Ba(CH_3COO)_2(aq) + 2H_2O(\ell)$

Plan: (1) Calculate the moles of $Ba(OH)_2$ used.
(2) Calculate the moles of CH_3COOH needed.
(3) Evaluate the volume of CH_3COOH needed

(1) ? mol $Ba(OH)_2$ = 0.0105 M $Ba(OH)_2$ × 0.0252 L = 2.65 × 10^{-4} mol $Ba(OH)_2$

(2) ? mol CH_3COOH = 2 × mol $Ba(OH)_2$ = 2 × 2.65 × 10^{-4} mol = 5.30 × 10^{-4} mol CH_3COOH

(3) ? L CH_3COOH soln = $\dfrac{5.30 \times 10^{-4} \text{ mol } CH_3COOH}{0.0150 \text{ } M \text{ } CH_3COOH}$ = **0.0353 L or 35.3 mL CH_3COOH soln**

11-14. *Refer to Section 3-6 and Example 3-19.*

Assume a 1 liter solution of 5.11% CH_3COOH.

$$? \text{ g } CH_3COOH \text{ soln in 1 L soln} = \frac{1.007 \text{ g soln}}{1 \text{ mL soln}} \times 1000 \text{ mL soln} = 1007 \text{ g soln}$$

$$? \text{ g } CH_3COOH \text{ in 1 L soln} = 1007 \text{ g soln} \times \frac{5.11 \text{ g } CH_3COOH}{100 \text{ g soln}} = 51.5 \text{ g } CH_3COOH$$

$$? \text{ mol } CH_3COOH \text{ in 1 L soln} = \frac{51.5 \text{ g } CH_3COOH}{60.1 \text{ g/mol}} = 0.857 \text{ mol } CH_3COOH$$

Therefore, $? M \ CH_3COOH = \mathbf{0.857\ M}$

11-16. *Refer to Section 11-1 and Example 11-4.*

Balanced equation: $3NaOH(aq) + H_3PO_4(aq) \rightarrow Na_3PO_4(aq) + 3H_2O(\ell)$

Plan: (1) Calculate the moles of NaOH and H_3PO_4 required to form 1 mole of Na_3PO_4.
          (2) Find the volumes of each solution.

(1) $? \text{ mol NaOH} = 3 \times \text{mol } Na_3PO_4 = 3 \times 1.00 \text{ mol} = 3.00 \text{ mol NaOH}$
      $? \text{ mol } H_3PO_4 = \text{mol } Na_3PO_4 = 1.00 \text{ mol } H_3PO_4$

(2) $? \text{ L NaOH soln} = \dfrac{3.00 \text{ mol NaOH}}{1.50 \ M \text{ NaOH}} = \mathbf{2.00 \text{ L NaOH soln}}$

      $? \text{ L } H_3PO_4 \text{ soln} = \dfrac{1.00 \text{ mol } H_3PO_4}{3.00 \ M \ H_3PO_4} = \mathbf{0.333 \text{ L } H_3PO_4 \text{ soln}}$

11-18. *Refer to Sections 11-2 and 3-9.*

A standard solution of NaOH cannot be prepared directly because the solid is hydroscopic and absorbs moisture and CO_2 from the air.

Step 1: Weigh out an amount of solid NaOH and dissolve it in water to obtain a solution with the approximate concentration.

Step 2: Weigh out an appropriate amount of an acidic material, suitable for use as a primary standard, such as potassium hydrogen phthalate (KHP).

Step 3: Titrate the KHP sample with the NaOH solution and calculate the molarity of the NaOH solution.

The NaOH solution thusly prepared is called a secondary standard because its concentration is determined by titration against a primary standard.

11-20. *Refer to Section 11-2.*

(a) Potassium hydrogen phthalate (KHP) is the acidic salt, $KC_6H_4(COO)(COOH)$.

(b) KHP is used as a primary standard for the standardization of strong bases.

11-22. *Refer to Section 11-2.*

Balanced equation: $2HNO_3(aq) + Na_2CO_3(s) \rightarrow 2NaNO_3(aq) + CO_2(g) + H_2O(\ell)$

Plan: $g\ Na_2CO_3 \overset{(1)}{\Longrightarrow} mol\ Na_2CO_3 \overset{(2)}{\Longrightarrow} mol\ HNO_3 \overset{(3)}{\Longrightarrow} M\ HNO_3\ soln$

(1) $?\ mol\ Na_2CO_3 = 0.364\ g\ Na_2CO_3 \times \dfrac{1\ mol\ Na_2CO_3}{106\ g\ Na_2CO_3} = 3.43 \times 10^{-3}\ mol\ Na_2CO_3$

(2) $?\ mol\ HNO_3 = 2 \times mol\ Na_2CO_3 = 2 \times 3.43 \times 10^{-3}\ mol = 6.86 \times 10^{-3}\ mol\ HNO_3$

(3) $?\ M\ HNO_3\ soln = \dfrac{6.86 \times 10^{-3}\ mol\ HNO_3}{0.03572\ L\ HNO_3} = \textbf{0.192 } \textbf{\textit{M}} \textbf{ HNO}_3$

Dimensional Analysis:

$?\ M\ HNO_3\ soln = \dfrac{0.364\ g\ Na_2CO_3}{0.03572\ L\ HNO_3\ soln} \times \dfrac{1\ mol\ Na_2CO_3}{106\ g\ Na_2CO_3} \times \dfrac{2\ mol\ HNO_3}{1\ mol\ Na_2CO_3} = \textbf{0.192 } \textbf{\textit{M}} \textbf{ HNO}_3$

11-24. *Refer to Section 11-2 and Example 11-6.*

Balanced equation: $NaOH + KHP \rightarrow NaKP + H_2O$

Plan: $g\ KHP \overset{(1)}{\Longrightarrow} mmol\ KHP \overset{(2)}{\Longrightarrow} mmol\ NaOH \overset{(3)}{\Longrightarrow} M\ NaOH$

(1) $?\ mmol\ KHP = \dfrac{0.8407\ g\ KHP}{204.2\ g/mol} \times \dfrac{1000\ mmol}{1\ mol} = 4.117\ mmol\ KHP$

(2) $?\ mmol\ NaOH = mmol\ KHP = 4.117\ mmol\ NaOH$

(3) $?\ M\ NaOH = \dfrac{4.117\ mmol\ NaOH}{(38.78\ mL - 0.23\ mL)} = \textbf{0.1068 } \textbf{\textit{M}} \textbf{ NaOH}$

11-26. *Refer to the Introduction to Section 11-2.*

(a) An ideal primary standard:

    (1) does not react with or absorb water vapor, oxygen or carbon dioxide,

    (2) reacts according to a single known reaction,

    (3) is available in high purity,

    (4) has a high formula weight,

    (5) is soluble in the solvent of interest,

    (6) is nontoxic,

    (7) is inexpensive, and

    (8) is environmentally friendly.

(b) The significance of each factor is given below.

    (1) The compound must be weighed accurately and must not undergo composition change due to reaction with atmospheric components.

    (2) The reaction must be one of known stoichiometry with no side reactions.

    (3) Solutions of precisely known concentration must be prepared by directly weighing the primary standard.

    (4) The high formula weight is necessary to minimize the effect of weighing errors.

    (5),(6),(7),(8) The significance is self-explanatory.

11-28. *Refer to Section 11-2 and Example 11-7.*

Balanced equation: $(COOH)_2 + 2NaOH \rightarrow Na_2(COO)_2 + 2H_2O$

Plan: M, mL NaOH $\overset{(1)}{\Longrightarrow}$ mol NaOH $\overset{(2)}{\Longrightarrow}$ mol $(COOH)_2 \overset{(3)}{\Longrightarrow}$ mol $(COOH)_2 \cdot 2H_2O \overset{(4)}{\Longrightarrow}$ g $(COOH)_2 \cdot 2H_2O \overset{(5)}{\Longrightarrow}$ % purity

(1) ? mol NaOH = 0.198 M NaOH $\times$ 0.03832 L = 0.00759 mol NaOH

(2) ? mol $(COOH)_2$ = 1/2 $\times$ mol NaOH = 1/2 $\times$ 0.00759 mol NaOH = 0.00379 mol $(COOH)_2$

(3) ? mol $(COOH)_2 \cdot 2H_2O$ = mol $(COOH)_2$ = 0.00379 mol $(COOH)_2 \cdot 2H_2O$

(4) ? g $(COOH)_2 \cdot 2H_2O$ = 0.00379 mol $(COOH)_2 \cdot 2H_2O \times$ 126 g/mol = 0.478 g $(COOH)_2 \cdot 2H_2O$

(5) ? % $(COOH)_2 \cdot 2H_2O = \dfrac{\text{g } (COOH)_2 \cdot 2H_2O}{\text{g sample}} \times 100 = \dfrac{0.478 \text{ g}}{2.00 \text{ g}} \times 100 =$ **23.9% $(COOH)_2 \cdot 2H_2O$**

11-30. *Refer to Section 11-2.*

Balanced equation: $2HCl + CaCO_3 \rightarrow CaCl_2 + CO_2 + H_2O$

Plan: M, mL HCl $\overset{(1)}{\Longrightarrow}$ mol HCl $\overset{(2)}{\Longrightarrow}$ mol $CaCO_3 \overset{(3)}{\Longrightarrow}$ g $CaCO_3$

(1) ? mol HCl = 0.0932 M HCl $\times$ 0.0245 L = 0.00228 mol HCl

(2) ? mol $CaCO_3$ = 1/2 $\times$ mol HCl = 0.00114 mol $CaCO_3$

(3) ? g $CaCO_3$ = 0.00114 mol $CaCO_3 \times$ 100. g/mol = **0.114 g $CaCO_3$**

Dimensional Analysis:

? g $CaCO_3$ = 0.0245 L HCl soln $\times \dfrac{0.0932 \text{ mol HCl}}{1 \text{ L HCl soln}} \times \dfrac{1 \text{ mol } CaCO_3}{2 \text{ mol HCl}} \times \dfrac{100. \text{ g } CaCO_3}{1 \text{ mol } CaCO_3} =$ **0.114 g $CaCO_3$**

11-32. *Refer to Section 11-3 and Example 11-8.*

For complete neutralization, 1 mole of H_3PO_4 yields 3 moles of H^+ ions.

Therefore, the equivalent weight of $H_3PO_4 = \dfrac{\text{molecular weight}}{3 \text{ eq/mol}} = \dfrac{98.0 \text{ g/mol}}{3 \text{ eq/mol}} = 32.7$ g/eq

? N $H_3PO_4 = \dfrac{\text{eq } H_3PO_4}{\text{L soln}} = \dfrac{(7.08 \text{ g } H_3PO_4)/(32.7 \text{ g/eq})}{0.185 \text{ L}} =$ **1.17 N H_3PO_4**

11-34. *Refer to Section 11-3 and Example 11-9.*

? M $H_3AsO_4 = \dfrac{(19.6 \text{ g } H_3AsO_4)/(142 \text{ g/mol})}{0.500 \text{ L}} =$ **0.276 M H_3AsO_4**

For complete neutralization, 1 mole of H_3AsO_4 yields 3 moles of H^+ ions.

Therefore,

? N H_3AsO_4 = 3 $\times$ 0.276 M H_3AsO_4 = **0.828 N H_3AsO_4**

11-36. *Refer to Section 11-3 and Example 11-10.*

At neutralization, $N_{acid} \times V_{acid} = N_{base} \times V_{base}$. Substituting and solving for N_{base},

$$? \ N \ Ba(OH)_2 = \frac{0.206 \ N \ HNO_3 \times 25.0 \ mL \ HNO_3}{35.2 \ mL \ Ba(OH)_2} = 0.146 \ N \ Ba(OH)_2$$

For complete neutralization, 1 mole of $Ba(OH)_2$ will yield 2 moles of OH^- ions.

Therefore,

$? \ M \ Ba(OH)_2 = 1/2 \times 0.146 \ N \ Ba(OH)_2 = \mathbf{0.0732 \ M \ Ba(OH)_2}$

11-38. *Refer to Section 11-3 and Example 11-11.*

Balanced equation: $2HCl + Na_2CO_3 \rightarrow 2NaCl + CO_2 + H_2O$

Plan: $g \ Na_2CO_3 \overset{(1)}{\Longrightarrow} eq \ Na_2CO_3 \overset{(2)}{\Longrightarrow} eq \ HCl \overset{(3)}{\Longrightarrow} N \ HCl \overset{(4)}{\Longrightarrow} M \ HCl$

For Na_2CO_3, equivalent weight (g/eq) $= \dfrac{MW \ (g/mol)}{2 \ eq/mol} = \dfrac{106.0 \ g/mol}{2 \ eq/mol} = 53.0 \ g/eq$

(1) $? \ eq \ Na_2CO_3 = \dfrac{0.318 \ g \ Na_2CO_3}{53.0 \ g/eq} = 6.00 \times 10^{-3} \ eq \ Na_2CO_3$

(2) $? \ eq \ HCl = eq \ Na_2CO_3 = 6.00 \times 10^{-3} \ eq \ HCl$

(3) $? \ N \ HCl = \dfrac{6.00 \times 10^{-3} \ eq \ HCl}{0.0431 \ L} = \mathbf{0.139 \ N \ HCl}$

(4) $? \ M \ HCl = N \ HCl = \mathbf{0.139 \ M \ HCl}$

11-40. *Refer to Sections 3-1 and 11-4, and Example 11-12.*

(a) $Fe(s) + 2HCl(aq) \rightarrow FeCl_2(aq) + H_2(g)$

(b) $2Cr(s) + 3H_2SO_4(aq) \rightarrow Cr_2(SO_4)_3(aq) + 3H_2(g)$

(c) $Sn(s) + 4HNO_3(aq) \rightarrow SnO_2(s) + 4NO_2(s) + 2H_2O(\ell)$

11-42. *Refer to Section 11-4 and Examples 11-13, 11-14 and 11-15.*

Using the Change-In-Oxidation-Number Method:

(a) $2MnO_4^-(aq) \ + \ 16H^+(aq) \ + \ 10Br^-(aq) \rightarrow 2Mn^{2+}(aq) \ + \ 5Br_2(\ell) \ + \ 8H_2O(\ell)$

$(+7)$ (-1) $(+2)$ (0) Oxidation Number

-5

$+1$

Oxidation Numbers	Change/Atom	Equalizing Changes Gives
$Br = -1 \rightarrow Br = 0$	$+1$	$5(+1) = +5$
$Mn = +7 \rightarrow Mn = +2$	-5	$1(-5) = -5$

Each change is multiplied by 2 since there are 2 Br atoms in Br_2.

(b) $Cr_2O_7^{2-}(aq) + 14H^+(aq) + 6I^-(aq) \rightarrow 2Cr^{3+}(aq) + 3I_2(s) + 7H_2O(\ell)$

Oxidation Numbers	Change/Atom	Equalizing Changes Gives
$I = -1 \rightarrow I = 0$	+1	$3(+1) = +3$
$Cr = +6 \rightarrow Cr = +3$	-3	$1(-3) = -3$

Each change is multiplied by 2 since there are 2 Cr atoms in $Cr_2O_7^{2-}$ and 2 I atoms in I_2.

(c) $2MnO_4^-(aq) + 5SO_3^{2-}(aq) + 6H^+(aq) \rightarrow 2Mn^{2+}(aq) + 5SO_4^{2-}(aq) + 3H_2O(\ell)$

Oxidation Numbers	Change/Atom	Equalizing Changes Gives
$S = +4 \rightarrow S = +6$	+2	$5(+2) = +10$
$Mn = +7 \rightarrow Mn = +2$	-5	$2(-5) = -10$

(d) $Cr_2O_7^{2-}(aq) + 6Fe^{2+}(aq) + 14H^+(aq) \rightarrow 2Cr^{3+}(aq) + 6Fe^{3+}(aq) + 7H_2O(\ell)$

Oxidation Numbers	Change/Atom	Equalizing Changes Gives
$Fe = +2 \rightarrow Fe = +3$	+1	$3(+1) = +3$
$Cr = +6 \rightarrow Cr = +3$	-3	$1(-3) = -3$

Each change is multiplied by 2 since there are 2 Cr atoms in $Cr_2O_7^{2-}$.

11-44. *Refer to Section 11-6, and Examples 11-16 and 11-17.*

Using the Half-Reaction Method:

(a) skeletal equation: $Al(s) + NO_3^-(aq) + OH^-(aq) + H_2O(\ell) \rightarrow Al(OH)_4^-(aq) + NH_3(g)$
    ox. half-rxn: $Al(s) \rightarrow Al(OH)_4^-(aq)$
    balanced ox. half-rxn: $Al(s) + 4OH^-(aq) \rightarrow Al(OH)_4^-(aq) + 3e^-$
    red. half-rxn: $NO_3^-(aq) \rightarrow NH_3(g)$
    balanced red. half-rxn: $8e^- + NO_3^-(aq) + 6H_2O(\ell) \rightarrow NH_3(g) + 9OH^-(aq)$

Now, we balance the electron transfer and add the half-reactions term-by-term and cancel electrons:

oxidation: $8[Al(s) + 4OH^-(aq) \rightarrow Al(OH)_4^-(aq) + 3e^-]$
reduction: $3[8e^- + NO_3^-(aq) + 6H_2O(\ell) \rightarrow NH_3(aq) + 9OH^-(aq)]$
balanced: $8Al(s) + 3NO_3^-(aq) + 5OH^-(aq) + 18H_2O(\ell) \rightarrow 8Al(OH)_4^-(aq) + 3NH_3(g)$

(b) skeletal equation: $NO_2(g) + OH^-(aq) \rightarrow NO_3^-(aq) + NO_2^-(aq) + H_2O(\ell)$
    ox. half-rxn: $NO_2(g) \rightarrow NO_3^-(aq)$
    balanced ox. half-rxn: $NO_2(g) + 2OH^-(aq) \rightarrow NO_3^-(aq) + H_2O(\ell) + e^-$
    red. half-rxn: $NO_2(g) \rightarrow NO_2^-(aq)$
    balanced red. half-rxn: $e^- + NO_2(g) \rightarrow NO_2^-(aq)$

The electron transfer is already balanced and we can write:

oxidation: $NO_2(g) + 2OH^-(aq) \rightarrow NO_3^-(aq) + H_2O(\ell) + e^-$
reduction: $e^- + NO_2(g) \rightarrow NO_2^-(aq)$
balanced: $2NO_2(g) + 2OH^-(aq) \rightarrow NO_3^-(aq) + NO_2^-(aq) + H_2O(\ell)$

(c) oxidation: $3[NO_2^-(aq) + 2OH^-(aq) \rightarrow NO_3^-(aq) + H_2O(\ell) + 2e^-]$
    reduction: $2[3e^- + MnO_4^-(aq) + 2H_2O(\ell) \rightarrow MnO_2 + 4OH^-(aq)]$
    balanced: $2MnO_4^-(aq) + 3NO_2^-(aq) + H_2O(\ell) \rightarrow 2MnO_2(s) + 3NO_3^-(aq) + 2OH^-(aq)$

(d) oxidation: $2I^-(aq) \rightarrow I_2(s) + 2e^-$
    reduction: $2[e^- + NO_2^-(aq) + 2H^+(aq) \rightarrow NO(g) + H_2O(\ell)]$
    balanced: $2I^-(aq) + 4H^+(aq) + 2NO_2^-(aq) \rightarrow 2NO(g) + 2H_2O(\ell) + I_2(s)$

(e) oxidation: $4NH_3(aq) + Hg_2Cl_2(s) \rightarrow 2HgNH_2Cl(s) + 2NH_4^+ + 2e^-$

    reduction: $2e^- + Hg_2Cl_2(s) \rightarrow 2Hg(\ell) + 2Cl^-(aq)$

    balanced: $2Hg_2Cl_2(s) + 4NH_3(aq) \rightarrow 2Hg(\ell) + 2HgNH_2Cl(s) + 2NH_4^+(aq) + 2Cl^-(aq)$

    or $Hg_2Cl_2(s) + 2NH_3(aq) \rightarrow Hg(\ell) + HgNH_2Cl(s) + NH_4^+(aq) + Cl^-(aq)$

11-46. *Refer to Section 11-6, and Examples 11-16 and 11-17.*

Using the Half-Reaction Method:

(a) skeletal equation: $Fe^{2+}(aq) + MnO_4^-(aq) \rightarrow Fe^{3+}(aq) + Mn^{2+}(aq)$

    ox. half-rxn: $Fe^{2+}(aq) \rightarrow Fe^{3+}(aq)$

    balanced ox. half-rxn: $Fe^{2+}(aq) \rightarrow Fe^{3+}(aq) + e^-$

    red. half-rxn: $MnO_4^-(aq) \rightarrow Mn^{2+}(aq)$

    balanced red. half-rxn: $5e^- + 8H^+(aq) + MnO_4^-(aq) \rightarrow Mn^{2+}(aq) + 4H_2O(\ell)$

Now, we balance the electron transfer and add the half-reactions term-by-term and cancel electrons:

    oxidation: $5[Fe^{2+}(aq) \rightarrow Fe^{3+}(aq) + e^-]$

    reduction: $5e^- + MnO_4^-(aq) + 8H^+(aq) \rightarrow Mn^{2+}(aq) + 4H_2O(\ell)$

    balanced: $5Fe^{2+}(aq) + MnO_4^-(aq) + 8H^+(aq) \rightarrow 5Fe^{3+}(aq) + Mn^{2+}(aq) + 4H_2O(\ell)$

(b) skeletal equation: $Br_2(\ell) + SO_2(g) \rightarrow Br^-(aq) + SO_4^{2-}(aq)$

    ox. half-rxn: $SO_2(g) \rightarrow SO_4^{2-}(aq)$

    balanced ox. half-rxn: $2H_2O(\ell) + SO_2(g) \rightarrow SO_4^{2-}(aq) + 4H^+(aq) + 2e^-$

    red. half-rxn: $Br_2(\ell) \rightarrow Br^-(aq)$

    balanced red. half-rxn: $2e^- + Br_2(\ell) \rightarrow 2Br^-(aq)$

The electron transfer is already balanced and we can write:

    oxidation: $2H_2O(\ell) + SO_2(g) \rightarrow SO_4^{2-}(aq) + 4H^+(aq) + 2e^-$

    reduction: $2e^- + Br_2(\ell) \rightarrow 2Br^-(aq)$

    balanced: $Br_2(\ell) + SO_2(g) + 2H_2O(\ell) \rightarrow 2Br^-(aq) + SO_4^{2-}(aq) + 4H^+(aq)$

(c) oxidation: $Cu(s) \rightarrow Cu^{2+}(aq) + 2e^-$

    reduction: $2[1e^- + 2H^+(aq) + NO_3^-(aq) \rightarrow NO_2(g) + H_2O(\ell)]$

    balanced: $Cu(s) + 2NO_3^-(aq) + 4H^+(aq) \rightarrow Cu^{2+}(aq) + 2NO_2(g) + 2H_2O(\ell)$

(d) oxidation: $2Cl^-(aq) \rightarrow Cl_2(g) + 2e^-$

    reduction: $2e^- + 4H^+(aq) + 2Cl^-(aq) + PbO_2(s) \rightarrow PbCl_2(s) + 2H_2O(\ell)$

    balanced: $PbO_2(s) + 4Cl^-(aq) + 4H^+(aq) \rightarrow PbCl_2(s) + Cl_2(g) + 2H_2O(\ell)$

(e) oxidation: $5[Zn(s) \rightarrow Zn^{2+}(aq) + 2e^-]$

    reduction: $10e^- + 12H^+(aq) + 2NO_3^-(aq) \rightarrow N_2(g) + 6H_2O(\ell)$

    balanced: $5Zn(s) + 2NO_3^-(aq) + 12H^+(aq) \rightarrow 5Zn^{2+}(aq) + N_2(g) + 6H_2O(\ell)$

11-48. *Refer to Sections 11-4, 11-5 and 11-6, and Example 11-18.*

Using the Change-In-Oxidation-Number Method:

(a) Balanced net ionic equation:

$$2MnO_4^-(aq) + 16H^+(aq) + 5C_2O_4^{2-}(aq) \rightarrow 2Mn^{2+}(aq) + 10CO_2(g) + 8H_2O(\ell)$$

    (+7)   (+3)   (+2)   (+4)   Oxidation Number

                −5   +1

Oxidation Numbers	Change/Atom	Equalizing Changes Gives
C = +3 → C = +4	+1	5(+1) = +5
Mn = +7 → Mn = +2	-5	1(-5) = -5

Each change is multiplied by 2 since there are 2 C atoms in $C_2O_4^{2-}$.

Balanced formula unit equation (cations = K^+, anions Cl^-):

$$2KMnO_4(aq) + 16HCl(aq) + 5K_2C_2O_4(aq) \rightarrow 2MnCl_2(aq) + 12KCl(aq) + 10\ CO_2(g) + 8H_2O(\ell)$$

(b) Balanced net ionic equation: $4Zn(s) + NO_3^-(aq) + 10\ H^+(aq) \rightarrow 4Zn^{2+}(aq) + NH_4^+(aq) + 3H_2O(\ell)$

Oxidation Numbers	Change/Atom	Equalizing Changes Gives
Zn = 0 → Zn = +2	+2	4(+2) = +8
N = +5 → N = -3	-8	1(-8) = -8

Balanced formula unit equation (cations = H^+, anions = unreacted NO_3^-):

$$4Zn(s) + 10\ HNO_3(aq) \rightarrow 4Zn(NO_3)_2(aq) + NH_4NO_3(aq) + 3H_2O(\ell)$$

11-50. *Refer to Sections 11-4, 11-5 and 11-6, and Example 11-18.*

Using the Half-Reaction Method:

(a) Balanced net ionic equation:

oxidation: $\qquad\qquad\qquad\qquad\qquad Zn(s) \rightarrow Zn^{2+}(aq) + 2e^-$

reduction: $\qquad\qquad\qquad 2e^- + Cu^{2+}(aq) \rightarrow Cu(s)$

balanced: $\qquad\qquad\qquad Zn(s) + Cu^{2+}(aq) \rightarrow Zn^{2+}(aq) + Cu(s)$

Balanced formula unit equation (anion = SO_4^{2-}): $Zn(s) + CuSO_4(aq) \rightarrow ZnSO_4(aq) + Cu(s)$

(c) Balanced net ionic equation:

oxidation: $\qquad\qquad\qquad\qquad\qquad 2[Cr(s) \rightarrow Cr^{3+}(aq) + 3e^-]$

reduction: $\qquad\qquad 3[2e^- + 2H^+(aq) \rightarrow H_2(g)]$

balanced: $\qquad\qquad 2Cr(s) + 6H^+(aq) \rightarrow 2Cr^{3+}(aq) + 3H_2(g)$

Balanced formula unit equation (anion = SO_4^{2-}): $2Cr(s) + 3H_2SO_4(aq) \rightarrow Cr_2(SO_4)_3(aq) + 3H_2(g)$

11-52. *Refer to Section 11-7 and Example 11-19.*

Balanced <u>net ionic</u> equation is:

$$5Fe^{2+}(aq) + MnO_4^-(aq) + 8H^+(aq) \rightarrow 5Fe^{3+}(aq) + Mn^{2+}(aq) + 4H_2O(\ell)$$

Note: This exercise uses $KMnO_4$ and $FeSO_4$. These are both soluble salts which dissociate into their ions. The K^+ and SO_4^{2-} ions are spectator ions and are omitted from the balanced net ionic equation.

Plan: M, mL $FeSO_4$ soln $\overset{(1)}{\Longrightarrow}$ mmol $FeSO_4$ $\overset{(2)}{\Longrightarrow}$ mmol $KMnO_4$ $\overset{(3)}{\Longrightarrow}$ mL $KMnO_4$

(1) ? mmol $FeSO_4$ = 0.100 M × 20.0 mL = 2.00 mmol $FeSO_4$

(2) ? mmol $KMnO_4$ = 1/5 × mmol $FeSO_4$ = 1/5 × 2.00 mmol = 0.400 mmol $KMnO_4$

(3) ? mL $KMnO_4 = \dfrac{0.400 \text{ mmol } KMnO_4}{0.150\ M\ KMnO_4}$ = **2.67 mL $KMnO_4$ soln**

Balanced <u>net</u> <u>ionic</u> equation:

$$2MnO_4^-(aq) + 16H^+(aq) + 10\,I^-(aq) \rightarrow 2Mn^{2+}(aq) + 5I_2(s) + 8H_2O(\ell)$$

Note: This exercise uses $KMnO_4$ and KI, which are both soluble salts that dissociate into their ions. The K^+ ions are spectator ions and are omitted from the balanced net ionic equation.

$$\text{Plan:}\quad M,\ \text{mL KI soln} \overset{(1)}{\Longrightarrow} \text{mol KI} \overset{(2)}{\Longrightarrow} \text{mol } KMnO_4 \overset{(3)}{\Longrightarrow} \text{L } KMnO_4 \text{ soln}$$

(1) ? mol KI $= 0.100\ M \times 0.0500\ L = 0.00500$ mol KI

(2) ? mol $KMnO_4 = 1/5 \times$ mol KI $= 0.00100$ mol $KMnO_4$

(3) ? L $KMnO_4$ soln $= \dfrac{0.00100 \text{ mol } KMnO_4}{0.200\ M\ KMnO_4} =$ **0.00500 L or 5.00 mL $KMnO_4$ soln**

(a) Balanced equation: $2Na_2S_2O_3 + I_2 \rightarrow Na_2S_4O_6 + 2NaI$

$$\text{Plan:}\quad M,\ \text{L } Na_2S_2O_3 \text{ soln} \overset{(1)}{\Longrightarrow} \text{mol } Na_2S_2O_3 \overset{(2)}{\Longrightarrow} \text{mol } I_2 \overset{(3)}{\Longrightarrow} M\ I_2 \text{ soln}$$

(1) ? mol $Na_2S_2O_3 = 0.1455\ M \times 0.04000\ L = 0.005820$ mol $Na_2S_2O_3$

(2) ? mol $I_2 = 1/2 \times$ mol $Na_2S_2O_3 = 0.002910$ mol I_2

(3) ? $M\ I_2 = \dfrac{\text{mol } I_2}{\text{L } I_2} = \dfrac{0.002910 \text{ mol } I_2}{0.02636 \text{ L } I_2} =$ **0.1104 $M\ I_2$**

(b) Balanced equation: $As_2O_3 + 5H_2O + 2I_2 \rightarrow 2H_3AsO_4 + 4HI$

$$\text{Plan:}\quad M,\ \text{L } I_2 \text{ soln} \overset{(1)}{\Longrightarrow} \text{mol } I_2 \overset{(2)}{\Longrightarrow} \text{mol } As_2O_3 \overset{(3)}{\Longrightarrow} \text{g } As_2O_3$$

(1) ? mol $I_2 = 0.1104\ M \times 0.02532\ L = 0.002795$ mol I_2

(2) ? mol $As_2O_3 = 1/2 \times$ moles $I_2 = 0.001398$ mol As_2O_3

(3) ? g $As_2O_3 = 0.001398$ mol $As_2O_3 \times 197.8$ g/mol $=$ **0.2765 g As_2O_3**

(a) Balanced equation: $HI + NaOH \rightarrow NaI + H_2O$

$$\text{Plan:}\quad M,\ \text{L NaOH soln} \overset{(1)}{\Longrightarrow} \text{mol NaOH} \overset{(2)}{\Longrightarrow} \text{mol HI} \overset{(3)}{\Longrightarrow} \text{L HI soln}$$

(1) ? mol NaOH $= 0.100\ M \times 0.0250\ L = 0.00250$ mol NaOH

(2) ? mol HI $=$ mol NaOH $= 0.00250$ mol HI

(3) ? L HI soln $= \dfrac{0.00250 \text{ mol HI}}{0.250\ M\ HI} =$ **0.0100 L HI soln**

Dimensional Analysis:

$$? \text{ L HI soln} = 0.0250 \text{ L NaOH soln} \times \frac{0.100 \text{ mol NaOH}}{1 \text{ L NaOH soln}} \times \frac{1 \text{ mol HI}}{1 \text{ mol NaOH}} \times \frac{1 \text{ L HI soln}}{0.250 \text{ mol HI}} = \textbf{0.0100 L HI soln}$$

(b) Balanced net ionic equation: $Ag^+ + I^- \rightarrow AgI$

Plan: $\text{g AgNO}_3 \overset{(1)}{\Longrightarrow} \text{mol AgNO}_3 \overset{(2)}{\Longrightarrow} \text{mol HI} \overset{(3)}{\Longrightarrow} \text{L HI soln}$

(1) $? \text{ mol AgNO}_3 = \dfrac{5.03 \text{ g AgNO}_3}{169.9 \text{ g/mol}} = 0.0296 \text{ mol AgNO}_3$

(2) $? \text{ mol HI} = \text{mol AgNO}_3 = 0.0296 \text{ mol HI}$

(3) $? \text{ L HI soln} = \dfrac{0.0296 \text{ mol HI}}{0.250 \text{ M HI}} = \textbf{0.118 L HI soln}$

(c) Balanced equation: $2Cu^{2+} + 4I^- \rightarrow 2CuI + I_2$

Plan: $\text{g CuSO}_4 \overset{(1)}{\Longrightarrow} \text{mol CuSO}_4 \overset{(2)}{\Longrightarrow} \text{mol HI} \overset{(3)}{\Longrightarrow} \text{L HI soln}$

(1) $? \text{ mol CuSO}_4 = \dfrac{0.621 \text{ g CuSO}_4}{159.6 \text{ g/mol}} = 0.00389 \text{ mol CuSO}_4$

(2) $? \text{ mol HI} = 4/2 \times \text{mol CuSO}_4 = 2 \times 0.00389 \text{ mol} = 0.00778 \text{ mol HI}$

(3) $? \text{ L HI} = \dfrac{0.00778 \text{ mol HI}}{0.250 \text{ M HI}} = \textbf{0.0311 L HI soln}$

11-60. *Refer to Sections 11-7 and 3-6.*

Plan: $\text{g KMnO}_4 \overset{(1)}{\Longrightarrow} \text{mol KMnO}_4 \overset{(2)}{\Longrightarrow} M \text{ KMnO}_4$

(1) $? \text{ mol KMnO}_4 = \dfrac{12.6 \text{ g KMnO}_4}{158 \text{ g/mol}} = 0.0797 \text{ mol KMnO}_4$

(2) $? \, M \text{ KMnO}_4 = \dfrac{0.0797 \text{ mol KMnO}_4}{0.500 \text{ L}} = \textbf{0.159} \, \boldsymbol{M} \textbf{ KMnO}_4$

The balanced half-reaction involving the reduction of MnO_4^- to MnO_4^{2-} requires 1 electron:

$$e^- + MnO_4^-(aq) \rightarrow MnO_4^{2-}(aq)$$

This fact is irrelevant since the molarity of a solution is *independent* of the number of electrons involved in the reaction. The molarity depends only on the moles of solute and the liters of solution.

11-62. *Refer to Sections 11-2 and 3-8.*

Balanced equation: $2HCl + Na_2CO_3 \rightarrow 2NaCl + H_2O + CO_2$

Plan: $\text{g Na}_2SO_4 \overset{(1)}{\Longrightarrow} \text{mol Na}_2SO_4 \overset{(2)}{\Longrightarrow} \text{mol HCl} \overset{(3)}{\Longrightarrow} M \text{ HCl}$

(1) $? \text{ mol Na}_2CO_3 = \dfrac{0.2013 \text{ g Na}_2CO_3}{106.0 \text{ g/mol}} = 0.001899 \text{ mol Na}_2CO_3$

(2) $? \text{ mol HCl} = 2 \times 0.001899 \text{ mol Na}_2CO_3 = 0.003798 \text{ mol HCl}$

(3) $? \, M \text{ HCl} = \dfrac{0.003798 \text{ mol HCl}}{0.03275 \text{ L soln}} = \textbf{0.1160} \, \boldsymbol{M} \textbf{ HCl}$

11-64. *Refer to Sections 11-2 and 3-8.*

Balanced equation: $HCl + NaOH \rightarrow NaCl + H_2O$

(1) Plan: M, mL NaOH $\overset{(1)}{\Longrightarrow}$ mmol NaOH $\overset{(2)}{\Longrightarrow}$ mmol HCl

$? \text{ mmol HCl} = 25.5 \text{ mL NaOH} \times \dfrac{0.110 \text{ mmol NaOH}}{1 \text{ mL NaOH}} \times \dfrac{1 \text{ mmol HCl}}{1 \text{ mmol NaOH}} = \textbf{2.80 mmol HCl}$

(2) $? \text{ mL HCl} = 2.80 \text{ mmol HCl} \times \dfrac{1 \text{ mL HCl}}{0.205 \text{ mmol HCl}} = \textbf{13.7 mL HCl}$

11-66. *Refer to Sections 11-2 and 3-8.*

Balanced equation: $2HCl + Ca(OH)_2 \rightarrow CaCl_2 + 2H_2O$

Plan: g $Ca(OH)_2 \overset{(1)}{\Longrightarrow}$ mol $Ca(OH)_2 \overset{(2)}{\Longrightarrow}$ mol HCl $\overset{(3)}{\Longrightarrow} V$ HCl

(1) $? \text{ mol } Ca(OH)_2 = \dfrac{1.58 \text{ g } Ca(OH)_2}{74.1 \text{ g/mol}} = 0.0213 \text{ mol } Ca(OH)_2$

(2) $? \text{ mol HCl} = 2 \times 0.0213 \text{ mol } Ca(OH)_2 = 0.0426 \text{ mol HCl}$

(3) $? \text{ L HCl} = \dfrac{0.0426 \text{ mol HCl}}{0.1123 \text{ } M \text{ HCl}} = \textbf{0.379 L} \text{ or } \textbf{379 mL HCl soln}$

Dimensional Analysis:

$? \text{ L HCl} = 1.58 \text{ g } Ca(OH)_2 \times \dfrac{1 \text{ mol } Ca(OH)_2}{74.1 \text{ g } Ca(OH)_2} \times \dfrac{2 \text{ mol HCl}}{1 \text{ mol } Ca(OH)_2} \times \dfrac{1 \text{ L HCl soln}}{0.1123 \text{ mol HCl}}$

$= \textbf{0.380 L} \text{ or } \textbf{380 mL HCl soln}$

11-68. *Refer to Sections 11-2 and 3-8.*

Balanced equation: $H_2SO_4 + 2KOH \rightarrow K_2SO_4 + 2H_2O$

Plan: M, L KOH $\overset{(1)}{\Longrightarrow}$ mol KOH $\overset{(2)}{\Longrightarrow}$ mol $H_2SO_4 \overset{(3)}{\Longrightarrow} V$ H_2SO_4

(1) $? \text{ mol KOH} = 0.302 \text{ } M \times 0.0394 \text{ L} = 0.0119 \text{ mol KOH}$

(2) $? \text{ mol } H_2SO_4 = 1/2 \times 0.0119 \text{ mol KOH} = 0.00595 \text{ mol } H_2SO_4$

(3) $? \text{ L } H_2SO_4 = \dfrac{0.00595 \text{ mol } H_2SO_4}{0.203 \text{ } M \text{ } H_2SO_4} = \textbf{0.0293 L} \text{ or } \textbf{29.3 mL } H_2SO_4 \textbf{ soln}$

Dimensional Analysis:

$? \text{ L } H_2SO_4 = 0.0394 \text{ L KOH soln} \times \dfrac{0.302 \text{ mol KOH}}{1 \text{ L KOH soln}} \times \dfrac{1 \text{ mol } H_2SO_4}{2 \text{ mol KOH}} \times \dfrac{1 \text{ L } H_2SO_4 \text{ soln}}{0.203 \text{ mol } H_2SO_4}$

$= \textbf{0.0293 L} \text{ or } \textbf{29.3 mL } H_2SO_4 \textbf{ soln}$

11-70. *Refer to Sections 11-3 and Example 11-10.*

At neutralization, $N_{acid} \times V_{acid} = N_{base} \times V_{base}$. Therefore,

$? \text{ } V \text{ NaOH} = \dfrac{0.1023 \text{ } N \text{ } H_2SO_4 \times 38.38 \text{ mL } H_2SO_4}{0.2045 \text{ } N \text{ NaOH}} = \textbf{19.20 mL NaOH soln}$

Balanced equation: $Cr_2O_7^{2-} + 6Fe^{2+} + 14H^+ \rightarrow 2Cr^{3+} + 6Fe^{3+} + 7H_2O$

Plan: M, L $Na_2Cr_2O_7 \overset{(1)}{\Longrightarrow}$ mol $Na_2Cr_2O_7$ (= mol $Cr_2O_7^{2-}$) $\overset{(2)}{\Longrightarrow}$ mol $Fe^{2+} \overset{(3)}{\Longrightarrow}$ g Fe^{2+} (= g Fe) $\overset{(4)}{\Longrightarrow}$ %Fe

(1) ? mol $Cr_2O_7^{2-}$ = 42.96 mL $Na_2Cr_2O_7 \times \dfrac{1.000 \text{ L } Na_2Cr_2O_7}{1000 \text{ mL } Na_2Cr_2O_7} \times \dfrac{0.02130 \text{ mol } Na_2Cr_2O_7}{1.000 \text{ L } Na_2Cr_2O_7} \times \dfrac{1 \text{ mol } Cr_2O_7^{2-}}{1 \text{ mol } Na_2Cr_2O_7}$

$= 9.150 \times 10^{-4}$ mol $Cr_2O_7^{2-}$

(2) ? mol Fe^{2+} = 9.150×10^{-4} mol $Cr_2O_7^{2-} \times \dfrac{6 \text{ mol } Fe^{2+}}{1 \text{ mol } Cr_2O_7^{2-}}$ = 5.490×10^{-3} mol Fe^{2+}

(3) ? g Fe = 5.490×10^{-3} mol $Fe^{2+} \times \dfrac{1 \text{ mol Fe}}{1 \text{ mol } Fe^{2+}} \times \dfrac{55.85 \text{ g Fe}}{1 \text{ mol Fe}}$ = 0.3066 g Fe

(4) ? %Fe = $\dfrac{0.3066 \text{ g Fe}}{0.5166 \text{ g sample}} \times 100$ = **59.36% Fe**

If the percentage of Fe in the limonite ore had been greater than 100%, one conclusion would be that there were other components in the dissolved ore solution in addition to Fe^{2+} that were capable of reducing $Cr_2O_7^{2-}$ to Cr^{3+}, (assuming of course that the analytical data were correct). Therefore, the volume of $Na_2Cr_2O_7$ necessary to reach the equivalence point would increase, and the amount of Fe present would appear to be larger than it really was.

Balanced equation: $NaAl(OH)_2CO_3 + 4HCl \rightarrow NaCl + AlCl_3 + CO_2 + 3H_2O$

Plan: (1) Calculate the mmoles of HCl in your stomach acid.
       (2) Calculate the mmoles of $NaAl(OH)_2CO_3$ in one Rolaids® antacid tablet.
       (3) Calculate the mmoles of HCl that can be neutralized by the antacid tablet.

(1) ? mmol HCl in stomach = 0.10 M HCl × 800 mL = **80 mmol HCl in stomach**

(2) ? mmol $NaAl(OH)_2CO_3$ = $\dfrac{334 \text{ mg } NaAl(OH)_2CO_3}{144 \text{ mg/mmol}}$ = 2.32 mmol $NaAl(OH)_2CO_3$

(3) ? mmol neutralized HCl = 2.32 mmol $NaAl(OH)_2CO_3 \times \dfrac{4 \text{ mmol HCl}}{1 \text{ mmol } NaAl(OH)_2CO_3}$ = **9.28 mmol HCl**

The number of **mmoles of HCl in your stomach** is roughly nine times greater than the number of mmoles of HCl that can be neutralized by a single antacid tablet. However, about 2 tablets are sufficient to neutralize the excess HCl in the stomach by reducing its concentration down to the normal 8.0×10^{-2} M level.

Balanced equation: $SiO_2(s) + 6HF(aq) \rightarrow H_2SiF_6(aq) + 2H_2O(\ell)$

The etching of glass, SiO_2, by hydrofluoric acid, HF, is **not** an oxidation-reduction reaction, since no element in the reaction is undergoing a change in oxidation number.

11-78. *Refer to Sections 4-6 and 4-7.*

An antioxidant is a compound that opposes oxidation or inhibits reactions promoted by oxygen or peroxides. Such a compound is ascorbic acid, $H_2C_6H_6O_6$, also called Vitamin C, which can undergo a decomposition reaction as follows:

Oxidation Numbers: $\overset{+1}{H_2}\overset{+\frac{2}{3}}{C_6}\overset{+1}{H_6}\overset{-2}{O_6} \rightarrow \overset{+1}{C_6}\overset{+1}{H_6}\overset{-2}{O_6} + \overset{0}{H_2}$

Vitamin C is both oxidized ($C = +2/3 \rightarrow C = +1$) and reduced ($H = +1 \rightarrow H = 0$) in this reaction.

11-80. *Refer to Sections 11-2 and 3-8.*

Balanced equation: $HCl + NaHCO_3 \rightarrow NaCl + CO_2 + H_2O$

Plan: M, L HCl $\overset{(1)}{\Longrightarrow}$ mol HCl $\overset{(2)}{\Longrightarrow}$ mol $NaHCO_3$ $\overset{(3)}{\Longrightarrow}$ g $NaHCO_3$

(1) $? \text{ mol HCl} = 0.085 \text{ L HCl} \times \frac{0.17 \text{ mol HCl}}{1 \text{ L HCl}} = 0.014 \text{ mol HCl}$

(2) $? \text{ mol NaHCO}_3 = 0.014 \text{ mol HCl} \times \frac{1 \text{ mol NaHCO}_3}{1 \text{ mol HCl}} = 0.014 \text{ mol NaHCO}_3$

(3) $? \text{ g NaHCO}_3 = 0.014 \text{ mol NaHCO}_3 \times \frac{84.01 \text{ g NaHCO}_3}{1 \text{ mol NaHCO}_3} = \mathbf{1.2 \text{ g NaHCO}_3}$

Dimensional Analysis:

$$? \text{ g NaHCO}_3 = 0.085 \text{ L HCl} \frac{0.17 \text{ mol HCl}}{1 \text{ L HCl}} \times \frac{1 \text{ mol NaHCO}_3}{1 \text{ mol HCl}} \times \frac{84.01 \text{ g NaHCO}_3}{1 \text{ mol NaHCO}_3} = \mathbf{1.2 \text{ g NaHCO}_3}$$

12 Gases and the Kinetic Molecular Theory

12-2. *Refer to Sections 12-1 and 12-2.*

All gases are (a) transparent to light. Some gases are (b) colorless and (e) odorless. However, no gas (c) is unable to pass through filter paper, (d) is more difficult to compress than water and (f) settles on standing.

12-4. *Refer to Sections 12-1 and 12-2.*

(a) The material is not a gas. If the container did hold a gas and was opened to the atmosphere, the material would expand without limit.

(b) The material discharging from the smokestack is not a gas, but a colloidal mixture that light cannot penetrate.

(c) The material is not a gas because its density, 8.2 g/mL, is far too great.

(d) The material is a gas for two reasons. (1) It is much less dense than seawater since it rises rapidly to the surface. (2) At 30 ft below the water's surface the material is exposed to 2 atm pressure: 1 atm (760 mm Hg) atmospheric pressure and 1 atm (76 cm Hg) of water pressure. As the pressure on the material decreased by a factor of 2 to 1 atm pressure at the surface, its volume increased by a factor of 2. This is an illustration of Boyle's Law (Section 12-4).

(e) The material may be a gas, but insufficient information is given.

(f) The material is definitely a gas.

12-6. *Refer to Section 12-3 and the inside back cover of the textbook.*

Some of the units of pressure include mm Hg, torr, atmosphere (atm), pascal (Pa), and pounds per square inch (psi).

$$1 \text{ atmosphere (atm)} = 760 \text{ mm Hg} = 760 \text{ torr} = 1.01325 \times 10^5 \text{ Pa} = 14.7 \text{ psi}$$

Different units are used for different purposes in different areas of endeavor. The English unit is psi, whereas the SI base unit is the pascal. Units of torr are commonly used when working at low pressures ($\leq$ 1 atm) and atmospheres at high pressures ($\geq$ 1 atm).

12-8. *Refer to Section 12-3 and Figure 12-1.*

A manometer is a device employing the change in liquid levels to measure gas pressure differences between a standard and an unknown system. For example, a typical mercury manometer consists of a glass tube partially filled with mercury. One arm is open to the atmosphere and the other is connected to a container of gas. When the pressure of the gas in the container is greater than atmospheric pressure, the level of the mercury in the open side will be higher and

$$P_{gas} = P_{atm} + \Delta h$$

where Δh is the difference in mercury levels

However, when the pressure of the gas is less than atmospheric pressure, the level of the mercury in the closed side will be higher, and

$$P_{gas} = P_{atm} - \Delta h$$

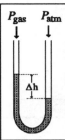

where Δh is the difference in mercury levels

12-10. *Refer to Section 12-3 and Appendix C.*

(a) ? psi $= 755$ torr $\times \dfrac{1 \text{ atm}}{760 \text{ torr}} \times \dfrac{14.70 \text{ psi}}{1 \text{ atm}} = $ **14.6 psi**

(b) ? cm Hg $= 755$ torr $\times \dfrac{1 \text{ mm Hg}}{1 \text{ torr}} \times \dfrac{1 \text{ cm Hg}}{10 \text{ mm Hg}} = $ **75.5 cm Hg**

(c) ? inches Hg $= 755$ torr $\times \dfrac{1 \text{ mm Hg}}{1 \text{ torr}} \times \dfrac{1 \text{ cm Hg}}{10 \text{ mm Hg}} \times \dfrac{1 \text{ in Hg}}{2.54 \text{ cm Hg}} = $ **29.7 in Hg**

(d) ? kPa $= 755$ torr $\times \dfrac{1 \text{ atm}}{760 \text{ torr}} \times \dfrac{1.013 \times 10^5 \text{ Pa}}{1 \text{ atm}} \times \dfrac{1 \text{ kPa}}{1000 \text{ Pa}} = $ **101 kPa**

(e) ? atm $= 755$ torr $\times \dfrac{1 \text{ atm}}{760 \text{ torr}} = $ **0.993 atm**

(f) ? ft H$_2$O $= 755$ torr $\times \dfrac{1 \text{ mm Hg}}{1 \text{ torr}} \times \dfrac{1 \text{ cm Hg}}{10 \text{ mm Hg}} \times \dfrac{1 \text{ in Hg}}{2.54 \text{ cm Hg}} \times \dfrac{1 \text{ ft Hg}}{12 \text{ in Hg}} \times \dfrac{13.59 \text{ ft H}_2\text{O}}{1.00 \text{ ft Hg}} = $ **33.7 ft H$_2$O**

Note: The final unit factor uses the relative densities of water and mercury at 25°C. Since the density of water is only 1/13.59 that of mercury, a 13.59 ft column of H$_2$O has the same mass as a 1.00 ft column of mercury.

12-12. *Refer to Section 12-3.*

Within a container of liquid mercury, the pressure at a certain depth, P_{depth} (torr) equals the sum of the atmospheric pressure (torr) and the depth (mm Hg $\equiv$ torr).

$$P_{depth} \text{ (torr)} = P_{atm} \text{ (torr)} + P_{Hg} \text{ (mm Hg)}$$

(a) at 100 mm, P (torr) $= 754$ torr $+ 100$ mm Hg $= $ **854 torr**

P (atm) $= 854$ torr $\times \dfrac{1 \text{ atm}}{760 \text{ torr}} = $ **1.12 atm**

(b) at 5.04 cm, P (torr) $= 754$ torr $+ \left[5.04 \text{ cm} \times \dfrac{10 \text{ mm}}{1 \text{ cm}} \right] = $ **804 torr**

P (atm) $= 804$ torr $\times \dfrac{1 \text{ atm}}{760 \text{ torr}} = $ **1.06 atm**

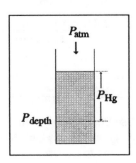

12-14. *Refer to Section 12-3.*

Since 1 atm = 14.7 psi,

$$? \text{ psi} = 150 \text{ atm} \times \frac{14.7 \text{ psi}}{1 \text{ atm}} = \textbf{2.20} \times \textbf{10}^\textbf{3} \textbf{ psi}$$

12-16. *Refer to Section 12-4 and Figures 12-3 and 12-4.*

(a) Boyle studied the effect of changing pressure on a volume of a known mass of gas at constant temperature. Boyle's Law states: at a given temperature, the product of pressure and volume of a definite mass of gas is constant.

(b) When the mathematical relationship, XY = constant, is plotted on the X-Y axes, a hyperbola results. Boyle's Law can be stated as

$$\text{pressure} \times \text{volume} = \text{constant} \quad \text{(at constant } n, T)$$

resulting in the graph shown in Figure 12-4. Since pressure and volume can never have negative values, the other branch of the hyperbola is omitted.

12-18. *Refer to Section 12-4 and Examples 12-1 and 12-2.*

Boyle's Law states: $P_1V_1 = P_2V_2$ at constant n and T

Substituting, $P_2 = \dfrac{P_1V_1}{V_2} = \dfrac{1.50 \text{ atm} \times 35 \text{ L}}{105 \text{ L}} = \textbf{0.50 atm}$

12-20. *Refer to Section 12-4 and Examples 12-1 and 12-2.*

Recall Boyle's Law: $P_1V_1 = P_2V_2$ at constant n and T

(a) Given: $P_1 = 18.3 \text{ torr} \times \dfrac{1 \text{ atm}}{760 \text{ torr}} = 2.41 \times 10^{-2} \text{ atm}$ $\qquad V_1 = 50 \text{ L}$

$\qquad\qquad\quad P_2 = ?$ $\qquad\qquad\qquad\qquad\qquad\qquad\qquad\quad V_2 = 150 \text{ mL} \times \dfrac{1 \text{ L}}{1000 \text{ mL}} = 0.150 \text{ L}$

$$P_2 = \frac{P_1V_1}{V_2} = \frac{2.41 \times 10^{-2} \text{ atm} \times 50 \text{ L}}{0.150 \text{ L}} = \textbf{8.0 atm}$$

(b) Given: $P_1 = 2.41 \times 10^{-2} \text{ atm}$ $\qquad\qquad\qquad V_1 = 50 \text{ L}$
$\qquad\qquad\quad P_2 = 10.0 \text{ atm}$ $\qquad\qquad\qquad\qquad V_2 = ?$

$$V_2 = \frac{P_1V_1}{P_2} = \frac{2.41 \times 10^{-2} \text{ atm} \times 50 \text{ L}}{10.0 \text{ atm}} = \textbf{0.12 L} \quad \text{(2 significant figures)}$$

12-22. *Refer to Section 12-4.*

Plan: (1) Use Boyle's Law to find the maximum volume occupied by the gas at 1.1 atm.
(2) After subtracting out the volume of the cylinder, divide the remaining volume by the volume of each balloon to get the number of balloons.

(1) Recall Boyle's Law: $P_1V_1 = P_2V_2$ at constant n and T

    Given: $P_1 = 165$ atm $V_1 = 29$ L

              $P_2 = 1.1$ atm $V_2 = ?$

    Solving, $V_2 = \dfrac{P_1V_1}{P_2} = \dfrac{165 \text{ atm} \times 29 \text{ L}}{1.1 \text{ atm}} = 4350$ L

(2) This volume of gas is distributed between the balloons and the "empty" cylinder.

$$N \times V_{balloon} = V_2 - V_{cylinder}$$

                      where N = number of balloons (a whole number)

$$N \times 2.0 \text{ L} = 4350 \text{ L} - 29 \text{ L}$$

$$N = \frac{4350 \text{ L} - 29 \text{ L}}{2.0 \text{ L}}$$

$$N = 2161 \text{ or } \textbf{2200 balloons} \text{ (to 2 significant figures)}$$

12-24. *Refer to Section 12-5 and Figure 12-5.*

(a) An "absolute temperature scale" is a scale in which properties such as gas volume change linearly with temperature while the origin of the scale is set at absolute zero. The Kelvin scale is a typical example of it.

(b) Boyle, in his experiments, noticed that temperature affected gas volume. About 1800, Charles and Gay-Lussac found that the rate of gas expansion with increased temperature was constant at constant pressure. Later, Lord Kelvin noticed that for a series of constant pressure systems, volume decreased as temperature decreased and the extrapolation of these different T-V lines back to zero volume yielded a common intercept, -273.15°C on the temperature axis. He defined this temperature as absolute zero. The relationship between the Celsius and Kelvin temperature scales is

$$K = °C + 273.15°.$$

(c) Absolute zero may be thought of as the limit of thermal contraction for an ideal gas. In other words, an ideal gas would have zero volume at absolute zero temperature. Theoretically, it is also the temperature at which molecular motion ceases.

12-26. *Refer to Section 12-5 and Figure 12-5.*

(a) Experiments have shown that at constant pressure, the volume of a definite mass of gas is directly proportional to its absolute temperature (in K).

(b) This is known as Charles' Law and is expressed as $V/T =$ constant at constant n and P. Therefore, for a sample of gas when volume is plotted against temperature, a straight line results.

12-28. *Refer to Section 12-5.*

This is a Charles' Law calculation: $\dfrac{V_1}{T_1} = \dfrac{V_2}{T_2}$ at constant n and P

Given: $V_1 = 90.0$ mL $T_1 = 25°C + 273° = 298$ K

         $V_2 = 30.0$ mL $T_2 = ?$

$$T_2 \text{ (K)} = \frac{V_2 T_1}{V_1} = \frac{30.0 \text{ mL} \times 298 \text{ K}}{90.0 \text{ mL}} = 99.3 \text{ K}$$

t_2 (°C) = 99.3 K - 273.15° = **-173.8°C**

(a) Recall Charles' Law: $\dfrac{V_1}{T_1} = \dfrac{V_2}{T_2}$ at constant n and P

Given: $V_1 = 1.400$ L $\qquad\qquad$ $T_1 = 0.0°C + 273.15° = 273.2$ K

$\qquad\quad$ $V_2 = ?$ $\qquad\qquad\qquad$ $T_2 = 8.0°C + 273.15° = 281.2$ K

$$V_2 = \frac{V_1 T_2}{T_1} = \frac{1.400 \text{ L} \times 281.2 \text{ K}}{273.2 \text{ K}} = \mathbf{1.441 \text{ L}}$$

(b) The volume change corresponding to the temperature change from 0.0°C to 8.0°C is (1.441 - 1.400) L = 0.041 L or 41 mL or 41 cm³. When the cross-sectional area of the graduated arm is 1.0 cm², the difference in height (cm) is equivalent to the difference in volume (cm³). Hence, the height will increase by **41 cm**.

(c) To improve the thermometer's sensitivity (measured in Δheight/°C) for the same volume change, the cross-sectional area of the graduated arm should be decreased. This will cause the height difference to increase. Also, a larger volume of gas could be used.

Recall Charles' Law: $\dfrac{V_1}{T_1} = \dfrac{V_2}{T_2}$ at constant n and P

volume at -78.5°C: $\quad V_2 = \dfrac{V_1 T_2}{T_1} = \dfrac{5.00 \text{ L} \times (-78.5°C + 273.15°)}{25.0°C + 273.15°} = \mathbf{3.26 \text{ L}}$

volume at -195.8°C: $\quad V_2 = \dfrac{V_1 T_2}{T_1} = \dfrac{5.00 \text{ L} \times (-195.8°C + 273.15°)}{25.0°C + 273.15°} = \mathbf{1.30 \text{ L}}$

volume at -268.9°C: $\quad V_2 = \dfrac{V_1 T_2}{T_1} = \dfrac{5.00 \text{ L} \times (-268.9°C + 273.15°)}{25.0°C + 273.15°} = \mathbf{0.0715 \text{ L}}$

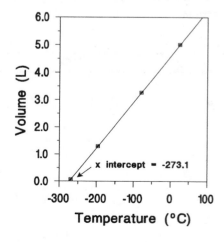

If the line is extrapolated to the x axis, the x intercept is the temperature at which zero volume is reached. That temperature is -273.1°C, also known as absolute zero, 0.0 K.

12-34. *Refer to Sections 12-4 and 12-5, and Figures 12-4 and 12-5.*

(a) $P \times V = \text{constant}$

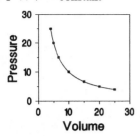

(b) $P = \text{constant} \times 1/V$

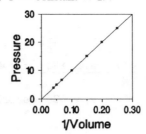

(c) $V = \text{constant} \times T$ or $V/T = \text{constant}$

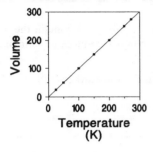

(d) $P = \text{constant} \times T$ or $P/T = \text{constant}$

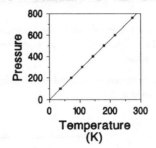

The above graphs were obtained by plotting the hypothetical data given below. It is assumed that for (a) and (b), n and T are constant; for (c), n and P are constant; and for (d), n and V are constant.

(a),(b)

P	V	$P \times V$	$1/V$
4.00	25	100	0.0400
5.00	20.0	100	0.0500
6.67	15.0	100	0.0667
10.0	10.0	100	0.100
15.0	6.67	100	0.150
20.0	5.00	100	0.200
25.0	4.00	100	0.250

(c)

V	$T(K)$	V/T
273	273	1.00
250	250	1.00
200	200	1.00
150	150	1.00
100	100	1.00
50	50	1.00
25	25	1.00

(d)

P	$T(K)$	P/T
760	273	2.784
600	216	2.784
500	180	2.784
400	144	2.784
300	108	2.784
200	71.8	2.784
100	35.9	2.784

12-36. *Refer to Section 12-7, and Examples 12-4 and 12-5.*

Recall: at STP, $T = 273.15$ K and $P = 1$ atm

Given: $T_1 = 273.15$ K $V_1 = 375$ mL $P_1 = 1$ atm
      $T_2 = 773°C + 273.15° = 1046$ K $V_2 = 375$ mL $P_2 = ?$

Combined Gas Law: $\dfrac{P_1 V_1}{T_1} = \dfrac{P_2 V_2}{T_2}$ at constant n

$$\frac{1 \text{ atm} \times 375 \text{ mL}}{273.15 \text{ K}} = \frac{P_2 \times 375 \text{ mL}}{1046 \text{ K}}$$

$$P_2 = 3.83 \text{ atm}$$

147

Refer to Section 12-7, and Examples 12-4 and 12-5.

Given: $T_1 = 26°C + 273.15° = 299$ K $V_1 = 280$ mL $P_1 = 660$ torr

   $T_2 = ?$ $V_2 = 440$ mL $P_2 = 880$ torr

Combined Gas Law: $\dfrac{P_1 V_1}{T_1} = \dfrac{P_2 V_2}{T_2}$ at constant n

$$\frac{660 \text{ torr} \times 280 \text{ mL}}{299 \text{ K}} = \frac{880 \text{ torr} \times 440 \text{ mL}}{T_2}$$

$$T_2 = \textbf{626 K or 353°C}$$

12-40. **Refer to Section 12-8.**

(a) Avogadro's Law states that, at the same temperature and pressure, equal volumes of all gases contain the same number of molecules. This means that equal number of moles of any gas take up equal volumes as long as the temperature and pressure are the same.

(b) The standard molar volume is the volume occupied by 1 mole of an ideal gas under standard conditions. It is taken to be 22.4 L/mol at STP.

12-42. **Refer to Section 12-8.**

A "1 ppb by volume limit of sensitivity" means that in a 1 liter sample of air, one-billionth of a liter $(1 \times 10^{-9}$ L$)$ of CO is the minimum volume of CO that can be detected.

Recall that at STP, 1 mole of gas occupies 22.4 L.

Plan: V_{STP} air $\overset{(1)}{\Longrightarrow} V_{\text{STP}}$ CO $\overset{(2)}{\Longrightarrow}$ mol CO $\overset{(3)}{\Longrightarrow}$ molecules CO

Method 1:

(1) ? V_{STP} CO $= 10$ L air $\times \dfrac{1 \text{ part CO}}{10^9 \text{ parts air}} = 1.0 \times 10^{-8}$ L CO

(2) ? mol CO $= \dfrac{V_{\text{STP}}}{22.4 \text{ L/mol}} = \dfrac{1 \times 10^{-8} \text{ L CO}}{22.4 \text{ L/mol}} = 4.5 \times 10^{-10}$ mol CO

(3) ? molecules CO $= N \times$ mol CO $= (6.0 \times 10^{23}$ molecules/mol$)(4.5 \times 10^{-10}$ mol CO$)$

$\qquad\qquad\qquad\qquad\qquad = 2.7 \times 10^{14}$ molecules CO

$\qquad\qquad\qquad\qquad\qquad = \textbf{3} \times \textbf{10}^{14}$ **molecules CO** (1 significant figure)

Method 2: Dimensional Analysis

? molecules CO $= 10$ L air $\times \dfrac{1 \times 10^{-9} \text{ L CO}}{1 \text{ L air}} \times \dfrac{1 \text{ mol CO}}{22.4 \text{ L CO}} \times \dfrac{6.0 \times 10^{23} \text{ molecules CO}}{1 \text{ mol CO}} = \textbf{3} \times \textbf{10}^{14}$ **molecules CO**

12-44. **Refer to Sections 12-8 and 12-9, and Examples 12-6 and 12-11.**

Plan: (1) Assume that we have 1 mole of ethylene dibromide (EDB) vapor (188 g) which occupies 22.4 L at STP. Using the Combined Gas Law, which simplifies to Charles' Law, calculate the volume that the vapor would occupy at 180°C and 1.00 atm.

   (2) Calculate the density, D (g/L), of the EDB vapor.

(1) $T_1 = 273.15$ K $\qquad\qquad$ $P_1 = 1.00$ atm $\qquad\qquad$ $V_1 = 22.4$ L

 $T_2 = 180°C + 273° = 453$ K $\qquad$ $P_2 = 1.00$ atm $\qquad\qquad$ $V_2 = ?$

Combined Gas Law: $\qquad\qquad \dfrac{P_1 V_1}{T_1} = \dfrac{P_2 V_2}{T_2}$ $\qquad\qquad$ at constant n

$$\frac{1.00 \text{ atm} \times 22.4 \text{ L}}{273.15 \text{ K}} = \frac{1.00 \text{ atm} \times V_2}{453 \text{ K}}$$

$$V_2 = 37.1 \text{ L}$$

(2) Density (g/L) = $\dfrac{\text{mass of vapor (g)}}{\text{volume of vapor (L)}} = \dfrac{188 \text{ g}}{37.1 \text{ L}} = \textbf{5.07 g/L}$

Note: It is standard procedure to solve this problem using the ideal gas law. See Example 12-11.

12-46. *Refer to Section 12-8 and Example 12-6.*

Recall: at STP, 1 mol of gas having a mass equal to its molecular weight occupies 22.4 L. Therefore at STP for a gas that behaves ideally,

$$\text{Density (g/L)} = \frac{\text{molecular weight (g/mol)}}{22.4 \text{ L/mol}}$$

Plan: Using the above formula, calculate the molecular weights of the 2 unknown gases and identify them.

Cylinder #1: $D = 3.74$ g/L

$\qquad$ MW (g/mol) = D (g/L) $\times$ 22.4 L/mol = 3.74 g/L $\times$ 22.4 L/mol = 83.8 g/mol

$\qquad$ Therefore, the gas must be **krypton, Kr**.

Cylinder #2: $D = 0.178$ g/L

$\qquad$ MW (g/mol) = D (g/L) $\times$ 22.4 L/mol = 0.178 g/L $\times$ 22.4 L/mol = 3.99 g/mol

$\qquad$ Therefore, the gas must be **helium, He**.

12-48. *Refer to Section 12-9.*

(a) An "ideal gas" is a hypothetical gas that follows all of the postulates of the kinetic molecular theory. It also obeys exactly all of the gas laws.

(b) The ideal gas equation, also called the ideal gas law, is the relationship, $PV = nRT$.

(c) The ideal gas law is derived by combining Boyle's Law, Charles' Law and Avogadro's Law, obtaining

$$V \propto \frac{nT}{P} \qquad \text{with no restrictions}$$

(d) The symbol for the proportionality constant for the conversion of the above proportion to an equality is R. The formula obtained is

$$V = R\left[\frac{nT}{P}\right]$$

which can be rearranged to give

$$PV = nRT.$$

The value of R is obtained by experimentally measuring a complete set of P, V, n and T values, then solving for R by substituting into the ideal gas law.

12-50. *Refer to Section 12-9.*

Recall the ideal gas law: $PV = nRT$

$$P = \frac{nRT}{V} = \frac{(2.44 \text{ mol})(0.0821 \text{ L·atm/mol·K})(45°C + 273°)}{3.45 \text{ L}} = \textbf{18.5 atm}$$

12-52. *Refer to Section 12-9 and Example 12-11.*

(a) Recall the ideal gas law: $PV = nRT$

$$T = \frac{PV}{nR} = \frac{(2.50 \text{ atm})(3.14 \text{ L})}{(5.00 \text{ mol})(0.0821 \text{ L·atm/mol·K})} = \textbf{19.1 K or -254.0°C}$$

(b) Density, D (g/L) $= \dfrac{(5.00 \text{ mol Ne})(20.18 \text{ g Ne/mol})}{3.14 \text{ L}} = \textbf{32.1 g Ne/L}$

12-54. *Refer to Section 12-9, Table 1-7, and Example 12-8.*

Plan: (1) Calculate the moles of Cl_2 involved.
(2) Determine the volume (in L and ft³) of Cl_2 at 750 torr and 18.0°C using $PV = nRT$.
(3) Determine the length (ft) of the Cl_2 cloud knowing that V (ft³) = length (ft) × width (ft) × depth (ft).

(1) ? mol Cl_2 = 580 tons × $\dfrac{2000 \text{ lb}}{1 \text{ ton}}$ × $\dfrac{453.6 \text{ g}}{1 \text{ lb}}$ × $\dfrac{1 \text{ mol}}{70.9 \text{ g}}$ = 7.42×10^6 mol Cl_2

(2) ? V Cl_2 (in L) $= \dfrac{nRT}{P} = \dfrac{(7.42 \times 10^6 \text{ mol})(0.0821 \text{ L·atm/mol·K})(18.0°C + 273°)}{(750/760) \text{ atm}} = \textbf{1.80} \times \textbf{10}^8 \textbf{ L Cl}_2$

? V Cl_2 (in ft³) $= 1.80 \times 10^8$ L × $\dfrac{1 \text{ ft}^3}{28.32 \text{ L}}$ = $\textbf{6.34} \times \textbf{10}^6 \textbf{ ft}^3 \textbf{ Cl}_2$

(3) ? length of Cl_2 cloud (ft) $= \dfrac{V \text{ (ft}^3)}{\text{width (ft)} \times \text{depth (ft)}} = \dfrac{6.34 \times 10^6 \text{ ft}^3}{(0.50 \text{ mi} \times 5280 \text{ ft/mi}) \times 60 \text{ ft}}$

$= \textbf{40 ft}$ (to 2 significant figure)

12-56. *Refer to Section 12-10 and Example 12-8.*

Plan: (1) Use the ideal gas law, $PV = nRT$, to calculate the number of moles of ethane in the container at STP.
(2) Determine the experimental molecular weight of ethane and compare it to the theoretical value.

(1) $n = \dfrac{PV}{RT} = \dfrac{(1 \text{ atm})(0.185 \text{ L})}{(0.0821 \text{ L·atm/mol·K})(273 \text{ K})} = 8.25 \times 10^{-3}$ mol C_2H_6

(2) MW $C_2H_6 = \dfrac{0.244 \text{ g } C_2H_6}{8.25 \times 10^{-3} \text{ mol}} = \textbf{29.6 g/mol}$

The actual molecular weight of ethane, C_2H_6, is 30.1 g/mol.

Percent error $= \dfrac{\text{actual MW - experimental MW}}{\text{actual MW}} \times 100 = \dfrac{30.1 - 29.6}{30.1} \times 100 = \textbf{2\%}$

Possible sources of error which would result in a slightly low experimental molecular weight include:
(a) the container volume is slightly less than 185 mL,
(b) the mass of ethane is slightly more than 0.244 g, and
(c) ethane deviates slightly from ideal behavior under STP conditions (Refer to methane, CH_4, in Table 12-5.)

12-58. *Refer to Section 12-10 and Example 12-12.*

Plan: (1) Use the ideal gas law, $PV = nRT$, to find the number of moles of gas.

(2) Calculate the molecular weight of the gas.

(1) $n = \dfrac{PV}{RT} = \dfrac{(365/760 \text{ atm})(0.590 \text{ L})}{(0.0821 \text{ L·atm/mol·K})(45°C + 273°)} = 0.0109 \text{ mol}$

(2) MW (g/mol) $= \dfrac{0.480 \text{ g gas}}{0.0109 \text{ mol}} = $ **44.0 g/mol**

12-60. *Refer to Section 12-10 and Example 12-14.*

This experiment illustrates the Dumas method for determining the molecular weight of a gas.

Plan: (1) Find the volume, V, of the container. In this case, you cannot assume that the flask is exactly 250 mL.

(2) Determine the number of moles of gas, using $PV = nRT$.

(3) Determine the mass of the gas in the flask.

(4) Calculate the molecular weight of the gas.

(1) ? V, volume of container = volume of water in container

$$= \frac{\text{mass of water (g)}}{\text{density of water (g/mL)}} \qquad \left(\text{since } D = \frac{\text{mass}}{\text{volume}} \right)$$

$$= \frac{\text{mass of flask filled with water - empty flask}}{\text{density of water}}$$

$$= \frac{327.4 \text{ g} - 65.347 \text{ g}}{0.997 \text{ g/mL}}$$

$$= 263 \text{ mL}$$

(2) $n = \dfrac{PV}{RT} = \dfrac{(743.3/760 \text{ atm})(0.263 \text{ L})}{(0.0821 \text{ L·atm/mol·K})(99.8°C + 273°)} = 0.00840 \text{ mol}$

(3) mass of gas = mass of condensed liquid = mass of flask and condensed liquid - mass of empty flask

$$= 65.739 \text{ g} - 65.347 \text{ g}$$
$$= 0.392 \text{ g}$$

(4) MW (g/mol) $= \dfrac{0.392 \text{ g gas}}{0.00840 \text{ mol}} = $ **46.7 g/mol**

12-62. *Refer to Section 12-11.*

(a) The partial pressure of a gas is the pressure it exerts in a mixture of gases. It is equal to the pressure the gas would exert if it were alone in the container at the same temperature.

(b) Dalton's Law states that the total pressure exerted by a mixture of ideal gases is the sum of the partial pressures of those gases.

$$P_{\text{total}} = P_A + P_B + P_C \ldots \qquad \text{at constant } V, T$$

12-64. *Refer to Section 12-11 and Example 12-16.*

From Dalton's Law of Partial Pressures,

$$P_{total} = \frac{n_{total}RT}{V} \qquad \text{where } n_{total} = n_{CHCl_3} + n_{C_2H_6} \quad = \frac{5.23 \text{ g CHCl}_3}{119.4 \text{ g/mol}} + \frac{1.66 \text{ g CH}_4}{16.04 \text{ g/mol}}$$

$$= 0.0438 \text{ mol} + 0.103 \text{ mol}$$

$$= 0.147 \text{ mol}$$

$$P_{total} = \frac{(0.147 \text{ mol gas})(0.0821 \text{ L·atm/mol·K})(345°C + 273°)}{0.0500 \text{ L}} = \textbf{149 atm}$$

$$P_{CHCl_3} = \left[\frac{n_{CHCl_3}}{n_{CHCl_3} + n_{C_2H_6}}\right] P_{total} = \left[\frac{0.0438 \text{ mol}}{0.0438 \text{ mol} + 0.103 \text{ mol}}\right] 149 \text{ atm} = \textbf{44.5 atm}$$

12-66. *Refer to Section 12-11 and Example 12-17.*

mole fraction of He $\quad X_{He} = \dfrac{P_{He}}{P_{total}} = \dfrac{0.267 \text{ atm He}}{0.267 \text{ atm He} + 0.317 \text{ atm Ar} + 0.277 \text{ atm Xe}} = \dfrac{0.267 \text{ atm}}{0.861 \text{ atm}} = \textbf{0.310}$

mole fraction of Ar $\quad X_{Ar} = \dfrac{P_{Ar}}{P_{total}} = \dfrac{0.317 \text{ atm}}{0.861 \text{ atm}} = \textbf{0.368}$

mole fraction of Xe $\quad X_{Xe} = \dfrac{P_{Xe}}{P_{total}} = \dfrac{0.277 \text{ atm}}{0.861 \text{ atm}} = \textbf{0.322}$

12-68. *Refer to Section 12-11 and Examples 12-16 and 12-18.*

(a) Boyle's Law states that $P_1 V_1 = P_2 V_2$.

For each gas, $P_2 = \dfrac{P_1 V_1}{V_2} = \dfrac{1.50 \text{ atm} \times 2.50 \text{ L}}{1.00 \text{ L}} = 3.75 \text{ atm}$

Dalton's Law of Partial Pressures states that $P_{total} = P_1 + P_2 + P_3 + \ldots$ at constant V, T

Therefore, $P_{total} = P_{O_2} + P_{N_2} + P_{He} = 3.75 \text{ atm} + 3.75 \text{ atm} + 3.75 \text{ atm} = \textbf{11.25 atm}$

(b) partial pressure of O_2, $P_{O_2} = \textbf{3.75 atm}$

(c) partial pressure of N_2, $P_{N_2} =$ partial pressure of He, $P_{He} = \textbf{3.75 atm}$

12-70. *Refer to Section 12-11, Figure 12-8, Table 12-4 and Example 12-20.*

Plan: (1) Calculate the partial pressure of nitrogen in the container at 26°C and 750 torr.
(2) Use the Combined Gas Law to calculate the volume of dry nitrogen at the new conditions.

(1) $P_{H_2} = P_{atm} - P_{H_2O} = 750 \text{ torr} - 25 \text{ torr} = 725 \text{ torr}$

(2) Combined Gas Law: $\qquad \dfrac{P_1 V_1}{T_1} = \dfrac{P_2 V_2}{T_2}$

$$\frac{760 \text{ torr} \times 447 \text{ mL}}{273 \text{ K}} = \frac{725 \text{ torr} \times V_2}{26°C + 273°}$$

$$V_2 = \textbf{513 mL}$$

12-72. *Refer to Section 12-11 and Example 12-19.*

(a) Recall Boyle's Law: $P_1V_1 = P_2V_2$ at constant n, T

for He: 6.00 atm × 6.00 L $= P_2$ × 9.00 L

$P_2 = \textbf{4.00 atm}$

for N_2: 3.00 atm × 3.00 L $= P_2$ × 9.00 L

$P_2 = \textbf{1.00 atm}$

(b) According to Dalton's Law of Partial Pressures, $P_{total} = P_1 + P_2 + \ldots$

Therefore,

$P_{total} = P_{He} + P_{N_2} = 4.00$ atm $+ 1.00$ atm $= \textbf{5.00 atm}$

(c) mole fraction of He, $X_{He} = \dfrac{P_{He}}{P_{total}} = \dfrac{4.00 \text{ atm}}{5.00 \text{ atm}} = \textbf{0.800}$

12-74. *Refer to Section 12-12.*

(a) Since 1 Å = 1×10^{-10} m = 1×10^{-8} cm; and 1 cm³ = 1 mL

Molecular radius: $r = 2.0 \text{ Å} \times \dfrac{1 \times 10^{-8} \text{ cm}}{1 \text{ Å}} = 2.0 \times 10^{-8}$ cm

Molecular volume: $V = \frac{4}{3}\pi r^3 = \frac{4}{3} \times 3.1416 \times (2.0 \times 10^{-8} \text{ cm})^3 = \textbf{3.4} \times \textbf{10}^{-23} \textbf{ cm}^3$ or $\textbf{3.4} \times \textbf{10}^{-23} \textbf{ mL}$

(b) Molecular volume for 1 mol molecules:

? $V = 1.00 \text{ mol} \times \dfrac{6.02 \times 10^{23} \text{ molecules}}{1.00 \text{ mol}} \times \dfrac{3.4 \times 10^{-23} \text{ mL}}{1 \text{ molecule}} = \textbf{20. mL}$ or $\textbf{0.020 L}$

(c) fraction of volume occupied by 1 mole of molecules $= \dfrac{0.020 \text{ L}}{22.4 \text{ L}} = \textbf{8.9} \times \textbf{10}^{-4}$ ($\equiv \textbf{0.089\%}$)

(d) This result verifies the first statement of the Kinetic Molecular Theory of Ideal Gases: the molecular volume of the gas molecules is indeed negligible in comparison to the volume occupied by them.

12-76. *Refer to Section 12-12 and Exercise 12-78 Solution.*

According to the Kinetic-Molecular Theory, all gas molecules have the same average kinetic energy ($= 1/2 \, m\bar{u}^2$) at the same temperature, where $\bar{u}$ is the average velocity. Hence, at the same T:

$$1/2(m_{SiH_4})(\bar{u}_{SiH_4})^2 = 1/2 \, (m_{CH_4})(\bar{u}_{CH_4})^2$$

$$\frac{\bar{u}_{CH_4}}{\bar{u}_{SiH_4}} = \sqrt{\frac{m_{SiH_4}}{m_{CH_4}}} = \sqrt{\frac{MW_{SiH_4}}{MW_{CH_4}}} = \sqrt{\frac{32}{16}} = 1.4$$

SiH_4 is heavier than CH_4; however, both molecules have the same average kinetic energy. This is due to the fact that methane molecules have an average speed which is 1.4 times faster than that of silane molecules.

12-78. *Refer to Section 12-12.*

(a) The third assumption of the Kinetic-Molecular Theory states that the average kinetic energy of gaseous molecules is directly proportional to the absolute temperature of the sample.

$$\text{kinetic energy} = 1/2 m\bar{u}^2 \propto T \qquad \text{where} \quad m = \text{mass (g)}$$
$$\bar{u} = \text{average molecular speed (m/s)}$$
$$T = \text{absolute temperature (K)}$$

We see that the average molecular speed is directly proportional to the square root of the absolute temperature.

(b) $\dfrac{\text{rms speed of N}_2 \text{ molecules at } 100°C}{\text{rms speed of N}_2 \text{ molecules at } 0°C} = \sqrt{\dfrac{100°C + 273°}{0°C + 273°}} = \mathbf{1.17}$

12-80. *Refer to Section 12-14.*

For H_2, F_2 and HF under the same conditions, H_2 would behave the most ideally, because for such small nonpolar molecules, the London forces would be small and therefore the intermolecular attractions would be negligible. The behavior of HF, on the other hand, would deviate the most from ideality, because even though HF is smaller than F_2, it is very polar and its molecules exhibit great attraction for one another.

12-82. *Refer to Section 12-14.*

(a) The effect of molecular volume on the properties of a gas becomes more important when a gas is compressed at constant volume.

(b) Molecular volume also becomes more important when more gas molecules are added to a system.

(c) When the temperature of the gas is raised at constant pressure, the volume expands. At a larger occupied volume, the effect of molecular volume on the properties of a gas becomes less significant.

12-84. *Refer to Section 12-14.*

The pressure and volume given are P_{real} and V_{real}. Therefore,

$$\text{compressibility factor} = \frac{(P_{real})(V_{real})}{RT} = \frac{(30.0 \text{ atm})(0.500 \text{ L})}{(0.0821 \text{ L·atm/mol·K})(-10°C + 273°)} = \mathbf{0.695}$$

If we assume that NH_3 is behaving like an ideal gas, the *ideal* pressure can be calculated from the ideal gas law.

$$P = \frac{nRT}{V} = \frac{(1.00 \text{ mol})(0.0821 \text{ L·atm/mol·K})(-10.0°C + 273°)}{0.500 \text{ L}} = \mathbf{43.2 \text{ atm}}$$

By comparison, the real pressure is 13.2 atm (30.6%) less than the ideally exerted pressure. Since the system is at relatively high pressure and low temperature, there are apparently attractive forces at work between the ammonia molecules to render the real pressure lower.

12-86. *Refer to Section 12-14 and Example 12-24.*

(a) Assuming CCl_4 obeys the ideal gas law: $PV = nRT$

$$P = \frac{nRT}{V} = \frac{(1.00 \text{ mol})(0.0821 \text{ L·atm/mol·K})(77.0°C + 273°)}{35.0 \text{ L}} = \textbf{0.821 atm}$$

(b) Assuming CCl_4 obeys the van der Waals equation: $\left[P + \frac{n^2a}{V^2}\right](V-nb) = nRT$

for CCl_4, $a = 20.39 \text{ L}^2\text{·atm/mol}^2$, $b = 0.1383 \text{ L/mol}$

$$\left[P + \frac{(1.00 \text{ mol})^2(20.39 \text{ L}^2\text{·atm/mol}^2)}{(35.0 \text{ L})^2}\right]\left[35.0 \text{ L} - (1.00 \text{ mol})\left(0.1383 \frac{\text{L}}{\text{mol}}\right)\right]$$

$$= (1.00 \text{ mol})(0.0821 \text{ L·atm/mol·K})(77°C + 273°)$$

$$[P + 0.0166 \text{ atm}][34.9 \text{ L}] = 28.7 \text{ L·atm}$$

$$P + 0.0166 \text{ atm} = 0.822 \text{ atm}$$

$$P = \textbf{0.805 atm}$$

12-88. *Refer to Section 12-15 and Example 12-23.*

Balanced equation: $2NaN_3(s) \rightarrow 2Na(s) + 3N_2(g)$

Plan: $V N_2 \overset{(1)}{\Longrightarrow} \text{mol } N_2 \overset{(2)}{\Longrightarrow} \text{mol } NaN_3 \overset{(3)}{\Longrightarrow} \text{g } NaN_3$

(1) $? \text{ mol } N_2 = n = \dfrac{PV}{RT} = \dfrac{(1.40 \text{ atm})(30.0 \text{ L})}{(0.0821 \text{ L·atm/mol·K})(25°C + 273°)} = 1.72 \text{ mol } N_2$

(2) $? \text{ mol } NaN_3 = \dfrac{2 \text{ mol } NaN_3}{3 \text{ mol } N_2} \times 1.72 \text{ mol } N_2 = 1.15 \text{ mol } NaN_3$

(3) $? \text{ g } NaN_3 = 1.15 \text{ mol } NaN_3 \times 65.0 \text{ g/mol} = \textbf{74.8 g } NaN_3$

12-90. *Refer to Sections 12-15 and 24-8.*

Let us assume that the balanced equation is: $S(g) + O_2(g) \rightarrow SO_2(g)$

In reality, above 444°C, sulfur boils to give a vapor containing a mixture of S_8, S_6, S_4 and S_2 molecules. The problem at this point becomes very complicated. To calculate the moles of SO_2 properly in Step 2, one would have to know the mole fraction and the moles of each of the different sulfur compounds that was present in the mixture. Then one would calculate the moles of SO_2 produced from each compound and finally add them together.

Plan: $V S \overset{(1)}{\Longrightarrow} \text{mol } S \overset{(2)}{\Longrightarrow} \text{mol } SO_2 \overset{(3)}{\Longrightarrow} \text{g } SO_2$

(1) $? \text{ mol } S = n = \dfrac{PV}{RT} = \dfrac{(1.00 \text{ atm})(1.00 \text{ L})}{(0.0821 \text{ L·atm/mol·K})(600°C + 273°)} = 0.0140 \text{ mol } S$

(2) $? \text{ mol } SO_2 = \dfrac{1 \text{ mol } SO_2}{1 \text{ mol } S} \times 0.0140 \text{ mol } S = 0.0140 \text{ mol } SO_2$

(3) $? \text{ g } SO_2 = 0.0140 \text{ mol } SO_2 \times 64.07 \text{ g/mol} = \textbf{0.897 g } SO_2$

Balanced equation: $2KClO_3(s) \rightarrow 2KCl(s) + 3O_2(g)$

$$\text{Plan:} \quad V_{\text{actual}} \, O_2 \overset{(1)}{\Longrightarrow} V_{\text{theoretical}} \, O_2 \overset{(2)}{\Longrightarrow} \text{mol } O_2 \overset{(3)}{\Longrightarrow} \text{mol } KClO_3 \overset{(4)}{\Longrightarrow} \text{g } KClO_3$$

(1) In order to fill four 250 mL bottles, 1.00 L O_2 is actually required. However, more than 1.00 L O_2 must be produced since some O_2 will be lost in the process. If 25% of O_2 will be wasted, the percentage yield of the process is 75%. The theoretical amount of O_2 that must be produced can be calculated:

Recall: $\text{percentage yield} = \dfrac{\text{actual yield}}{\text{theoretical yield}} \times 100\%$.   Therefore,

$$\text{theoretical volume of } O_2 \text{ needed} = \frac{\text{actual volume of } O_2 \text{ needed}}{75\%} \times 100\% = \frac{1.00 \text{ L}}{75\%} \times 100\% = 1.33 \text{ L } O_2$$

(2) $? \text{ mol } O_2 = n = \dfrac{PV}{RT} = \dfrac{(741/760 \text{ atm})(1.33 \text{ L})}{(0.0821 \text{ L·atm/mol·K})(25°C + 273°)} = 0.0529 \text{ mol } O_2$

(3) $? \text{ mol } KClO_3 = 2/3 \times 0.0529 \text{ mol } O_2 = 0.0353 \text{ mol } KClO_3$

(4) $? \text{ g } KClO_3 = 0.0353 \text{ mol } KClO_3 \times 122.6 \text{ g/mol} = \textbf{4.33 g } KClO_3$

Balanced equation: $N_2(g) + 3H_2(g) \rightarrow 2NH_3(g)$

This is a limiting reactant problem. Due to Gay-Lussac's Law, we can work directly in volumes instead of moles.

(1) Compare the required ratio to the available ratio of reactants to find the limiting reactant.

Required ratio $= \dfrac{1 \text{ volume } N_2}{3 \text{ volumes } H_2} = 0.333$ Available ratio $= \dfrac{1.67 \text{ L } N_2}{4.42 \text{ L } H_2} = 0.378$

Available ratio > required ratio; H_2 is the limiting reactant.

(2) $? \text{ L } NH_3 = 2/3 \times 4.42 \text{ L } H_2 = 2.95 \text{ L } NH_3$

(3) $? \text{ g } NH_3 \text{ at STP} = 2.95 \text{ L } NH_3 \times \dfrac{1 \text{ mol } NH_3}{22.4 \text{ L } NH_3} \times \dfrac{17.04 \text{ g } NH_3}{1 \text{ mol } NH_3} = \textbf{2.24 g } NH_3$

Balanced equation: $2KNO_3(s) \overset{\Delta}{\rightarrow} 2KNO_2(s) + O_2(g)$

Recall that 1 mole of ideal gas at STP occupies 22.4 L.

$$\text{Plan:} \quad V_{\text{STP}} \, O_2 \overset{(1)}{\Longrightarrow} \text{mol } O_2 \overset{(2)}{\Longrightarrow} \text{mol } KNO_3 \overset{(3)}{\Longrightarrow} \text{g } KNO_3$$

Method 1:

(1) $? \text{ mol } O_2 = \dfrac{21.1 \text{ L}_{\text{STP}} \, O_2}{22.4 \text{ L}_{\text{STP}}/\text{mol}} = 0.942 \text{ mol } O_2$

(2) $? \text{ mol } KNO_3 = 0.942 \text{ mol } O_2 \times 2/1 = 1.88 \text{ mol } KNO_3$

(3) $? \text{ g } KNO_3 = 1.88 \text{ mol } KNO_3 \times 101 \text{ g/mol} = \textbf{190. g } KNO_3$

Method 2: Dimensional Analysis

$? \text{ g } KNO_3 = 21.1 \text{ L}_{\text{STP}} \, O_2 \times \dfrac{1 \text{ mol } O_2}{22.4 \text{ L}_{\text{STP}} \, O_2} \times \dfrac{2 \text{ mol } KNO_3}{1 \text{ mol } O_2} \times \dfrac{101 \text{ g } KNO_3}{1 \text{ mol } KNO_3} = \textbf{190. g } KNO_3$

Plan: $V\,SO_2 \overset{(1)}{\Longrightarrow} mol\,SO_2 \overset{(2)}{\Longrightarrow} mol\,S \overset{(3)}{\Longrightarrow} g\,S \overset{(4)}{\Longrightarrow} \%S\ by\ mass$

(1) ? mol $SO_2 = n = \dfrac{PV}{RT} = \dfrac{(755/760\ atm)(1.177\ L)}{(0.0821\ L\cdot atm/mol\cdot K)(35.0°C + 273°)} = 0.0462\ mol\ SO_2$

(2) ? mol S = mol SO_2 = 0.0462 mol S

(3) ? g S = 0.0462 mol S $\times$ 32.066 g/mol = 1.48 g S

(4) ? %S by mass $= \dfrac{g\,S}{g\ sample} \times 100 = \dfrac{1.48\ g}{5.913\ g} \times 100 =$ **25.1% S by mass**

Balanced equations: A: $2C_8H_{18} + 25O_2 \rightarrow 16CO_2 + 18H_2O$

                           B: $2C_8H_{18} + 17O_2 \rightarrow 16CO + 18H_2O$

(a) Plan: CO concentration $\overset{(1)}{\Longrightarrow} g\,CO \overset{(2)}{\Longrightarrow} mol\,CO \overset{(3)}{\Longrightarrow} mol\,C_8H_{18}$ (from Reaction B) $\overset{(4)}{\Longrightarrow} mol\,C_8H_{18}$ (total)

              $\overset{(5)}{\Longrightarrow} g\,C_8H_{18}$ (total) $\overset{(6)}{\Longrightarrow} V\,C_8H_{18}$

Method 1:

(1) ? g CO = concentration (g/m³) $\times$ volume (m³) = 2.00 g/m³ $\times$ 97.5 m³ = 195 g CO produced

(2) ? mol CO $= \dfrac{195\ g\ CO}{28.0\ g/mol} = 6.96$ mol CO

(3) ? mol C_8H_{18} (Reaction B) = 2/16 $\times$ 6.96 mol CO = 0.871 mol C_8H_{18} (Reaction B)

(4) ? mol C_8H_{18} (total) $= \dfrac{0.871\ mol\ C_8H_{18}\ (Reaction\ B)}{0.050} = 17.4$ mol C_8H_{18} (total)

        since only 5.0% of the total amount of C_8H_{18} burned in the engine produced CO.

(5) ? g C_8H_{18} (total) = 17.4 mol C_8H_{18} $\times$ 114 g/mol = 1980 g C_8H_{18} (total)

(6) ? V $C_8H_{18} = \dfrac{mass\ (g)}{Density\ (g/mL)} = \dfrac{1980\ g\ C_8H_{18}}{0.702\ g/mL} = 2830$ mL or **2.83 L C_8H_{18}** $D\ (g/mL) = \dfrac{mass\ (g)}{volume\ (mL)}$

Method 2: Dimensional Analysis

? L C_8H_{18} = 97.5 m³ $\times \dfrac{2.00\ g\ CO}{1\ m^3} \times \dfrac{1\ mol\ CO}{28.0\ g\ CO} \times \dfrac{2\ mol\ C_8H_{18}\ (Reaction\ B)}{16\ mol\ CO} \times \dfrac{1.00\ mol\ C_8H_{18}\ (total)}{0.05\ mol\ C_8H_{18}\ (Reaction\ B)}$

                $\times \dfrac{114\ g\ C_8H_{18}\ (total)}{1\ mol\ C_8H_{18}\ (total)} \times \dfrac{1\ mL\ C_8H_{18}}{0.702\ g\ C_8H_{18}} \times \dfrac{1\ L\ C_8H_{18}}{1000\ mL\ C_8H_{18}} =$ **2.83 L C_8H_{18}**

(b) fuel rate $\left[\dfrac{L}{min}\right] = \dfrac{volume\ of\ fuel\ burned\ (L)}{time\ (min)}$

    therefore,   time (min) $= \dfrac{volume\ of\ fuel\ burned\ (L)}{fuel\ rate\ (L/min)} = \dfrac{2.83\ L}{0.0631\ L/min} =$ **44.8 min**

12-102. *Refer to Section 12-5.*

Recall Charles' Law: $\dfrac{V_1}{T_1} = \dfrac{V_2}{T_2}$ at constant n and P

Given: $V_1 = 150 \text{ m}^3$ $\quad T_1 = 10°C + 273° = 283 \text{ K}$
$\quad\quad\ V_2 = ?$ $\quad\quad\quad T_2 = 18°C + 273° = 291 \text{ K}$

$V_2 = \dfrac{V_1 T_2}{T_1} = \dfrac{150 \text{ m}^3 \times 291 \text{ K}}{283 \text{ K}} = 154 \text{ m}^3$

Therefore, **4m³** (= 154 m³ - 150 m³) of air had been forced out of the cabin.

? L air forced from cabin $= 4 \text{ m}^3 \times \dfrac{(100 \text{ cm})^3}{(1 \text{ m})^3} \times \dfrac{1 \text{ mL}}{1 \text{ cm}^3} \times \dfrac{1 \text{ L}}{1000 \text{ mL}} = \textbf{4000 L air}$

12-104. *Refer to Section 12-10 and Example 12-12.*

Plan: (1) Use the ideal gas law, $PV = nRT$, to find the moles of Freon-12.
(2) Calculate the molecular weight of Freon-12.

(1) $n = \dfrac{PV}{RT} = \dfrac{(790/760 \text{ atm})(8.29 \text{ L})}{(0.0821 \text{ L·atm/mol·K})(200°C + 273°)} = 0.222 \text{ mol Freon-12}$

(2) MW (g/mol) $= \dfrac{26.8 \text{ g Freon-12}}{0.222 \text{ mol}} = \textbf{121 g/mol}$

12-106. *Refer to Section 12-9 and Appendix E.*

(a) Plan: (1) Determine the actual partial pressure of H_2O vapor before and after air conditioning.
(2) Calculate the moles and mass of water present before and after air conditioning using the ideal gas law, $PV = nRT$.
(3) Determine the mass of water removed by the air conditioning process.

(1) From Appendix E, $\quad$ vapor pressure of water at 33°C = 37.7 torr
$\quad\quad\quad\quad\quad\quad\quad\quad\quad$ vapor pressure of water at 25°C = 23.8 torr

Given: relative humidity $= \dfrac{\text{actual partial pressure of } H_2O \text{ vapor}}{\text{partial pressure of } H_2O \text{ vapor if saturated}}$

before air conditioning at 33.0°C: $P_{H_2O,\text{actual}} = $ relative humidity $\times P_{H_2O,\text{sat}}$
$\quad\quad\quad\quad\quad\quad\quad\quad\quad\quad\quad\quad\quad = 0.800 \times 37.7 \text{ torr}$
$\quad\quad\quad\quad\quad\quad\quad\quad\quad\quad\quad\quad\quad = 30.2 \text{ torr}$

after air conditioning at 25.0°C: $P_{H_2O,\text{actual}} = $ relative humidity $\times P_{H_2O,\text{sat}}$
$\quad\quad\quad\quad\quad\quad\quad\quad\quad\quad\quad\quad\quad = 0.150 \times 23.8 \text{ torr}$
$\quad\quad\quad\quad\quad\quad\quad\quad\quad\quad\quad\quad\quad = 3.57 \text{ torr}$

(2) ? V_{house} (L) $= 245 \text{ m}^3 \times \dfrac{(100 \text{ cm})^3}{(1 \text{ m})^3} \times \dfrac{1 \text{ mL}}{1 \text{ cm}^3} \times \dfrac{1 \text{ L}}{1000 \text{ mL}} = 2.45 \times 10^5 \text{ L}$

before air conditioning: $n = \dfrac{PV}{RT} = \dfrac{(30.2/760 \text{ atm})(2.45 \times 10^5 \text{ L})}{(0.0821 \text{ L·atm/mol·K})(33.0°C + 273°)} = 388 \text{ mol } H_2O$

$\quad\quad\quad\quad\quad\quad\quad$? g $H_2O = 388 \text{ mol} \times 18.0 \text{ g/mol} = 6980 \text{ g } H_2O$

after air conditioning: $n = \dfrac{PV}{RT} = \dfrac{(3.57/760 \text{ atm})(2.45 \times 10^5 \text{ L})}{(0.0821 \text{ L·atm/mol·K})(25.0°C + 273°)} = 47.0 \text{ mol } H_2O$

$\quad\quad\quad\quad\quad\quad\quad$? g $H_2O = 47.0 \text{ mol} \times 18.0 \text{ g/mol} = 846 \text{ g } H_2O$

(3) The mass of water removed = 6980 g - 846 g = **6130 g H$_2$O**

(b) ? mL H$_2$O at 25°C = $\dfrac{6130 \text{ g H}_2\text{O}}{0.997 \text{ g/mL}}$ = **6150 mL H$_2$O**   since Density (g/mL) = $\dfrac{\text{mass (g)}}{\text{volume (mL)}}$

12-108. *Refer to Section 12-14, Example 12-22 and Table 12-5.*

(1) Assuming NH$_3$ obeys the ideal gas law: $PV = nRT$

$$P = \frac{nRT}{V} = \frac{(10.0 \text{ mol})(0.0821 \text{ L·atm/mol·K})(100°\text{C} + 273°)}{60.0 \text{ L}} = \textbf{5.10 atm}$$

(2) Assuming NH$_3$ obeys the van der Waals equation: $\left[P + \dfrac{n^2a}{V^2} \right](V\text{-}nb) = nRT$

for NH$_3$, $a = 4.17$ L^2·atm/mol^2, $b = 0.0371$ L/mol

$$\left[P + \frac{(10.0 \text{ mol})^2(4.17 \text{ L}^2\text{·atm/mol}^2)}{(60.0 \text{ L})^2} \right]\left[60.0 \text{ L} - (10.0 \text{ mol})\left(0.0371 \frac{\text{L}}{\text{mol}} \right) \right]$$

$$= (10.0 \text{ mol})(0.0821 \text{ L·atm/mol·K})(100°\text{C} + 273°)$$

Simplifying,
$$[P + 0.116 \text{ atm}][59.6 \text{ L}] = 306 \text{ L·atm}$$

$$P + 0.116 \text{ atm} = 5.13 \text{ atm}$$

$$P = \textbf{5.01 atm}$$

(3) % difference = $\dfrac{P_{ideal} - P_{real}}{P_{ideal}} \times 100 = \dfrac{5.10 - 5.02}{5.10} \times 100 = 1.57\%$ or **2%** (1 significant figure)

12-110. *Refer to Section 12-10 and Examples 12-14 and 12-15.*

Plan: (1) Find the empirical formula for cyanogen.
     (2) Calculate the molecular weight of cyanogen, using the ideal gas law, $PV = nRT$.
     (3) Determine the molecular formula.

(1) Assume 100 g of cyanogen.

? mol C = $\dfrac{46.2 \text{ g C}}{12.0 \text{ g/mol}}$ = 3.85 mol C   Ratio = $\dfrac{3.85}{3.84}$ = 1.00

? mol N = $\dfrac{53.8 \text{ g N}}{14.0 \text{ g/mol}}$ = 3.84 mol N   Ratio = $\dfrac{3.84}{3.84}$ = 1.00

The empirical formula for cyanogen is C$_1$N$_1$ or CN (formula weight = 26.0 g/mol)

(2) ? mol cyanogen = $n = \dfrac{PV}{RT} = \dfrac{(750/760 \text{ atm})(0.476 \text{ L})}{(0.0821 \text{ L·atm/mol·K})(25°\text{C} + 273°)}$ = 0.0192 mol

MW (g/mol) = $\dfrac{1.00 \text{ g}}{0.0192 \text{ mol}}$ = 52.1 g/mol

(3) let n = $\dfrac{\text{molecular weight}}{\text{simplest formula weight}} = \dfrac{52.1 \text{ g/mol}}{26.0 \text{ g/mol}}$ = 2

Therefore, the true molecular formula for cyanogen is **C$_2$N$_2$**.

12-112. *Refer to Section 12-9.*

Pressure is directly proportional to temperature, when moles and volume are kept constant, as shown here:

The Combined Gas Law states: $\dfrac{P_1 V_1}{T_1} = \dfrac{P_2 V_2}{T_2}$ at constant n

And, assuming the tire volume is constant, we have: $\dfrac{P_1}{T_1} = \dfrac{P_2}{T_2}$ at constant n and V

Tire manufacturers test tire pressures at room temperatures. Therefore, tires should be filled at room temperatures, and not at the higher temperatures that result from friction when the car is driven at high speeds. If the tires were filled at higher temperatures to the specified pressure, they would actually be underinflated when the tires cooled.

12-114. *Refer to Sections 12-9 and 12-15.*

Since $PV = nRT$; $P = \dfrac{nRT}{V}$. Therefore, at constant T and V, $P \propto n$.

As a result, at constant T and V, pressure can be used to measure the relative amount of compounds.

Balanced equation: $2H_2(g) + O_2(g) \rightarrow 2H_2O(g)$

Compare the required ratio of reactants (using moles) to the available ratio of reactants using partial pressures to find the limiting reactant.

$$\text{Required ratio} = \frac{2 \text{ moles } H_2}{1 \text{ mole } O_2} = 2.00 \quad \text{Available ratio} = \frac{0.588 \text{ atm } H_2}{0.302 \text{ atm } O_2} = 1.95$$

Available ratio < required ratio; **H_2 is the limiting reactant.**

12-116. *Refer to Section 12-15.*

Balanced equation: $C_5H_{12}(\ell) + 8O_2(g) \rightarrow 5CO_2(g) + 6H_2O(g)$

Plan: $\quad$ g $C_5H_{12} \overset{(1)}{\Longrightarrow}$ mol $C_5H_{12} \overset{(2)}{\Longrightarrow}$ mol CO_2 or $H_2O \overset{(3)}{\Longrightarrow} P_{CO_2}$ or P_{H_2O}

(1) $\ ? \text{ mol } C_5H_{12} = \dfrac{0.361 \text{ g } C_5H_{12}}{72.15 \text{ g/mol}} = 5.00 \times 10^{-3} \text{ mol } C_5H_{12}$

(2) $\ ? \text{ mol } CO_2 = 5/1 \times 5.00 \times 10^{-3} \text{ mol } C_5H_{12} = 2.50 \times 10^{-2} \text{ mol } CO_2$
$\quad\ ? \text{ mol } H_2O = 6/1 \times 5.00 \times 10^{-3} \text{ mol } C_5H_{12} = 3.00 \times 10^{-2} \text{ mol } H_2O$

(3) $\ P_{CO_2} = \dfrac{nRT}{V} = \dfrac{(2.50 \times 10^{-2} \text{ mol})(0.0821 \text{ L·atm/mol·K})(300°C + 273°)}{4.00 \text{ L}} = \textbf{0.294 atm}$

$\quad\ P_{H_2O} = \dfrac{nRT}{V} = \dfrac{(3.00 \times 10^{-2} \text{ mol})(0.0821 \text{ L·atm/mol·K})(300°C + 273°)}{4.00 \text{ L}} = \textbf{0.353 atm}$

12-118. *Refer to Section 12-15.*

Balanced equation: $Mg^{2+}(aq) + SiO_2(s, dispersed) + 2HCO_3^-(aq) \rightarrow MgSiO_3(s) + 2CO_2(g) + H_2O(\ell)$

Plan: L $CO_2 \overset{(1)}{\Longrightarrow}$ mol $CO_2 \overset{(2)}{\Longrightarrow}$ mol $MgSiO_3 \overset{(3)}{\Longrightarrow}$ g $MgSiO_3$

(1) $\ ? \text{ mol } CO_2 = n = \dfrac{PV}{RT} = \dfrac{(775/760 \text{ atm})(100 \text{ L})}{(0.0821 \text{ L·atm/mol·K})(30.°C + 273°)} = 4.10 \text{ mol } CO_2$

(2) $\ ? \text{ mol } MgSiO_3 = 1/2 \times 4.10 \text{ mol } CO_2 = 2.05 \text{ mol } MgSiO_3$

(3) $\ ? \text{ g } MgSiO_3 = 2.05 \text{ mol } MgSiO_3 \times 100.4 \text{ g/mol} = \textbf{206 g } MgSiO_3$

13 Liquids and Solids

13-2. *Refer to Section 13-1.*

A gas completely fills its container because the strength of the intermolecular forces at work between its molecules are weak. The average molecular kinetic energy is great enough to overcome the forces of attraction and allow the molecules to travel until they collide with the wall. As a result, the volume that the molecules occupy expands and fills the container.

In a liquid, the intermolecular forces acting between the molecules are much stronger than those of a gas. As a result, the molecules are always in close contact with each other, to effect a constant volume for the liquid. They do have enough kinetic energy to partially overcome the attractive forces; the molecules can glide past each other and assume the shape of the container.

In a solid, the molecular motion has decreased even more significantly. Strong attractive interactions which operate over very short distances restrict the particles' movements and lock the particles into a fixed three-dimensional structure. Therefore, the solid retains its shape.

13-4. *Refer to Sections 13-2 and the Key Terms for Chapter 13.*

Hydrogen bonding is an especially strong dipole-dipole interaction between molecules in which one contains H in a highly polarized bond and the other contains a lone pair of electrons. The energy of a hydrogen bond is 4 - 5 times larger than a normal dipole-dipole interaction and roughly 10% of a covalent bond. It only occurs in systems where a hydrogen atom is directly bonded to a small, highly electronegative atom, such as N, O or F.

13-6. *Refer to Section 13-2.*

Permanent dipole-dipole forces can be found acting between the polar molecules of:

(c) NO and

(d) SeF_4.

13-8. *Refer to Section 13-2 and Exercise 13-6.*

London forces are the only important intermolecular forces of attraction operating between the nonpolar molecules of:

(a) molecular $AlCl_3$ and

(b) SF_6.

13-10. *Refer to Section 13-2 and Exercise 13-4 Solution.*

(a) ammonia, NH_3

H-N̈-H
|
H

NH_3 has stronger hydrogen bonds operating between its molecules than PH_3. NH_3 contains a hydrogen atom directly bonded to the small, highly electronegative atom, N.

phosphine, PH_3

H-P̈-H
|
H

(b)   ethylene, C_2H_4

$$\begin{array}{ccc} H & & H \\ & C=C & \\ H & & H \end{array}$$

hydrazine, N_2H_4

$$\begin{array}{c} H-\overset{..}{N}-\overset{..}{N}-H \\ \;\;\;\; H \;\;\; H \end{array}$$

Hydrazine, N_2H_4, has stronger hydrogen bonding than ethylene, C_2H_4, since it contains the small highly electronegative element, N, directly bonded to hydrogen atoms and 2 lone pairs of electrons.

(c)   hydrogen fluoride, HF

$$H-\overset{..}{\underset{..}{F}}:$$

hydrogen chloride, HCl

$$H-\overset{..}{\underset{..}{Cl}}:$$

Hydrogen fluoride, HF, has stronger hydrogen bonding than hydrogen chloride, HCl, since in HF, the hydrogen atom is directly bonded to the small highly electronegative element, F.

13-12.   *Refer to Section 13-2.*

(a) The physical properties of ethyl alcohol (ethanol), $C_2H_6O \equiv CH_3CH_2OH$, are influenced mainly by hydrogen bonding since there is an H atom directly bonded to an O atom, but are also affected by London forces like any other molecule.

(b) Dipole-dipole interactions are the primary intermolecular forces operating between the molecules of phosphine, PH_3, since it is polar. Refer to Exercise 13-10a solution for its structure. London forces also exist in this system since it is a molecule.

(c) The properties of sulfur hexafluoride, SF_6, are only influenced by London forces, because the molecules are nonpolar.

13-14.   *Refer to Sections 13-2, 13-8, 13-10 and 13-12.*

(a) liquid (T = 0°C at STP)   (b) liquid; freezes; crystalline solid   (c) sublimation; gas; 24°C

13-16.   *Refer to Section 13-2.*

Acetic acid dimer
    (with 2 hydrogen bonds)

13-18.   *Refer to Section 13-6 and 13-8.*

Evaporation is the process by which molecules at the *surface* of the liquid escape and go into the gas phase at a temperature below its boiling point. During boiling, bubbles of vapor begin to form beneath the surface in the *bulk* of the liquid, rise to the surface and burst, releasing vapor into the air. At the boiling point, the vapor pressure equals the external pressure.

As temperature increases, a greater fraction of molecules have more kinetic energy and move more rapidly. They can then more readily overcome the attractions of neighboring molecules to escape into the vapor phase, so the rate of evaporation increases.

13-20. *Refer to Section 13-2 and Figure 13-5.*

(a) Ne, Ar, and Kr are nonpolar noble gases. Their boiling points are determined solely by London forces, which in turn are dependent on an atom's size and polarizability. In order of increasing boiling point (i.e., increasing size),

$$Ne\ (-246°C)\ <\ Ar\ (-186°C)\ <\ Kr\ (-152°C)$$

(b) All three compounds, NH_3, H_2O and HF exhibit hydrogen bonding. Figure 13-5 gives the order of increasing boiling points as:

$$NH_3\ (-33°C)\ <\ HF\ (20°C)\ <\ H_2O\ (100°C)$$

Since the electronegativities of the elements follow the order, F > O > N, the charge separation for the bonds in these three molecules should follow the order, F-H > O-H > N-H. The same order also should be followed when *each* of these hydrogen atoms forms hydrogen bonding with a lone pair of electrons on a neighboring molecule.

The reason why H_2O has a higher boiling point than HF is because there is a larger number of H-bonds in the H_2O system. The limiting reactant for the formation of H-bonds in the NH_3 system is the number of lone pairs of electrons, while that in the HF system is the number of H atoms. In the H_2O system, all of the H atoms and lone pairs can participate in H-bond formation. In either the NH_3 or the HF system, 1 mol of the molecules could only give a maximum of 1 mole of H-bonds because of the limiting reactants. But, in 1 mole of H_2O, a maximum of 2 moles of H-bonds could be obtained. This explains why the *total* H-bonding force is much higher in H_2O than in HF, as shown by the higher boiling point.

13-22. *Refer to Section 13-8.*

(a) The normal boiling point is the temperature at which the vapor pressure of a liquid is exactly equal to one atmosphere (760 torr) pressure.

(b) When the boiling point of a liquid is measured, the atmospheric pressure must be specified since the boiling point of a liquid is the temperature at which its vapor pressure is exactly equal to the applied pressure. As the atmospheric pressure decreases, so does the boiling point of a liquid.

13-24. *Refer to Section 13-4 and Figure 13-8.*

Surface tension is a measure of the inward intermolecular forces of attraction among liquid particles that must be overcome to expand the surface area. A measurable tension in the surface of the liquid is caused by an imbalance set up between the intermolecular forces operating at the surface and those operating within the liquid. As the temperature increases, the increased kinetic energy and greater movement of the particles in the liquid tend to counteract the intermolecular forces, resulting in the lowering of surface tension.

13-26. *Refer to Sections 13-2 and 13-7, and Table 13-3.*

The weaker the intermolecular forces between molecules in the liquid phase, the higher is the vapor pressure. Generally, hydrogen bonding and the cumulative London forces in larger molecules are the most significant factors (see Table 13-3). In order of decreasing vapor pressure,

(a) $BiCl_3 > BiBr_3$ $BiCl_3$ is smaller than $BiBr_3$, has weaker London forces and thus has a higher vapor pressure at the same temperature.

(b) $CO > CO_2$ — CO is smaller, has weaker London forces and a higher vapor pressure than CO_2. Note from Table 13-3 that even though CO is polar, its dipole moment is extremely low and its dipole-dipole interaction energy is nearly zero.

(c) $N_2 > NO$ — Both are small molecules of comparable size, and therefore have comparable London forces. NO however is polar and has permanent dipole-dipole interactions, whereas N_2 is nonpolar and has only London forces. Therefore N_2 has the higher vapor pressure.

(d) $HCOOCH_3 > CH_3COOH$ — Methyl formate ($HCOOCH_3$) and acetic acid (CH_3COOH) have the same molecular formula, $C_2H_4O_2$, and are about the same size, so both have similar London forces. CH_3COOH can form hydrogen bonds, whereas $HCOOCH_3$ cannot. Therefore $HCOOCH_3$ has weaker intermolecular interactions and has higher vapor pressure than CH_3COOH.

To confirm the above reasoning, use boiling points given below to indicate strengths of intermolecular forces. Remember, the lower the boiling point is, the weaker the intermolecular forces present in the liquid, and the higher the vapor pressure at a specific temperature.

Compound	B.P. (°C)	Compound	B.P. (°C)
$BiCl_3$	447	N_2	-195.8
$BiBr_3$	453	NO	-151.8
CO	-191.5	CH_3COOH	117.9
CO_2	-78.5 (sublimes)	$HCOOCH_3$	31.5

13-28. Refer to Section 13-8.

The order of increasing boiling points corresponds with the order of increasing temperatures when their vapor pressures are constant, e.g. 100 torr:

$$\text{normal butane} < \text{diethyl ether} < \text{1-butanol}$$

13-30. Refer to Sections 13-7 and 12-11.

Using Dalton's Law of Partial Pressures, $P_{total} = P_{air} + P_{water\ vapor}$

Therefore, $P_{air} = P_{total} - P_{water\ vapor} = 633.5$ torr - 289.1 torr = **344.4 torr**

13-32. Refer to Section 13-9.

(a) at 37°C, the heat of vaporization of water = 2.41 kJ/g

at 37°C, ΔH°_{vap} (kJ/mol) $= 2.41 \dfrac{kJ}{g} \times \dfrac{18.0\ g}{1\ mol} = $ **43.4 kJ/mol**

(b) The heat of vaporization at a certain temperature is the amount of heat required to change 1 gram of liquid to 1 gram of vapor at that temperature. The heat of vaporization is greater at 37°C than at 100°C because the average kinetic energy of the molecules is lower at lower temperatures. Therefore, more energy must be added per unit mass of the liquid to break the intermolecular forces between the molecules at the lower temperature.

13-34. _Refer to Section 13-8._

Vapor pressure curve for $C_2H_4F_2$

The boiling point of $C_2H_4F_2$ at 200 torr from the graph is about -1°C.

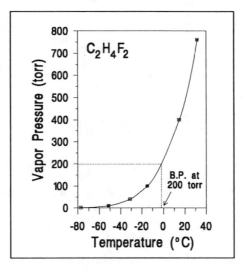

t (°C)	VP (torr)
-77.2	1
-51.2	10
-31.1	40
-15.0	100
14.8	400
31.7	760

13-36. _Refer to Section 13-9._

We must use the Clausius-Clapeyron equation to find the vapor pressure at 80°C:

$$\log\left(\frac{P_2}{P_1}\right) = \frac{\Delta H_{vap}}{2.303R}\left[\frac{1}{T_1} - \frac{1}{T_2}\right]$$

where ΔH_{vap} = molar heat of vaporization (J/mol)
P_1 = 760.00 torr
P_2 = vapor pressure at 80°C
R = 8.314 J/mol·K
T_1 = normal boiling point of liquid (K)
T_2 = 80.00°C + 273.15° = 353.15 K

(1) for water, H_2O: ΔH_{vap} = 40,656 J/mol
T_1 = 100.00°C + 273.15K = 373.15 K

Substituting,

$$\log\left(\frac{P_2}{760.00\ torr}\right) = \frac{40656\ J/mol}{(2.303)(8.314\ J/mol\cdot K)}\left[\frac{1}{373.15\ K} - \frac{1}{353.15\ K}\right]$$

$$= (2123)(0.0026799 - 0.0028317)$$

$$= -0.3223$$

$$\frac{P_2}{760.00\ torr} = 0.4761$$

$$P_2 = \textbf{361.8 torr}$$

(2) for heavy water, D_2O: ΔH_{vap} = 41,606 J/mol
T_1 = 101.41°C + 273.15K = 374.56 K

Substituting,

$$\log\left(\frac{P_2}{760.00\ torr}\right) = \frac{41606\ J/mol}{(2.303)(8.314\ J/mol\cdot K)}\left[\frac{1}{374.56\ K} - \frac{1}{353.15\ K}\right]$$

$$= (2173)(0.0026698 - 0.0028317)$$

$$= -0.3518$$

$$\frac{P_2}{760.00\ torr} = 0.4448$$

$$P_2 = \textbf{338.1 torr}$$

(a) The Clausius-Clapeyron equation is

$$\log\left(\frac{P_2}{P_1}\right) = \frac{\Delta H_{vap}}{2.303R}\left(\frac{1}{T_1} - \frac{1}{T_2}\right)$$

where
ΔH_{vap} = molar heat of vaporization (J/mol)
P_1 = vapor pressure at temperature, T_1 (K)
P_2 = vapor pressure at temperature, T_2 (K)
R = 8.314 J/mol·K

Expanding the equation we obtain:

$$\log P_2 - \log P_1 = \frac{\Delta H_{vap}}{2.303R}\left(\frac{1}{T_1}\right) - \frac{\Delta H_{vap}}{2.303R}\left(\frac{1}{T_2}\right)$$

$$\log P_2 = \frac{-\Delta H_{vap}}{2.303RT_2} + \left[\frac{\Delta H_{vap}}{2.303R}\left(\frac{1}{T_1}\right) + \log P_1\right]$$

If we let P_1 be a known vapor pressure of the substance at a particular temperature, T_1, and simplify by letting B stand for all the terms in the square brackets, we have

$$\log P = \frac{-\Delta H_{vap}}{2.303RT} + B$$

When $\log P$ is plotted against $1/T$, we obtain a straight line with a slope of $-\Delta H_{vap}/2.303R$ and y-intercept of B.

(b) Vapor pressure data for ethyl acetate, $CH_3COOC_2H_5$:

t (°C)	T (K)	$1/T$ (K^{-1})	P (torr)	$\log P$
-43.4	229.8	4.35×10^{-3}	1	0.00
-23.5	249.7	4.00×10^{-3}	5	0.70
-13.5	259.7	3.85×10^{-3}	10	1.00
-3.0	270.2	3.70×10^{-3}	20	1.30
9.1	282.3	3.54×10^{-3}	40	1.60
16.6	289.8	3.45×10^{-3}	60	1.78
27.0	300.2	3.33×10^{-3}	100	2.00
42.0	315.2	3.17×10^{-3}	200	2.30
59.3	332.5	3.01×10^{-3}	400	2.60
?	?	?	760	2.88

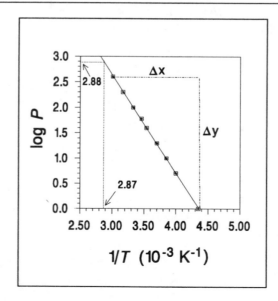

(c) The molar heat of vaporization of ethyl acetate is determined from the slope of the line. A linear regression fit to the data gives a slope of -1.940 with a coefficient of determination, $r^2 = 0.99954$, which indicates a very good fit. If you do not have access to a line-fitting program, the slope can be estimated by using two data points that are far apart as shown in the graph. This will work because the line fits the data well.

$$\text{slope} = \frac{-\Delta H_{vap}}{2.303R} = \frac{\Delta y}{\Delta x} = \frac{\Delta \log P}{\Delta\, 1/T} = \frac{(0.00 - 2.60)}{(4.35 \times 10^{-3} - 3.01 \times 10^{-3})} = -1.94 \times 10^3 \text{ K}$$

Therefore, $\Delta H_{vap} = \text{-slope} \times 2.303R = -(-1.94 \times 10^3 \text{ K})(2.303)(8.314 \text{ J/mol·K}) = \mathbf{3.71 \times 10^4 \text{ J/mol}}$

(d) The normal boiling point is the temperature at which the vapor pressure of ethyl acetate is 760 torr. From the graph, when $\log P = \log 760 = 2.88$,

$1/T = 2.87 \times 10^{-3} \text{ K}^{-1}$ and solving, $T = \mathbf{3.48 \times 10^2 \text{ K} \text{ or } 75°C}$

13-40. *Refer to Section 13-6 and the Key Terms for Chapter 13.*

A dynamic equilibrium is a situation in which two (or more) processes occur at the same rate so that no net change occurs. This is the kind of equilibrium that is established between two physical states of matter, e.g., between a liquid and its vapor, in which the rate of evaporation is equal to the rate of condensation in a closed container:

$$\text{liquid} \rightleftarrows \text{vapor}.$$

On a molecular level at equilibrium, molecules of liquid are escaping into the vapor and vapor molecules are condensing into the liquid. This is not a static situation. No net change occurs because the rates are the same.

13-42. *Refer to Section 13-11.*

Plan: (1) Calculate the amount of heat (in J) required to melt the zinc sample.
(2) Determine ΔH°_{fusion} at its melting point.

(1) ? heat absorbed = Rate (J/s) × time (s) = 9.84 J/s × (3.6 min × 60 s/min) = 2.13×10^3 J

(2) $\Delta H^\circ_{fusion} \left(\dfrac{J}{mol}\right) = \dfrac{\text{heat required to melt Zn sample at m.p.}}{\text{mol Zn}} = \dfrac{2.13 \times 10^3 \text{ J}}{21.8 \text{ g}/65.4 \text{ g/mol}} = \mathbf{6.38 \times 10^3 \text{ J/mol}}$

13-44. *Refer to Sections 13-9 and 13-11.*

This exercise requires 3 separate calculations:

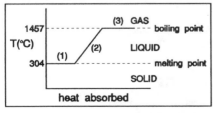

$$\begin{array}{ccccccc}
 & (1) & & (2) & & (3) & \\
\text{solid} & \Longrightarrow & \text{liquid} & \Longrightarrow & \text{liquid} & \Longrightarrow & \text{gas} \\
304°C & & 304°C & & 1457°C & & 1457°C
\end{array}$$

(1) heat required = mass × heat of fusion = (225 g)(21 J/g) = 4.7×10^3 J

(2) heat required = mass × specific heat(ℓ) × Δt = (225 g)(0.13 J/g·°C)(1457°C - 304°C) = 3.4×10^4 J

(3) heat required = mass × heat of vaporization = (225 g)(795 J/g) = 1.79×10^5 J

Therefore, the total heat required = (1) + (2) + (3) = $\mathbf{2.18 \times 10^5 \text{ J}}$

This exercise involves 5 separate calculations:

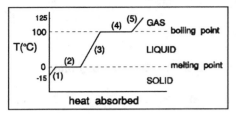

$$
\begin{array}{ccccccccccc}
& (1) & & (2) & & (3) & & (4) & & (5) & \\
\text{ice} & \Longrightarrow & \text{ice} & \Longrightarrow & \text{water} & \Longrightarrow & \text{water} & \Longrightarrow & \text{steam} & \Longrightarrow & \text{steam} \\
-15.0°C & & 0.0°C & & 0.0°C & & 100.0°C & & 100.0°C & & 125.0°C
\end{array}
$$

(1) heat required = mass × specific heat (s) × Δt = (75.0 g)(2.09 J/g·°C)(0.0°C - (-15.0°C)) = 2.35×10^3 J

(2) heat required = mass × heat of fusion = (75.0 g)(334 J/g) = 2.51×10^4 J

(3) heat required = mass × specific heat (ℓ) × Δt = (75.0 g)(4.18 J/g·°C)(100.0°C - 0.0°C) = 3.14×10^4 J

(4) heat required = mass × heat of vaporization = (75.0 g)(2260 J/g) = 1.70×10^5 J

(5) heat required = mass × specific heat (g) × Δt = (75.0 g)(2.03 J/g·°C)(125.0°C - 100.0°C) = 3.81×10^3 J

Therefore, the total heat required = (1) + (2) + (3) + (4) + (5) = **2.33×10^5 J**

Recall: When two substances are brought into contact with each other, the heat lost by one substance is equal in absolute value to the heat gained by the other.

$$| \text{heat gained by cold water} | = | \text{heat lost by hot water} |$$
$$| \text{mass} \times \text{specific heat}(\ell) \times \Delta t |_{\text{cold}} = | \text{mass} \times \text{specific heat}(\ell) \times \Delta t |_{\text{hot}}$$
$$| \text{mass} \times \Delta t |_{\text{cold}} = | \text{mass} \times \Delta t |_{\text{hot}}$$
$$(475 \text{ g})(t_{\text{final}} - 30.0°C) = (275 \text{ g})(100°C - t_{\text{final}})$$
$$(475 \times t_{\text{final}}) - 14250 = 27500 - (275 \times t_{\text{final}})$$
$$750 \times t_{\text{final}} = 41750$$
$$t_{\text{final}} = \textbf{55.7°C}$$

Plan: (1) Determine if the final phase will be liquid or gas.

(2) Calculate the final temperature, t_{final} (°C).

(1) The amount of heat required to change the liquid water to steam at 100°C is

heat required = $| \text{mass} \times \text{specific heat } (\ell) \times \Delta t |_{\text{water}} + | \text{mass} \times \text{heat of vaporization} |$
= (175 g)(4.18 J/g·°C)(100.0°C - 0.0°C) + (175 g)(2.26 × 10^3 J/g)
= 4.69×10^5 J

The amount of heat released when the steam changes to liquid water at 100°C is

heat released = $| \text{mass} \times \text{specific heat } (g) \times \Delta t |_{\text{steam}} + | \text{mass} \times \text{heat of vaporization} |$
= (17.5 g)(2.03 J/g·°C)(110.0°C - 100.0°C) + (17.5 g)(2.26 × 10^3 J/g)
= 3.99×10^4 J

The heat required to convert water to steam is greater than the heat released when the steam condenses to water. Therefore, when the two systems are mixed, the liquid water will cause all of the steam to condense and the final temperature will be between 0°C and 100°C.

(2) | amount of heat gained by water | = | amount of heat lost by steam |

| mass × specific heat (ℓ) × Δt | = | mass × specific heat (g) × Δt | + | mass × heat of vaporization |

+ | mass × specific heat (ℓ) × Δt | $_{\text{water from steam}}$

$$(175 \text{ g})(4.18 \text{ J/g} \cdot {}^{\circ}\text{C})(t_{\text{final}} - 0.0{}^{\circ}\text{C}) = (17.5 \text{ g})(2.03 \text{ J/g} \cdot {}^{\circ}\text{C})(110.0{}^{\circ}\text{C} - 100.0{}^{\circ}\text{C}) + (17.5 \text{ g})(2.26 \times 10^3 \text{ J/g})$$

$$+ (17.5 \text{ g})(4.18 \text{ J/g} \cdot {}^{\circ}\text{C})(100.0{}^{\circ}\text{C} - t_{\text{final}})$$

$$(732 \times t_{\text{final}}) - 0 = (355 \text{ J}) + (3.96 \times 10^4 \text{ J}) + (7.32 \times 10^3 \text{ J}) - (73.2 \times t_{\text{final}})$$

$$805 \times t_{\text{final}} = 4.73 \times 10^4$$

$$t_{\text{final}} = \mathbf{58.8{}^{\circ}C}$$

13-52.   Refer to Sections 13-9 and 13-11.

(a) This part of the exercise requires 2 separate calculations:

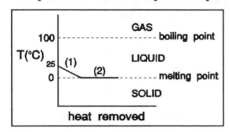

$$
\begin{array}{ccccc}
 & (1) & & (2) & \\
\text{water} & \Longrightarrow & \text{water} & \Longrightarrow & \text{ice} \\
25.0{}^{\circ}\text{C} & & 0.0{}^{\circ}\text{C} & & 0.0{}^{\circ}\text{C}
\end{array}
$$

(1) heat removed = mass × specific heat (ℓ) × Δt = (15.0 g)(4.18 J/g·°C)(25.0°C - 0.0°C) = 1.57 × 10³ J

(2) heat removed = mass × heat of fusion = (15.0 g)(334 J/g) = 5.01 × 10³ J

Therefore, the total heat removed = (1) + (2) = **6.58 × 10³ J**

(b) | heat removed from water | = | heat gained by Freon-12 |

$$6.58 \times 10^3 \text{ J} = | \text{mass}_{\text{Freon-12}} \times \text{heat of vaporization} |$$

$$= \text{mass}_{\text{Freon-12}} \times 165.1 \text{ J/g}$$

$$\text{mass}_{\text{Freon-12}} = \mathbf{39.9 \text{ g}}$$

13-54.   Refer to Section 13-13 and the Key Terms for Chapter 13.

The critical point is the combination of critical temperature and critical pressure of a substance. The critical temperature is the temperature above which a gas cannot be liquefied, i.e., above this temperature, one cannot distinguish between a liquid and a gas. The critical pressure is the pressure required to liquefy a gas at its critical temperature. If the temperature is less than the critical temperature, the substance can be either a gas, liquid or solid, depending on the pressure.

13-56.   Refer to Section 13-13 and Figure 13-17b.

(a) This point lies on the sublimation curve where the solid and gas phases are in equilibrium. So, both the solid and gaseous phases are present.

(b) This point is called the triple point where all three phases are in equilibrium with each other. So, the solid, liquid and gaseous phases are all present.

The melting point of carbon dioxide increases with increasing pressure, since the solid-liquid equilibrium line on its phase diagram slopes up and to the right. If the pressure on a sample of liquid carbon dioxide is increased at constant temperature, causing the molecules to get closer together, the liquid will solidify. This indicates that solid carbon dioxide has a higher density than the liquid phase. This is true for most substances. The notable exception is water.

13-60. *Refer to Section 13-13.*

(a) Phase diagram for butane:
    (not to scale)

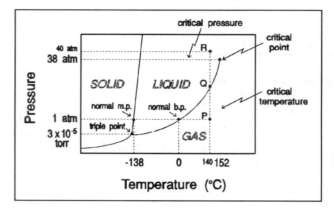

(b) As butane is compressed from 1 atm (point P) to 40 atm (point R) at 140°C, butane is converted from a gas to a liquid. Both phases are present simultaneously where the vapor pressure curve intersects the vertical isothermal line: $T = 140$°C indicated by point Q. At point Q, both phases are in equilibrium.

(c) Since 200°C is a temperature greater than the critical temperature of 152°C, there is no pressure at which two phases exist. At all pressures the liquid phase cannot be distinguished from the gas phase.

13-62. *Refer to Section 13-13 and the phase diagram for sulfur.*

(a) liquid   (b) solid (rhombic)   (c) solid (rhombic)   (d) solid (monoclinic)   (e) vapor   (f) liquid

13-64. *Refer to Section 13-13.*

Ice, i.e., solid water, floats in liquid water because the solid state is less dense than the liquid state. However, like most other substances, solid mercury is more dense than liquid mercury and therefore, solid mercury sinks when placed in liquid mercury.

13-66. *Refer to Section 13-16.*

SiO_2 covalent (network) solid Na_2S ionic solid $Cr(CO)_6$ molecular solid   Ti metallic solid

13-68. *Refer to Section 13-16.*

(a) SO_2F molecular solid   (c) W   metallic solid   (e) PF_5 molecular solid

(b) MgF_2 ionic solid   (d) Pb   metallic solid

13-70. *Refer to Sections 13-2 and 13-16.*

Melting points of ionic compounds increase with increasing ion-ion interactions which are functions of d, the distance between the ions, and q, the charge on the ions:

$$F \propto \frac{q^+q^-}{d^2}$$

Due to the differences in the number of charges on the cations (Na^+, Mg^{2+} and Al^{3+}), the following order of increasing melting points is predicted:

$$NaF \ < \ MgF_2 \ < \ AlF_3$$

13-72. *Refer to Section 13-15, and Figures 13-22 and 13-28.*

Simple cubic lattice:
   (Example: CsCl)

The eight corners of a cubic unit cell are occupied by one kind of atom, while the center of the unit cell is occupied by another kind of atom. Note that this is *not* the same as a body-centered cubic arrangement, because the atom at the center of the cell is *not* the same as those at the corners.

Body-centered cubic lattice (bcc):
   (Example: Na)

All eight corners *and* the point at the center of a cubic unit cell are occupied by a single kind of atom.

Face-centered cubic lattice (fcc):
   (Example: Ni)

The eight corners of a cubic unit cell as well as the central points on each of the six faces of the cube are occupied by the same kind of atom.

CsCl 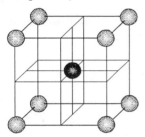   Na 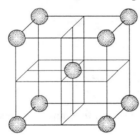   Ni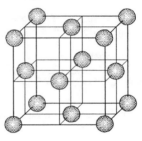

Unit cell 1: Cs^+ ⬤ Cl^- ⬤   Na ⬤   Ni ⬤
Unit cell 2: Cs^+ ⬤ Cl^- ⬤

13-74. *Refer to Section 13-15 and Figures 13-22 and 13-28b, c.*

(a) 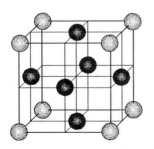   cation: ⬤

anion:

(b) The cations are in the corners: 8 cations $\times$ 1/8 = 1 cation in the unit cell.

(c) The anions are in the faces: 6 anions $\times$ 1/2 = 3 anions in the unit cell.

(d) The simplest formula for the ionic compound is AB_3, where A^{3+} represents the cation and B^- represents the anion.

13-76. *Refer to Section 13-15.*

Consider the two-dimensional lattice:

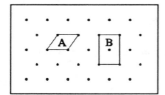

Shown are the simplest unit cell, labeled A, and the orthogonal unit cell (in which all the angles are right angles), labeled B. Obviously, it is easier to calculate areas, etc., for the rectangular unit cell B than for unit cell A, a rhombus.

13-78. *Refer to Section 13-15 and Figure 13-28b.*

Consider the NaCl face-centered cubic structure with a unit cell edge represented as a, shown here.

(a) The distance from Na^+ to its nearest neighbor is $a/2$.

(b) Each Na^+ ion has **6** equidistant nearest neighbors. They are Cl^- ions.

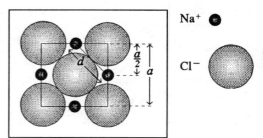

(c) The distance, d, from Na^+ to nearest Na^+ is the length of a hypotenuse of an isosceles right triangle, with sides equal to $a/2$.

$$d^2 = (a/2)^2 + (a/2)^2$$

$$d = \sqrt{(a/2)^2 + (a/2)^2} = \sqrt{2(a/2)^2} = \sqrt{a^2/2} = \frac{a}{\sqrt{2}} \text{ or } \frac{a\sqrt{2}}{2}$$

(d) Each Cl^- ion has **6** equidistant nearest neighbors; they are Na^+ ions.

13-80. *Refer to Section 13-16 and Examples 13-9 and 13-10.*

Plan:　(1) Calculate the volume of the unit cell, V.
　　　　(2) Calculate the mass, m, of Na atoms in the unit cell.
　　　　(3) Determine the density of Na:

$$\text{Density} = \frac{m_{\text{unit cell}}}{V_{\text{unit cell}}}$$

(1) let the length of the cube edge = a = 4.24 Å

$V_{\text{unit cell}} = a^3 = [4.24 \text{ Å} \times (1 \times 10^{-8} \text{ cm/Å})]^3 = 7.62 \times 10^{-23} \text{ cm}^3$

(2) A body-centered cubic unit cell contains 2 atoms in total:

　　　　in the corners:　8 Na atoms $\times$ 1/8 = 1 Na atom
　　　　in the center:　1 Na atom $\times$ 1　= 1 Na atom

$? \ m_{\text{unit cell}} = 2 \text{ Na atoms} \times \dfrac{1 \text{ mol Na}}{6.02 \times 10^{23} \text{ atoms Na}} \times \dfrac{22.99 \text{ g Na}}{1 \text{ mol Na}} = 7.64 \times 10^{-23} \text{ g}$

(3) Density of Na = $\dfrac{m}{V} = \dfrac{7.64 \times 10^{-23} \text{ g Na/unit cell}}{7.62 \times 10^{-23} \text{ cm}^3/\text{unit cell}}$ = **1.00 g/cm³**

The density of sodium as given by the *Handbook of Chemistry and Physics* is 0.97 g/cm³.

Plan: (1) Calculate the volume of the unit cell, V.
   (2) Determine the mass of the unit cell, m, given the density.
   (3) Knowing the number of atoms in a face-centered cube, find the moles of atoms in the unit cell.
   (4) Determine the atomic weight of the unknown and its identity.

(1) let the length of the cube edge $= a = 4.95$ Å

$V_{\text{unit cell}} = a^3 = [4.95$ Å $\times (1 \times 10^{-8}$ cm/Å$)]^3 = 1.21 \times 10^{-22}$ cm^3

(2) $m_{\text{unit cell}} = D \times V_{\text{unit cell}} = (11.35$ g/cm$^3)(1.21 \times 10^{-22}$ cm$^3) = 1.37 \times 10^{-21}$ g   since $D = \dfrac{m}{V}$

(3) A face-centered cubic unit cell contains 4 atoms:

   in the corners   8 atoms $\times$ 1/8 = 1 atom
   on the faces   6 atoms $\times$ 1/2 = 3 atoms

? mol element in unit cell $= 4$ atoms $\times \dfrac{1 \text{ mol atoms}}{6.02 \times 10^{23} \text{ atoms}} = 6.64 \times 10^{-24}$ mol

(4) $\text{AW} = \dfrac{\text{g element}}{\text{mol element}} = \dfrac{m_{\text{unit cell}}}{\text{mol}_{\text{unit cell}}} = \dfrac{1.37 \times 10^{-21} \text{ g}}{6.64 \times 10^{-24} \text{ mol}} = \textbf{206 g/mol}$

Therefore, the IVA element is **Pb (207.2 g/mol)**

(a) The cubic unit cell of diamond:

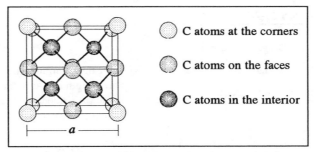

   C atoms at the corners

   C atoms on the faces

   C atoms in the interior

The unit cell contains 8 C atoms: 8 atoms at the corners $\times$ 1/8 = 1 atom
            6 atoms on the faces $\times$ 1/2 = 3 atoms
            4 atoms in the interior $\times$ 1 = 4 atoms

(b) Each C atom is at the center of a **tetrahedron** and hence has 4 nearest neighbors.

(c) Consider the lower right hand corner of diamond's unit cell:

Triangle BCD is in the plane of the base of the unit cell.
Line BC = line CD = a/4.
So, (line BD)^2 = (line BC)^2 + (line CD)^2

   line BD $= \sqrt{(a/4)^2 + (a/4)^2} = \dfrac{a}{2\sqrt{2}}$

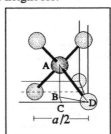

Triangle ABD is perpendicular to the base of the unit cell. Line AD is the distance from any carbon atom to its nearest neighbor.

   (line AD)^2 = (line AB)^2 + (line BD)^2

   line AD $= \sqrt{\left[\dfrac{a}{4}\right]^2 + \left[\dfrac{a}{2\sqrt{2}}\right]^2} = \sqrt{3}\left[\dfrac{a}{4}\right]$

(d) The unit cell edge is 3.567 Å. Therefore,

$$\text{C-C bond length} = \text{line AD} = \sqrt{3}(3.567 \text{ Å}/4) = \textbf{1.545 Å}$$

(e) $V_{\text{unit cell}} = a^3 = [3.567 \text{ Å} \times (1 \times 10^{-8} \text{ cm/Å})]^3 = 4.538 \times 10^{-23} \text{ cm}^3$

$$m_{\text{unit cell}} = 8 \text{ C atoms} \times \frac{1 \text{ mol}}{6.022 \times 10^{23} \text{ atoms}} \times \frac{12.01 \text{ g}}{1 \text{ mol}} = 1.596 \times 10^{-22} \text{ g}$$

$$\text{Density of C}_{\text{diamond}} = \frac{m}{V} = \frac{1.596 \times 10^{-22} \text{ g}}{4.538 \times 10^{-23} \text{ cm}^3} = \textbf{3.517 g/cm}^3$$

The density of C_{diamond} as given by the *Handbook of Chemistry and Physics* is 3.51 g/cm³.

13-86. *Refer to Section 13-14 and Figure 13-18.*

(a) Crystal diffraction studies use the x-ray region of the electromagnetic radiation.

(b) In the x-ray diffraction experiment, a monochromatic x-ray beam is attenuated by a system of slits and aimed at a crystal. The crystal is rotated to vary the angle of incidence. At certain angles which depend on the crystal's unit cell, x-rays are deflected and hit a photographic plate. After development the plate shows a set of symmetrically arranged spots. From the arrangement of the spots, the crystal structure can be determined.

(c) In order for diffraction to occur, the wavelength of the in-coming radiation must be about the same as the inter-nuclear separations in the crystal.

13-88. *Refer to Section 13-14.*

In x-ray diffraction, the Bragg equation is used:

$$n\lambda = 2d\sin\theta \qquad \text{where} \qquad \begin{array}{l} n = 1 \text{ for the minimum diffraction angle} \\ \lambda = \text{wavelength of Cu radiation} \\ \theta = \text{angle of incidence, } 19.98° \\ d = \text{spacing between parallel layers of Pt atoms, } 2.256 \text{ Å} \end{array}$$

Solving for λ,

$$\lambda = \frac{2d\sin\theta}{n} = \frac{2 \times 2.256 \text{ Å} \times \sin 19.98°}{1} = \textbf{1.542 Å}$$

13-90. *Refer to Section 13-17.*

Metallic bonding which occurs between the atoms in metals results from the electrical attractions among positively charged metal ions and mobile, delocalized electrons belonging to the entire crystal. Nonmetals do not form metallic bonds. These elements have very positive first ionization energies and do not wish to lose electrons to form positive ions.

13-92. *Refer to Section 13-16.*

When a metal is distorted (e.g., rolled into sheets or drawn into wire), new metallic bonds are formed and the environment around each atom is essentially unchanged. This can happen because the valence electrons of bonded metal atoms are only loosely associated with individual atoms, as though metal cations exist in a "cloud of electrons." In ionic solids, the lattice arrangements of cations and anions are more rigid. When an ionic solid is distorted, it is possible for cation-cation and anion-anion alignments to occur. But this will cause the solid to shatter due to electrostatic repulsions between ions of like charge.

The first gas to evaporate would be the gas with the weakest intermolecular forces acting between the molecules. Since a lower boiling point is an indicator of weaker intermolecular forces of attraction at work, N_2 (b.p. -196°C) would be the first gas to evaporate, then Ar (b.p. -186°C), and lastly, O_2 (b.p. -183°C).

13-96. *Refer to Section 12-14 and Example 12-22.*

(1) Assuming $H_2O(g)$ obeys the ideal gas law: $PV = nRT$

$$P = \frac{nRT}{V} = \frac{(1.00 \text{ mol})(0.0821 \text{ L·atm/mol·K})(100.00°C + 273.15°)}{31.0 \text{ L}} = \textbf{0.988 atm}$$

(2) Assuming $H_2O(g)$ obeys the van der Waals equation: $\left(P + \frac{n^2a}{V^2}\right)(V-Nb) = nRT$

for H_2O, $\quad a = 5.464$ L²·atm/mol², $b = 0.03049$ L/mol

$$\left[P + \frac{(1.00 \text{ mol})^2(5.464 \text{ L}^2\text{·atm/mol}^2)}{(31.0 \text{ L})^2}\right]\left[31.0 \text{ L} - (1.00 \text{ mol})\left(0.03049 \frac{\text{L}}{\text{mol}}\right)\right]$$

$$= (1.00 \text{ mol})(0.0821 \text{ L·atm/mol·K})(100.00°C + 273.15°)$$

$$[P + 0.00569 \text{ atm}][30.97 \text{ L}] = 30.6 \text{ L·atm}$$

$$P + 0.00569 \text{ atm} = 0.988 \text{ atm}$$

$$P = \textbf{0.982 atm}$$

(3) % difference $= \dfrac{P_{\text{ideal}} - P_{\text{real}}}{P_{\text{ideal}}} \times 100 = \dfrac{0.988 - 0.982}{0.988} \times 100 = \textbf{0.6\%}$

Therefore, steam does not deviate significantly from ideality at these conditions. This is expected because 100°C is a rather high temperature, and 1 atm is in the intermediate pressure range.

13-98. *Refer to Sections 13-2 and 13-8.*

Consider the following three molecules with formula $C_2H_2Cl_2$:

All of the compounds have London forces of attraction operating between their molecules. Since the compounds are approximately the same size, their London forces are about the same. Compounds (1) and (3) are polar and have permanent dipole-dipole interactions. The compound with the lowest boiling point is the one with the weakest intermolecular forces of attraction. Compound (2) which is nonpolar should therefore have the lowest boiling point.

13-100. *Refer to Section 13-15.*

Plan: (1) Calculate the mass of tantalum in a unit cell.
(2) Determine the number of atoms in a unit cell.
(3) Evaluate the type of cubic crystal lattice.

(1) Since Density $= \dfrac{m}{V}$,

$m_{\text{unit cell}} = D \times V_{\text{unit cell}} = 16.7 \text{ g/cm}^3 \times [3.32 \text{ Å} \times (1 \times 10^{-8} \text{ cm/Å})]^3 = \textbf{6.11} \times \textbf{10}^{-22} \textbf{ g}$

(2) number of atoms/unit cell $= \dfrac{\text{mass of unit cell}}{\text{mass of 1 atom}} = \dfrac{6.11 \times 10^{-22}\ g}{(181\ g/mol)(1\ mol/6.02 \times 10^{23}\ atoms)}$

$$= \dfrac{6.11 \times 10^{-22}\ g}{3.01 \times 10^{-22}\ g/atom}$$

$$= 2\ atoms$$

(3) Tantalum must crystallize in a body-centered cubic lattice which has 2 atoms/unit cell.

13-102. *Refer to Section 13-9.*

? time (min) $= 1.00\ mol \times \dfrac{84.2\ g}{1.00\ mol} \times \dfrac{390\ J}{1.00\ g} \times \dfrac{1.00\ s}{10.0\ J} \times \dfrac{1.00\ min}{60.0\ s} = $ **54.7 min**

13-104. *Refer to Section 13-1.*

(a) The volume occupied by 1 mole of Pb in the solid, liquid and gaseous states:

? $V_{Pb(s)} = 1.000\ mol \times \dfrac{207.2\ g}{1\ mol} \times \dfrac{1\ cm^3}{11.288\ g} = $ **18.36 cm³**

? $V_{Pb(\ell)} = 1.000\ mol \times \dfrac{207.2\ g}{1\ mol} \times \dfrac{1\ cm^3}{10.43\ g} = $ **19.87 cm³**

? $V_{Pb(g)} = 1.000\ mol \times \dfrac{207.2\ g}{1\ mol} \times \dfrac{1\ L}{1.110\ g} \times \dfrac{1000\ cm^3}{1\ L} = $ **1.867 × 10⁵ cm³**

(b) The volume actually occupied by 1 mole of Pb atoms:

$V_{atom\ Pb} = \frac{4}{3}\pi r^3 = \frac{4}{3}\pi(0.175 \times 10^{-9}\ m \times 100\ cm/m)^3 = 2.24 \times 10^{-23}\ cm^3$

$V_{mol\ Pb} = V_{atom\ Pb} \times N = (2.24 \times 10^{-23}\ cm^3/atom)(6.02 \times 10^{23}\ atom/mol) = $ **13.5 cm³**

(c) fraction of volume occupied by Pb atoms $\quad$ in Pb(s) $= \dfrac{13.5\ cm^3}{18.36\ cm^3} = $ **0.735**

$\qquad\qquad\qquad\qquad\qquad\qquad\qquad\qquad$ in Pb(ℓ) $= \dfrac{13.5\ cm^3}{19.87\ cm^3} = $ **0.679**

$\qquad\qquad\qquad\qquad\qquad\qquad\qquad\qquad$ in Pb(g) $= \dfrac{13.5\ cm^3}{1.867 \times 10^5\ cm^3} = $ **7.23 × 10⁻⁵**

13-106. *Refer to Sections 13-13 and 13-14.*

(a) gas $\qquad\qquad$ (b) condense, liquid $\qquad\qquad$ (c) freeze, crystalline solid $\qquad\qquad$ (d) sublimation, gas

---

13-108. *Refer to Section 13-7 .*

The vapor pressure of a liquid in equilibrium with its vapor cannot be treated like an ideal gas that obeys the gas laws; the equilibrium (liquid ⇌ vapor) controls the vapor pressure. As conditions are changed, the system adjusts itself until the system reaches equilibrium again; either the liquid which is present evaporates, or the vapor condenses.

In particular, if the temperature increases, more liquid evaporates becoming vapor to increase the vapor pressure. Mathematically, the saturated vapor pressure of a liquid increases exponentially instead of linearly with increasing temperature. Vapor pressure cannot be calculated from the ideal gas law: $P_1/T_1 = P_2/T_2$ since n is not a constant but increases greatly with temperature.

(a)   CH₃COOH

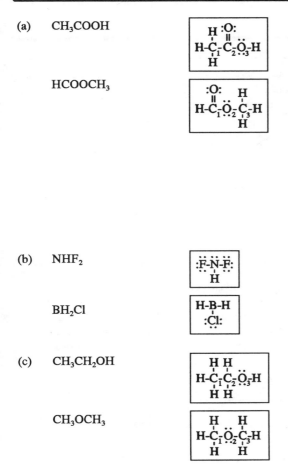

CH₃COOH   C₁   tetrahedral (*sp³*)
           C₂   trigonal planar (*sp²*)
           O₃   bent (*sp³* with 2 lone pairs)

HCOOCH₃   C₁   trigonal planar (*sp²*)
           O₂   bent (*sp³* with 2 lone pairs)
           C₃   tetrahedral (*sp³*)

Both molecules are about the same size and are polar. Both molecules have London and dipole-dipole intermolecular forces operating between their molecules. However, CH₃COOH has hydrogen bonds operating between its molecules, since CH₃COOH contains a hydrogen atom directly bonded to the small, highly electronegative atom, O; HCOOCH₃ does not. Therefore, CH₃COOH will have the higher boiling point.

HCOOCH₃

(b)   NHF₂

       BH₂Cl

NHF₂ is pyramidal (*sp³* with 1 lone pair) and BH₂Cl is trigonal planar (*sp²*). Both molecules are polar and have London and dipole-dipole intermolecular forces operating between their molecules. However, NHF₂ has hydrogen bonds operating between its molecules, and BH₂Cl does not. Therefore, NHF₂ will have the higher boiling point.

(c)   CH₃CH₂OH

CH₃CH₂OH   C₁   tetrahedral (*sp³*)
            C₂   tetrahedral (*sp³*)
            O₃   bent (*sp³* with 2 lone pairs)

CH₃OCH₃   C₁   tetrahedral (*sp³*)
            O₂   bent (*sp³* with 2 lone pairs)
            C₃   tetrahedral (*sp³*)

Both molecules are about the same size and are polar. They have London and dipole-dipole intermolecular forces operating between their molecules. However, CH₃CH₂OH has hydrogen bonds operating between its molecules, and CH₃OCH₃ does not. Therefore, CH₃CH₂OH will have the higher boiling point.

CH₃OCH₃

13-112. *Refer to Sections 1-13, 13-9 and 13-11.*

Plan: (1) Find out the specific heat of the unknown metal in J/g·°C.
         (2) Using the Law of Dulong and Petit, find out the formula weight of the metal and then identify it.

(1) | heat lost by the metal | = | heat gained by the water |
                | mass × specific heat × Δ*t* | metal = | mass × specific heat(ℓ) × Δ*t* | water
         (100.2 g)(specific heat of metal)(99.9°C - 36.6°C) = (50.6 g)(4.18 J/g·°C)(36.6°C - 24.8°C)

         Solving, the specific heat of the metal = 0.393 J/g·°C

(2) ? FW = $\dfrac{25 \text{ J/mol·°C}}{0.393 \text{ J/g·°C}}$ ≈ 64 g/mol

     Therefore, the metal is likely to be **Cu**.

14 Solutions

14-2.	*Refer to the Introduction to Chapter 14.*

	Type of Solution	Example	Solute	Solvent
(a)	solid dissolved in liquid	salt water	$NaCl(s)$	$H_2O(\ell)$
(b)	gas dissolved in gas	air (major components)	$O_2(g)$	$N_2(g)$
(c)	gas dissolved in liquid	$HCl(aq)$	$HCl(g)$	$H_2O(\ell)$
(d)	liquid dissolved in liquid	$CH_3COOH(aq)$	$CH_3COOH(\ell)$	$H_2O(\ell)$
(e)	solid dissolved in solid	brass	$Zn(s)$	$Cu(s)$

14-4.	*Refer to Section 14-1 and Figure 14-1.*

Dissolution is favored when (a) solute-solute attractions and (b) solvent-solvent attractions are relatively small and (c) solvent-solute attractions are relatively large. Both processes (a) and (b) require energy, first to separate the solute particles from each other, then to separate the solvent molecules from each other. Process (c) releases energy as solute particles and solvent molecules interact. If the absolute value of heat absorbed in processes (a) and (b) is less than the absolute value of heat released in process (c), then the dissolving process is favored and releases heat.

14-6.	*Refer to Sections 14-1, 14-2 and 14-3.*

The dissolution of many solids in liquids is endothermic (requires heat) due to the large solute-solute attractions that must be overcome relative to the solvent-solute attractions.

The mixing of two miscible liquids is exothermic (releases heat) since solute-solute attractions are less than the solvent-solute attractions.

14-8.	*Refer to Section 14-1.*

There are two factors which control the spontaneity of a dissolution process: (1) the amount of heat absorbed or released and (2) the amount of increase in the disorder, or randomness of the system. All dissolution processes are accompanied by an increase in the disorder of both solvent and solute. Thus, their disorder factor is invariably favorable to solubility. Dissolution will always occur if the dissolution process is exothermic and the disorder term increases. Dissolution will occur when the dissolution process is endothermic if the disorder term is large enough to overcome the endothermicity, which opposes dissolution.

14-10.	*Refer to Section 14-3.*

When two completely miscible, nonreactive liquids, A and B, are mixed in any proportions, the molecules of A and B will intermingle, and one phase only is always produced. When two completely immiscible liquids, C and D, are mixed in any proportions, two separate phases are always produced; one is pure C and the other is pure D. However, when liquid E is slowly added to liquid F, with which it is only partially miscible, at first only one phase is present ($V_E \ll V_F$). When the solubility of liquid E in liquid F is exceeded, two phases result. After a very large quantity of liquid E is added to liquid F ($V_E \gg V_F$), one phase again is present.

14-12. *Refer to Sections 14-2, 14-3 and 14-4.*

(a) HCl in H_2O — high solubility since both are polar covalent molecules.

(b) HF in H_2O — high solubility since both are polar covalent molecules which are capable of forming hydrogen bonds.

(c) Al_2O_3 in H_2O — low solubility since both Al^{3+} and O^{2-} in the ionic solid have high charge-to-size ratios and therefore have a larger lattice energy. The dissolution of Al_2O_3 is very endothermic and is not favored.

(d) S_8 in H_2O — low solubility since S_8 is a nonpolar covalent molecule and H_2O is very polar.

(e) $NaNO_3$ in C_6H_{14} — low solubility since $NaNO_3$ is an ionic solid and C_6H_{14} is a nonpolar solvent.

14-14. *Refer to Section 4-2 and Exercise 14-12.*

(a) HCl in H_2O — strong electrolyte

(b) HF in H_2O — weak electrolyte

(c) Al_2O_3 in H_2O — cannot be prepared in "reasonable" concentration

(d) S_8 in H_2O — cannot be prepared in "reasonable" concentration

(e) $NaNO_3$ in C_6H_{14} — cannot be prepared in "reasonable" concentration

---

14-16. *Refer to Sections 14-2 and 14-3.*

(a) The solubility of a solid in a liquid does not depend appreciably on pressure.

(b) The solubility of a liquid in a liquid also is essentially independent of pressure.

14-18. *Refer to Sections 14-4 and 14-6.*

Most gases undergo exothermic dissolution in water:

$$\text{gas} \rightleftarrows \text{dissolved gas} + \text{heat}.$$

Applying LeChatelier's Principle to this process, we see that at higher temperatures, more heat is available; the dissolution process will reverse itself to minimize the heat gain. Therefore at higher temperatures, the solubilities of most gases decrease.

14-20. *Refer to Section 14-7 and Exercise 14-19.*

Henry's Law states: $P_{gas} = kX_{gas}$ where P_{gas} = partial pressure of gas above the solution
k = Henry's Law constant
X_{gas} = mole fraction of gas at a certain temperature

For CH_4 at 25°C:

$$P_{CH_4} = k\,X_{CH_4}$$
$$10 \text{ atm} = (4.13 \times 10^4 \text{ atm})(X_{CH_4})$$
$$X_{CH_4} = 2.4 \times 10^{-4}$$

For CH_4 at 50°C:

$$P_{CH_4} = k\,X_{CH_4}$$
$$10 \text{ atm} = (5.77 \times 10^4 \text{ atm})(X_{CH_4})$$
$$X_{CH_4} = 1.7 \times 10^{-4}$$

Therefore, the solubility of $CH_4(g)$ **decreases** with increasing temperature.

14-22. *Refer to Section 14-2 and Table 14-1.*

The hydration energy of ions generally increases with increasing charge and decreasing size. The greater the hydration energy for an ion, the more strongly hydrated it is. In order of decreasing hydration energy:

(a) $Na^+ > Rb^+$ due to smaller size   (c) $Fe^{3+} > Fe^{2+}$ due to higher charge

(b) $Cl^- > Br^-$ due to smaller size   (d) $Mg^{2+} > Na^+$ due to higher charge and smaller size

---

14-24. *Refer to Sections 14-6 and 13-6.*

LeChatelier's Principle states that when a stress is applied to a system, the system will react in such a way as to minimize the stress.

(a) Consider the process for a solid dissolving by an *exothermic* change:

$$Substance(s) \rightarrow Substance(aq) + heat$$

When the temperature of the system is increased, heat is added to the system. To remove some of this heat, the above process will reverse, producing more solid; the substance's solubility will decrease. When the temperature is decreased, heat is removed from the system. The system will adjust itself to replace some of the lost heat. The forward process is favored, enhancing the concentration of the substance in the aqueous phase and the substance's solubility will increase.

(b) Consider the process for a solid dissolving by an *endothermic* change:

$$Substance(s) + heat \rightarrow Substance(aq)$$

When the temperature is increased and heat is added to the system, it will adjust itself to remove some of the heat. The forward process is favored, dissolving some of the solid and producing more Substance(aq). The solubility of the substance increases. When the temperature is decreased and heat is removed from the system, the reverse process is favored and the solubility of the substance decreases.

14-26. *Refer to Section 3-6.*

If s = solubility of A in $\left(\dfrac{g\ A}{100\ g\ H_2O} \right)$, then the maximum mass of A that will dissolve in 100 g of H_2O = s.

The solubility of A as a mass percent:

$$\text{solubility of A in } \left[\frac{g\ A}{100\ g\ solution} \right] = \frac{g\ A}{g\ A\ +\ 100\ g\ H_2O} \times 100 = \frac{s}{s + 100} \times 100$$

14-28. *Refer to Section 14-8 and Example 14-1.*

Plan: (1) Determine the mass of solution for 1 kg of water in 0.225 m NaCl solution.
         (2) Calculate the mass of NaCl in 1 liter of solution.

(1) We know that a 0.225 m NaCl solution contains 0.225 moles of NaCl in 1000 g of water.

? g NaCl in 1000 g water = 0.225 mol $\times$ 58.44 g/mol = 13.2 g NaCl

? g solution = g solute + g solvent = 13.2 g NaCl + 1000 g water = 1013.2 g soln

(2) ? g NaCl in 1000 mL soln = $1000\ \text{mL soln} \times \dfrac{1.01\ \text{g soln}}{1\ \text{mL soln}} \times \dfrac{1000\ g\ H_2O}{1013.2\ \text{g soln}} \times \dfrac{0.225\ \text{mol NaCl}}{1000\ g\ H_2O} \times \dfrac{58.44\ \text{g NaCl}}{1\ \text{mol NaCl}}$

            = 13.1 g NaCl

Therefore, to prepare 1.000 L of a 0.225 m NaCl soln, add 13.1 g of NaCl to a one-liter volumetric flask. Add enough distilled water to dissolve the NaCl(s), then continue to add water until the volume of the solution is exactly one liter and mix thoroughly.

14-30. *Refer to Section 3-6.*

(1) Recall: mass % = $\dfrac{\text{g solute}}{\text{g soln}} \times 100$

Substituting,

$$15.0\% \text{ NaCl soln} = \dfrac{?\text{ g NaCl}}{210\text{ g soln}} \times 100$$

$$?\text{ g NaCl} = \dfrac{15.0\% \times 210\text{ g}}{100} = \textbf{31.5 g NaCl}$$

(2) Since g soln = g solute + g solvent, then

$$?\text{ g H}_2\text{O} = 210\text{ g soln} - 31.5\text{ g NaCl} = \textbf{178 g H}_2\textbf{O}$$

14-32. *Refer to Section 14-8, and Examples 14-1, 14-2 and 14-3.*

(a) mass percent $K_2ZnF_6 = \dfrac{25\text{ g K}_2\text{ZnF}_6}{25\text{ g K}_2\text{ZnF}_6 + 100\text{ g H}_2\text{O}} \times 100\% = \textbf{20.\%}$

(b) $X_{K_2ZnF_6} = \dfrac{\text{mol K}_2\text{ZnF}_6}{\text{mol K}_2\text{ZnF}_6 + \text{mol H}_2\text{O}}$

$\qquad = \dfrac{(25\text{g}/258\text{ g/mol})}{(25\text{ g}/258\text{ g/mol}) + (100\text{ g}/18\text{ g/mol})} = \dfrac{0.097\text{ mol}}{0.097\text{ mol} + 5.56\text{ mol}} = \textbf{0.017}$

(c) $m\ K_2ZnF_6 = \dfrac{\text{mol K}_2\text{ZnF}_6}{\text{kg H}_2\text{O}} = \dfrac{25\text{ g}/258\text{ g/mol}}{0.100\text{ kg H}_2\text{O}} = \textbf{0.97 } \boldsymbol{m} \textbf{ K}_2\textbf{ZnF}_6$

14-34. *Refer to Section 12-8.*

We know: $m = \dfrac{\text{mol solute}}{\text{kg solvent}}$

$?\text{ mol C}_6\text{H}_5\text{COOH} = \dfrac{90.0\text{ g C}_6\text{H}_5\text{COOH}}{122\text{ g/mol}} = 0.738\text{ mol C}_6\text{H}_5\text{COOH}$

$?\text{ kg C}_2\text{H}_5\text{OH} = 350\text{ mL C}_2\text{H}_5\text{OH} \times \dfrac{0.789\text{ g C}_2\text{H}_5\text{OH}}{1\text{ mL C}_2\text{H}_5\text{OH}} \times \dfrac{1\text{ kg}}{1000\text{ g}} = 0.276\text{ kg C}_2\text{H}_5\text{OH}$

Therefore, $m = \dfrac{0.738\text{ mol C}_6\text{H}_5\text{COOH}}{0.276\text{ kg C}_2\text{H}_5\text{OH}} = \textbf{2.67 } \boldsymbol{m} \textbf{ C}_6\textbf{H}_5\textbf{COOH in C}_2\textbf{H}_5\textbf{OH}$

14-36. *Refer to Section 14-8.*

(a) Plan: (1) Assuming 100 g of solution, calculate the moles of $C_6H_{12}O_6$.

              (2) Calculate the mass of water.

              (3) Determine the molality of $C_6H_{12}O_6$.

(1) Recall: % by mass = $\dfrac{\text{g solute}}{\text{g solution}} \times 100$

$?\text{ g C}_6\text{H}_{12}\text{O}_6 \text{ in 100 g soln} = \dfrac{24.0\%\text{ C}_6\text{H}_{12}\text{O}_6 \times 100\text{ g soln}}{100} = 24.0\text{ g C}_6\text{H}_{12}\text{O}_6$

$?\text{ mol C}_6\text{H}_{12}\text{O}_6 = \dfrac{24.0\text{ g}}{180\text{ g/mol}} = 0.133\text{ mol C}_6\text{H}_{12}\text{O}_6$

(2) $? \text{ g } H_2O = \text{g soln} - \text{g } C_6H_{12}O_6 = 100.0 \text{ g} - 24.0 \text{ g} = 76.0 \text{ g}$

(3) $m \text{ } C_6H_{12}O_6 = \dfrac{\text{mol } C_6H_{12}O_6}{\text{kg } H_2O} = \dfrac{0.133 \text{ mol}}{0.0760 \text{ kg}} = \textbf{1.75 } \boldsymbol{m} \textbf{ } C_6H_{12}O_6$

(b) Molality is independent of density and temperature since it is a measure of the number of moles of substance dissolved in 1 kilogram of solvent. Only concentration units that involve volume, e.g., molarity, are dependent on density and temperature. Therefore, the molality at a higher temperature would be the **same** as the molality at 20°C.

14-38. *Refer to Section 14-8 and Example 14-3.*

Plan: (1) Calculate the moles of each component.
(2) Calculate the mole fraction.

(1) $? \text{ mol } C_2H_5OH = \dfrac{75.0 \text{ g}}{46.1 \text{ g/mol}} = 1.63 \text{ mol } C_2H_5OH$

$? \text{ mol } H_2O = \dfrac{35.0 \text{ g}}{18.0 \text{ g/mol}} = 1.94 \text{ mol } H_2O$

(2) $X_{C_2H_5OH} = \dfrac{\text{mol } C_2H_5OH}{\text{mol } C_2H_5OH + \text{mol } H_2O} = \dfrac{1.63 \text{ mol}}{1.63 \text{ mol} + 1.94 \text{ mol}} = \textbf{0.457}$

$X_{H_2O} = \dfrac{\text{mol } H_2O}{\text{mol } H_2O + \text{mol } C_2H_5OH} = \dfrac{1.94 \text{ mol}}{1.94 \text{ mol} + 1.63 \text{ mol}} = \textbf{0.543}$

Alternative method: $X_{H_2O} = 1 - X_{C_2H_5OH} = 1.000 - 0.457 = \textbf{0.543}$

14-40. *Refer to Sections 3-6 and 14-8.*

(1) $? \text{ mol } K_2SO_4 = \dfrac{12.50 \text{g}}{174.3 \text{ g/mol}} = 0.07172 \text{ mol } K_2SO_4$

$? \text{ mL soln} = \dfrac{\text{g soln}}{\text{density}} = \dfrac{100.00 \text{ g}}{1.083 \text{ g/mL}} = 92.34 \text{ mL}$

Therefore,

$M \text{ } K_2SO_4 = \dfrac{\text{mol } K_2SO_4}{\text{L soln}} = \dfrac{0.07172 \text{ mol}}{0.09234 \text{ L}} = \textbf{0.7767 } \boldsymbol{M} \textbf{ } K_2SO_4$

(2) $m \text{ } K_2SO_4 = \dfrac{\text{mol } K_2SO_4}{\text{kg } H_2O} = \dfrac{0.07172 \text{ mol}}{0.08750 \text{ kg}} = \textbf{0.8197 } \boldsymbol{m} \textbf{ } K_2SO_4$

since $\text{kg } H_2O = \text{kg soln} - \text{kg } K_2SO_4 = 0.10000 \text{ kg} - 0.01250 \text{ kg} = 0.08750 \text{ kg}$

(3) $\text{mass \% } K_2SO_4 = \dfrac{\text{g } K_2SO_4}{\text{g soln}} \times 100 = \dfrac{12.50 \text{ g}}{100.00 \text{g}} \times 100 = \textbf{12.50\% } K_2SO_4$

(4) $X_{H_2O} = \dfrac{\text{mol } H_2O}{\text{mol } H_2O + \text{mol } K_2SO_4} = \dfrac{\left[\dfrac{87.50 \text{ g}}{18.02 \text{ g/mol}}\right]}{\left[\dfrac{87.50 \text{ g}}{18.02 \text{ g/mol}} + \dfrac{12.50 \text{ g}}{174.3 \text{ g/mol}}\right]} = \dfrac{4.856 \text{ mol}}{4.856 \text{ mol} + 0.07172 \text{ mol}} = \textbf{0.9854}$

The vapor pressure of a liquid depends upon the ease with which the molecules are able to escape from the surface of the liquid. The vapor pressure of a liquid always decreases when nonvolatile solutes (ions or molecules) are dissolved in it, since after dissolution there are fewer solvent molecules at the surface to vaporize.

14-44. *Refer to Section 14-9 and Example 14-4.*

(a) From Raoult's Law,

$$\Delta P_{benzene} = X_{naphthalene}\, P^o_{benzene}$$

where $\Delta P_{benzene}$ = vapor pressure lowering of benzene, C_6H_6
$P^o_{benzene}$ = vapor pressure of pure benzene, 74.6 torr at 20°C
$X_{naphthalene}$ = mole fraction of naphthalene, $C_{10}H_8$

$$? \text{ mol benzene} = \frac{150.0 \text{ g } C_6H_6}{78.1 \text{ g/mol}} = 1.92 \text{ mol } C_6H_6$$

$$? \text{ mol naphthalene} = \frac{35.5 \text{ g } C_{10}H_8}{128 \text{ g/mol}} = 0.277 \text{ mol } C_{10}H_8$$

$$X_{naphthalene} = \frac{\text{mol } C_{10}H_8}{\text{mol } C_{10}H_8 + \text{mol } C_6H_6} = \frac{0.277 \text{ mol}}{0.277 \text{ mol} + 1.92 \text{ mol}} = 0.126$$

Substituting into Raoult's Law, $\Delta P_{benzene} = 0.126 \times 74.6 \text{ torr} = \mathbf{9.40 \text{ torr}}$

(b) $P_{benzene} = P^o_{benzene} - \Delta P_{benzene} = 74.6 \text{ torr} - 9.40 \text{ torr} = \mathbf{65.2 \text{ torr}}$

14-46. *Refer to Section 14-9 and Example 14-5.*

Plan: (1) Calculate the mole fraction of each component in the solution.
(2) By assuming the solution to be ideal, apply Raoult's Law to calculate the partial pressures of acetone and chloroform.

(1) Let X = mole fraction of the components

$$X_{acetone} = \frac{0.250 \text{ mol acetone}}{(0.250 \text{ mol acetone}) + (0.300 \text{ mol chloroform})} = 0.455$$

$$X_{chloroform} = \frac{0.300 \text{ mol chloroform}}{(0.250 \text{ mol acetone}) + (0.300 \text{ mol chloroform})} = 0.545 \ (= 1 - X_{acetone})$$

(2) From Raoult's Law:

$$P_{acetone} = X_{acetone}\, P^o_{acetone} = (0.455)(345 \text{ torr}) = \mathbf{157 \text{ torr}}$$
$$P_{chloroform} = X_{chloroform}\, P^o_{chloroform} = (0.545)(295 \text{ torr}) = \mathbf{161 \text{ torr}}$$

Since the acetone-chloroform system is expected to show negative deviation from an ideal solution, $P_{acetone}$ should be less that 157 torr and $P_{chloroform}$ should be less than 161 torr.

14-48. *Refer to Sections 14-9 and 12-11, Examples 14-5 and 14-6, and Exercise 14-46 Solution.*

Plan: (1) Calculate the total pressure using Exercise 14-46 solution: $P_{acetone} = 157 \text{ torr}$, $P_{chloroform} = 161 \text{ torr}$.
(2) Since the mole fraction of a component in a gaseous mixture equals the ratio of its partial pressure to the total pressure, the composition of the solution can be calculated.

(1) $P_{total} = P_{acetone} + P_{chloroform} = 157 \text{ torr} + 161 \text{ torr} = \mathbf{318 \text{ torr}}$

(2) In the vapor, $X_{acetone} = \dfrac{P_{acetone}}{P_{total}} = \dfrac{157\ torr}{318\ torr} = \mathbf{0.494}$

$X_{chloroform} = \dfrac{P_{chloroform}}{P_{total}} = \dfrac{161\ torr}{318\ torr} = \mathbf{0.506}$

14-50. *Refer to Section 14-9 and Exercise 14-49.*

Assuming real behavior for a chloroform/acetone solution in which the mole fraction of chloroform, $CHCl_3$, is 0.3 and using the dashed (curved) lines on the diagram in Exercise 14-49,

(a) $P_{chloroform} = \mathbf{55\ torr}$

(b) $P_{acetone} = \mathbf{210\ torr}$

(c) $P_{total} = \mathbf{280\ torr}$

14-52. *Refer to Section 14-11, Example 14-7 and Table 14-2.*

Plan: (1) Find ΔT_b.
  (2) Determine T_b for the ethylene glycol solution.

(1) From Table 14-2, K_b for H_2O = 0.512 °C/m; B.P. = 100.00°C.

$\Delta T_b = K_b\,m = (0.512\ °C/m)(2.15\ m) = 1.10°C$

(2) Boiling point of the ethylene glycol solution, $T_{b(soln)} = T_{b(solvent)} + \Delta T_b = 100.00°C + 1.10°C = \mathbf{101.10°C}$

14-54. *Refer to Section 14-12, Example 14-9, Exercise 14-52 Solution and Table 14-2.*

Plan: (1) Find ΔT_f.
  (2) Determine T_f for the ethylene glycol solution.

(1) From Table 14-2, K_f for H_2O = 1.86 °C/m; F.P. = 0.00°C.

$\Delta T_f = K_f\,m = (1.86\ °C/m)(2.15\ m) = 4.00°C$

(2) Freezing point of the ethylene glycol solution, $T_{f(soln)} = T_{f(solvent)} - \Delta T_f = 0.00°C - 4.00°C = \mathbf{-4.00°C}$

14-56. *Refer to Sections 14-11 and 14-12, and Table 14-2.*

Consider a 0.150 m solution of a non-electrolyte in the solvents listed in Table 14-2.

(a) The greatest freezing point depression, ΔT_f, occurs in a **camphor** solution since $\Delta T_f = K_f\,m$ and this solvent has the largest value of K_f.

(b) The lowest freezing point, $T_{f(solution)}$, occurs in a solution of the non-electrolyte in **water**. Pure water freezes at 0°C, the lowest freezing point of all the listed solvents. The effect of freezing point depression on the freezing point of a 0.150 m solution is minor compared to the actual freezing point of the pure solvents.

(c) The greatest boiling point elevation, ΔT_b, occurs in a solution of **camphor** since $\Delta T_b = K_b m$ and this solvent has the largest value of K_b.

(d) The highest boiling point, $T_{b(solution)}$, for all the listed solvents occurs in a solution of **nitrobenzene** since it has the highest boiling point.

14-58. *Refer to Section 14-12 and Example 14-9.*

Plan: (1) Determine the molality of the solute, Zn, in the solid solution.
(2) Calculate the melting point (freezing point) of brass using $\Delta T_f = K_f m$.

(1) Assume 100. g of brass containing 10. g of Zn and 90. g of Cu.

$$m\ Zn = \frac{\text{mol solute}}{\text{kg solvent}} = \frac{\text{mol Zn}}{\text{kg Cu}} = \frac{10.g\ Zn/65.4\ g/mol}{0.090\ kg\ Zn} = 1.7\ m$$

(2) $\Delta T_f = K_f m = 23\ °C/m \times 1.7\ m = 39°C$

$T_{f(brass)} = T_{f(Cu)} - \Delta T_f = 1083°C - 39°C = \mathbf{1044°C}$

14-60. *Refer to Section 14-11 and Table 14-2.*

Plan: (1) Solve for the molality of the solution using $\Delta T_b = K_b m$
(2) Calculate the mass of $C_{10}H_8$.

From Table 14-2, for nitrobenzene: $T_b = 210.88°C$ $\qquad K_b = 5.24\ °C/m$

(1) $m = \dfrac{\Delta T_b}{K_b} = \dfrac{214.20°C - 210.88°C}{5.24\ °C/m} = 0.634\ m\ C_{10}H_8$

(2) Recall: $m = \dfrac{\text{mol solute}}{\text{kg solvent}} = \dfrac{\text{g solute/MW solute}}{\text{kg solvent}}$

Solving,

$?\ g\ C_{10}H_8 = (m)(MW\ C_{10}H_8)(kg\ C_6H_5NO_2) = (0.634\ m)(128\ g/mol)(0.400\ kg) = \mathbf{32.5\ g\ C_{10}H_8}$

14-62. *Refer to Sections 14-12 and 2-9, Exercise 2-48 Solution, and Table 14-2.*

Plan: (1) Solve for the molality and the approximate formula weight of the nonelectrolyte, using $\Delta T_f = K_f m$.
(2) Determine the simplest (empirical) formula from the % composition data.
(3) Using the approximate formula weight of the simplest formula, determine the true molecular formula and the exact molecular weight.

From Table 14-2, for benzene: $T_f = 5.48°C$ $\qquad K_f = 5.12\ °C/m$

(1) $m = \dfrac{\Delta T_f}{K_f} = \dfrac{0.42°C}{5.12\ °C/m} = 0.082\ m$

Recall: $m = \dfrac{\text{mol solute}}{\text{kg solvent}} = \dfrac{\text{g solute/MW solute}}{\text{kg solvent}}$

Solving,

$?\ MW\ solute = \dfrac{\text{g solute}}{m \times \text{kg benzene}} = \dfrac{0.500\ g}{0.082\ m \times 0.050\ kg} = 120\ g/mol$ (to 2 significant figures)

(2) Assume 100 g of sample containing 40.0 g C, 6.67 g H and 53.3 g O.

$?\ mol\ C = \dfrac{40.0\ g}{12.0\ g/mol} = 3.33\ mol$ $\qquad\qquad$ Ratio $= \dfrac{3.33}{3.33} = 1$

$?\ mol\ H = \dfrac{6.67\ g}{1.008\ g/mol} = 6.62\ mol$ $\qquad\qquad$ Ratio $= \dfrac{6.62}{3.33} = 2$

$?\ mol\ O = \dfrac{53.3\ g}{16.0\ g/mol} = 3.33\ mol$ $\qquad\qquad$ Ratio $= \dfrac{3.33}{3.33} = 1$

Therefore, the simplest formula is CH_2O (FW $= 30.03\ g/mol$)

(3) $n = \dfrac{\text{molecular weight}}{\text{simplest formula weight}} = \dfrac{120 \text{ g/mol}}{30 \text{ g/mol}} = 4$

The true molecular formula for the organic compound is $(CH_2O)_4 = C_4H_8O_4$ with a molecular weight (to 5 significant figures) of **120.10 g/mol**.

14-64. *Refer to Section 14-12 and Fundamental Algebra.*

(a) Plan: (1) Calculate the molality of the solution using $\Delta T_f = K_f m$
            (2) Calculate the total number of moles of solutes.
            (3) Calculate the masses of $C_{10}H_8$ and $C_{14}H_{10}$ and the % composition of each in the sample.

(1) From Table 14-2, for benzene: $T_b = 80.1°C$ $K_b = 2.53 \text{ °C/}m$
                                     $T_f = 5.48°C$ $K_f = 5.12 \text{ °C/}m$

$$m = \frac{\Delta T_f}{K_f} = \frac{5.48°C - 4.85°C}{5.12 \text{ °C/}m} = 0.12 \text{ } m \text{ solutes}$$

(2) Recall: $m = \dfrac{\text{mol solute}}{\text{kg solvent}}$

    ? mol solutes $= 0.12 \text{ } m \times 0.360 \text{ kg} = 0.043 \text{ mol solutes}$

(3) Let $x = \text{g } C_{10}H_8$ and $\dfrac{x}{128 \text{ g/mol}} = \text{mol } C_{10}H_8$

                $6.00 \text{ g} - x = \text{g } C_{14}H_{10}$ $\dfrac{6.00 \text{ g} - x}{178 \text{ g/mol}} = \text{mol } C_{14}H_{10}$

Therefore,

        the total moles of solute $= \text{mol } C_{10}H_8 + \text{mol } C_{14}H_{10} = \dfrac{x}{128} + \dfrac{6.00 - x}{178}$

    So,   total moles $= 0.043 \text{ mol} = \dfrac{x}{128} + \dfrac{6.00 - x}{178}$

When we multiply both sides of the equation by (128)(178), we obtain,

$$980 = (178x) + (768 - 128x)$$
$$50x = 212$$
$$x = 4.2 \text{ g } C_{10}H_8$$
$$6.00 - x = 1.8 \text{ g } C_{14}H_{10}$$

And, $\% \text{ } C_{10}H_8 = \dfrac{4.2 \text{ g } C_{10}H_8}{6.00 \text{ g sample}} \times 100 = \mathbf{70\%}$

        $\% \text{ } C_{14}H_{10} = \dfrac{1.8 \text{ g } C_{14}H_{10}}{6.00 \text{ g sample}} \times 100 = \mathbf{30\%}$

(b) $\Delta T_b = K_b m_{total} = (2.53 \text{ °C/}m)(0.12 \text{ } m) = 0.30°C$

   $T_{b(solution)} = T_{b(benzene)} + \Delta T_b = 80.1°C + 0.30°C = \mathbf{80.4°C}$

14-66. *Refer to Section 14-14.*

In an ionic solution, the electrical interactions between dissolved ions increase with (1) increasing charge on the ions and (2) increasing concentration. The weaker the electrical attraction between the ions of a dissolved salt, the more completely dissociated is the salt. A 0.10 M $LiNO_3$ solution is more completely dissociated than either a 0.10 M $Ca(NO_3)_2$ or a 0.10 M $Al(NO_3)_3$ solution since the charge on the lithium ion is +1, compared to the +2 charge on the calcium ion and the +3 charge on the aluminum ion.

The solution with the greatest number of ions will conduct electricity the strongest. If completely dissociated, the total number of ions expected from $LiNO_3$, $Ca(NO_3)_2$ and $Al(NO_3)_3$ would be 2, 3 and 4, respectively. Therefore, $Al(NO_3)_3$ is expected to be the best conductor of electricity among the three, even though there is some degree of association of ions.

14-68. *Refer to Section 14-14 and Table 14-3.*

The ideal value for the van't Hoff factor, i, for strong electrolytes at infinite dilution is the total number of ions present in a formula unit.

	ideal value of i
(a) Na_2SO_4	3
(b) KOH	2
(c) $Al_2(SO_4)_3$	5
(d) $SrSO_4$	2

14-70. *Refer to Section 14-14.*

We know that $\Delta T_f = iK_f m$. For solutions of the same solvent at the same molality, the solute with the largest van't Hoff factor, i, has the lowest freezing point. Since $CaCl_2(aq)$ will dissociate into 3 ions ($i_{ideal} = 3$) and the other compounds have a van't Hoff factor that is less than 3, the solution with the lowest freezing point is **0.010 m $CaCl_2$**.

14-72. *Refer to Sections 14-9 and 14-14.*

Note: The van't Hoff factor, i, must be included in all calculations involving colligative properties, including vapor pressure lowering.

Plan: (1) Determine the total number of moles of ions in the solution.
(2) Determine the mole fraction of water in the solution.
(3) Calculate the vapor pressure above the solution.

(1) For NaCl, NaBr and NaI, the ideal van't Hoff factor, i_{ideal}, is 2.

? mol ions $= (2 \times$ mol NaCl$) + (2 \times$ mol NaBr$) + (2 \times$ mol NaI$)$

$$= \left[2 \times \frac{1.00 \text{ g NaCl}}{58.4 \text{ g/mol}}\right] + \left[2 \times \frac{1.00 \text{ g NaBr}}{103 \text{ g/mol}}\right] + \left[2 \times \frac{1.00 \text{ g NaI}}{150 \text{ g/mol}}\right]$$

$$= 0.0669 \text{ mol ions}$$

(2) $X_{water} = \dfrac{\text{mol } H_2O}{\text{mol ions} + \text{mol } H_2O} = \dfrac{(100 \text{ g}/18.0 \text{ g/mol})}{0.0669 \text{ mol} + (100 \text{ g}/18.0 \text{ g/mol})} = 0.988$

(3) At 100°C, the boiling point of water, the vapor pressure of pure water, $P^\circ_{water} = 760$ torr.

$P_{water} = X_{water} P^\circ_{water} = 0.988 \times 760$ torr $=$ **751 torr**

14-74. *Refer to Section 14-14 and Example 14-12.*

Plan: (1) Calculate $m_{effective}$.
(2) Determine the % ionization of C_3COOH.

(1) From Table 14-2, for water: $T_f = 0$°C, $K_f = 1.86$ °C/m

$$m_{effective} = \frac{\Delta T_f}{K_f} = \frac{0.0000°C - (-0.1884°C)}{1.86 \text{ °C}/m} = 0.101 \text{ } m$$

(2) Let $\quad$ x = molality of CH_3COOH that ionizes

Then $\quad$ x = molality of H^+ and CH_3COO^- that formed

Consider:	CH_3COOH	$\rightarrow$	H^+	$+$	CH_3COO^-
Start	0.100 m		$\approx 0\ m$		0 m
Change	- x m		+ x m		+ x m
Final	(0.100 - x) m		x m		x m

$m_{effective}$ $\quad = m_{CH_3COOH} + m_{H^+} + m_{CH_3COO^-}$

$\quad 0.101 = (0.100 - x) + x + x$

$\quad 0.101 = 0.100 + x$

$\qquad x = 0.001\ m$

% ionization $= \dfrac{m_{ionized}}{m_{original}} \times 100 = \dfrac{0.001\ m}{0.100\ m} \times 100 = \mathbf{1\%}$

14-76. **Refer to Section 14-14 and Example 14-12.**

Plan: $\quad$ (1) Find ΔT_f for the solution if CsCl had been a nonelectrolyte.

$\qquad$ (2) Determine the van't Hoff factor, i.

$\qquad$ (3) Calculate $m_{effective}$ for CsCl.

$\qquad$ (4) Calculate the % dissociation.

(1) If CsCl were a nonelectrolyte

$\qquad \Delta T_f = K_f m = (1.86\ °C/m)(0.121\ m) = 0.225°C$ $\qquad\qquad$ for water, $K_f = 1.86\ °C/m$

(2) The van't Hoff factor, $i = \dfrac{\Delta T_{f(actual)}}{\Delta T_{f(nonelectrolyte)}} = \dfrac{0°C - (-0.403°C)}{0.225°C} = \mathbf{1.79}$

(3) $m_{effective} = i \times m_{original} = 1.79 \times 0.121\ m = 0.217\ m$

(4) Let $\quad$ x = molality of CsCl that apparently dissociated

Then $\quad$ x = molality of Cs^+ and Cl^- that formed

Consider:	CsCl	$\rightarrow$	Cs^+	$+$	Cl^-
Start	0.121 m		0 m		0 m
Change	- x m		+ x m		+ x m
Final	(0.121 - x) m		x m		x m

$m_{effective}$ $\quad = m_{CsCl} + m_{Cs^+} + m_{Cl^-}$

$\quad 0.217 = (0.121 - x) + x + x$

$\quad 0.217 = 0.121 + x$

$\qquad x = 0.096\ m$

% ionization $= \dfrac{m_{dissociated}}{m_{original}} \times 100 = \dfrac{0.096\ m}{0.121\ m} \times 100 = \mathbf{79\%}$

14-78. Refer to Section 14-8.

Assume 1 liter of solution.

? g NaCl in 1 L soln = 1.00×10^{-4} mol NaCl $\times$ 58.4 g/mol = 5.84×10^{-3} g NaCl

The contribution of NaCl to the mass of 1 liter of water is nearly negligible. Since the density of water (1.00 g/mL) is essentially the same as the density of the solution:

$$M = \frac{1.00 \times 10^{-4} \text{ mol NaCl}}{1 \text{ L soln}} \cong \frac{1.00 \times 10^{-4} \text{ mol NaCl}}{1 \text{ kg H}_2\text{O}} = m$$

However, if acetonitrile (density at 20°C = 0.786 g/mL) were the solvent, 1 liter of solution which would be essentially pure solvent would have a mass of only 786 g. We see that molarity of NaCl in acetonitrile would *not* be equivalent to the molality, but would be less.

14-80. Refer to Section 14-15 and Example 14-13.

osmotic pressure, $\pi = iMRT$ = (1)(0.0111 M)(0.0821 L·atm/mol·K)(75°C + 273°) = **0.317 atm**

14-82. Refer to Sections 14-11, 14-12 and 14-15, and Example 14-81.

Plan: (1) Determine the molality of the aqueous solution.
(2) Calculate ΔT_f and ΔT_b.

(1) Recall, osmotic pressure, $\pi = iMRT$

$$M = \frac{\pi}{iRT} = \frac{1.17 \text{ atm}}{(1)(0.0821 \text{ L·atm/mol·K})(273 \text{ K})} = 0.0522 \ M$$

For dilute aqueous solution, M $\cong$ m (See Exercise 14-78 Solution).
Therefore, ? m = 0.0522 m

(2) From Table 14-2, for water: K_f = 1.86 °C/m $\qquad$ K_b = 0.512 °C/m

$\Delta T_f = iK_f m$ = (1)(1.86 °C/m)(0.0522 m) = **0.0971°C**

$\Delta T_b = iK_b m$ = (1)(0.512 °C/m)(0.0522 m) = **0.0267°C**

14-84. Refer to Section 14-15 and Table 14-3.

From Table 14-3, the van't Hoff factor, i, for 0.10 m K_2CrO_4 is 2.39. Assume i for 0.12 m K_2CrO_4 is the same.

$m_{effective} = i \times m_{stated}$ = 2.39 $\times$ 0.12 m = 0.29 m

Therefore, using the approximation equation for dilute solutions, $M \cong m$:

osmotic pressure, $\pi = m_{effective}RT$ = (0.29 m)(0.0821 L·atm/mol·K)(25°C + 273°) = **7.1 atm**

14-86. Refer to Sections 14-4 and 14-15, Table 14-2, and Example 14-14.

From Table 14-2, for water: K_f = 1.86 °C/m

(a) $m = \dfrac{\text{mol substance}}{\text{kg solvent}} = \dfrac{0.0110 \text{ g}/2.00 \times 10^4 \text{ g/mol}}{0.0100 \text{ kg}} = 5.50 \times 10^{-5} \ m$

$\Delta T_f = iK_f m$ = (1)(1.86 °C/m)(5.50 $\times$ 10^{-5} m) = 1.02×10^{-4} °C

$T_{f(soln)} = T_{f(water)} - \Delta T_f$ = 0°C - 1.02 $\times$ 10^{-4} °C = **-1.02 × 10^{-4} °C**

(b) For dilute solutions, $M \cong m$,

$$\pi = MRT \cong mRT = (5.50 \times 10^{-5}\, m)(0.0821 \text{ L·atm/mol·K})(25°C + 273°) = \mathbf{1.35 \times 10^{-3}\ atm\ or\ 1.02\ torr}$$

(c) From the equation given in (a),

% error in $T_{f(\text{soln})}$ = % error in ΔT_f = % error in MW

$$?\ \%\text{ error in } T_{f(\text{soln})} = \frac{\text{error in } T_f}{T_f} \times 100 = \frac{0.001°C}{1.02 \times 10^{-4}\ °C} \times 100 = \mathbf{1000\%}$$

Therefore, an error of only $0.001°C$ in the freezing point temperature corresponds to a 1000% error in the macromolecule's molecular weight.

(d) Since the osmotic pressure, $\pi = MRT = \dfrac{nRT}{V} = (\text{mol/MW})\dfrac{RT}{V}$, % error in π = % error in MW

$$?\ \text{error in } \pi = \frac{\text{error in } \pi}{\pi} \times 100 = \frac{0.1\ \text{torr}}{1.02\ \text{torr}} \times 100 = \mathbf{10\%\ error}$$

Therefore, an error of 0.1 torr in osmotic pressure gives only a 10% error in molecular weight.

14-88. *Refer to Table 14-4 and the Key Terms for Chapter 14.*

(a) sol — a colloidal suspension of a solid dispersed in a liquid, e.g., detergents in water

(b) gel — a colloidal suspension of a solid dispersed in a liquid; a semirigid sol, e.g., jelly

(c) emulsion — a colloidal suspension of a liquid in a liquid, e.g., some cough medicines

(d) foam — a colloidal suspension of a gas in a liquid, e.g., bubbles in a bubble bath

(e) solid sol — a colloidal suspension of a solid in a solid, e.g., dirty ice

(f) solid emulsion — a colloidal suspension of a liquid dispersed in a solid, e.g., some kinds of sea ice containing pockets of brine

(g) solid foam — a colloidal suspension of a gas dispersed in a solid, e.g., marshmallows

(h) solid aerosol — a colloidal suspension of a solid in a gas, e.g., fine dust

(i) liquid aerosol — a colloidal suspension of a liquid in gas, e.g., insect spray

---

14-90. *Refer to Section 14-18 and the Key Terms for Chapter 14.*

Hydrophilic colloids are colloidal particles that attract water molecules, whereas hydrophobic colloids are colloidal particles that repel water molecules.

14-92. *Refer to Section 14-18.*

Soaps and detergents are both emulsifying agents. Solid soaps are usually sodium salts of long chain organic acids called fatty acids with the general formula, $R\text{-COO}^- Na^+$. On the other hand, synthetic detergents contain sulfonate, $-SO_3^-$, sulfate, $-SO_4^-$, or phosphate groups instead of carboxylate groups, $-COO^-$.

"Hard" water contains Fe^{3+}, Ca^{2+} and/or Mg^{2+} ions, all of which displace Na^+ from soap molecules and give an undesirable precipitate coating. However, detergents do not form precipitates with the ions of "hard" water.

A typical equation between soap and hard water that contains Ca^{2+} ions is shown below:

$$Ca^{2+}(aq) + 2RCOO^-Na^+(aq) \rightarrow (RCOO^-)_2Ca^{2+}(s) + 2Na^+(aq)$$

For a solution at infinite dilution at 25°C,

      the overall heat of solution = hydration energy + ($-\Delta H_{xtal}$)

The reason for using $-\Delta H_{xtal}$ is as follows.

The crystal lattice energy, ΔH_{xtal} = -825.9 kJ/mol. This means that 825.9 kJ of heat is released in the reaction:

      $K^+(g) + F^-(g) \rightarrow KF(s) + 825.9$ kJ

However, we are interested in the reverse reaction:

      $KF(s) + 825.9$ kJ $\rightarrow K^+(g) + F^-(g)$ $\Delta H_{rxn} = -\Delta H_{xtal} = +825.9$ kJ/mol of KF(s).

We will discuss sign conventions more thoroughly in Chapter 15, the thermodynamics chapter.

Rearranging the above equation, we have

      hydration energy of KF = heat of solution + ΔH_{xtal} = -17.7 kJ/mol - 825.0 kJ/mol = **-843.6 kJ/mol**

14-96. *Refer to Sections 14-12 and 14-13, and Exercise 13-16 Solution.*

From Table 14-2,   for water: $T_f = 0°C$ $K_f = 1.86 °C/m$
                for benzene: $T_f = 5.48°C$ $K_f = 5.12 °C/m$

Plan: (1) Solve for the molality of CH_3COOH in water or benzene using $\Delta T_f = K_f m$ (assume $i = 1$).
         (2) Calculate the apparent molecular weight of CH_3COOH in each solvent.

(a) in aqueous solution:

   (1) $m_{apparent} = \dfrac{\Delta T_f}{K_f} = \dfrac{0.310°C}{1.86 °C/m} = 0.167\ m\ CH_3COOH$ in H_2O

   (2) Recall: $m = \dfrac{\text{mol solute}}{\text{kg solvent}} = \dfrac{\text{g solute/MW}}{\text{kg solvent}}$

       Assume 100.00 g of solution containing 1.00 g of CH_3COOH in 99.00 g of water.

       $?\ MW_{apparent} = \dfrac{\text{g } CH_3COOH}{m_{apparent} \times \text{kg } H_2O} = \dfrac{1.00\ \text{g}}{0.167\ m \times 0.09900\ \text{kg}} = $ **60.5 g/mol**

The true molecular weight of acetic acid is 60.05 g/mol. Acetic acid is a weak acid and dissociates very slightly in water; the van't Hoff factor, i, is then only slightly larger than 1. The behavior of acetic acid approximates that of a nonelectrolyte in water.

(b) in benzene solution:

   (1) $m_{apparent} = \dfrac{\Delta T_f}{K_f} = \dfrac{0.441°C}{5.12 °C/m} = 0.0861\ m\ CH_3COOH$ in benzene

   (2) Assume 100.00 g of solution containing 1.00 g of CH_3COOH in 99.00 g of benzene.

       $?\ MW_{apparent} = \dfrac{\text{g } CH_3COOH}{m_{apparent} \times \text{kg benzene}} = \dfrac{1.00\ \text{g}}{0.0861\ m \times 0.09900\ \text{kg}} = $ **117 g/mol**

The apparent molecular weight, determined in benzene is almost twice as large as the true molecular weight. The reason is that in benzene, the acetic acid dimerizes due to hydrogen bonding (refer to the figure in Exercise 13-16 Solution). Most of the acetic acid exists as 2 acetic acid molecules held together by hydrogen bonds.

14-98. *Refer to Section 14-18.*

(a) hydrophobic   hydrocarbons   (b) hydrophilic   starch

(c) The hydrophilic dispersion of starch in water is much easier to make and maintain. Using hydrogen bonding, the very polar water molecules surround and isolate each starch molecule from one another. The starch molecules cannot coalesce and therefore remain dispersed.

14-100. Refer to Section 14-15.

Plan: (1) Determine the total molarity of the drug-sugar mixture dissolved in water, using $\pi = MRT$ ($i = 1$).
(2) Determine the mass of lactose in the solution.
(3) Calculate the % lactose present.

(1) $M_{lactose} + M_{drug} = M_{total} = \dfrac{\pi}{RT} = \dfrac{(527/760) \text{ atm}}{(0.0821 \text{ L·atm/mol·K})(25°C + 273)} = 0.0283\ M_{total}$

(2) Let x = g lactose, $C_{12}H_{22}O_{11}$ (MW: 342 g/mol)
$1 - x$ = g drug, $C_{21}H_{23}O_5N$ (MW: 369 g/mol)

$M_{total} = \dfrac{\text{mol lactose + mol drug}}{\text{L solution}}$

Substituting,

$$0.0283\ M = \dfrac{\left[\dfrac{x}{342 \text{ g/mol}} + \dfrac{1.00 - x}{369 \text{ g/mol}} \right]}{0.100 \text{ L}}$$

$$0.00283 = \dfrac{x}{342} + \dfrac{1.00 - x}{369}$$

Multiplying both sides by the product, (342)(369), we obtain

$$357 = 369x + 342(1.00 - x)$$
$$15 = 27x$$
$$x = 0.56 \text{ g lactose}$$

(3) ? lactose by mass = $\dfrac{0.56 \text{ g lactose}}{1.00 \text{ g mixture}} \times 100 = $ **56% lactose by mass**

14-102. Refer to Section 14-1.

To have an ideal solution, the solvent-solvent, solute-solute and solvent-solute intermolecular forces should be as identical as possible. Solution (c) consisting of $CH_4(\ell)$ dissolved in $CH_3CH_3(\ell)$ would be the most ideal.

14-104. Refer to Section 14-9 and Figure 14-17b.

When the system reaches equilibrium, the mole fractions of solute in both solutions are equal.

$$X_{solute(unknown)} = X_{solute(known)}$$

$$\left(\dfrac{\text{mol solute}}{\text{mol solute + mol solvent}} \right)_{unknown} = \left(\dfrac{\text{mol solute}}{\text{mol solute + mol solvent}} \right)_{known}$$

Assuming that the moles of each solute is negligible compared to the moles of solvent: $mol_{solute} << mol_{solvent}$,

$$\left(\dfrac{\text{mol solute}}{\text{mol solvent}} \right)_{unknown} = \left(\dfrac{\text{mol solute}}{\text{mol solvent}} \right)_{known}$$

Since the solvents are the same in both the unknown and known solution and the moles of solvent are directly proportional to the volume of solvent, we can write

$$\left(\frac{\text{mol solute}}{V_{\text{solvent}}}\right)_{\text{unknown}} = \left(\frac{\text{mol solute}}{V_{\text{solvent}}}\right)_{\text{known}}$$

Since the volume of each solute is negligible compared to the solvent's volume: $V_{\text{solute}} << V_{\text{solvent}}$, then the $V_{\text{solvent}} \cong V_{\text{solution}}$.

$$\left(\frac{\text{mol solute}}{V_{\text{solution}}}\right)_{\text{unknown}} = \left(\frac{\text{mol solute}}{V_{\text{solution}}}\right)_{\text{known}}$$

or

$$\left(\frac{\text{mass/MW}}{V_{\text{solution}}}\right)_{\text{unknown}} = \left(\frac{\text{mass/MW}}{V_{\text{solution}}}\right)_{\text{known}}$$

Rearranging,

$$MW_{\text{unknown}} = MW_{\text{known}} \times \frac{\text{mass}_{\text{unknown}}}{\text{mass}_{\text{known}}} \times \frac{V_{\text{known}}}{V_{\text{unknown}}}$$

$$= 168.2 \text{ g/mol} \times \frac{40.6 \text{ mg}}{45.1 \text{ mg}} \times \frac{1.80 \text{ mL}}{1.41 \text{ mL}}$$

$$= \textbf{193 g/mol}$$

14-106. *Refer to Section 14-15 and Example 14-13.*

Plan: (1) Determine the osmotic pressure (in torr) required to force the sap to the top of the tree.
(2) Calculate the molarity of the sugar solution (sap).

(1) ? height of the tree (mm) = $45 \text{ ft} \times \dfrac{12 \text{ in}}{1 \text{ ft}} \times \dfrac{2.54 \text{ cm}}{1 \text{ in}} \times \dfrac{10 \text{ mm}}{1 \text{ cm}} = 1.4 \times 10^4 \text{ mm}$

The pressure exerted by a column of sugar solution in the tree can be stated as 1.4×10^4 mm sugar solution. However, pressure is more often given in units of the height of an equivalent column of mercury in millimeters, called torr, which can be easily converted to atmospheres. The unit factor relating a column of mercury to a column of a liquid exerting the same pressure involves their densities. In this case since the density of the solution is only 1/13.6 that of mercury as determined from the relative densities, a 13.6 mm column of solution exerts the same pressure as a 1 mm column of mercury. Thus,

? pressure (atm) = 1.4×10^4 mm sugar soln $\times \dfrac{1.00 \text{ mm Hg}}{13.6 \text{ mm sugar soln}} \times \dfrac{1 \text{ torr}}{1 \text{ mm Hg}} \times \dfrac{1 \text{ atm}}{760 \text{ torr}} = \textbf{1.4 atm}$

(2) Recall, osmotic pressure, $\pi = iMRT$

$$M = \frac{\pi}{iRT} = \frac{1.4 \text{ atm}}{(1)(0.0821 \text{ L·atm/mol·K})(273 \text{ K})} = \textbf{0.062 } \textit{M}$$

Note: If the numbers are not rounded off to 2 significant figures until the end, the answer is 0.059 M.

15 Chemical Thermodynamics

15-2. *Refer to the Key Terms for Chapters 1 and 15.*

(a) Heat is a form of energy that flows between two samples of matter due to their differences in temperature.

(b) Temperature is a measure of the intensity of heat, i.e., the hotness or coldness of an object.

(c) The system refers to the substances of interest in a process, i.e., it is the part of the universe that is under investigation.

(d) The surroundings refer to everything in the environment of the system of interest.

(e) The thermodynamic state of a system refers to a set of conditions that completely specifies all of the thermodynamic properties of the system.

(f) Work is the application of a force through a distance. For physical or chemical changes that occur at constant pressure, the work done by the system is $-P\Delta V$.

15-4. *Refer to Sections 1-1 and 15-1.*

An endothermic process absorbs heat energy from its surroundings; an exothermic process releases heat energy to its surroundings. If a reaction is endothermic in one direction, it is exothermic in the opposite direction. For example, the melting of 1 mole of ice water is an endothermic process requiring 6.02 kJ of heat:

$$H_2O(s) + 6.02 \text{ kJ} \rightarrow H_2O(\ell)$$

The reverse process, the freezing of 1 mole of liquid water, releasing 6.02 kJ of heat, is an exothermic process:

$$H_2O(\ell) \rightarrow H_2O(s) + 6.02 \text{ kJ}$$

15-6. *Refer to Sections 15-5 and 15-6.*

(a) ΔH, the enthalpy change or heat of reaction, is the heat change of a reaction occurring at some constant pressure.

ΔH° is the standard enthalpy change of a reaction that occurs at 1 atm pressure. Unless otherwise stated, the reaction temperature is 25°C.

(b) As stated in (a), ΔH°_{rxn} is the standard enthalpy change of a reaction occurring at 1 atm pressure.

ΔH°_f, the standard molar enthalpy of formation of a substance, is the amount of heat absorbed in a reaction in which 1 mole of the substance in a specific state is formed from its elements in their standard states.

15-8. *Refer to Sections 15-4 and 15-5.*

(1) Since the reaction is endothermic,

(a) enthalpy increases, (b) $H_{product} > H_{reactant}$ and (c) ΔH is positive.

(2) Since the reaction is exothermic,

(a) enthalpy decreases, (b) $H_{reactant} > H_{product}$ and (c) ΔH is negative.

15-10. *Refer to Section 15-5 and Example 15-4.*

Balanced equation: $CH_3OH(g) + 1\frac{1}{2}O_2(g) \rightarrow CO_2(g) + 2H_2O(\ell)$ $\qquad\qquad \Delta H = -762$ kJ/mol rxn

(a) Plan: heat evolved/mol rxn $\overset{(1)}{\Longrightarrow}$ heat evolved/mol CH_3OH $\overset{(2)}{\Longrightarrow}$ heat evolved/g CH_3OH $\overset{(3)}{\Longrightarrow}$ heat evolved

? heat evolved (kJ) $= \dfrac{762\text{ kJ}}{\text{mol rxn}} \times \dfrac{1\text{ mol rxn}}{1\text{ mol }CH_3OH} \times \dfrac{1\text{ mol }CH_3OH}{32.0\text{ g }CH_3OH} \times 90.0\text{ g }CH_3OH =$ **2140 kJ evolved**

(b) Plan: heat evolved $\overset{(1)}{\Longrightarrow}$ mol reaction $\overset{(2)}{\Longrightarrow}$ mol O_2 $\overset{(3)}{\Longrightarrow}$ g O_2

? g $O_2 = 825$ kJ $\times \dfrac{1\text{ mol rxn}}{762\text{ kJ}} \times \dfrac{1.5\text{ mol }O_2}{1\text{ mol rxn}} \times \dfrac{32.0\text{ g }O_2}{1\text{ mol }O_2} =$ **52.0 g O_2**

15-12. *Refer to Section 15-5 and Example 15-4.*

Balanced equation: $PbO(s) + C(s) \rightarrow Pb(s) + CO(g)$

Since the equation involves one mole of PbO, ΔH can be expressed in the units of kJ/mol PbO.

? heat supplied to the reaction $= \dfrac{23.8\text{ kJ}}{49.7\text{ g PbO}} \times \dfrac{223.2\text{ g PbO}}{1\text{ mol PbO}} = 107$ kJ/mol PbO

Therefore, since the heat is being added to the reaction, $\Delta H = +$**107 kJ/mol rxn**

15-14. *Refer to Sections 15-1 and 15-5.*

Consider the balanced reactions:
(1) $CH_4(g) + 2O_2(g) \rightarrow CO_2(g) + 2H_2O(\ell)$ $\qquad \Delta H_1 = (-)$
(2) $CH_4(g) + 2O_2(g) \rightarrow CO_2(g) + 2H_2O(g)$ $\qquad \Delta H_2 = (-)$

The only difference between them is that Reaction (1) involves water in the liquid phase and Reaction (2) involves water as water vapor. Since more heat is released when $H_2O(g) \rightarrow H_2O(\ell)$, as shown in the adjacent diagram, Reaction (1) is more exothermic than Reaction (2).

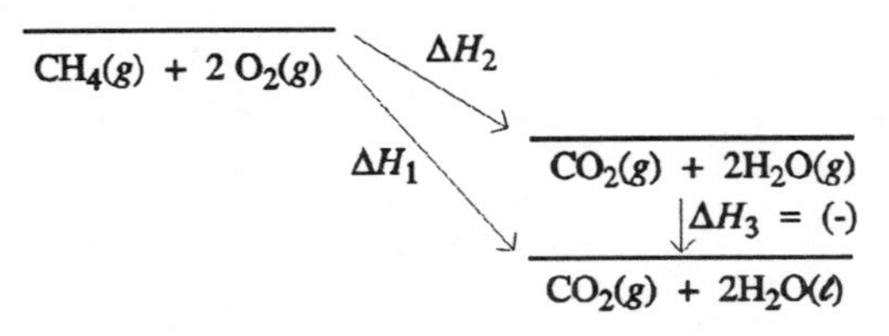

15-16. *Refer to Section 15-6.*

The thermochemical standard state of a substance is its most stable state under standard pressure (1 atm) and at some specific temperature (usually 25°C). Examples of elements in their standard states are the following:

Element	hydrogen	chlorine	bromine	helium	copper
Standard State	$H_2(g)$	$Cl_2(g)$	$Br_2(\ell)$	$He(g)$	$Cu(s)$

15-18. *Refer to Section 15-7 and Appendix K.*

The standard molar enthalpy of formation, ΔH_f°, of elements in their standard states is zero. From the tabulated values of standard molar enthalpies in Appendix K, we can identify the standard states of elements.

(a) chlorine $\quad Cl_2(g)$ (c) nitrogen $\quad N(g)$ (e) sulfur $\quad S(s,\text{rhombic})$

(b) iron $\quad Fe(s)$ (d) iodine $\quad I_2(s)$

15-20. *Refer to Section 15-7, Example 15-5 and Appendix K.*

Hint: Use Appendix K to identify an element's standard state since its ΔH_f° value is equal to zero.

(a) $Ca(s) + O_2(g) + H_2(g) \rightarrow Ca(OH)_2(s)$ (e) $\frac{1}{4} P_4(s,\text{white}) + 1\frac{1}{2} H_2(g) \rightarrow PH_3(g)$

(b) $2C(s, \text{graphite}) + 3H_2(g) \rightarrow C_2H_6(g)$ (f) $3C(s,\text{graphite}) + 4H_2(g) \rightarrow C_3H_8(g)$

(c) $Na(s) + C(s,\text{graphite}) + 1\frac{1}{2} O_2(g) \rightarrow Na_2CO_3(s)$ (g) $S(s,\text{rhombic}) \rightarrow S(g)$

(d) $Ca(s) + F_2(g) \rightarrow CaF_2(s)$

15-22. *Refer to Section 15-5.*

The balanced equation for the standard molar enthalpy of formation of $Li_2O(s)$ is: $\quad 2Li(s) + \frac{1}{2} O_2(g) \rightarrow Li_2O(s)$

$$? \text{ kJ/mol } Li_2O(s) = \frac{1029 \text{ kJ}}{24.5 \text{ g Li}} \times \frac{6.94 \text{ g Li}}{1 \text{ mol Li}} \times \frac{2 \text{ mol Li}}{1 \text{ mol } Li_2O} = 583 \text{ kJ/mol}$$

And so, ΔH_f° $_{Li_2O(s)}$ = **-583 kJ/mol** since the reaction is exothermic.

15-24. *Refer to Section 15-8 and Examples 15-6 and 15-7.*

To obtain the desired equation,

(1) divide the first equation by 2 to give 2 moles of HCl on the reactant side,
(2) multiply the second equation by 2, giving 2 moles of HF on the product side. Then,
(3) reverse the third equation, so that H_2O, H_2 and $\frac{1}{2} O_2$ are eliminated when the modified equations are added together.

$$\Delta H^\circ$$

$2HCl(g) + \frac{1}{2} O_2(g) \rightarrow H_2O(\ell) + Cl_2(g)$ -74.2 kJ/mol rxn

$H_2(g) + F_2(g) \rightarrow 2HF(\ell)$ -1200.0 kJ/mol rxn

$H_2O(\ell) \rightarrow H_2(g) + \frac{1}{2} O_2(g)$ +285.8 kJ/mol rxn

---

$2HCl(g) + F_2(g) \rightarrow 2HF(\ell) + Cl_2(g)$ **-988.4 kJ/mol rxn**

15-26. *Refer to Section 15-8 and Examples 15-6 and 15-7.*

To obtain the desired equation,

(1) multiply the first equation by 2 to give 2 moles of SO_2 on the product side, then
(2) reverse the second equation and multiply by 2, giving 2 moles of SO_3 on the reactant side.

$$\Delta H^\circ$$

$2S(s) + 2O_2(g) \rightarrow 2SO_2(g)$ -593.6 kJ/mol rxn

$2SO_3(g) \rightarrow 2S(s) + 3O_2(g)$ +791.4 kJ/mol rxn

$2SO_3(g) \rightarrow 2SO_2(g) + O_2(g)$ **+197.8 kJ/mol rxn**

15-28. *Refer to Section 15-8 and Examples 15-6 and 15-7.*

To obtain the desired hydrogenation equation,

(1) use the first equation as it is to give 2 moles of H_2 on the reactant side,
(2) use the second equation as it is to give 1 mole of C_3H_4 on the reactant side, then
(3) reverse the third equation to give 1 mole of C_3H_8 on the product side.

$$\Delta H°$$

$$2H_2(g) + O_2(g) \rightarrow 2H_2O(\ell) \qquad \text{-571.7 kJ/mol rxn}$$
$$C_3H_4(g) + 4O_2(g) \rightarrow 3\,CO_2(g) + 2H_2O(\ell) \quad \text{-1941 kJ/mol rxn}$$
$$\underline{3CO_2(g) + 4H_2O(\ell) \rightarrow C_3H_8(g) + 5O_2(g) \qquad \text{+2220 kJ/mol rxn}}$$
$$C_3H_4(g) + 2H_2(g) \rightarrow C_3H_8(g) \qquad \textbf{-293 kJ/mol rxn}$$

15-30. *Refer to Section 15-8, Example 15-8 and Appendix K.*

(a) Balanced equation: $NH_4NO_3(s) \rightarrow N_2O(g) + 2H_2O(g)$

$\Delta H°_{rxn} = [\Delta H°_f\ _{N_2O(g)} + 2\Delta H°_f\ _{H_2O(g)}] - [\Delta H°_f\ _{NH_4NO_3(s)}]$

$\quad = [(1\ mol)(82.05\ kJ/mol) + (2\ mol)(-241.8\ kJ/mol)] - [(1\ mol)(-365.6\ kJ/mol)]$

$\quad = \textbf{-36.0 kJ/mol rxn}$

(b) Balanced equation: $2FeS_2(s) + \frac{11}{2}O_2(g) \rightarrow Fe_2O_3(s) + 4SO_2(g)$

$\Delta H°_{rxn} = [\Delta H°_f\ _{Fe_2O_3(s)} + 4\Delta H°_f\ _{SO_2(g)}] - [2\Delta H°_f\ _{FeS_2(s)} + \frac{11}{2}\Delta H°_f\ _{O_2(g)}]$

$\quad = [(1\ mol)(-824.2\ kJ/mol) + (4\ mol)(-296.8\ kJ/mol)] - [(2\ mol)(-177.5\ kJ/mol) + (\frac{11}{2}\ mol)(0\ kJ/mol)]$

$\quad = \textbf{-1656 kJ/mol rxn}$

(c) Balanced equation: $SiO_2(s) + 3C(s,graphite) \rightarrow SiC(s) + 2CO(g)$

$\Delta H°_{rxn} = [\Delta H°_f\ _{SiC(s)} + 2\Delta H°_f\ _{CO(g)}] - [\Delta H°_f\ _{SiO2(s)} + 3\Delta H°_f\ _{C(s,graphite)}]$

$\quad = [(1\ mol)(-65.3\ kJ/mol) + (2\ mol)(-110.5\ kJ/mol)] - [(1\ mol)(-910.9\ kJ/mol) + (3\ mol)(0\ kJ/mol)]$

$\quad = \textbf{+624.6 kJ/mol rxn}$

15-32. *Refer to Section 15-5 and Example 15-3.*

Balanced equation: $8Al(s) + 3Fe_3O_4(s) \rightarrow 4Al_2O_3(s) + 9Fe(s)$ $\qquad \Delta H° = \text{-3347.6 kJ/mol rxn}$

Plan: (1) Determine the limiting reactant.
$\qquad$ (2) Calculate the heat released based on the limiting reactant.

(1) $? \text{ mol Al} = \dfrac{19.2 \text{ g Al}}{27.0 \text{ g/mol}} = 0.711 \text{ mol Al}$

$\quad ? \text{ mol Fe}_3O_4 = \dfrac{48.0 \text{ g Fe}_3O_4}{231.6 \text{ g/mol}} = 0.207 \text{ mol Fe}_3O_4$

$\quad \text{Required ratio} = \dfrac{8 \text{ mol Al}}{3 \text{ mol Fe}_3O_4} = 2.67 \qquad \text{Available ratio} = \dfrac{0.711 \text{ mol Al}}{0.207 \text{ mol Fe}_3O_4} = 3.43$

Available ratio > Required ratio; Fe_3O_4 is the limiting reactant.

(2) $\Delta H° = 48.0 \text{ g Fe}_3O_4 \times \dfrac{1 \text{ mol Fe}_3O_4}{231.6 \text{ g Fe}_3O_4} \times \dfrac{\text{-3347.6 kJ}}{3 \text{ mol Fe}_3O_4} = \text{-231 kJ}$

Therefore, there are **231 kJ** of heat released.

15-34. *Refer to Sections 15-5 and 15-8, and Example 15-9.*

Plan: Use Hess' Law to solve for ΔH°_{rxn} of silicon carbide. Assume that the ΔH°_{rxn} of coke is the same as that of graphite.

Balanced equation: $SiO_2(s) + 3C(s) \rightarrow SiC(s) + 2CO(g)$

$$\Delta H^{\circ}_{rxn} = [\Delta H^{\circ}_{f\ SiC(s)} + 2\Delta H^{\circ}_{f\ CO(g)}] - [\Delta H^{\circ}_{f\ SiO_2(s)} + 3\Delta H^{\circ}_{f\ C(s)}]$$

624.6 kJ = $[(1\ mol)(\Delta H^{\circ}_{f\ SiC(s)}) + (2\ mol)(-110.5\ kJ/mol)] - [(1\ mol)(-910.9\ kJ/mol) + (3\ mol)(0\ kJ/mol)]$

624.6 kJ = $(1\ mol)(\Delta H^{\circ}_{f\ SiC(s)}) + 689.9\ kJ$

$\Delta H^{\circ}_{f\ SiC(s)}$ = **-65.3 kJ/mol SiC(s)**

15-36. *Refer to Sections 15-7 and 15-9.*

(a) For a reaction occurring in the gaseous phase, the net enthalpy change, ΔH°_{rxn}, equals the sum of the bond energies in the reactants minus the sum of the bond energies in the products:

$$\Delta H^{\circ}_{rxn} = \Sigma\ B.E._{reactants} - \Sigma\ B.E._{products}$$

If the products have higher bond energy and are therefore more stable than the reactants, the reaction is exothermic. If the opposite is true, the reaction is endothermic.

(b) Analogy: 100 = 500 - 400
  100 = 900 - 800 This does not imply that 500 is the same as 900.

Bond energies involve the breaking of one mole of bonds in a gaseous substance to form gaseous atoms of the elements. The value of ΔH°_f is for the formation of the substance from its elements in their standard states. They differ in two major aspects: (1) In bond energy considerations, all the bonds are broken to give free atoms, while in ΔH°_f, some bonds may still be maintained as diatomic or polyatomic free elements (e.g., $O_2(g)$ or $P_4(s)$.) (2) The standard states of the elements are not necessarily the gaseous state. Therefore, one cannot say that $\Delta H^{\circ}_{f\ substance} = -\Sigma\ B.E._{substance}$. Moreover, the ΔH°_f equation is an exact calculation, but the bond energy equation is only an estimation of ΔH°_f.

15-38. *Refer to Section 15-9, Tables 15-2 and 15-3, and Examples 15-10 and 15-11.*

(a) Balanced equation in terms of Lewis structures of the reactants and products:

$$\begin{array}{ccccc}
\underset{H}{\overset{H}{C}}=\underset{H}{\overset{H}{C}}\ (g) & + & :\ddot{Br}-\ddot{Br}:\ (g) & \rightarrow & :\ddot{Br}-\underset{H\,H}{\overset{H\,H}{C-C}}-\ddot{Br}:\ (g)
\end{array}$$

$\Delta H^{\circ}_{rxn} = \Sigma\ B.E._{reactants} - \Sigma\ B.E._{products}$ in the gas phase

  = $[B.E._{C=C} + 4B.E._{C-H} + B.E._{Br-Br}] - [B.E._{C-C} + 4B.E._{C-H} + 2B.E._{C-Br}]$

  = $[(1\ mol)(611\ kJ/mol) + (4\ mol)(414\ kJ/mol) + (1\ mol)(192\ kJ/mol)]$
   $- [(1\ mol)(347\ kJ/mol) + (4\ mol)(414\ kJ/mol) + (2\ mol)(276\ kJ/mol)]$

  = **-96 kJ/mol rxn**

(b) Balanced equation in terms of Lewis structures of the reactants and products:

$$H-\ddot{O}-\ddot{O}-H\ (g) \rightarrow H-\ddot{O}-H\ (g) + \frac{1}{2}\ :\ddot{O}=\ddot{O}:\ (g)$$

$\Delta H^{\circ}_{rxn} = \Sigma\ B.E._{reactants} - \Sigma\ B.E._{products}$ in the gas phase

  = $[2B.E._{O-H} + B.E._{O-O}] - [2B.E._{O-H} + 1/2\ B.E._{O=O}]$

  = $[(2\ mol)(464\ kJ/mol) + (1\ mol)(138\ kJ/mol)] - [(2\ mol)(464\ kJ/mol) + (1/2\ mol)(498\ kJ/mol)]$

  = **-111 kJ/mol rxn**

15-40. *Refer to Section 15-9, Table 15-2, and Examples 15-10 and 15-11.*

Balanced equation: $CCl_2F_2(g) + F_2(g) \rightarrow CF_4(g) + Cl_2(g)$

$\Delta H^\circ_{rxn} = \Sigma \text{ B.E.}_{reactants} - \Sigma \text{ B.E.}_{products}$ in the gas phase

$\phantom{\Delta H^\circ_{rxn}} = [2\text{B.E.}_{C-Cl} + 2\text{B.E.}_{C-F} + \text{B.E.}_{F-F}] - [4\text{B.E.}_{C-F} + \text{B.E.}_{Cl-Cl}]$

$\phantom{\Delta H^\circ_{rxn}} = [(2 \text{ mol})(330 \text{ kJ/mol}) + (2 \text{ mol})(439 \text{ kJ/mol}) + (1 \text{ mol})(159 \text{ kJ/mol})]$

$\phantom{\Delta H^\circ_{rxn} = } - [(4 \text{ mol})(439 \text{ kJ/mol}) + (1 \text{ mol})(243 \text{ kJ/mol})]$

$\phantom{\Delta H^\circ_{rxn}} = \textbf{-302 kJ/mol rxn}$

15-42. *Refer to Section 15-9 and Appendix K.*

The ΔH°_{rxn} of the reaction: $OF_2(g) \rightarrow O(g) + 2F(g)$
is equal to 2 times the average O-F bond energy in OF_2, since this reaction involves the breaking of 2 O-F bonds.

$\Delta H^\circ_{rxn} = [\Delta H^\circ_{f\ O(g)} + 2\Delta H^\circ_{f\ F(g)}] - [\Delta H^\circ_{f\ OF_2(g)}]$

$\phantom{\Delta H^\circ_{rxn}} = [(1 \text{ mol})(249.2 \text{ kJ/mol}) + (2 \text{ mol})(78.99 \text{ kJ/mol})] - [(1 \text{ mol})(23 \text{ kJ/mol})]$

$\phantom{\Delta H^\circ_{rxn}} = 384 \text{ kJ/mol rxn}$

Therefore, the average bond energy of a O-F bond in $OF_2(g)$ is (384/2) kJ or **192 kJ**.

15-44. *Refer to Section 15-9 and Appendix K.*

The ΔH°_{rxn} of this reaction: $SF_6(g) \rightarrow S(g) + 6F(g)$
is equal to 6 times the average S-F bond energy in $SF_6(g)$ since this reaction involves the breaking of 6 S-F bonds.

$\Delta H^\circ_{rxn} = [\Delta H^\circ_{f\ S(g)} + 6\Delta H^\circ_{f\ F(g)}] - [\Delta H^\circ_{f\ SF_6(g)}]$

$\phantom{\Delta H^\circ_{rxn}} = [(1 \text{ mol})(278.8 \text{ kJ/mol}) + (6 \text{ mol})(78.99 \text{ kJ/mol})] - [(1 \text{ mol})(-1209 \text{ kJ/mol})]$

$\phantom{\Delta H^\circ_{rxn}} = 1962 \text{ kJ/mol rxn}$

Therefore, the average bond energy of an S-F bond in $SF_6(g)$ is (1962/6) kJ or **327.0 kJ**.

15-46. *Refer to Section 15-9, Table 15-2 and Example 15-11.*

$\Delta H^\circ_{rxn} = [5\text{B.E.}_{C-H} + \text{B.E.}_{C-C} + \text{B.E.}_{C-N} + 2\text{B.E.}_{N-H}] - [4\text{B.E.}_{C-H} + \text{B.E.}_{C=C} + 3\text{B.E.}_{N-H}]$

Substituting,

$54.68 \text{ kJ} = [(5 \text{ mol})(414 \text{ kJ/mol}) + (1 \text{ mol})(347 \text{ kJ/mol}) + (1 \text{ mol})(\text{B.E.}_{C-N}) + (2 \text{ mol})(389 \text{ kJ/mol})]$

$\phantom{54.68 \text{ kJ} = } - [(4 \text{ mol})(414 \text{ kJ/mol}) + (1 \text{ mol})(611 \text{ kJ/mol}) + (3 \text{ mol})(389 \text{ kJ/mol})]$

$54.68 \text{ kJ} = (1 \text{ mol})(\text{B.E.}_{C-N}) - 239 \text{ kJ}$

$\text{B.E.}_{C-N} = \textbf{294 kJ/mol}$

Table 15-2 gives the bond energy for an average C-N bond as 293 kJ/mol. The two values agree very well.

15-48. *Refer to Section 15-10 and Figure 15-7.*

A bomb calorimeter is a mechanical device used to measure the heat transfer between a system and its surroundings at constant volume. Generally, the systems are chemical reactions. No work is done in a bomb calorimeter since $\Delta V = 0$. Therefore,

$$\Delta E = q_v + w = q_v - P\Delta V = q_v - 0 = q_v$$

and we see that bomb calorimeters measure the change in internal energy, ΔE, for a reaction directly.

15-50. *Refer to Sections 1-13 and 15-4, Example 15-1 and Exercise 1-64 Solution.*

Plan: (1) Determine the heat gained by the calorimeter.
(2) Find the heat capacity of the calorimeter (calorimeter constant).

(1) | heat lost |$_{iron}$ = | heat gained |$_{water}$ + | heat gained |$_{calorimeter}$

| specific heat × mass × Δt |$_{iron}$ = | specific heat × mass × Δt |$_{water}$ + | heat gained |$_{calorimeter}$

$(0.444 \text{ J/g·°C})(93.3 \text{ g})(65.58°C - 19.68°C) = (4.184 \text{ J/g·°C})(75.0 \text{ g})(19.68°C - 16.95°C)$

$+ \text{ | heat gained |}_{calorimeter}$

$1.90 \times 10^3 \text{ J} = 8.57 \times 10^2 \text{ J} + \text{ | heat gained |}_{calorimeter}$

Therefore, | heat gained |$_{calorimeter}$ = $1.90 \times 10^3 \text{ J} - 857 \text{ J} = 1.04 \times 10^3 \text{ J}$

(2) heat capacity of calorimeter (J/°C) = $\dfrac{\text{| heat gained |}_{calorimeter}}{\Delta T} = \dfrac{1.04 \times 10^3 \text{ J}}{19.68°C - 16.95°C}$ = **381 J/°C**

15-52. *Refer to Sections 15-4 and 15-5, Examples 15-2 and 15-3, and Exercise 1-64 Solution.*

Balanced equation: $Pb(NO_3)_2(aq) + 2NaI(aq) \rightarrow PbI_2(s) + 2NaNO_3(aq)$

(a) | heat released | = | heat gained |$_{soln}$ + | heat gained |$_{calorimeter}$

= | specific heat × mass × Δt |$_{soln}$ + | heat capacity × Δt |$_{calorimeter}$

= $(4.18 \text{ J/g·°C})(200 \text{ g})(24.2°C - 22.6°C) + (472 \text{ J/°C})(24.2°C - 22.6°C)$

= $1.3 \times 10^3 \text{ J} + 7.6 \times 10^2 \text{ J}$

= **2.1 × 10³ J**

(b) Plan: (1) Determine if the reaction has a limiting reagent and how many moles of $Pb(NO_3)_2$ are actually consumed in the reaction.
(2) Find ΔH_{rxn} per mole of $Pb(NO_3)_2$.

(1) ? mol $Pb(NO_3)_2$ = $\dfrac{6.62 \text{ g Pb(NO}_3)_2}{331 \text{ g/mol}}$ = 0.0200 mol $Pb(NO_3)_2$

? mol NaI = $\dfrac{6.00 \text{ g NaI}}{150. \text{ g/mol}}$ = 0.0400 mol NaI

Required ratio = $\dfrac{1 \text{ mol Pb(NO}_3)_2}{2 \text{ mol NaI}} = \dfrac{1}{2}$ Available ratio = $\dfrac{0.0200 \text{ mol Pb(NO}_3)_2}{0.0400 \text{ mol NaI}} = \dfrac{1}{2}$

Therefore, we have stoichiometric amounts of both reactants and 0.0200 mol of $Pb(NO_3)_2$ are consumed in the reaction.

(2) $\Delta H_{rxn}\left(\dfrac{\text{J}}{\text{mol Pb(NO}_3)_2}\right) = -\dfrac{2.1 \times 10^3 \text{ J}}{0.0200 \text{ mol Pb(NO}_3)_2}$ = **-1.0 × 10⁵ J/mol Pb(NO₃)₂**

ΔH_{rxn} is a negative number because heat was released during the reaction. We know this because the temperature of the solution increased as the reaction proceeded.

15-54. *Refer to Sections 15-4 and 15-10, and Examples 15-2 and 15-13.*

(a) $2C_6H_6(\ell) + 15\,O_2(g) \rightarrow 12CO_2(g) + 6H_2O(\ell)$

(b) | heat released | = | heat gained | $_{water}$ + | heat gained | $_{calorimeter}$

$\qquad\qquad$ = | specific heat × mass × Δt | $_{water}$ + | heat capacity × Δt | $_{calorimeter}$

$\qquad\qquad$ = (4.18 J/g·°C)(945 g)(32.692°C - 23.640°C) + (891 J/°C)(32.692°C - 23.640°C)

$\qquad\qquad$ = 3.58×10^4 J + 8.07×10^3 J

$\qquad\qquad$ = 4.38×10^4 J or 43.8 kJ

Since heat is released in this reaction (the temperature of the water increased), ΔE is a negative quantity.

$$\Delta E = -\frac{43.8\ kJ}{1.048\ g\ C_6H_6(\ell)} = \textbf{-41.8 kJ/g } C_6H_6(\ell)$$

$$\Delta E = -\frac{43.8\ kJ}{1.048\ g\ C_6H_6(\ell)} \times \frac{78.11\ g}{1\ mol} = \textbf{-3260 kJ/mol } C_6H_6(\ell) \quad \text{(to 3 significant figures)}$$

15-56. *Refer to Sections 15-10.*

(a) When heat is added to a system, q is "+." When heat is removed from a system, q is "−."

(b) When work is done on a system, w is "+." When work is done by a system, w is "−."

15-58. *Refer to Section 15-10.*

We know: $\Delta E = q + w$ $\quad$ where ΔE = change in internal energy
$\qquad\qquad\qquad\qquad\qquad\qquad q$ = heat absorbed by the system from the surroundings
$\qquad\qquad\qquad\qquad\qquad\qquad w$ = work done on the system by the surroundings

(a) $q > 0$ and $w > 0$: $\quad \Delta E = +$ $\qquad$ (internal energy increases)

(b) $q = w = 0$: $\qquad\quad \Delta E = 0$ $\qquad$ (no change in internal energy)

(c) $q < 0$ and $w > 0$: $\quad \Delta E = +$ or $-$, depending on the relative values of q and w

---

15-60. *Refer to Section 15-10.*

For the system: q = -212 J, $w_{electrical}$ = +73 J and w_{PV} = -227 J

$\Delta E = q + w_{total} = q + (w_{electrical} + w_{PV}) = $ -212 J + [+73 J + (-227 J)] = **-366 J**

15-62. *Refer to Section 15-10 and Example 15-12.*

Plan: $\quad$ Evaluate $\Delta n_{gas} = n_{gaseous\ products} - n_{gaseous\ reactants}$. The sign of the work term is opposite that of Δn_{gas} since $w = -P\Delta V = -\Delta nRT$ at constant P and T.

(a) $2SO_2(g) + O_2(g) \rightarrow 2SO_3(g)$

Δn_{gas} = 2 mol - 3 mol = -1 mol. $\quad$ Therefore, $w > 0$ and the work is done by the surroundings on the system.

(b) $CaCO_3(s) \rightarrow CaO(s) + CO_2(g)$

$\Delta n_{gas} = 1$ mol - 0 mol = +1 mol. Therefore, work < 0, and the work is done by the system on the surroundings.

(c) $CO_2(g) + H_2O(\ell) + CaCO_3(s) \rightarrow Ca^{2+}(aq) + 2HCO_3^-(aq)$

$\Delta n_{gas} = 0$ mol - 1 mol = -1 mol. Therefore, work > 0 and work is done on the system by the surroundings.

15-64. *Refer to Sections 15-10 and 15-11.*

(a) The balanced equation for the oxidation of 1 mole of HCl: $HCl(g) + 1/4\ O_2(g) \rightarrow 1/2\ Cl_2(g) + 1/2\ H_2O(g)$

$$\begin{aligned} \text{work} = -P\Delta V = -\Delta n_{gas}RT &= -(n_{gaseous\ products} - n_{gaseous\ reactants})RT \\ &= -(1\ mol - 5/4\ mol)(8.314\ J/mol \cdot K)(200°C + 273°) \\ &= +983\ J \end{aligned}$$

Work is a positive number, therefore, work is done on the system by the surroundings. As the system "shrinks" from 5/4 mole of gas to 1 mole of gas, work is done on the system by the surroundings to decrease the volume. (Recall that $V \propto n$ at constant T and P.)

(b) The balanced reaction for the decomposition of 1 mole of NO: $NO(g) \rightarrow 1/2\ N_2(g) + 1/2\ O_2(g)$

$$\text{work} = -P\Delta V = -\Delta n_{gas}RT = -(1\ mol - 1\ mol)RT = \mathbf{0\ J}$$

There is no work done since the number of moles of gas, and hence the volume of the system, remains constant.

15-66. *Refer to Section 15-14 and the Key Terms for Chapter 15.*

Entropy is a thermodynamic state function that measures the degree of disorder or randomness of a system. The greater the disorder of a system, the higher (more positive) is its entropy.

15-68. *Refer to Section 15-14.*

The Third Law of Thermodynamics states that the entropy of a pure, perfect crystalline substance is zero at 0 K. This means that all substances have some disorder except when the substance is a pure, perfect, motionless crystal at absolute zero Kelvin. This also implies that the entropy of a substance can be expressed on an absolute basis.

15-70. *Refer to Section 15-14.*

When the volume occupied by one mole of Ar at 0°C is decreased by one-half, there is a *decrease* in disorder or randomness, as signified by the negative sign of the entropy change, -5.76 J/(mol rxn)·K. In the smaller volume there are less positions available for the argon molecules to occupy and so, there is more order in the smaller volume.

15-72. *Refer to Sections 15-13 and 15-14, and Table 15-4.*

(a) decrease in entropy   The pennies are placed in a definite pattern (separated and heads up) and are more ordered on the table.

(b) increase in entropy   The pennies become disordered when put back into the bag.

(c) decrease in entropy   The solid phase is always more ordered than the liquid phase of a substance.

(d) increase in entropy   The gas phase is always more disordered than the liquid phase of a substance.

(e) increase in entropy   The reaction is producing 2 moles of gas from 1 mole of gas. A system with 2 moles of gas is more disordered than a system with only 1 mole of gas.

(f) decrease in entropy   The reaction is the opposite of (e).

15-74. *Refer to Sections 15-13 and 15-14.*

Consider the boiling of a pure liquid at constant pressure.

(a) $\Delta S_{system} > 0$ (b) $\Delta H_{system} > 0$ (c) $\Delta T_{system} = 0$

15-76. *Refer to Section 15-13 and Example 15-15.*

(a) Balanced equation: $4HCl(g) + O_2(g) \rightarrow 2Cl_2(g) + 2H_2O(g)$

$\Delta S^\circ_{rxn} = [2S^\circ_{Cl_2(g)} + 2S^\circ_{H_2O(g)}] - [4S^\circ_{HCl(g)} + S^\circ_{O_2(g)}]$
$= [(2\ mol)(223.0\ J/mol\cdot K) + (2\ mol)(188.7\ J/mol\cdot K)] - [(4\ mol)(+186.8\ J/mol\cdot K) + (1\ mol)(+205.0\ J/mol\cdot K)]$
$= $ **-128.8 J/(mol rxn)·K**

The reaction is producing 4 moles of gas from 5 moles of gas. The system is becoming more ordered as the number of moles of gas decreases; entropy is decreasing and the change in entropy is expected to be negative.

(b) Balanced equation: $PCl_3(g) + Cl_2(g) \rightarrow PCl_5(g)$

$\Delta S^\circ_{rxn} = [S^\circ_{PCl_5(g)}] - [S^\circ_{PCl_3(g)} + S^\circ_{Cl_2(g)}]$
$= [(1\ mol)(353\ J/mol\cdot K)] - [(1\ mol)(311.7\ J/mol\cdot K) + (1\ mol)(223.0\ J/mol\cdot K)]$
$= $ **-182 J/(mol rxn)·K**

The reaction is producing 1 mole of gas from 2 moles of gas. For the same reasoning as shown in (a), the entropy is decreasing and the change in entropy is expected to be negative.

(c) Balanced equation: $2NO(g) \rightarrow N_2(g) + O_2(g)$

$\Delta S^\circ_{rxn} = [S^\circ_{N_2(g)} + S^\circ_{O_2(g)}] - [2S^\circ_{NO(g)}]$
$= [(1\ mol)(191.5\ J/mol\cdot K) + (1\ mol)(205.0\ J/mol\cdot K)] - [(2\ mol)(210.7\ J/mol\cdot K)]$
$= $ **-24.9 J/(mol rxn)·K**

The number of moles of gas in the system is not changing, i.e., $\Delta n_{gas} = 0$; the entropy change is expected to be small. The value is negative indicating a decrease in randomness because homonuclear diatomic molecules such as N_2 and O_2 are more ordered than heteronuclear diatomic molecules such as NO.

15-78. *Refer to Sections 15-15 and 15-16, and Table 15-7.*

(a) always spontaneous: (iii) $\Delta H < 0,\ \Delta S > 0$

(b) always nonspontaneous: (ii) $\Delta H > 0,\ \Delta S < 0$

(c) spontaneous or nonspontaneous, depending on T and the magnitudes of ΔH and ΔS: (i) $\Delta H > 0,\ \Delta S > 0$
                                                                                          (iv) $\Delta H < 0,\ \Delta S < 0$

15-80. *Refer to Section 15-15 and Example 15-16.*

Plan: Use the Gibbs-Helmholtz equation to calculate ΔG_{rxn} from ΔH_{rxn} and ΔS_{rxn}. Remember to change the units of ΔS to kJ/K.

$$\Delta G^\circ_{rxn} = \Delta H^\circ_{rxn} - T\Delta S^\circ_{rxn}$$
$$= -57.20 \text{ kJ} - (25.00°C + 273.15°)(-0.17583 \text{ kJ/K})$$
$$= \textbf{-4.78 kJ}$$

Since $\Delta G^\circ_{rxn} < 0$, the reaction is spontaneous. Its exothermicity ($\Delta H < 0$) is the driving force for spontaneity.

15-82. *Refer to Sections 15-8 and 15-15.*

Since ΔG° is a state function like ΔH°, we can use Hess' Law type of manipulations to determine the ΔG°_{rxn}.

The balanced equation representing the ΔG° of formation of HBr(g) is: $\frac{1}{2}H_2(g) + \frac{1}{2}Br_2(\ell) \rightarrow HBr(g)$

	ΔG°
$\frac{1}{2}Br_2(\ell) \rightarrow \frac{1}{2}Br_2(g)$	1.555 kJ
$H(g) + Br(g) \rightarrow HBr(g)$	-339.09 kJ
$\frac{1}{2}Br_2(g) \rightarrow Br(g)$	82.396 kJ
$\frac{1}{2}H_2(g) \rightarrow H(g)$	203.247 kJ
$\frac{1}{2}H_2(g) + \frac{1}{2}Br_2(\ell) \rightarrow HBr(g)$	**-51.89 kJ/mol rxn**

15-84. *Refer to Section 15-15, Example 15-16 and Appendix K.*

Plan: Calculate ΔH°_{rxn} and ΔS°_{rxn}, then use the Gibbs-Helmholtz equation to determine ΔG°_{rxn}.

(a) Balanced equation: $3NO_2(g) + H_2O(\ell) \rightarrow 2HNO_3(\ell) + NO(g)$

$\Delta H^\circ_{rxn} = [2\Delta H^\circ_{f\ HNO_3(\ell)} + \Delta H^\circ_{f\ NO(g)}] - [3\Delta H^\circ_{f\ NO_2(g)} + \Delta H^\circ_{f\ H_2O(\ell)}]$
$= [(2 \text{ mol})(-174.1 \text{ kJ/mol}) + (1 \text{ mol})(90.25 \text{ kJ/mol})] - [(3 \text{ mol})(33.2 \text{ kJ/mol}) + (1 \text{ mol})(-285.8 \text{ kJ/mol})]$
$= \textbf{-71.75 kJ/mol rxn}$

$\Delta S^\circ_{rxn} = [2S^\circ_{HNO_3(\ell)} + S^\circ_{NO(g)}] - [3S^\circ_{NO_2(g)} + S^\circ_{H_2O(\ell)}]$
$= [(2 \text{ mol})(155.6 \text{ J/mol·K}) + (1 \text{ mol})(210.7 \text{ J/mol·K})]$
$\qquad - [(3 \text{ mol})(240.0 \text{ J/mol·K}) + (1 \text{ mol})(69.91 \text{ J/mol·K})]$
$= \textbf{-268.0 J/(mol rxn)·K}$

$\Delta G^\circ_{rxn} = \Delta H^\circ_{rxn} - T\Delta S^\circ_{rxn} = -71.75 \text{ kJ} - (298.15 \text{ K})(-0.268 \text{ kJ/K}) = \textbf{8.15 kJ/mol rxn}$

(b) Balanced equation: $SnO_2(s) + 2CO(g) \rightarrow 2CO_2(g) + Sn(s,white)$

$\Delta H^\circ_{rxn} = [2\Delta H^\circ_{f\ CO_2(g)} + \Delta H^\circ_{f\ Sn(s)}] - [\Delta H^\circ_{f\ SnO_2(s)} + 2\Delta H^\circ_{f\ CO(g)}]$
$= [(2 \text{ mol})(-393.5 \text{ J/mol}) + (1 \text{ mol})(0 \text{ kJ/mol})] - [(1 \text{ mol})(-580.7 \text{ kJ/mol}) + (2 \text{ mol})(-110.5 \text{ kJ/mol})]$
$= \textbf{14.7 kJ/mol rxn}$

$\Delta S^\circ_{rxn} = [2S^\circ_{CO_2(g)} + S^\circ_{Sn(s)}] - [S^\circ_{SnO_2(s)} + 2S^\circ_{CO(g)}]$
$= [(2 \text{ mol})(213.6 \text{ J/mol·K}) + (1 \text{ mol})(51.55 \text{ J/mol·K})]$
$\qquad - [(1 \text{ mol})(52.3 \text{ J/mol·K}) + (2 \text{ mol})(197.6 \text{ J/mol·K})]$
$= \textbf{31.2 J/(mol rxn)·K}$

$\Delta G^\circ_{rxn} = \Delta H^\circ_{rxn} - T\Delta S^\circ_{rxn} = 14.7 \text{ kJ} - (298.15 \text{ K})(0.0312 \text{ kJ/K}) = \textbf{5.4 kJ/mol rxn}$

(c) Balanced equation: $2Na(s) + 2H_2O(\ell) \rightarrow 2NaOH(aq) + H_2(g)$

$$\Delta H^\circ_{rxn} = [2\Delta H^\circ_{f\ NaOH(aq)} + \Delta H^\circ_{f\ H_2(g)}] - [2\Delta H^\circ_{f\ Na(s)} + 2\Delta H^\circ_{f\ H_2O(\ell)}]$$
$$= [(2\ mol)(-469.6\ kJ/mol) + (1\ mol)(0\ kJ/mol)] - [(2\ mol)(0\ kJ/mol) + (2\ mol)(-285.8\ kJ/mol)]$$
$$= \textbf{-367.6 kJ/mol rxn}$$

$$\Delta S^\circ_{rxn} = [2S^\circ_{NaOH(s)} + S^\circ_{H_2(g)}] - [2S^\circ_{Na(s)} + 2S^\circ_{H_2O(\ell)}]$$
$$= [(2\ mol)(49.8\ J/mol\cdot K) + (1\ mol)(130.6\ J/mol\cdot K)]$$
$$- [(2\ mol)(51.0\ J/mol\cdot K) + (2\ mol)(69.91\ J/mol\cdot K)]$$
$$= \textbf{-11.62 J/(mol rxn)}\cdot K$$

$$\Delta G^\circ_{rxn} = \Delta H^\circ_{rxn} - T\Delta S^\circ_{rxn} = -367.6\ kJ - (298.15\ K)(-0.01162\ kJ/K) = \textbf{-364.1 kJ/mol rxn}$$

15-86. *Refer to Sections 15-15 and 15-16.*

Recall: Gibbs-Helmholtz equation: $\Delta G = \Delta H - T\Delta S$

(a) false   An exothermic reaction ($\Delta H < 0$) will be spontaneous ($\Delta G < 0$) only if either ΔS is positive, or, in the event ΔS is negative, the absolute value of $T\Delta S$ is smaller than that of ΔH.

(b) true   from the Gibbs-Helmholtz equation; the $T\Delta S$ term has a negative sign in front

(c) false   A reaction with $\Delta S_{sys} > 0$ will be spontaneous ($\Delta G < 0$) only if either ΔH is negative, or, in the event ΔH is positive, its absolute value is smaller than that of $T\Delta S$.

15-88. *Refer to Section 15-16.*

Plan: (1) Use the Gibbs-Helmholtz equation, $\Delta G^\circ = \Delta H^\circ - T\Delta S^\circ$, to calculate ΔS° at 25°C.
        (2) Assuming that ΔH° and ΔS° do not change with temperature, calculate ΔG° at 95°C.

(1) $\Delta S^\circ = \dfrac{\Delta H^\circ - \Delta G^\circ}{T} = \dfrac{-91.34\ kJ/mol\ rxn - (-103.75\ kJ/mol\ rxn)}{(25.00°C + 273.15°)} = +0.04162\ kJ/(mol\ rxn)\cdot K$

(2) ΔG° at 95°C $= \Delta H^\circ - T\Delta S^\circ = -91.34\ kJ/mol\ rxn - (95.00°C + 273.15°)(0.04162\ kJ/(mol\ rxn)\cdot K)$
$$= \textbf{-106.66 kJ}$$

ΔG° becomes more negative in going from 25°C to 95°C. This is in agreement with the prediction that since ΔS° is positive, an increase in temperature will make ΔG° more negative.

15-90. *Refer to Section 15-16 and Appendix K.*

Balanced equation: $2H_2O_2(\ell) \rightarrow 2H_2O(\ell) + O_2(g)$

(a) $\Delta H^\circ_{rxn} = [2\Delta H^\circ_{f\ H_2O(\ell)} + \Delta H^\circ_{f\ O_2(g)}] - [2\Delta H^\circ_{f\ H_2O_2(\ell)}]$
$$= [(2\ mol)(-285.8\ kJ/mol) + (1\ mol)(0\ kJ/mol)] - [(2\ mol)(-187.8\ kJ/mol)]$$
$$= \textbf{-196.0 kJ/mol rxn}$$

$\Delta G^\circ_{rxn} = [2\Delta G^\circ_{f\ H_2O(\ell)} + \Delta G^\circ_{f\ O_2(g)}] - [2\Delta G^\circ_{f\ H_2O_2(\ell)}]$
$$= [(2\ mol)(-237.2\ kJ/mol) + (1\ mol)(0\ kJ/mol)] - [(2\ mol)(-120.4\ kJ/mol)]$$
$$= \textbf{-233.6 kJ/mol rxn}$$

$\Delta S^\circ_{rxn} = [2S^\circ_{H_2O(\ell)} + S^\circ_{O_2(g)}] - [2S^\circ_{H_2O_2(\ell)}]$
$$= [(2\ mol)(69.91\ J/mol\cdot K) + (1\ mol)(205.0\ J/mol\cdot K)] - [(2\ mol)(109.6\ J/mol\cdot K)]$$
$$= \textbf{+125.6 J/(mol rxn)}\cdot K$$

(b) Hydrogen peroxide, $H_2O_2(\ell)$, will be stable if $\Delta G° > 0$ for the above balanced reaction at some temperature, i.e., if the above reaction is non-spontaneous. However, $\Delta H°_{rxn} < 0$ and $\Delta S°_{rxn} > 0$ for the decomposition of $H_2O_2(\ell)$ and the reaction is spontaneous ($\Delta G° < 0$) for all temperatures. Hence, there is no temperature at which $H_2O_2(\ell)$ is stable at 1 atm.

15-92. *Refer to Section 15-16, Appendix K and Example 15-17.*

(a) The process is: $H_2O(\ell) \rightarrow H_2O(g)$

$\Delta H°_{rxn} = \Delta H°_{f\ H_2O(g)} - \Delta H°_{f\ H_2O(\ell)} = (1\ mol)(-241.8\ kJ/mol) - (1\ mol)(-285.8\ kJ/mol) = 44.0\ kJ$

$\Delta S°_{rxn} = S°_{H_2O(g)} - S°_{H_2O(\ell)} = (1\ mol)(188.7\ J/mol \cdot K) - (1\ mol)(69.91\ J/mol \cdot K) = 118.8\ J/K$

$T_{eq} = \dfrac{\Delta H_{rxn}}{\Delta S_{rxn}} = \dfrac{44.0\ kJ}{0.1188\ kJ/K} = 370\ K\ or\ \mathbf{97°C}$

(b) The known boiling point of water is, of course, 100°C. The discrepancy is because we assumed that the standard values of enthalpy of formation and entropy in Appendix K are independent of temperature. However, these tabulated values were determined at 25°C; we are using them to solve a problem at 100°C. Nevertheless, this assumption allows us to estimate the boiling point of water with reasonable accuracy.

15-94. *Refer to Section 15-16 and Example 15-15.*

Balanced equation: $C(s,\ diamond) \rightarrow C(s,\ graphite)$

(a) $\Delta G°_{rxn} = \Delta G°_{f\ C(s,\ graphite)} - \Delta G°_{f\ C(s,\ diamond)} = (1\ mol)(0\ kJ/mol) - (1\ mol)(2.900\ kJ/mol) = -2.900\ kJ$

Since $\Delta G°_{rxn} < 0$, the reaction is spontaneous at standard conditions.

(b) From common experience, we know that diamonds do not readily change to graphite. Thermodynamics tells us if a reaction will occur but it says nothing about the *rate* at which a reaction will occur. The rate at which diamonds change to graphite is very, very slow.

(c) $\Delta H°_{rxn} = \Delta H°_{f\ C(s,\ graphite)} - \Delta H°_{f\ C(s,\ diamond)} = (1\ mol)(0\ kJ/mol) - (1\ mol)(1.897\ kJ/mol) = -1.897\ kJ$

$\Delta S°_{rxn} = S°_{C(s,\ graphite)} - S°_{C(s,\ diamond)} = (1\ mol)(5.740\ J/mol \cdot K) - (1\ mol)(2.38\ J/mol \cdot K) = 3.36\ J/K$

Since ΔH_{rxn} is negative and ΔS_{rxn} is positive, the reaction is spontaneous at all temperatures, i.e., there is no temperature at which diamond and graphite are in equilibrium.

(d) Diamonds form from graphite under conditions of high pressure. At higher pressures, carbon as graphite transforms itself into a higher density crystalline form in which the carbon atoms are closer together, called diamond.

15-96. *Refer to Section 15-10.*

Calculation: Activity Time Equivalent (min) $= \dfrac{\text{Food Fuel Value (kcal)}}{\text{Energy Output (kcal/min)}}$

Food	Fuel Value (kcal)	Activity Time Equivalent (min)				
		Sitting (1.7 kcal/min)	Walking (5.5 kcal/min)	Cycling (10 kcal/min)	Swimming (8.4 kcal/min)	Running (19 kcal/min)
Apple	100	59	18	10	12	5.3
Cola	105	62	19	11	13	5.5
Malted milk	500	294	91	50	60	26
Pasta	195	115	35	20	23	10
Hamburger	350	206	64	35	42	18
Steak	1000	588	182	100	119	53

15-98. *Refer to Sections 15-7 and 15-8, Example 15-9 and Appendix K.*

Plan: Use Hess' Law and solve for ΔH_f° of the organic compound.

(a) Balanced equation: $C_6H_{12}(\ell) + 9O_2(g) \rightarrow 6CO_2(g) + 6H_2O(\ell)$

$\Delta H^\circ_{\text{combustion}} = [6\Delta H_f^\circ {}_{CO_2(g)} + 6\Delta H_f^\circ {}_{H_2O(\ell)}] - [\Delta H_f^\circ {}_{C_6H_{12}(\ell)} + 9\Delta H_f^\circ {}_{O_2(g)}]$

$-3920 \text{ kJ} = [(6 \text{ mol})(-393.5 \text{ kJ/mol}) + (6 \text{ mol})(-285.8 \text{ kJ/mol})]$
$\qquad\qquad - [(1 \text{ mol})\Delta H_f^\circ {}_{C_6H_{12}(\ell)} + (9 \text{ mol})(0 \text{ kJ/mol})]$

$-3920 \text{ kJ} = -4075.8 \text{ kJ} - (1 \text{ mol})\Delta H_f^\circ {}_{C_6H_{12}(\ell)}$

$\Delta H_f^\circ {}_{C_6H_{12}(\ell)} = \textbf{-156 kJ/mol } \mathbf{C_6H_{12}(\ell)}$

(b) Balanced equation: $C_6H_5OH(s) + 7O_2(g) \rightarrow 6CO_2(g) + 3H_2O(\ell)$

$\Delta H^\circ_{\text{combustion}} = [6\Delta H_f^\circ {}_{CO_2(g)} + 3\Delta H_f^\circ {}_{H_2O(\ell)}] - [\Delta H_f^\circ {}_{C_6H_5OH(s)} + 7\Delta H_f^\circ {}_{O_2(g)}]$

$-3053 \text{ kJ} = [(6 \text{ mol})(-393.5 \text{ kJ/mol}) + (3 \text{ mol})(-285.8 \text{ kJ/mol})]$
$\qquad\qquad - [(1 \text{ mol})\Delta H_f^\circ {}_{C_6H_5OH(s)} + (7 \text{ mol})(0 \text{ kJ/mol})]$

$-3053 \text{ kJ} = -3218.4 \text{ kJ} - (1 \text{ mol})\Delta H_f^\circ {}_{C_6H_5OH(s)}$

$\Delta H_f^\circ {}_{C_6H_5OH(s)} = \textbf{-165 kJ/mol } \mathbf{C_6H_5OH(s)}$

15-100. *Refer to Sections 1-13 and 15-4.*

(a)
$$| \text{heat lost} |_{\text{metal}} = | \text{heat gained} |_{\text{water}}$$
$$| \text{specific heat} \times \text{mass} \times \Delta t |_{\text{metal}} = | \text{specific heat} \times \text{mass} \times \Delta t |_{\text{water}}$$
$$(\text{specific heat of metal})(32.6 \text{ g})(99.83°C - 24.41°C) = (4.184 \text{ J/g·°C})(100.0 \text{ g})(24.41°C - 23.62°C)$$
$$(\text{specific heat of metal})(2.46 \times 10^3) = 330$$
$$\text{specific heat of metal} = \textbf{0.13 J/g·°C}$$

Therefore, according to this calculation, the metal is **tungsten, W** (specific heat = 0.135 J/g·°C).

(b)
$$| \text{heat lost} |_{\text{metal}} = | \text{heat gained} |_{\text{water}} + | \text{heat gained} |_{\text{calorimeter}}$$
$$| \text{specific heat} \times \text{mass} \times \Delta t |_{\text{metal}} = | \text{specific heat} \times \text{mass} \times \Delta t |_{\text{water}}$$
$$+ | \text{heat capacity} \times \Delta t |_{\text{calorimeter}}$$
$$(\text{specific heat of metal})(32.6 \text{ g})(99.83°C - 24.41°C) = (4.184 \text{ J/g·°C})(100.0 \text{ g})(24.41°C - 23.62°C)$$
$$+ (410 \text{ J/°C})(24.41°C - 23.62°C)$$
$$(\text{specific heat of metal})(2.46 \times 10^3) = 330 + 320$$
$$\text{specific heat of metal} = 0.26 \text{ J/g·°C}$$

Yes, the identification of the metal was different. When the heat capacity of the calorimeter is taken into account, the specific heat of the metal is 0.26 J/g·°C and the metal is identified as molybdenum, Mo (specific heat = 0.250 J/g·°C).

15-102. *Refer to Section 15-5.*

The vaporization process is: ethanol(ℓ) → ethanol(g)

$\Delta E = q + w$ where ΔE = change in internal energy
q = heat absorbed by the system
w = work done on the system

(1) The heat absorbed by the system,

$q = \Delta H_{vap} \times$ g ethanol = +854 J/g $\times$ 10.0 g = **+8540 J**

(2) The work done on the system in going from a liquid to a gas,

$w = -P\Delta V = -P(V_{gas} - V_{liquid})$

where $V_{gas} = \dfrac{nRT}{P} = \dfrac{(10.0 \text{ g}/46.0 \text{ g/mol})(0.0821 \text{ L·atm/mol·K})(78.0°C + 273°)}{1.00 \text{ atm}} = 6.26 \text{ L}$

$V_{liquid} = 10.0$ g ethanol $\times \dfrac{1.00 \text{ mL ethanol}}{0.789 \text{ g ethanol}} = 12.7$ mL or 0.0127 L

Therefore,

$w = -P\Delta V = -$ (1 atm)(6.26 L - 0.01 L) = -6.25 L·atm (the negative value means the system is doing the work)

To find a factor to convert L·atm to J, we can equate two values of the molar gas constant, R

0.0821 L·atm/mol·K = 8.314 J/mol·K
1 L·atm = 101 J

And so, $w = -6.25$ L·atm $\times \dfrac{101 \text{ J}}{1 \text{ L·atm}} = $ **-631 J**

(3) Finally, $\Delta E = q + w = 8540 \text{ J} + (-631 \text{ J}) = $ **7910 J**

15-104. *Refer to Section 15-4.*

(a) heat gained by calorimeter = 0.01520 g $C_{10}H_8 \times \dfrac{1 \text{ mol } C_{10}H_8}{128.16 \text{ g } C_{10}H_8} \times \dfrac{5156.8 \text{ kJ}}{1 \text{ mol } C_{10}H_8} = 0.6116 \text{ kJ}$

We know: | heat gained by calorimeter | = | heat capacity $\times \Delta t$ |

Therefore, heat capacity = $\dfrac{|\text{ heat gained by calorimeter }|}{|\Delta t|} = \dfrac{0.6116 \text{ kJ}}{0.212°C} = $ **2.88 kJ/°C**

(b) | heat released in the reaction | = 0.1040 g $C_8H_{18} \times \dfrac{1 \text{ mol } C_8H_{18}}{114.22 \text{ g/mol}} \times \dfrac{5450 \text{ kJ}}{1 \text{ mol } C_8H_{18}} = 4.962 \text{ kJ}$

We also know:

| heat released in the reaction | = | heat gained by calorimeter |

Substituting,

4.962 kJ = | heat capacity $\times \Delta t$ |
= | 2.88 kJ/°C $\times \Delta t$ |
Δt = 1.72°C

Therefore, $t_{final} = t_{initial} + \Delta t = 22.102°C + 1.72°C = $ **23.82°C**.

15-106. *Refer to Sections 15-14, 15-15 and 15-16.*

When a rubber band is stretched: $\Delta H < 0$, since heat is released

$\Delta S < 0$, since the rubber band is becoming more ordered (more linear); therefore

$\Delta G > 0$, since the process does not occur spontaneously

When the stretched rubber band is relaxed, the signs of the thermodynamic state functions change:

$\Delta H > 0$, since heat is absorbed (that's why your hand feels colder)

$\Delta S > 0$, since the rubber band is becoming more disordered; therefore

$\Delta G < 0$, since the process occurs spontaneously

The spontaneous process that occurs when the stretched rubber band is allowed to return to its original, random arrangement of polymer molecules must be driven by the increase in the disorder of the system, since the reaction is endothermic ($\Delta H > 0$).

15-108. *Refer to Section 15-3 and Fundamental Algebra.*

? hrs to utilize 100 g protein while resting $= 100 \text{ g protein} \times \dfrac{17 \text{ kJ}}{1 \text{ g protein}} \times \dfrac{1 \text{ hr}}{335 \text{ kJ}} = 5.1 \text{ hrs}$

? rate at which fat burns while walking (g/hr) $= \dfrac{1 \text{ g}}{39 \text{ kJ}} \times \dfrac{1250 \text{ kJ}}{1 \text{ hr}} = 32 \text{ g/hr}$

? rate at which fat burns while resting (g/hr) $= \dfrac{1 \text{ g}}{39 \text{ kJ}} \times \dfrac{335 \text{ kJ}}{1 \text{ hr}} = 8.6 \text{ g/hr}$

Let x = time spent walking (in hours)

5.1 - x = time spent resting (in hours)

? g of fat utilized = (walking rate for fat × walking time) + (resting rate for fat × resting time)

100 g fat = (32 g/hr × x) + (8.6 g/hr × (5.1 - x))

100 = 32x + 44 - 8.6x

56 = 23x

x = 2.4 hr spent walking

Therefore, in order to digest 100 g of fat in the same time it takes to digest 100 g of protein while resting, i.e., 5.1 hr, it is necessary to walk 2.3 hours out of the total 5.1 hours.

15-110. *Refer to Sections 1-13 and 15-4.*

$| \text{ heat lost } |_{\text{lead}} = | \text{ heat gained } |_{\text{water}}$

$| \text{ specific heat} \times \text{mass} \times \Delta t \, |_{\text{lead}} = | \text{ specific heat} \times \text{mass} \times \Delta t \, |_{\text{water}}$

(Specific heat of Pb)(436 g)(100.0°C - 40.8°C) = (4.184 J/g·°C)(50.0 g)(40.8°C - 25.0°C)

(Specific heat of Pb)(2.58 × 10⁴) = 3.31 × 10³

Specific heat of Pb = **0.128 J/g·°C**

Molar heat capacity of Pb = 0.128 J/g·°C × 207.2 g/mol = **26.5 J/mol·°C**

15-112. *Refer to Section 15-4.*

(a) Heat gain by calorimeter = (4572 J/°C)(27.93°C - 24.76°C) = 1.449 × 10⁴ J or 14.49 kJ

Fuel value of butter $= \dfrac{14.49 \text{ kJ}}{0.483 \text{ g}} = $ **30.0 kJ/g**

(b) Nutritional Calories/g butter $= \dfrac{30.0 \text{ kJ/g}}{4.184 \text{ kJ/kilocalorie}} = $ **7.17 kilocalorie/g**

(c) Nutritional Calories/5.00 g pat of butter = 7.17 kilocalorie/g × 5.00 g = **35.9 kilocalorie**

16 Chemical Kinetics

16-2. *Refer to Sections 16-5 and 16-6.*

The collision theory of reaction rates states that molecules, atoms or ions must collide effectively in order to react. For an effective collision to occur, the reacting species must have (1) at least a minimum amount of energy in order to break old bonds and make new ones, and (2) the proper orientation toward each other.

Transition state theory complements collision theory. When particles collide with enough energy to react, called the activation energy, E_a, the reactants form a short-lived, high energy activated complex, or transition state, before forming the products. The transition state also could revert back to the reactants.

16-4. *Refer to the Introduction to Chapter 16.*

In Chapter 15, we learned that reactions which are thermodynamically favorable have negative ΔG values and occur spontaneously as written. However, thermodynamics cannot be used to determine the rate of a reaction. Kinetically favorable reactions must be thermodynamically favorable *and* have a low enough activation energy to occur at a reasonable rate at a certain temperature.

16-6. *Refer to Section 16-3.*

The coefficients of the balanced overall equation bear no necessary relationship to the exponents to which the concentrations are raised in the rate law expression. The exponents are determined experimentally and describe how the concentrations of each reactant affect the reaction rate. The exponents are related to the rate-determining (slow) step in a sequence of mainly unimolecular and bimolecular reactions called the mechanism of the reaction. It is the mechanism which lays out exactly the order in which bonds are broken and made as the reactants are transformed into the products of the reaction.

16-8. *Refer to Section 16-1.*

(a) $3ClO^-(aq) \rightarrow ClO_3^-(aq) + 2Cl^-(aq)$ $\qquad$ rate of reaction $= -\dfrac{\Delta[ClO^-]}{3\Delta t} = \dfrac{\Delta[ClO_3^-]}{\Delta t} = \dfrac{\Delta[Cl^-]}{2\Delta t}$

(b) $2SO_2(g) + O_2(g) \rightarrow 2SO_3(g)$ $\qquad$ rate of reaction $= -\dfrac{\Delta[SO_2]}{2\Delta t} = -\dfrac{\Delta[O_2]}{\Delta t} = \dfrac{\Delta[SO_3]}{2\Delta t}$

(c) $C_2H_4(g) + Br_2(g) \rightarrow C_2H_4Br_2(g)$ $\qquad$ rate of reaction $= -\dfrac{\Delta[C_2H_4]}{\Delta t} = -\dfrac{\Delta[Br_2]}{\Delta t} = \dfrac{\Delta[C_2H_4Br_2]}{\Delta t}$

16-10. *Refer to Section 16-1.*

Balanced reaction: $4NH_3 + 5O_2 \rightarrow 4NO + 6H_2O$ $\qquad$ rate of reaction $= -\dfrac{\Delta[NH_3]}{4\Delta t} = -\dfrac{\Delta[O_2]}{5\Delta t} = \dfrac{\Delta[NO]}{4\Delta t} = \dfrac{\Delta[H_2O]}{6\Delta t}$

Substituting,

$\qquad$ rate of reaction $= -\dfrac{\Delta[NH_3]}{4\Delta t} = \dfrac{1.20\ M\ NH_3}{4 \times 1\ min} = 0.300\ M/min$

$\qquad$ Therefore,

rate of disappearance of $O_2 = -\dfrac{\Delta[O_2]}{\Delta t} = 5 \times$ rate of reaction $= 5 \times 0.300\ M/\text{min} = \textbf{1.50 } \textit{M}\textbf{/min}$

rate of appearance of NO $= \dfrac{\Delta[NO]}{\Delta t} = 4 \times$ rate of reaction $= 4 \times 0.300\ M/\text{min} = \textbf{1.20 } \textit{M}\textbf{/min}$

rate of appearance of $H_2O = \dfrac{\Delta[H_2O]}{\Delta t} = 6 \times$ rate of reaction $= 6 \times 0.300\ M/\text{min} = \textbf{1.80 } \textit{M}\textbf{/min}$

16-12. *Refer to Section 16-3.*

Plan: Use dimensional analysis and the rate-law expression to determine the units of k, the rate constant, in the following general equation:

$$\text{Rate } (M/s) = k[A]^x \qquad\qquad \text{where} \qquad x = \text{the overall order of the reaction}$$
$$[A] = \text{the reactant concentration } (M)$$

	Overall Reaction Order	Example	Units of k
(a)	1	Rate $= k[A]$	$(M/s)/M = s^{-1}$
(b)	2	Rate $= k[A]^2$	$(M/s)/M^2 = M^{-1}\cdot s^{-1}$
(c)	3	Rate $= k[A]^3$	$(M/s)/M^3 = M^{-2}\cdot s^{-1}$
(d)	1.5	Rate $= k[A]^{1.5}$	$(M/s)/M^{1.5} = M^{-0.5}\cdot s^{-1}$

16-14. *Refer to Section 16-3 and Examples 16-1 and 16-2.*

The form of the rate-law expression: Rate $= k[A]^x[B]^y[C]^z$

Step 1: Rate dependence on [A]. Consider Experiments 1 and 3:

Method 1: By observation, [B] and [C] do not change; [A] increases by a factor of 3. However, the reaction rate does not change. Therefore, changing [A] does not affect reaction rate and the reaction is zero order with respect to A. In all subsequent determinations, the effect of A can be ignored.

Method 2: A mathematical solution is obtained by substituting the experimental values of Experiments 1 and 3 into rate-law expressions and dividing the latter by the former. Note: the calculations are easier when the experiment with the larger rate is in the numerator.

$$\begin{array}{l} \text{Expt 3} \\ \text{Expt 1} \end{array} \qquad \frac{5.0 \times 10^{-4}\ M/\text{min}}{5.0 \times 10^{-4}\ M/\text{min}} = \frac{k(0.30\ M)^x(0.20\ M)^y(0.10\ M)^z}{k(0.10\ M)^x(0.20\ M)^y(0.10\ M)^z}$$

$$1 = 3^x$$
$$x = 0$$

Step 2: Rate dependence on [C]. Consider Experiments 1 and 2:

Method 1: [B] does not change; [C] changes by a factor of 3; the reaction rate also changes by a factor of 3 $(= 1.5 \times 10^{-3}/5.0 \times 10^{-4})$. The reaction rate is directly proportional to [C] and z must be equal to 1. The reaction is first order with respect to C.

Method 2:

$$\begin{array}{l} \text{Expt 2} \\ \text{Expt 1} \end{array} \qquad \frac{1.5 \times 10^{-3}\ M/\text{min}}{5.0 \times 10^{-4}\ M/\text{min}} = \frac{k(0.20\ M)^0(0.20\ M)^y(0.30\ M)^z}{k(0.10\ M)^0(0.20\ M)^y(0.10\ M)^z}$$

$$3 = 3^z$$
$$z = 1$$

Step 3: Rate dependence on [B]. Consider Experiments 2 and 4:

Method 1: [C] does not change; [B] changes by a factor of 3; the reaction rate changes by a factor of 3 ($= 4.5 \times 10^{-3}/1.5 \times 10^{-3}$). The reaction rate is directly proportional to [B] and y must be equal to 1. The reaction is first order with respect to B.

Method 2:

Expt 4
Expt 2
$$\frac{4.5 \times 10^{-3}\ M/min}{1.5 \times 10^{-3}\ M/min} = \frac{k(0.40\ M)^0(0.60\ M)^y(0.30\ M)^1}{k(0.20\ M)^0(0.20\ M)^y(0.30\ M)^1}$$

$$3 = 3^y$$
$$y = 1$$

The rate-law expression is: Rate $= k[A]^0[B]^1[C]^1 = k[B][C]$. To calculate the value of k, substitute the values from any one of the experiments into the rate-law expression and solve for k. If we use the data from Experiment 1,

$$5.0 \times 10^{-4}\ M/min = k(0.20\ M)(0.10\ M)$$
$$k = 2.5 \times 10^{-2}\ M^{-1}\cdot min^{-1}$$

The rate-law expression is now: **Rate $= (2.5 \times 10^{-2}\ M^{-1}\cdot min^{-1})[B][C]$**

16-16. *Refer to Section 16-3, Examples 16-1 and 16-2, and Exercise 16-14 Solution.*

Balanced equation: $2NO + O_2 \rightarrow 2NO_2$

(a) The form of the rate-law expression: Rate $= k[NO]^x[O_2]^y$

Step 1: Rate dependence on [NO]. Consider Experiments 1 and 2:

Method 1: By observation, $[O_2]$ does not change. [NO] increases by a factor of 2 ($= 0.04/0.02$) and the rate increases by a factor of 4 ($= 6.0 \times 10^{-4}/1.5 \times 10^{-4}$). The reaction rate increases as the square of [NO]; x must be 2. The reaction is second order with respect to NO.

Method 2:

Expt 2
Expt 1
$$\frac{6.0 \times 10^{-4}\ M/min}{1.5 \times 10^{-4}\ M/min} = \frac{k(0.040\ M)^x(0.010\ M)^y}{k(0.020\ M)^x(0.010\ M)^y}$$

$$4 = 2^x$$
$$x = 2$$

Step 2: Rate dependence on $[O_2]$. Consider Experiments 1 and 3:

Method 1: By observation, [NO] does not change. $[O_2]$ increases by a factor of 4 ($= 0.040/0.010$) and the rate increases by a factor of 4 ($= 6.0 \times 10^{-4}/1.5 \times 10^{-4}$). The reaction rate is directly proportional to $[O_2]$ and y must be equal to 1. The reaction is first order with respect to O_2.

Method 2:

Expt 3
Expt 1
$$\frac{6.0 \times 10^{-4}\ M/min}{1.5 \times 10^{-4}\ M/min} = \frac{k(0.020\ M)^2(0.040\ M)^y}{k(0.020\ M)^2(0.010\ M)^y}$$

$$4 = 4^y$$
$$y = 1$$

The rate-law expression is: Rate $= k[NO]^2[O_2]^1 = k[NO]^2[O_2]$

Substituting the data from Experiment 1 into the rate law expression to calculate k,

$$1.5 \times 10^{-4}\ M/s = k(0.020\ M)^2(0.010\ M)$$
$$k = 38\ M^{-2}\cdot s^{-1}$$

The rate-law expression is now: Rate of disappearance of NO $= (38\ M^{-2} \cdot s^{-1})[NO]^2[O_2]$

We know: rate of reaction $= -\dfrac{\Delta[NO]}{2\Delta t} = -\dfrac{\Delta[O_2]}{\Delta t} = \dfrac{\Delta[NO_2]}{2\Delta t}$

Therefore: **rate of reaction $= (19\ M^{-2} \cdot s^{-1})[NO]^2[O_2]$**

(b) Substituting into the rate law expression,

$$\text{Rate} = (19\ M^{-2} \cdot s^{-1})[NO]^2[O_2] = (19\ M^{-2} \cdot s^{-1})(0.055\ M)^2(0.035\ M) = \mathbf{2.0 \times 10^{-3}\ M/s}$$

(c) rate of disappearance of NO $= -\dfrac{\Delta[NO]}{\Delta t} = 2 \times$ rate of reaction $= \mathbf{4.0 \times 10^{-3}\ M/s}$

rate of disappearance of $O_2 = -\dfrac{\Delta[O_2]}{\Delta t} =$ rate of reaction $= \mathbf{2.0 \times 10^{-3}\ M/s}$

rate of formation of $NO_2 = \dfrac{\Delta[NO_2]}{\Delta t} = 2 \times$ rate of reaction $= \mathbf{4.0 \times 10^{-3}\ M/s}$

16-18.   *Refer to Section 16-3.*

The simplest approach to this problem is to assume that the initial concentrations of A and B_2 are each 1 M. Then the final concentrations of A and B_2 are each 1/2 M. Let us substitute these values into the rate-law expression,

$$\text{Rate} = k[A]^2[B_2]$$

Initial: Rate $= k(1\ M)^2(1\ M) = k$
Final: Rate $= k(1/2\ M)^2(1/2\ M) = 1/8\ k$

Therefore, the rate of reaction would **decrease** by a factor of **8**.

16-20.   *Refer to Section 16-3, Examples 16-1 and 16-2, and Exercise 16-14 Solution.*

Balanced equation: $2ClO_2(aq) + 2OH^-(aq) \rightarrow ClO_3^-(aq) + ClO_2^-(aq) + H_2O(\ell)$

(a) The form of the rate-law expression: Rate $= k[ClO_2]^x[OH^-]^y$

Step 1: Rate dependence on $[ClO_2]$. Consider Experiments 1 and 2.

Method 1: By observation, $[OH^-]$ is constant and $[ClO_2]$ increases by a factor of 2 ($= 0.024/0.012$). The rate of reaction increases by a factor of 4 ($= 8.28 \times 10^{-4}/2.07 \times 10^{-4}$). The reaction rate increases as the square of $[ClO_2]$ and x equals 2. The reaction is second order with respect to ClO_2.

Method 2:

$\dfrac{\text{Expt 2}}{\text{Expt 1}}$ $\dfrac{8.28 \times 10^{-4}\ M/s}{2.07 \times 10^{-4}\ M/s} = \dfrac{k(0.024\ M)^x(0.012\ M)^y}{k(0.012\ M)^x(0.012\ M)^y}$

$$4 = 2^x$$
$$x = 2$$

Step 2: Rate dependence on $[OH^-]$. Consider Experiments 2 and 4:

Method 1: By observation, $[ClO_2]$ is constant and $[OH^-]$ increases by a factor of 2 ($= 0.024/0.012$). The rate of reaction increases by a factor of 2 ($= 1.66 \times 10^{-3}/8.28 \times 10^{-4}$). The reaction rate is directly proportional to $[OH^-]$ and y equals 1. The reaction is first order with respect to OH^-.

Method 2:

Expt 4	$\dfrac{1.66 \times 10^{-3} \, M/s}{8.28 \times 10^{-4} \, M/s} = \dfrac{k(0.024 \, M)^2(0.024 \, M)^y}{k(0.024 \, M)^2(0.012 \, M)^y}$
Expt 2	

$$2 = 2^y$$
$$y = 1$$

The rate-law expression is: Rate $= k[ClO_2]^2[OH^-]^1 = k[ClO_2]^2[OH^-]$

Using the data from Experiment 1 to calculate k, we have

$$2.07 \times 10^{-4} \, M/s = k(0.012 \, M)^2(0.012 \, M)$$
$$k = 1.2 \times 10^2 \, M^{-2} \cdot s^{-1}$$

The rate-law expression is now: **Rate $= (1.2 \times 10^2 \, M^{-2} \cdot s^{-1})[ClO_2]^2[OH^-]$**

(b) The reaction is second order with respect to ClO_2, first order with respect to OH^- and third order overall.

16-22. *Refer to Section 16-3, Examples 16-1 and 16-2, and Exercise 16-14 Solution.*

Balanced equation: A + B → C

The form of the rate-law expression: Rate $= k[A]^x[B]^y$

Step 1: Rate dependence on [A]. Consider Experiments 1 and 2:

Method 1: [B] does not change; [A] changes by a factor of 1.5 (= 0.30/0.20); reaction rate changes also by a factor of 1.5 (= $7.5 \times 10^{-6}/5.0 \times 10^{-6}$). The reaction rate is directly proportional to [A] and x equals 1. The reaction is first order with respect to A.

Method 2:

Expt 2	$\dfrac{7.5 \times 10^{-6} \, M/s}{5.0 \times 10^{-6} \, M/s} = \dfrac{k(0.30 \, M)^x(0.10 \, M)^y}{k(0.20 \, M)^x(0.10 \, M)^y}$
Expt 1	

$$1.5 = 1.5^x$$
$$x = 1$$

Step 2: Rate dependence on [B].

There is no pair of experiments in which [B] is changing and [A] is constant. Therefore, one may choose any 2 experiments in which [B] is varying and use Method 2. If we choose Experiments 1 and 3:

Expt 3	$\dfrac{4.0 \times 10^{-5} \, M/s}{5.0 \times 10^{-6} \, M/s} = \dfrac{k(0.40 \, M)^1(0.20 \, M)^y}{k(0.20 \, M)^1(0.10 \, M)^y}$
Expt 1	

$$8 = (2)^1(2)^y$$
$$4 = 2^y$$
$$y = 2$$

Therefore, the reaction is second order with respect to B.

The rate-law expression is: Rate $= k[A]^1[B]^2 = k[A][B]^2$

Using the data from Experiment 1 to calculate k, we have

$$5.0 \times 10^{-6} \, M/s = k(0.20 \, M)(0.10 \, M)^2$$
$$k = 2.5 \times 10^{-3} \, M^{-2} \cdot s^{-1}$$

The rate-law expression is now: **Rate $= (2.5 \times 10^{-3} \, M^{-2} \cdot s^{-1})[A][B]^2$**

16-24. Refer to Section 16-3.

The rate-law expression: Rate = $k[A][B]^2$

Plan: (1) Use the data for Experiment 1 and the rate-law expression to calculate the rate constant, k.
(2) Substitute the given values into the complete rate-law expression to determine the reaction rate.

(1) Substituting,
$$0.150 \ M/s = k(1.00 \ M)(0.200 \ M)^2$$
$$k = 3.75 \ M^{-2}\cdot s^{-1}$$

(2) Expt 2: Rate = $(3.75 \ M^{-2}\cdot s^{-1})(2.00 \ M)(0.200 \ M)^2 = \mathbf{0.300 \ M/s}$
Expt 3: Rate = $(3.75 \ M^{-2}\cdot s^{-1})(2.00 \ M)(0.400 \ M)^2 = \mathbf{1.20 \ M/s}$

16-26. Refer to Section 16-4.

The half-life of a reactant is the time required for half of that reactant to be converted into products. For a first order reaction, the half-life is independent of concentration so that the same time is required to consume half of any starting amount or concentration of the reactant. On the other hand, the half-life of a second-order reaction *does* depend on the starting amount of the reactant.

16-28. Refer to Section 16-4 and Example 16-6.

For the reaction, $NO_2 \rightarrow NO + 1/2 \ O_2$, the units of the rate constant, $M^{-1}\cdot min^{-1}$, tell us that the reaction is second order overall. Since the reaction has only one reactant, NO_2, the reaction is second order with respect to NO_2.

The integrated rate equation for a reaction that is second order with respect to NO_2 as the only reactant:

$$\frac{1}{[NO_2]} - \frac{1}{[NO_2]_o} = akt \qquad \text{where} \quad a = \text{stoichiometric coefficient of } NO_2$$

Substituting,
$$\frac{1}{1.25 \ M} - \frac{1}{2.00 \ M} = (1)(1.70 \ L\cdot mol^{-1}\cdot min^{-1})t$$

$$t = \frac{0.800 \ M^{-1} - 0.500 \ M^{-1}}{1.70 \ L\cdot mol^{-1}\cdot min^{-1}}$$

$$t = \mathbf{0.176 \ min \ or \ 10.6 \ s}$$

16-30. Refer to Section 16-4 and Examples 16-3 and 16-4.

For the reaction, $CH_3N{=}NCH_3 \rightarrow N_2 + C_2H_6$, the units of the rate constant, min^{-1}, tell us that the reaction is first order overall. Since the reaction has only one reactant, $CH_3N{=}NCH_3$, the reaction is first order with respect to $CH_3N{=}NCH_3$.

The integrated rate equation for a reaction that is first order with respect to $CH_3N{=}NCH_3$ as the only reactant:

$$\log\left(\frac{[CH_3N{=}NCH_3]_o}{[CH_3N{=}NCH_3]_t}\right) = \frac{akt}{2.303} \qquad \text{where} \quad a = \text{stoichiometric coefficient of } CH_3N{=}NCH_3$$

(a) Substituting to find the number of moles of $CH_3N{=}NCH_3$ remaining:

$$\log\left(\frac{2.00 \ g/58.1 \ g/mol}{? \ mol \ CH_3N{=}NCH_3 \ at \ time \ t}\right) = \frac{(1)(40.8 \ min^{-1})(0.0500 \ min)}{2.303}$$

$$\log\left(\frac{0.0344 \ mol}{? \ mol \ CH_3N{=}NCH_3}\right) = 0.886$$

$$\frac{0.0344 \ mol}{? \ mol \ CH_3N{=}NCH_3} = 7.69$$

$$? \ mol \ CH_3N{=}NCH_3 = \mathbf{4.48 \times 10^{-3} \ mol \ CH_3N{=}NCH_3}$$

$$? \text{ mol } N_2 \text{ formed} = \text{mol } CH_3N{=}NCH_3 \text{ reacted}$$
$$= \text{mol } CH_3N{=}NCH_3 \text{ initial - mol } CH_3N{=}NCH_3 \text{ at time } t$$
$$= 0.0344 \text{ mol} - 0.00448 \text{ mol}$$
$$= \textbf{0.0299 mol } N_2$$

(b) Substituting,

$$\log\left(\frac{2.00 \text{ g}}{\text{g } CH_3N{=}NCH_3 \text{ at time } t}\right) = \frac{(1)(40.8 \text{ min}^{-1})(12.0 \text{ s} \times 1 \text{ min/60 s})}{2.303}$$

$$\log\left(\frac{2.00 \text{ g}}{\text{g } CH_3N{=}NCH_3}\right) = 3.54$$

$$\frac{2.00 \text{ g}}{\text{g } CH_3N{=}NCH_3} = 3490 \text{ (to 3 significant figures)}$$

$$\text{g } CH_3N{=}NCH_3 = \textbf{5.7} \times \textbf{10}^{-4} \textbf{ g } \textbf{CH}_3\textbf{N}{=}\textbf{NCH}_3 \text{ (to 2 significant figures)}$$

16-32.   *Refer to Section 16-4, and Examples 16-5, 16-6 and 16-7.*

The rate equation, Rate $= (1.4 \times 10^{-10} \ M^{-1}{\cdot}s^{-1})([NO_2]^2$, tells us that the reaction is second order with respect to NO_2 as the only reactant.

(a) For the second order reaction, $2NO_2 \rightarrow 2NO + O_2$:

$$t_{1/2} = \frac{1}{ak[NO_2]_o} = \frac{1}{(2)(1.4 \times 10^{-10} \ M^{-1}{\cdot}s^{-1})(3.00 \text{ mol/2.00 L})} = \textbf{2.4} \times \textbf{10}^9 \textbf{ s } \text{ or } \textbf{76 yrs}$$

(b) The integrated second order rate equation:

$$\frac{1}{[NO_2]} - \frac{1}{[NO_2]_o} = akt \qquad\qquad \text{where } \quad a = \text{stoichiometric coefficient of } NO_2$$

Substituting,

$$\frac{1}{[NO_2]} - \frac{1}{(1.50 \ M)} = (2)(1.4 \times 10^{-10} \ M^{-1}{\cdot}s^{-1})(115 \text{ yr} \times 3.15 \times 10^7 \text{ s/yr})$$

$$\frac{1}{[NO_2]} = 1.0 \ M^{-1} + 0.667 \ M^{-1} = 1.7 \ M^{-1}$$

$$[NO_2] = \textbf{0.59 } \textbf{\textit{M}}$$

$$? \text{ g } NO_2 \text{ remaining} = 2.00 \text{ L} \times \frac{0.59 \text{ mol}}{1.00 \text{ L}} \times \frac{46.0 \text{ g}}{1.0 \text{ mol}} = \textbf{54 g } \textbf{NO}_2$$

(c) $[NO_2]_{reacted} = [NO_2]_o - [NO_2] = 1.50 \ M - 0.59 \ M = \textbf{0.91 } \textbf{\textit{M}}$

$[NO]_{produced} = [NO_2]_{reacted} = \textbf{0.91 } \textbf{\textit{M}}$

16-34.   *Refer to Section 16-4 and Example 16-4.*

The integrated first order rate equation: $\log\left(\dfrac{A_o}{A_t}\right) = \dfrac{akt}{2.303}$

If 99.0% of the cyclopropane disappeared, then 1.0% of it remains. It is not necessary to know the actual starting concentration to do this calculation.

Substituting, $\log\left(\dfrac{100.0\%}{1.0\%}\right) = \dfrac{(1)(2.74 \times 10^{-3} \ s^{-1})t}{2.303}$

$$2.00 = (1.19 \times 10^{-3})t$$

$$t = \textbf{1680 s } \text{ or } \textbf{28.0 min}$$

216

16-36. *Refer to Section 16.4.*

The integrated first order rate equation: $\log\left[\dfrac{[A]_o}{[A]_t}\right] = \dfrac{akt}{2.303}$

We have: $[A]_o = 0.75\ M$ $\qquad [A]_t = 0.25\ M$ $\qquad t = 47.0$ min

Substituting, $\qquad \log\left[\dfrac{0.75\ M}{0.25\ M}\right] = \dfrac{k(47.0\ \text{min})}{2.303}$ $\qquad$ where $a = 1$

Solving, $\qquad\qquad\qquad k = \mathbf{0.023\ min^{-1}}$

16-38. *Refer to Sections 16-5 and 16-6, and Example 16-8.*

Consider the decomposition reaction: $2HI \rightarrow H_2 + I_2$

(a)

t (s)	[HI] (mmol/L)	ln[HI]	1/[HI]
0	5.46	1.697	0.183
250	4.10	1.411	0.244
500	2.73	1.004	0.366
750	1.37	0.3148	0.730

(1) For a zero order reaction, a plot of [HI] vs. t gives a straight line.
(2) For a first order reaction, a plot of ln[HI] vs. t gives a straight line.
(3) For a second order reaction, a plot of 1/[HI] vs. t gives a straight line.

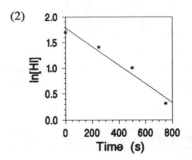
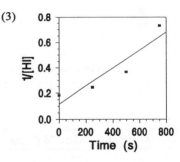

The data lay on a straight line only for Plot (1), the graph of [HI] vs. t. Therefore, the reaction is **zero order** with respect to HI. The slope of the line $= -0.00546\ mM\cdot s^{-1}$, using a least mean square regression fitting program. However, the slope can be estimated from any two points on the line. If we use the first and last points:

$$\text{slope} = \frac{\Delta y}{\Delta x} = \frac{\Delta[HI]}{\Delta t} = \frac{1.37\ mM - 5.46\ mM}{750\ s - 0\ s} = -5.45 \times 10^{-3}\ mM\cdot s^{-1}.$$

Table 16-2 summarizes: for a zero order reaction,

$$\text{slope} = -ak \qquad\qquad \text{where} \qquad a = \text{stoichiometric coefficient}$$
$$k = \text{rate constant}$$

Rearranging, $k = \dfrac{\text{slope}}{-a} = \dfrac{-0.00546\ mM\cdot s^{-1}}{-2} = 0.00273\ mM\cdot s^{-1}$

The rate equation is: **Rate** $= k[HI]^0 = k = \mathbf{0.00273\ mM\cdot s^{-1}}$

The integrated rate equation is: $[HI] = [HI]_o - akt = \mathbf{5.46\ mM - (2)(2.73 \times 10^{-3}\ mM\cdot s^{-1})t}$

(b) At 900 s, $[HI] = 5.46\ mM - (2)(2.73 \times 10^{-3}\ mM\cdot s^{-1})(600\ s) = 2.18\ mM$ or **2.18 mmoles/L**

16-40. *Refer to Section 16-4 and Exercise 16-38 Solution.*

(a) The units of the rate constant, 0.080 M·s^{-1}, relate that the reaction is **zeroth order**.

(b) For zeroth order:
$$[HI]_o - [HI] = akt$$
$$1.50 \, M - 0.15 \, M = (2)(0.080 \, M\text{·s}^{-1})t$$
$$t = \textbf{8.4 s}$$

16-42. *Refer to Section 16-9 and Figures 16-15 and 16-16.*

Catalysts are substances which increase the rate of reaction when added to a system by providing an alternative mechanism with a lower activation energy. Although a catalyst may enter into a reaction, it would not appear in the overall balanced equation, because if it is a reactant in one step, it is a product in another. It is not consumed during a reaction.

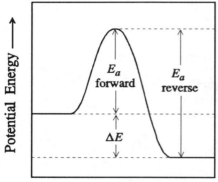

Reaction Coordinate for
Uncatalyzed Reaction

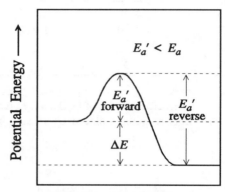

Reaction Coordinate for
Catalyzed Reaction

16-44. *Refer to Sections 16-6 and 16-9, and Figures 16-10 and 16-15.*

For the hypothetical reaction: $A + B \rightarrow C$

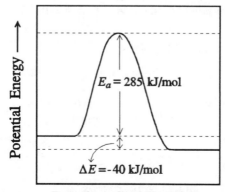

Reaction Coordinate for
Uncatalyzed Reaction

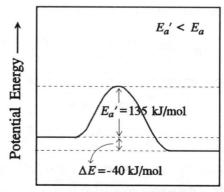

Reaction Coordinate for
Catalyzed Reaction

16-46. *Refer to Section 16-8 and Example 16-10.*

The Arrhenius equation can be presented as:

$$\log\left(\frac{k_2}{k_1}\right) = \frac{E_a}{2.303R}\left(\frac{T_2 - T_1}{T_1 T_2}\right)$$

where

k_2/k_1 = ratio of rate constants = 3.000
E_a = activation energy (J/mol)
R = 8.314 J/mol·K
T_1 = 298.0 K
T_2 = 308.0 K

Substituting,

$$\log 3.000 = \frac{E_a}{(2.303)(8.314\ \text{J/mol·K})}\left[\frac{308.0\ \text{K} - 298.0\ \text{K}}{(298.0\ \text{K})(308.0\ \text{K})}\right]$$

$$0.477 = \frac{E_a}{19.15\ \text{J/mol·K}}(1.09 \times 10^{-4}\ \text{K}^{-1})$$

$$E_a = \mathbf{8.38 \times 10^4\ \text{J/mol rxn}\ \ or\ \ 83.8\ \text{kJ/mol rxn}}$$

Note on significant figures and logarithms: A logarithm consists of 2 parts: an integer called the characteristic and a decimal called the mantissa. When working with base 10 logarithms, it is the number of significant digits in the mantissa that give the number of significant figures in the antilogarithm.

16-48. *Refer to Sections 16-6 and 16-8, Figure 16-10 and Example 16-46 Solution.*

For a particular reaction: $\Delta E° = 51.51$ kJ/mol reaction

(1) From the Arrhenius equation:

$$\log\left(\frac{k_2}{k_1}\right) = \frac{E_a}{(2.303)(8.314\ \text{J/mol·K})}\left[\frac{323.15\ \text{K} - 273.15\ \text{K}}{273.15\ \text{K} \times 323.15\ \text{K}}\right]$$

$$\log\left(\frac{8.9 \times 10^{-4}\ \text{s}^{-1}}{8.0 \times 10^{-7}\ \text{s}^{-1}}\right) = \frac{E_a}{19.15\ \text{J/mol·K}}(5.664 \times 10^{-4}\ \text{K}^{-1})$$

$$E_a = \mathbf{1.03 \times 10^5\ \text{J/mol rxn}\ \ or\ \ 103\ \text{kJ/mol rxn}}$$

(2) The reaction coordinate diagram for the reaction is:

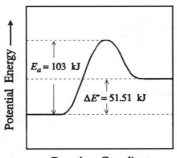

$E_a = 103$ kJ

$\Delta E = 51.51$ kJ

Potential Energy →

Reaction Coordinate

16-50. *Refer to Section 16-8.*

Plan: The Arrhenius equation can be rearranged: $\ln k = -\left(\dfrac{E_a}{R}\right)\left(\dfrac{1}{T}\right) + \ln A$.

Plot $\ln k$ against $1/T$. The slope of the line = $-E_a/R$ and solve for E_a.

T (K)	$1/T$ (K^{-1})	k (s^{-1})	ln k
600	1.67×10^{-3}	3.30×10^{-9}	-19.53
650	1.54×10^{-3}	2.19×10^{-7}	-15.33
700	1.43×10^{-3}	7.96×10^{-6}	-11.74
750	1.33×10^{-3}	1.80×10^{-4}	-8.623
800	1.25×10^{-3}	2.74×10^{-3}	-5.900
850	1.18×10^{-3}	3.04×10^{-2}	-3.493
900	1.11×10^{-3}	2.58×10^{-1}	-1.355

(a) By plotting the data as shown, we obtained
    slope $= -E_a/R = -3.25 \times 10^4$ K

Therefore,

$E_a = -\text{(slope)} \times R$
$= -(-3.25 \times 10^4 \text{ K}) \times 8.314 \text{ J/mol·K}$
$= \mathbf{2.70 \times 10^5 \text{ J/mol} \text{ or } 270. \text{ kJ/mol}}$

(b) On the x axis, $1/T = 1/500$ K $= 0.00200$ K^{-1}

From the graph, we can estimate:
    $y = \ln k = -30.2$

Therefore, $k = 7.7 \times 10^{-14}$ s^{-1} at 500 K

(c) On the y axis, $\ln k = \ln (5.00 \times 10^{-5}) = -9.9$

From the graph, we can estimate:
    $x = 1/T = 0.00137$ K^{-1}

Therefore, $T = \mathbf{730 \text{ K}}$ when $k = 5.00 \times 10^{-5}$ s^{-1}

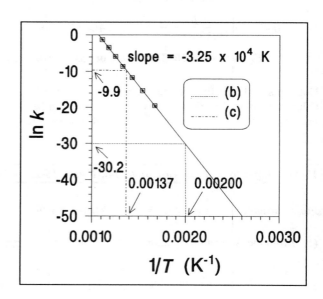

16-52. *Refer to Sections 16-4 and 16-9.*

The hydration reaction of CO_2, $CO_2 + H_2O \rightarrow H_2CO_3$, is enzyme catalyzed. The rate of reaction does not depend on [CO_2] or [H_2O]. This is deduced from the fact that it only takes 1 molecule of enzyme to react with 10^6 molecules of CO_2. Therefore, the reaction is zero-order with respect to CO_2.: Rate $= k[CO_2]^0 = k$.

Plan: mol enzyme/L $\overset{(1)}{\Longrightarrow}$ molecules enzyme/L $\overset{(2)}{\Longrightarrow}$ hydration rate (molecules CO_2/L·s)
$\overset{(3)}{\Longrightarrow}$ hydration rate (molecules CO_2/L·hr) $\overset{(4)}{\Longrightarrow}$ hydration rate (mol CO_2/L·hr)
$\overset{(5)}{\Longrightarrow}$ hydration rate (g CO_2/L·hr) $\overset{(6)}{\Longrightarrow}$ hydration rate (kg CO_2/L·hr)

? kg CO_2 hydrated/L·hr $= \dfrac{1.0 \times 10^{-6} \text{ mol enzyme}}{1 \text{ liter}} \times \dfrac{6.02 \times 10^{23} \text{ molecules enzyme}}{1 \text{ mol enzyme}} \times \dfrac{10^6 \text{ molecules } CO_2}{1 \text{ molecule enzyme} \times 1\text{s}}$

$\times \dfrac{3600 \text{ s}}{1 \text{ hr}} \times \dfrac{1 \text{ mol } CO_2}{6.02 \times 10^{23} \text{ molecules } CO_2} \times \dfrac{44 \text{ g } CO_2}{1 \text{ mol } CO_2} \times \dfrac{1 \text{ kg } CO_2}{1000 \text{ g } CO_2}$

$= \mathbf{160 \text{ kg } CO_2/\text{L·hr}}$

(a) From the Arrhenius equation: $\log\left(\dfrac{k_2}{k_1}\right) = \dfrac{E_a}{2.303R}\left[\dfrac{T_2 - T_1}{T_1 T_2}\right]$

Substituting,

$\log\dfrac{k_2}{9.16 \times 10^{-3} \text{ s}^{-1}} = \dfrac{88 \times 10^3 \text{ J/mol}}{(2.303)(8.314 \text{ J/mol·K})}\left[\dfrac{298 \text{ K} - 273 \text{ K}}{(273 \text{ K})(298 \text{ K})}\right] = 1.4$

$\dfrac{k_2}{9.16 \times 10^{-3} \text{ s}^{-1}} = 25$

$k_2 = 0.23 \text{ s}^{-1}$

(b) Substituting into the following version of the Arrhenius equation and solving for T_2,

$$\log\dfrac{k_2}{k_1} = \dfrac{E_a}{2.303R}\left[\dfrac{1}{T_1} - \dfrac{1}{T_2}\right]$$

$\log\left[\dfrac{3.00 \times 10^{-2} \text{ s}^{-1}}{9.16 \times 10^{-3} \text{ s}^{-1}}\right] = \dfrac{88 \times 10^3 \text{ J/mol}}{(2.303)(8.314 \text{ J/mol·K})}\left[\dfrac{1}{273 \text{ K}} - \dfrac{1}{T_2}\right]$

$0.515 = (4.60 \times 10^3 \text{ K})(3.66 \times 10^{-3} \text{ K}^{-1} - 1/T_2)$

$1/T_2 = 3.66 \times 10^{-3} - 1.12 \times 10^{-4}$

$1/T_2 = 3.55 \times 10^{-3}$

$T_2 = 282 \text{ K} \text{ or } 9°\text{C}$

The rate-law expression for the reaction is: Rate $= k[Cl_2][H_2S]$. If the rate-law expression derived from a proposed mechanism is different from this expression, the mechanism cannot be the correct one.

(a) The rate-law expression consistent with the slow step of this mechanism is: Rate $= k[Cl_2]$ and this cannot be a mechanism for the reaction.

(b) The rate-law expression consistent with the slow step of this mechanism is: Rate $= k[Cl_2][H_2S]$ and this is a possible mechanism for the reaction.

(c) The rate-law expression consistent with this mechanism is more complicated. It can be determined by following the procedure presented in Section 16-7:

From Step 3 (the slow step),   Rate $= k[HS][Cl]$ where HS and Cl are intermediates

From Step 1, $[Cl] = \left(\dfrac{k_{1f}}{k_{1r}}\right)^{1/2}[Cl_2]^{1/2}$

From Step 2, $[HS] = \left(\dfrac{k_{2f}}{k_{2r}}\right)\dfrac{[Cl][H_2S]}{[HCl]} = \left(\dfrac{k_{2f}}{k_{2r}}\right)\left(\dfrac{k_{1f}}{k_{1r}}\right)^{1/2}\dfrac{[Cl_2]^{1/2}[H_2S]}{[HCl]}$

Substituting   Rate $= k\left(\dfrac{k_{2f}}{k_{2r}}\right)\left(\dfrac{k_{1f}}{k_{1r}}\right)^{1/2}\dfrac{[Cl_2]^{1/2}[H_2S]}{[HCl]} \times \left(\dfrac{k_{1f}}{k_{1r}}\right)^{1/2}[Cl_2]^{1/2}$

            $= k'\dfrac{[Cl_2][H_2S]}{[HCl]}$

And so, this **cannot** be a mechanism for this reaction.

16-58. *Refer to Section 16-7.*

The reaction mechanism:

$$O_3 \underset{k_{1r}}{\overset{k_{1f}}{\rightleftarrows}} O_2 + O \qquad \text{(fast, equilibrium)}$$

$$O + O_3 \underset{k_2}{\rightarrow} 2O_2 \qquad \text{(slow)}$$

From Step 2 (the slow step), $\qquad$ Rate $= k_2[O][O_3]$ $\qquad$ where O is an intermediate

From the slow step, Rate $= k_2[O][O_3]$. However, O is an intermediate and its concentration must be expressed in terms of O_3. For a fast, equilibrium step, we know:

$$Rate_{1f} = Rate_{1r}$$

$$k_{1f}[O_3] = k_{1r}[O_2][O]$$

$$[O] = \frac{k_{1f}}{k_{1r}}\frac{[O_3]}{[O_2]}$$

Substituting for [O] in the original rate equation, we have

$$Rate = k_2[O][O_3] = k_2 \left(\frac{k_{1f}}{k_{1r}} \frac{[O_3]}{[O_2]} \right) [O_3] \quad \text{or} \quad \textbf{Rate} = k \frac{[O_3]^2}{[O_2]}$$

16-60. *Refer to Section 16-7.*

The reaction mechanism:

$$N_2 + Cl \underset{k_{1r}}{\overset{k_{1f}}{\rightleftarrows}} N_2Cl \qquad \text{(fast, equilibrium)}$$

$$N_2Cl + Cl \underset{k_2}{\rightarrow} Cl_2 + N_2 \qquad \text{(slow)}$$

(a) The intermediate species is N_2Cl.

(b) From the slow step, Rate $= k_2[N_2Cl][Cl]$. However, N_2Cl is an intermediate and its concentration must be expressed in terms of N_2 and Cl. For a fast, equilibrium step, we know:

$$Rate_{1f} = Rate_{1r}$$

$$k_{1f}[N_2][Cl] = k_{1r}[N_2Cl]$$

$$[N_2Cl] = \frac{k_{1f}}{k_{1r}}[N_2][Cl]$$

Substituting for [N_2Cl] in the original rate equation, we have

$$Rate = k_2[N_2Cl][Cl] = k_2 \left(\frac{k_{1f}}{k_{1r}}[N_2][Cl] \right) [Cl] \quad \text{or} \quad Rate = k[N_2][Cl]^2$$

Yes, the mechanism is consistent with the experimental rate law, Rate $= k[N_2][Cl]^2$.

16-62. *Refer to Section 16-7.*

The reaction mechanism:

$$H_2 \underset{k_{1r}}{\overset{k_{1f}}{\rightleftarrows}} 2H \qquad \text{(fast, equilibrium)}$$

$$H + CO \underset{k_2}{\rightarrow} HCO \qquad \text{(slow)}$$

$$H + HCO \underset{k_3}{\rightarrow} H_2CO \qquad \text{(fast)}$$

(a) Balanced equation: $H_2 + CO \rightarrow H_2CO$

(b) From the slow step, Rate = k_2[H][CO]. However, H is an intermediate and its concentration must be expressed in terms of H_2. For a fast, equilibrium step, we know:

$$Rate_{1f} = Rate_{1r}$$

$$k_{1f}[H_2] = k_{1r}[H]^2$$

$$[H] = \left(\frac{k_{1f}}{k_{1r}}\right)^{1/2}[H_2]^{1/2}$$

Substituting for [H] in the original rate equation, we have

$$Rate = k_2[H][CO] = k_2\left(\frac{k_{1f}}{k_{1r}}\right)^{1/2}[H_2]^{1/2}[CO] = k[H_2]^{1/2}[CO]$$

Yes, the mechanism is consistent with the observed rate dependence.

16-64. *Refer to Sections 13-9 and 16-8.*

Plan: (1) Use the Clausius-Clapeyron equation to calculate the steam temperature in the pressure cooker. Assume that ΔH_{vap} for H_2O is independent of temperature.

(2) Use the Arrhenius equation to calculate the activation energy for the process of steaming vegetables.

(1) From the Clausius-Clapeyron equation: $\log\left(\frac{P_2}{P_1}\right) = \frac{\Delta H_{vap}}{2.303R}\left(\frac{1}{T_1} - \frac{1}{T_2}\right)$

where
ΔH_{vap} = molar heat of vaporization for H_2O, 40.7 kJ/mol
P_1 = atmospheric pressure, 15 psi
P_2 = cooker pressure = P_1 + gauge pressure = (15 + 15) psi
T_1 = boiling point of water at 1 atm, 100.0°C
T_2 = steam temperature in the pressure cooker

Substituting,

$$\log\left(\frac{30 \text{ psi}}{15 \text{ psi}}\right) = \frac{40.7 \times 10^3 \text{ J/mol}}{(2.303)(8.314 \text{ J/mol·K})}\left(\frac{1}{373 \text{ K}} - \frac{1}{T_2}\right)$$

$$1.4 \times 10^{-4} \text{ K}^{-1} = (2.68 \times 10^{-3} \text{ K}^{-1} - 1/T_2)$$

$$T_2 = 394 \text{ K}$$

(2) From the Arrhenius equation: $\log\left(\frac{k_2}{k_1}\right) = \frac{E_a}{2.303R}\left[\frac{T_2 - T_1}{T_1 T_2}\right]$

where
T_1 = 373 K
T_2 = 394 K
k_1 = rate constant for cooking vegetables at atmospheric pressure
k_2 = rate constant for cooking vegetables in the pressure cooker
k_2/k_1 = 3 since the cooking process is 3 times faster in the pressure cooker

Substituting,

$$\log 3 = \frac{E_a}{(2.303)(8.314 \text{ J/mol·K})}\left[\frac{394 \text{ K} - 373 \text{ K}}{(373 \text{ K})(394 \text{ K})}\right]$$

$$E_a = 6.4 \times 10^4 \text{ J/mol or } 64 \text{ kJ/mol}$$

16-66. *Refer to Section 16-3.*

Balanced equation: $2N_2O_5(aq) \rightarrow 4NO_2(aq) + O_2(g)$

The form of the rate law expression: Rate $= k[N_2O_5]^x$

Consider Experiments 1 and 2:

Method 1: By observation, $[N_2O_5]$ decreases by a factor of 2 (= 0.900/0.450) and the initial rate decreases also by a factor of 2 (= $5.58 \times 10^{-4}/2.79 \times 10^{-4}$). The reaction rate is directly proportional to $[N_2O_5]$ and x equals 1. The reaction is first order with respect to N_2O_5.

Method 2:

$$\begin{array}{ll}\text{Expt 1} \\ \text{Expt 2}\end{array} \qquad \frac{5.58 \times 10^{-4}\ M/\text{s}}{2.79 \times 10^{-4}\ M/\text{s}} = \frac{k(0.900\ M)^x}{k(0.450\ M)^x}$$

$$2 = (2)^x$$

$$x = 1$$

The rate-law expression is: Rate $= k[N_2O_5]^1 = k[N_2O_5]$

Substituting the data from Experiment 1 into the rate-law expression to calculate k, we have

$$5.58 \times 10^{-4}\ M/\text{s} = k(0.900\ M)$$

$$k = 6.20 \times 10^{-4}\ \text{s}^{-1}$$

The rate law expression is now: **Rate = $(6.20 \times 10^{-4}\ \text{s}^{-1})[N_2O_5]$**

16-68. *Refer to Section 16-4 and Exercise 16-54 Solution.*

From Exercise 16-54, we know that the reaction, $N_2O_5 \rightarrow NO_2 + NO_3$, is first order with specific rate constant, $k = 0.23\ \text{s}^{-1}$ at 25°C.

(a) The integrated rate equation: $\log \frac{A_o}{A} = \frac{akt}{2.303}$

where $A_o = 3.60$ mol/3.00 L

a = stoichiometric coefficient of N_2O_5

Substituting,

$$\log\left(\frac{3.60\ \text{mol}}{?\ \text{mol}\ N_2O_5}\right) = \frac{(0.23\ \text{s}^{-1})(1.00\ \text{min} \times 60\ \text{s/min})}{2.303} = 6.0$$

? mol N_2O_5 = **3.7×10^{-6} mol N_2O_5 remaining after 1.00 min**

Note: Any small change in the value of k can greatly alter the answer.

(b) If 99.0% of N_2O_5 have decomposed, then 1.0% of N_2O_5 remains.

Substituting into the integrated rate equation, we have $\log\left(\frac{100.0\%}{1.0\%}\right) = \frac{(0.23\ \text{s}^{-1})t}{2.303}$ Solving, t = **20 s**

16-70. *Refer to Section 16-3.*

Balanced equation: $4Hb + 3CO \rightarrow Hb_4(CO)_3$ where Hb is hemoglobin

(a) The form of the rate-law expression: Rate = $k[Hb]^x[CO]^y$

Step 1: Rate dependence on [Hb]. Consider Experiments 1 and 2:

Method 1: By observation, [CO] does not change. [Hb] increases by a factor of 2 (= 6.72/3.36). The rate of disappearance of Hb also increases by a factor of 2 (= 1.88/0.941). The rate is directly proportional to [Hb] and x is equal to 1. The reaction is first order with respect to Hb.

Method 2:

$$\frac{\text{Expt 2}}{\text{Expt 1}} \qquad \frac{1.88\ \mu\text{mol/L}\cdot\text{s}}{0.941\ \mu\text{mol/L}\cdot\text{s}} = \frac{k(6.72\ \mu\text{mol/L})^x(1.00\ \mu\text{mol/L})^y}{k(3.36\ \mu\text{mol/L})^x(1.00\ \mu\text{mol/L})^y}$$

$$2 = (2)^x$$

$$x = 1$$

Step 2: Rate dependence on [CO]. Consider Experiments 2 and 3.

Method 1: By observation, [Hb] does not change. [CO] increases by a factor of 3 (= 3.00/1.00). The rate of disappearance of Hb also increases by a factor of 3 (= 5.64/1.88). The rate is directly proportional to [CO] and y is equal to 1. The reaction is first order with respect to CO.

Method 2:

$$\frac{\text{Expt 3}}{\text{Expt 2}} \qquad \frac{5.64\ \mu\text{mol/L}\cdot\text{s}}{1.88\ \mu\text{mol/L}\cdot\text{s}} = \frac{k(6.72\ \mu\text{mol/L})^x(3.00\ \mu\text{mol/L})^y}{k(6.72\ \mu\text{mol/L})^x(1.00\ \mu\text{mol/L})^y}$$

$$3 = (3)^x$$

$$x = 1$$

The rate-law expression is: **Rate** = $k[Hb]^1[CO]^1$ = $\mathbf{k[Hb][CO]}$

(b) Substituting, the data from Experiment 1 into the rate-law expression to calculate k, we have,

$$0.941\ \mu\text{mol/L}\cdot\text{s} = k(3.36\ \mu\text{mol/L})(1.00\ \mu\text{mol/L})$$

$$k = \mathbf{0.280\ L/\mu mol\cdot s}$$

(c) Substituting into the complete rate-law expression, gives

$$\text{Rate} = (0.280\ \text{L}/\mu\text{mol}\cdot\text{s})[\text{Hb}][\text{CO}]$$

$$= (0.280\ \text{L}/\mu\text{mol}\cdot\text{s})(1.50\ \mu\text{mol/L})(0.600\ \mu\text{mol/L})$$

$$= \mathbf{0.252\ \mu mol/L\cdot s}$$

16-72. *Refer to Section 16-3.*

Plan: Use dimensional analysis and the rate-law expression to determine the units of k, the rate constant, in the following general equation:

Rate (M/s) = $k[A]^x$ where x = the overall order of the reaction; [A] = the reactant concentration (M)

	Overall Reaction Order	Example	Units of k
(a)	1	Rate = $k[A]$	$(M/\text{s})/M = \text{s}^{-1}$
(b)	2	Rate = $k[A]^2$	$(M/\text{s})/M^2 = M^{-1}\cdot\text{s}^{-1}$
(c)	n	Rate = $k[A]^n$	$(M/\text{s})/M^n = M^{1-n}\cdot\text{s}^{-1}$

16-74. *Refer to Section 16-7 and Figure 16-12.*

The overall reaction: $H_2(g) + I_2(g) \rightarrow 2HI(g)$ rate $= k[H_2][I_2]$

The reaction mechanism: (1) $I_2 \rightleftarrows 2I$ (fast, equilibrium)
                         (2) $I + H_2 \rightleftarrows H_2I$ (fast, equilibrium)
                         <u>(3) $H_2I + I \rightarrow 2HI$ (slow)</u>
                              $H_2 + I_2 \rightarrow 2HI$ (overall)

Step 1	Reactant	Excited State (first peak)	Products (first trough)
	$\ddot{:}\ddot{I}\text{-}\ddot{I}\ddot{:}$ (H-H present)	$\ddot{:}\ddot{I}\text{—}\ddot{I}\ddot{:}$ (H-H present)	$2 \; \ddot{:}\ddot{I}\cdot$ (H-H present)

Step 2	Reactants	Excited State (second peak)	Product (second trough)
	H-H and $\ddot{:}\ddot{I}\cdot$ ($:\ddot{I}\cdot$ present)	$H\text{–}H\text{---}\ddot{I}\ddot{:}$ ($:\ddot{I}\cdot$ present)	$H\text{-}H\text{-}\ddot{I}\ddot{:}$ ($:\ddot{I}\cdot$ present)
		weaker bond	(9 e^- present) *

Step 3	Reactants	Excited State (third peak)	Products (final)
	$\ddot{:}\ddot{I}\cdot$ and $H\text{-}H\text{-}\ddot{I}\ddot{:}$ *	$\ddot{:}\ddot{I}\text{---}H\text{---}H\text{---}\ddot{I}\ddot{:}$	$2 \; H\text{-}\ddot{I}\ddot{:}$

* The extra weak H-H bond contains only one electron.

16-76. *Refer to Sections 15-3, 15-8 and 16-6, Appendix K and Figure 16-10.*

Balanced equation: $O_3(g) + NO(g) \rightarrow NO_2(g) + O_2(g)$ $E_a = 9.6$ kJ/mol reaction

(a) $\Delta H^\circ_{rxn} = [\Delta H^\circ_f \, {}_{NO_2(g)} + \Delta H^\circ_f \, {}_{O_2(g)}] - [\Delta H^\circ_f \, {}_{O_3(g)} + \Delta H^\circ_f \, {}_{NO(g)}]$

          $= [(1 \text{ mol})(33.2 \text{ kJ/mol}) + (1 \text{ mol})(0 \text{ kJ/mol})] - [(1 \text{ mol})(143 \text{ kJ/mol}) + (1 \text{ mol})(90.25 \text{ kJ/mol})]$

          $= $ **-200. kJ/mol reaction**

(b)   For the reaction, $\Delta E^\circ = \Delta H^\circ$ since $\Delta n_{gas} = 0$.

    Therefore, the activation energy plot for this reaction is:

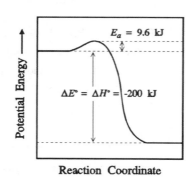

17 Chemical Equilibrium

17-2. *Refer to Section 17-2.*

Consider the equilibrium: $A(g) \rightleftarrows B(g)$ $\qquad\qquad K_c = \dfrac{[B]}{[A]}$

The magnitude of the equilibrium constant, K_c, is a measure of the extent to which a reaction occurs. If the equilibrium lies far to the right, then this means that at equilibrium most of the reactants would be converted into products and the value of K_c would be much greater than one. If the equilibrium lies far to the left, then at equilibrium, most of the reactants remain unreacted and there are very little products formed. The value of K_c would be a very small fraction.

17-4. *Refer to Sections 17-1 and 17-2.*

The magnitude of an equilibrium constant tells us *nothing* about how fast the system will reach equilibrium. Equilibrium constants are thermodynamic quantities, whereas the speed of a reaction is a kinetic quantity. The two are not related. Rather, an equilibrium constant is a measure of the extent to which a reaction occurs.

17-6. *Refer to Section 17-2 and 17-10.*

(a) $K_c = \dfrac{[CO][H_2O]}{[CO_2][H_2]}$ $\qquad$ (b) $K_c = \dfrac{[NO]^2[O_2]}{[NO_2]^2}$ $\qquad$ (c) $K_c = [CO_2]$

(d) $K_c = \dfrac{[H_2]}{[HBr]^2}$ $\qquad$ (e) $K_c = \dfrac{1}{[P_4][O_2]^3}$ $\qquad$ (f) $K_c = \dfrac{[CO_2]^2}{[CO]^2[O_2]}$

17-8. *Refer to Sections 17-2 and 17-7.*

Balanced equation: $Cl_2(aq) + 2Br^-(aq) \rightleftarrows 2Cl^-(aq) + Br_2(aq)$ $\qquad K_c = \dfrac{[Br_2][Cl^-]^2}{[Cl_2][Br^-]^2}$

When additional Br^- (from solid NaBr) is added to the above system at equilibrium, the system is no longer at equilibrium. Some Br^- ions will react with molecules of Cl_2 to form extra Cl^- ions and molecules of Br_2. The system readjusts itself so that when dynamic equilibrium is regained, the above equilibrium expression is still valid; the new concentrations of Br_2, Cl^-, Cl_2 and Br^- will give the same K_c value as long as the temperature remains constant.

17-10. *Refer to Section 17-2.*

The products are favored in those reactions in which $K > 1$: (b), (c) and (d).

17-12. *Refer to Section 17-2 and Example 17-1.*

Balanced equation: $N_2(g) + O_2(g) \rightleftarrows 2NO(g)$ $\qquad K_c = \dfrac{[NO]^2}{[N_2][O_2]} = \dfrac{(1.1 \times 10^{-5})^2}{(6.4 \times 10^{-3})(1.7 \times 10^{-3})} = \mathbf{1.1 \times 10^{-5}}$

17-14. *Refer to Section 17-2 and Example 17-1.*

Balanced equation: $PCl_3(g) + Cl_2(g) \rightleftarrows PCl_5(g)$

$$K_c = \frac{[PCl_5]}{[PCl_3][Cl_2]} = \frac{(12)}{(10)(9)} = \textbf{0.13}$$

17-16. *Refer to Sections 17-2 and 17-3, and Example 17-2.*

Balanced equation: $H_2(g) + Br_2(g) \rightleftarrows 2HBr(g)$

$$K_c = \frac{[HBr]^2}{[H_2][Br_2]} = 7.9 \times 10^{11}$$

(a) $1/2\ H_2(g) + 1/2\ Br_2(g) \rightleftarrows HBr(g)$

$$K_c' = \frac{[HBr]}{[H_2]^{1/2}[Br_2]^{1/2}} = \sqrt{K_c} = \textbf{8.9} \times \textbf{10}^5$$

(b) $2HBr(g) \rightleftarrows H_2(g) + Br_2(g)$

$$K_c'' = \frac{[H_2][Br_2]}{[HBr]^2} = \frac{1}{K_c} = \textbf{1.3} \times \textbf{10}^{-12}$$

(c) $4HBr(g) \rightleftarrows 2H_2(g) + 2Br_2(g)$

$$K_c''' = \frac{[H_2]^2[Br_2]^2}{[HBr]^4} = \frac{1}{K_c^2} = \textbf{1.6} \times \textbf{10}^{-24}$$

17-18. *Refer to Sections 17-1 and 17-2.*

Preparation (b) is the more suitable reaction for the preparation of HCl(g) since the equilibrium constant for reaction (b) is roughly 10,000 times greater than the equilibrium constant for reaction (a). A larger value of K_c indicates that more products will be formed when a system reaches equilibrium.

17-20. *Refer to Sections 17-2 and 17-5, and Examples 17-1, 17-5 and 17-6.*

Balanced equation: $H_2(g) + I_2(g) \rightleftarrows 2HI(g)$

(a) Let   x = moles of H_2 or I_2 that react.   Then
2x = moles of HI that are produced.

	H_2	+	I_2	$\rightleftarrows$	2HI
initial	9.84×10^{-4} mol		1.38×10^{-3} mol		0 mol
change	- x mol		- x mol		+ 2x mol
at equilibrium	$(9.84 \times 10^{-4}$ - x) mol		$(1.38 \times 10^{-3}$ - x) mol		2x mol

At equilibrium, mol $I_2 = (1.38 \times 10^{-3}$ - x) mol $= 4.73 \times 10^{-4}$ mol I_2
$$x = 9.1 \times 10^{-4}\ \text{mol}$$

Therefore,   mol $H_2 = 9.84 \times 10^{-4}$ - x $= 9.84 \times 10^{-4}$ mol - 9.1×10^{-4} mol $= \textbf{7.4} \times \textbf{10}^{-5}$ **mol** H_2
mol HI $= 2x = 2(9.1 \times 10^{-4}$ mol$) = \textbf{1.8} \times \textbf{10}^{-3}$ **mol HI**

(b) Let V = volume of container in liters. Then, at equilibrium,

$$[H_2] = \frac{7.4 \times 10^{-5}\ \text{mol}}{V} \qquad [I_2] = \frac{4.73 \times 10^{-4}\ \text{mol}}{V} \qquad [HI] = \frac{1.8 \times 10^{-3}\ \text{mol}}{V}$$

$$K_c = \frac{[HI]^2}{[H_2][I_2]} = \frac{\left(\dfrac{1.8 \times 10^{-3}}{V}\right)^2}{\left(\dfrac{7.4 \times 10^{-5}}{V}\right)\left(\dfrac{4.73 \times 10^{-4}}{V}\right)} = \textbf{93}$$

Note: In this exercise, the volume of the container cancels out in the equilibrium expression. This does not often happen, so you must work in units of molarity, not moles, in the equilibrium expression.

Balanced equation: $A(g) + B(g) \rightleftarrows C(g) + 2D(g)$

Plan: (1) Determine the concentrations of the species of interest.

(2) Determine the concentrations of all species after equilibrium is reached.

(3) Calculate K_c.

(1) $[A]_{initial} = [B]_{initial} = \dfrac{1.00 \text{ mol}}{0.400 \text{ L}} = 2.50 \text{ M}$

$[C]_{equil} = \dfrac{0.20 \text{ mol}}{0.400 \text{ L}} = 0.50 \text{ M}$

(2)

	A	+	B	$\rightleftarrows$	C	+	2D
initial	2.50 M		2.50 M		0 M		0 M
change	- 0.50 M		- 0.50 M		+ 0.50 M		+ 1.00 M
at equilibrium	2.00 M		2.00 M		0.50 M		1.00 M

(3) $K_c = \dfrac{[C][D]^2}{[A][B]} = \dfrac{(0.50)(1.00)^2}{(2.00)(2.00)} = 0.12$

17-24. *Refer to Section 17-4.*

Many systems are not at equilibrium. The mass action expression, also called the reaction quotient, Q, is a measure of how far a system is from equilibrium and in what direction the system must go to get to equilibrium. The reaction quotient has the same form as the equilibrium constant, K, but the concentration values put into Q are the actual values found in the system at that given moment.

(a) If $Q = K$, the system is at equilibrium.

(b) If $Q < K$, the system has greater concentrations of reactants than it would have if it were at equilibrium. The forward reaction will dominate until equilibrium is established.

(c) If $Q > K$, the system has greater concentrations of products than it would have if it were at equilibrium. The reverse reaction will dominate until equilibrium is reached.

17-26. *Refer to Section 17-4.*

If $Q > K$ for a reversible reaction, the reverse reaction occurs to a greater extent than the forward reaction until equilibrium is reached. If $Q < K$, the forward reaction occurs to a greater extent than the reverse reaction until equilibrium is reached.

17-28. *Refer to Section 17-4 and Example 17-3.*

Balanced equation: $H_2CO \rightleftarrows H_2 + CO$ $K_c = 0.50$

The reaction quotient, Q, for this reaction at the given moment is: $Q = \dfrac{[H_2][CO]}{[H_2CO]} = \dfrac{(1.50)(0.25)}{(0.50)} = 0.75$

Under the given conditions, $Q > K_c$.

(a) false   The reaction is not at equilibrium since $Q \neq K_c$.

(b) false   The reaction is not at equilibrium. However, the reaction will continue to proceed until equilibrium is reached.

(c) false   When $Q > K_c$, the system has more products and less reactants than it would have at equilibrium. Equilibrium will be reached by forming more H_2CO.

(d) false   The forward rate of the reaction is less than that of the reverse reaction since $Q > K_c$. The forward and reverse rates are equal only at equilibrium.

17-30.   *Refer to Section 17-5 and Example 17-6.*

Balanced equation: $N_2(g) + C_2H_2(g) \rightleftharpoons 2HCN(g)$ $K_c = 2.3 \times 10^{-4}$

Plan: Determine the equilibrium concentration of HCN using the K_c expression.

Let   x = moles per liter of N_2 that react. Then
         x = moles per liter of C_2H_2 that react, and
        2x = moles per liter of HCN that are formed.

	N_2	$+$	C_2H_2	$\rightleftharpoons$	2HCN
initial	2.5 *M*		1.0 *M*		0 *M*
change	- x *M*		- x *M*		+ 2x *M*
at equilibrium	(2.5 - x) *M*		(1.0 - x) *M*		2x *M*

$$K_c = \frac{[HCN]^2}{[N_2][C_2H_2]} = \frac{(2x)^2}{(2.5 - x)(1.0 - x)} = 2.3 \times 10^{-4}$$

Rearranging into a quadratic equation: $4.0x^2 + (8.1 \times 10^{-4})x - 5.8 \times 10^{-4} = 0$
Solving,

$$x = \frac{-8.1 \times 10^{-4} \pm \sqrt{(8.1 \times 10^{-4})^2 - 4(4.0)(-5.8 \times 10^{-4})}}{(2)(4.0)} = \frac{-8.1 \times 10^{-4} \pm 0.096}{8.00} = 0.012 \text{ or } -0.012 \text{ (discard)}$$

Note: There are always two solutions when solving quadratic equations, but only one is meaningful. The other solution, -0.012, is discarded because a negative value for concentration of HCN has no physical meaning in this problem.

Therefore, at equilibrium: [HCN] = 2x = **0.024 *M***

17-32.   *Refer to Section 17-2 and Example 17-1.*

Balanced equation: $SbCl_5(g) \rightleftharpoons SbCl_3(g) + Cl_2(g)$

Plan: (1) Determine the equilibrium concentrations of $SbCl_5$, $SbCl_3$ and Cl_2 at 448°C.
        (2) Evaluate K_c at 448°C by substituting the equilibrium concentrations into the K_c expression.

(1) $[SbCl_5] = \dfrac{(3.84 \text{ g}/299 \text{ g/mol})}{5.00 \text{ L}} = 2.57 \times 10^{-3} M$

    $[SbCl_3] = \dfrac{(9.14 \text{ g}/228 \text{ g/mol})}{5.00 \text{ L}} = 8.02 \times 10^{-3} M$

    $[Cl_2] = \dfrac{(2.84 \text{ g}/70.9 \text{ g/mol})}{5.00 \text{ L}} = 8.01 \times 10^{-3} M$

(2) $K_c = \dfrac{[SbCl_3][Cl_2]}{[SbCl_5]} = \dfrac{(8.02 \times 10^{-3})(8.01 \times 10^{-3})}{(2.57 \times 10^{-3})} = \mathbf{2.50 \times 10^{-2}}$

17-34. _Refer to Section 17-5 and Example 17-6._

Balanced equation: $PCl_3(g) + Cl_2(g) \rightleftarrows PCl_5(g)$ $K_c = 96.2$

Plan: Determine the equilibrium concentration of Cl_2 using the equilibrium expression.

Let   x = moles per liter of PCl_3 that react = moles per liter of Cl_2 that react. Then
       x = moles per liter of PCl_5 that are formed.

	PCl_3	+	Cl_2	$\rightleftarrows$	PCl_5
initial	0.20 M		7.0 M		0 M
change	- x M		- x M		+ x M
at equilibrium	(0.20 - x) M		(7.0 - x) M		x M

$$K_c = \frac{[PCl_5]}{[PCl_3][Cl_2]} = \frac{(x)}{(0.20 - x)(7.0 - x)} = 96.2$$

Rearranging into a quadratic equation: $96.2x^2 - 694x + 135 = 0$

Solving, $x = \dfrac{694 \pm \sqrt{(694)^2 - 4(96.2)(135)}}{(2)(96.2)} = \dfrac{694 \pm 656}{192} = 0.2 \text{ or } 7.0 \text{ (discard)}$

Note: The solution, x = 7.0 is meaningless since it would result in a negative equilibrium concentration for PCl_3.

Therefore, at equilibrium: $[Cl_2] = 7.0 - x = \textbf{6.8 } \textbf{\textit{M}}$

Note: The equilibrium concentration of PCl_3 is not zero, but a small positive number. Solve for $[PCl_3]$ by plugging the calculated equilibrium concentrations of Cl_2 and PCl_5 into the equilibrium expression.

$$K_c = \frac{[PCl_5]}{[PCl_3][Cl_2]} = \frac{x}{[PCl_3](7.0 - x)} = \frac{0.2}{[PCl_3](6.8)} = 96.2 \qquad \text{Solving, } [PCl_3] = 0.0003 \text{ } M$$

17-36. _Refer to Section 17-5, Example 7-6, and Exercise 17-32._

Balanced equation: $SbCl_5(g) \rightleftarrows SbCl_3(g) + Cl_2(g)$ $K_c = 2.51 \times 10^{-2}$

Plan: (1) Determine the initial concentrations of $SbCl_5$ and $SbCl_3$.
        (2) Determine the equilibrium concentrations of $SbCl_5$, $SbCl_3$ and Cl_2 using the K_c expression.

(1) $[SbCl_5] = \dfrac{(7.50 \text{ g}/299 \text{ g/mol})}{7.52 \text{ L}} = 3.34 \times 10^{-3} \text{ } M$

     $[SbCl_3] = \dfrac{(5.00 \text{ g}/228 \text{ g/mol})}{7.52 \text{ L}} = 2.92 \times 10^{-3} \text{ } M$

(2) Since $[Cl_2] = 0$, the forward reaction will predominate. Some $SbCl_5$ will react and some $SbCl_3$ and Cl_2 will be produced.

Let x = moles per liter of $SbCl_5$ that react. Then
     x = moles per liter of $SbCl_3$ produced = moles per liter of Cl_2 produced.

	$SbCl_5$	$\rightleftarrows$	$SbCl_3$	+	Cl_2
initial	3.34×10^{-3} M		2.92×10^{-3} M		0 M
change	- x M		+ x M		+ x M
at equilibrium	$(3.34 \times 10^{-3}$ - x$)$ M		$(2.92 \times 10^{-3}$ + x$)M$		x M

$$K_c = \frac{[SbCl_3][Cl_2]}{[SbCl_5]} = \frac{(x)(2.92 \times 10^{-3} + x)}{(3.34 \times 10^{-3} - x)} = 0.0251$$

The quadratic equation: $x^2 + 0.0280x - 8.38 \times 10^{-5} = 0$

Solving,

$$x = \frac{-0.0280 \pm \sqrt{(0.0280)^2 - 4(1)(-8.38 \times 10^{-5})}}{2(1)} = \frac{-0.0280 \pm 0.0335}{2} = 2.8 \times 10^{-3} \text{ or } -0.0308 \text{ (discard)}$$

Therefore, at equilibrium: $[SbCl_5] = 3.34 \times 10^{-3} - x = \mathbf{6 \times 10^{-4}\ M}$
$[SbCl_3] = 2.92 \times 10^{-3} + x = \mathbf{5.7 \times 10^{-3}\ M}$
$[Cl_2] = x = \mathbf{2.8 \times 10^{-3}\ M}$

17-38. *Refer to Sections 17-2 and 17-5.*

Balanced equation: $N_2(g) + 3H_2(g) \rightleftharpoons 2NH_3(g)$ $K_c = \dfrac{[NH_3]^2}{[N_2][H_2]^3} = 1$

(a) false   The coefficients in a balanced equation represent the relative number of moles of reactants that are consumed or products that are formed. They have no bearing on the relative concentrations of the species in the equilibrium condition. If we assume $[N_2] = 1.0\ M$, $[H_2] = 3.0\ M$ and $[NH_3] = 6.0\ M$, the corresponding K_c value is 0.15 instead of 1.

$$K_c = \frac{(6.0)^2}{(1.0)(3.0)^3} = 1.3$$

(b) false   If the reaction started with only NH_3 and went to equilibrium, then $[H_2]$ is three times that of $[N_2]$, but there are many other possibilities.

(c) true   Let $x = [N_2]$
$3x = [H_2]$
$2x = [NH_3]$

$$K_c = \frac{[NH_3]^2}{[N_2][H_2]^3} = \frac{(2x)^2}{(x)(3x)^3} = \frac{4x^2}{27x^4} = \frac{0.148}{x^2} = 1$$

Solving, $x = 0.385$

Therefore, $[N_2] = x = 0.385\ M$
$[H_2] = 3x = 1.15\ M$
$[NH_3] = 2x = 0.770\ M$

(d) true   Substituting, $K_c = \dfrac{(1)^2}{(1)(1)^3} = 1$

(e) false   The equilibrium expression is not satisfied if the concentrations of all reactants and products are equal and have any value other than $1\ M$.

(f) false   An equilibrium mixture does not require all reactants and products to have the same concentration. It only requires that the concentrations of the components fulfill the K_c expression.

17-40. *Refer to Section 17-5 and Example 7-6.*

Balanced equation: $POCl_3(g) \rightleftharpoons POCl(g) + Cl_2(g)$ $K_c = 0.450$

Plan: (1) Determine the initial concentration of $POCl_3$.
(2) Calculate the concentration of $POCl_3$ that dissociated, using the K_c expression.
(3) Determine the percent dissociation of $POCl_3$ at equilibrium.

(1) $[POCl_3] = \dfrac{0.600 \text{ mol}}{2.00 \text{ L}} = 0.300 \ M$

(2) Let x = moles per liter of $POCl_3$ that react (dissociate). Then
   x = moles per liter of POCl produced = moles per liter of Cl_2 produced.

	$POCl_3$	$\rightleftarrows$	POCl	+	Cl_2
initial	0.300 M		0 M		0 M
change	- x M		+ x M		+ x M
at equilibrium	(0.300 - x) M		x M		x M

$K_c = \dfrac{[POCl][Cl_2]}{[POCl_3]} = \dfrac{x^2}{0.300 - x} = 0.450$

The quadratic equation: $x^2 + 0.450x - 0.135 = 0$
Solving,

$x = \dfrac{-0.450 \pm \sqrt{(0.450)^2 - 4(1)(-0.135)}}{2(1)} = \dfrac{-0.450 \pm 0.862}{2} = 0.206 \text{ or } -0.656 \text{ (discard)}$

Therefore, the concentration of $POCl_3$ that reacted = $x = 0.206 \ M$

(3) % PCl_5 dissociated $= \dfrac{[POCl_3]_{reacted}}{[POCl_3]_{initial}} \times 100\% = \dfrac{0.206 \ M}{0.300 \ M} \times 100\% = \mathbf{68.7\%}$

17-42. _Refer to Section 17-6 and Example 17-7._

When an equilibrium system involving gases is subjected to an increase in pressure resulting from a decrease in volume, the concentrations of the gases increase and there may or may not be a shift in the equilibrium. If there is the same number of moles of gas on each side of the equation, equilibrium is not affected. If the number of moles of gas on each side of the equation is different, the general rule is that such an increase in pressure shifts a system in the direction that produces the smaller number of moles of gas.

(a) shift to left   (c) equilibrium is unaffected

(b) shift to right   (d) equilibrium is unaffected

17-44. _Refer to Section 17-6, Example 17-7 and Exercise 17-42 Solution._

Balanced equation: $2C(s) + O_2(g) \rightleftarrows 2CO(g)$

If the total pressure were increased, the equilibrium would **shift to the left** to create fewer gas molecules.

17-46. _Refer to Section 17-6 and Example 17-7._

Balanced equation: $CaCO_3(s) \rightleftarrows CaO(s) + CO_2(g)$ $K_c = [CO_2]$

(a) When CO_2 is added, $[CO_2]$ is too high, and (i) the mass of $CaCO_3$ increases since the equilibrium shifts to the left until $[CO_2] = K_c$ once again.

(b) When the pressure is decreased by increasing the volume of the container, $[CO_2]$ is too low, and (ii) the mass of $CaCO_3$ decreases, since the equilibrium shifts to the right until $[CO_2] = K_c$.

(c) When $CaO(s)$ is removed, (iii) the mass of $CaCO_3$ remains the same. The equilibrium is not affected because CaO as a solid is considered to have a constant concentration (has an activity of unity) and does not appear in the equilibrium expression.

17-48. *Refer to Section 17-6 and Example 17-7.*

Balanced equation: $6CO_2(g) + 6H_2O(\ell) \rightleftarrows C_6H_{12}O_6(s) + 6O_2(g)$ $\Delta H° = 2801.69$ kJ/mol rxn

(a) If $[CO_2]$ is increased, (i) the equilibrium will shift to the right.

(b) If P_{O_2} is increased, (ii) the equilibrium will shift to the left.

(c) If one-half of $C_6H_{12}O_6(s)$ is removed, (iii) the equilibrium is unaffected since $C_6H_{12}O_6$ is a solid and does not appear in the equilibrium expression.

(d) If the total pressure is decreased, the equilibrium is unaffected since the total number of moles of gas is the same on both sides of the equation.

(e) If the temperature is increased, (i) the equilibrium will shift to the right since the forward reaction is endothermic ($\Delta H° > 0$) and the forward reaction will absorb more heat to minimize the effect of raising the temperature and adding heat.

(f) If a catalyst is added, (iii) the equilibrium is unaffected. The reaction would simply reach equilibrium at a faster rate.

17-50. *Refer to Section 17-6, Example 17-7 and Exercise 17-42 Solution.*

When the pressure is decreased by increasing the volume, the equilibrium in question :

(a) shifts to left   (b) is not affected   (c) shifts to right   (d) shifts to left   (e) shifts to left

---

17-52. *Refer to Sections 17-4 and 17-7, and Example 17-8.*

Balanced equation: $A(g) + B(g) \rightleftarrows C(g) + D(g)$

(a) $K_c = \dfrac{[C][D]}{[A][B]} = \dfrac{(1.60 \text{ mol}/1.00 \text{ L})^2}{(0.40 \text{ mol}/1.00 \text{ L})^2} = \textbf{16}$

(b) Since we are adding both reactant and product to the system, the value of Q_c must be evaluated to determine the direction of the reaction.

New $[B] = 0.40\ M + 0.20\ M = 0.60\ M$
New $[C] = 1.60\ M + 0.20\ M = 1.80\ M$ $Q_c = \dfrac{[C][D]}{[A][B]} = \dfrac{(1.80\ M)(1.60\ M)}{(0.40\ M)(0.60\ M)} = 12$

Since $Q_c < K_c$, the forward reaction proceeds.

Let x = moles per liter of A or B that react *after* the addition of 0.20 moles per liter of A and C. Then
x = moles per liter of C produced = moles per liter of D produced.

	A	+	B	$\rightleftarrows$	C	+	D
initial	0.40 *M*		0.40 *M*		1.60 *M*		1.60 *M*
mol/L added	0 *M*		+ 0.20 *M*		+ 0.20 *M*		0 *M*
new system	0.40 *M*		0.60 *M*		1.80 *M*		1.60 *M*
change	- x *M*		- x *M*		+ x *M*		+ x *M*
at equil	(0.40 - x) *M*		(0.60 - x) *M*		(1.80 + x) *M*		(1.60 + x) *M*

$K_c = \dfrac{[C][D]}{[A][B]} = \dfrac{(1.80 + x)(1.60 + x)}{(0.40 - x)(0.60 - x)} = \textbf{16}$

The quadratic equation: $15x^2 - 19.4x + 0.96 = 0$

Solving, $x = \dfrac{19.4 \pm \sqrt{(19.4)^2 - 4(15)(0.96)}}{2(15)} = \dfrac{19.4 \pm 17.9}{30} = 0.052$ or 1.24 (discard)

Therefore, the new equilibrium concentration of A is (0.40 - x) or **0.35 *M***

Balanced equation: $A(g) \rightleftarrows B(g) + C(g)$

(a) $K_c = \dfrac{[B][C]}{[A]} = \dfrac{(0.20)^2}{0.30} = \mathbf{0.13}$

(b) If the volume is suddenly doubled, the initial concentrations will be halved and the system is no longer at equilibrium. We learned in Section 17-5 that the equilibrium will then shift to the side with the greater number of moles of gas, i.e., the right side.

Let x = number of moles per liter of A that react *after* the volume is doubled.

   x = number of moles per liter of B produced = number of moles per liter of C produced.

	A	$\rightleftarrows$	B	+	C
initial	0.30 *M*		0.20 *M*		0.20 *M*
new system	0.15 *M*		0.10 *M*		0.10 *M*
change	- x *M*		+ x *M*		+ x *M*
at equilibrium	(0.15 - x) *M*		(0.10 + x) *M*		(0.10 + x) *M*

$K_c = \dfrac{[B][C]}{[A]} = \dfrac{(0.10 + x)^2}{(0.15 - x)} = \mathbf{0.13}$

The quadratic equation: $x^2 + 0.33x - 0.0095 = 0$

Solving, $x = \dfrac{-0.33 \pm \sqrt{(0.33)^2 - 4(1)(-0.0095)}}{2(1)} = \dfrac{-0.33 \pm 0.38}{2} = 0.02$ or -0.36 (discard)

Therefore,   [A] = 0.15 - x = **0.13 *M***
       [B] = [C] = 0.10 + x = **0.12 *M***

(c) If the volume is suddenly halved, the initial equilibrium concentrations will be doubled and this system is no longer at equilibrium. The equilibrium will shift to the left side, the side with the lesser number of moles of gas.

Let x = number of moles per liter of B that react *after* the volume is halved, and

   x = number of moles per liter of C that react after the volume is halved. Then

   x = number of moles of A that are produced.

	A	$\rightleftarrows$	B	+	C
initial	0.30 *M*		0.20 *M*		0.20 *M*
new system	0.60 *M*		0.40 *M*		0.40 *M*
change	+ x *M*		- x *M*		- x *M*
at equilibrium	(0.60 + x) *M*		(0.40 - x) *M*		(0.40 - x) *M*

$K_c = \dfrac{[B][C]}{[A]} = \dfrac{(0.40 - x)^2}{(0.60 + x)} = \mathbf{0.13}$

The quadratic equation: $x^2 - 0.93x - 0.082 = 0$

Solving, $x = \dfrac{0.93 \pm \sqrt{(-0.93)^2 - 4(1)(0.082)}}{2(1)} = \dfrac{0.93 \pm 0.73}{2} = 0.10$ or 0.83 (discard)

Therefore,   [A] = 0.60 + x = **0.70 *M***
       [B] = [C] = 0.40 - x = **0.30 *M***

Balanced equation: $N_2O_4(g) \rightleftarrows 2NO_2(g)$ $K_c = 5.84 \times 10^{-3}$

(a) $[N_2O_4]_{initial} = \dfrac{3.50 \text{ g}/92.0 \text{ g/mol}}{2.00 \text{ L}} = 0.0190 \ M$

Let x = number of moles per liter of N_2O_4 that react. Then
   2x = number of moles per liter of NO_2 that are produced.

	N_2O_4	$\rightleftarrows$	$2NO_2$
initial	0.0190 M		0 M
change	- x M		+ 2x M
at equilibrium	(0.0190 - x) M		2x M

$$K_c = \frac{[NO_2]^2}{[N_2O_4]} = \frac{(2x)^2}{(0.0190 - x)} = 5.84 \times 10^{-3}$$

The quadratic equation: $4x^2 + (5.84 \times 10^{-3})x - 1.11 \times 10^{-4} = 0$

Solving, $x = \dfrac{-5.84 \times 10^{-3} \pm \sqrt{(5.84 \times 10^{-3})^2 - 4(4)(-1.11 \times 10^{-4})}}{2(4)}$

$= \dfrac{-5.84 \times 10^{-3} \pm 4.25 \times 10^{-2}}{8} = 4.59 \times 10^{-3} \text{ or } -6.04 \times 10^{-3} \text{ (discard)}$

Therefore, at equilibrium $[N_2O_4] = 0.0190 - x = \mathbf{0.0144 \ M}$
                $[NO_2] = 2x = \mathbf{9.18 \times 10^{-3} \ M}$

(b) When the volume is suddenly doubled (2L → 4L), the concentrations of N_2O_4 and NO_2 are halved and the equilibrium shifts to the right.

Let x = number of moles per liter of N_2O_4 that react *after* the volume is doubled. Then
   2x = number of moles of NO_2 that are produced.

	N_2O_4	$\rightleftarrows$	$2 NO_2$
initial	$1.44 \times 10^{-2} \ M$		$9.18 \times 10^{-3} \ M$
new system	$7.2 \times 10^{-3} \ M$		$4.59 \times 10^{-3} \ M$
change	- x M		+ 2x M
at equilibrium	$(7.2 \times 10^{-3} - x) \ M$		$(4.59 \times 10^{-3} + 2x) \ M$

$$K_c = \frac{[NO_2]^2}{[N_2O_4]} = \frac{(4.59 \times 10^{-3} + 2x)^2}{(7.2 \times 10^{-3} - x)} = 5.84 \times 10^{-3}$$

The quadratic equation: $4x^2 + 0.0242x - 2.1 \times 10^{-5} = 0$

Solving, $x = \dfrac{-0.0242 \pm \sqrt{(0.0242)^2 - 4(4)(-2.1 \times 10^{-5})}}{2(4)}$

$= \dfrac{-0.0242 \pm 0.0303}{8} = 7.67 \times 10^{-4} \text{ or } -6.81 \times 10^{-3} \text{ (discard)}$

Therefore, at equilibrium $[N_2O_4] = 7.2 \times 10^{-3} - x = \mathbf{6.4 \times 10^{-3} \ M}$
                $[NO_2] = 4.59 \times 10^{-3} + 2x = \mathbf{6.12 \times 10^{-3} \ M}$

(c) When the volume in (a) is suddenly halved (2L → 1L), the concentrations of N_2O_4 and NO_2 are doubled and the equilibrium shifts to the left side.

Let x = number of moles per liter of N_2O_4 that react *after* the volume is halved. Then
   2x = number of moles of NO_2 that are consumed.

	N_2O_4	⇌	2 NO_2
initial	1.44×10^{-2} M		9.18×10^{-3} M
new system	0.029 M		0.0184 M
change	+ x M		- 2x M
at equilibrium	(0.029 + x) M		(0.0184 - 2x) M

$$K_c = \frac{[NO_2]^2}{[N_2O_4]} = \frac{(0.0184 - 2x)^2}{(0.029 + x)} = 5.84 \times 10^{-3}$$

The quadratic equation: $4x^2 - 0.0794x + 1.7 \times 10^{-4} = 0$

Solving, $x = \dfrac{0.0794 \pm \sqrt{(-0.0794)^2 - 4(4)(1.7 \times 10^{-4})}}{2(4)}$

$$= \frac{0.0794 \pm 0.060}{8} = 2.4 \times 10^{-3} \text{ or } 0.0174 \text{ (discard)}$$

Therefore, at equilibrium $[N_2O_4] = 0.029 + x = \mathbf{0.031\ M}$
                 $[NO_2] = 0.0184 - 2x = \mathbf{0.0136\ M}$

17-58.   *Refer to Sections 17-8 and 17-9.*

The values of K_p and K_c are numerically equal for reactions in which there are equal numbers of moles of gases on both sides of the equation, i.e., $\Delta n_{gas} = 0$.

17-60.   *Refer to Section 17-8 and Example 17-10.*

Balanced equation: C(graphite) + $CO_2(g)$ ⇌ 2CO(g)

Plan: (1) Calculate the partial pressures of CO and CO_2.
     (2) Determine K_p.

(1) Since the CO_2 gas stream contains 4.0×10^{-3} mol percent CO, the mole fraction of CO is 4.0×10^{-5}.

   P_{CO} = mole fraction CO $\times P_{total}$ = $(4.0 \times 10^{-5})(1.00 \text{ atm})$ = 4.0×10^{-5} atm
   P_{CO_2} = mole fraction $CO_2 \times P_{total}$ = 1.00 (to 3 significant figures) $\times$ 1 atm = 1.00 atm

(2) $K_p = \dfrac{(P_{CO})^2}{P_{CO_2}} = \dfrac{(4.0 \times 10^{-5} \text{ atm})^2}{1.00 \text{ atm}} = \mathbf{1.6 \times 10^{-9}}$

17-62.   *Refer to Section 17-8 and Example 17-10.*

Balanced equation: $H_2(g)$ + $CO_2(g)$ ⇌ CO(g) + $H_2O(g)$

$$K_p = \frac{P_{CO}\, P_{H_2O}}{P_{H_2}\, P_{CO_2}} = \frac{(0.180 \text{ atm})(0.252 \text{ atm})}{(0.387 \text{ atm})(0.152 \text{ atm})} = \mathbf{0.771}$$

Balanced equation: $Br_2(g) \rightleftharpoons 2Br(g)$ $\qquad\qquad$ $K_p = 2550$ at 4000 K

$K_c = K_p(RT)^{-\Delta n} = 2550(0.0821 \times 4000)^{-(2-1)} = \mathbf{7.76}$

Balanced equation: $NH_4NH_2CO_2(s) \rightleftharpoons 2NH_3(g) + CO_2(g)$

If we start with pure $NH_4NH_2CO_2(s)$, the equilibrium partial pressure of NH_3 would be twice as great as that of CO_2.

Since at equilibrium: $\qquad P_{CO_2} = 0.300$ atm
Then: $\qquad\qquad\qquad\quad P_{NH_3} = 2 \times 0.300$ atm $= 0.600$ atm

$K_p = (P_{NH_3})^2(P_{CO_2}) = (0.600)^2(0.300) = \mathbf{0.108}$

Balanced equation: $Fe_2O_3(s) + 3H_2(g) \rightleftharpoons 2Fe(s) + 3H_2O(g)$ $\qquad K_c = 8.11$ at 1000 K $\qquad \Delta H = 96$ kJ/mol rxn

(a) Let $\quad$ x = moles per liter of H_2 present initially
$\qquad\qquad$ y = moles per liter of H_2 that react

	$Fe_2O_3(s)$	+	$3H_2$	$\rightleftharpoons$	$2Fe(s)$	+	$3H_2O$
initial	-		x *M*		-		0 *M*
change			- y *M*				+ y *M*
at equilibrium			(x - y) *M*				y *M*

$K_c = \dfrac{[H_2O]^3}{[H_2]^3} = \dfrac{y^3}{(x-y)^3} = 8.11$ $\qquad$ Taking the cube root of both sides: $\qquad \dfrac{y}{x-y} = 2.01$

$$y = (x-y)2.01$$
$$y = 2.01x - 2.01y$$
$$3.01y = 2.01x$$
$$y = 0.668x$$

Therefore, $\quad$ the percentage of H_2 that reacts $= 66.8\%$
$\qquad\qquad$ the percentage of H_2 that remains unreacted $= (100 - 66.8)\% = \mathbf{33.2\%}$

(b) At lower temperatures, the equilibrium will shift to the left since the reaction is endothermic ($\Delta H > 0$), and the percentage of H_2 that remains unreacted will be **greater**.

Balanced equation: $NH_4Cl(s) \rightleftharpoons NH_3(g) + HCl(g)$ $\qquad\qquad \Delta H = 176$ kJ/mol rxn

(a) As the temperature decreases, the mass of NH_3 decreases because the reaction is endothermic ($\Delta H > 0$).

(b) When more NH_3 is added, the total mass of NH_3 is initially the sum of the original equilibrium amount plus the additional NH_3. After equilibrium is reestablished by shifting to the left, the final mass of NH_3 is greater than the original equilibrium amount but less than the total mass present immediately after the additional NH_3 was added.

(c) When more HCl is added, the equilibrium shifts to the left and the mass of NH_3 decreases.

(d) When more solid NH_4Cl is added, the equilibrium is unaffected if the total gas volume is unchanged.

(e) When more solid NH_4Cl is added and the total gas volume decreases, the concentration of the gases will increase and the equilibrium will then shift to the left, thereby decreasing the concentrations of NH_3 and HCl back to their original values. However, although there is no change in the concentration of NH_3, the mass of NH_3 decreases due to the volume shrinkage.

17-72. *Refer to Section 17-8.*

Balanced equation: $C(s) + CO_2(g) \rightleftarrows 2CO(g)$

Given: $P_{CO} + P_{CO_2} = 1.00$ atm; $K_c = 1.50$

Let $\quad$ x = partial pressure of CO = P_{CO}. Then,
$\qquad$ 1.00 - x = partial pressure of CO_2 = P_{CO_2}

$$K_p = \frac{(P_{CO})^2}{P_{CO_2}} = \frac{x^2}{1.00 - x} = \mathbf{1.50}$$

The quadratic equation: $x^2 + 1.50x - 1.50 = 0$

$$\text{Solving, } x = \frac{-1.50 \pm \sqrt{(1.50)^2 - 4(1)(-1.50)}}{2(1)}$$

$$= \frac{-1.50 \pm 2.87}{2} = 0.685 \text{ or } -2.19 \text{ (discard)}$$

Therefore, at equilibrium $\qquad P_{CO} = x = \mathbf{0.685 \text{ atm}}$
$\qquad\qquad\qquad\qquad\qquad\qquad P_{CO_2} = 1.00 - x = \mathbf{0.315 \text{ atm}}$

17-74. *Refer to Section 17-11.*

(a) If $K \gg 1$, the forward reaction is likely to be spontaneous, and ΔG°_{rxn} must be negative.

(b) If $K = 1$, then ΔG°_{rxn} is 0, and the reaction is at equilibrium when the concentrations of aqueous species are 1 M and the partial pressures of gaseous species are 1 atm.

(c) If $K \ll 1$, the forward reaction is likely to be nonspontaneous and ΔG°_{rxn} must be positive.

17-76. *Refer to Section 17-11, Examples 17-14 and 17-15, and Appendix K.*

Balanced equation: $2SO_2(g) + O_2(g) \rightleftarrows 2SO_3(g)$

$$\begin{aligned}
\Delta G^\circ_{rxn} &= [2\Delta G^\circ_f \, _{SO_3(g)}] - [2\Delta G^\circ_f \, _{SO_2(g)} + \Delta G^\circ_f \, _{O_2(g)}] \\
&= [(2 \text{ mol})(-371.1 \text{ kJ/mol})] - [(2 \text{ mol})(-300.2 \text{ kJ/mol}) + (1 \text{ mol})(0 \text{ kJ/mol})] \\
&= -141.8 \text{ kJ/mol rxn}
\end{aligned}$$

We know $\qquad\qquad\qquad\qquad \Delta G^\circ_{rxn} = -2.303RT \log K_p \qquad$ for a gas phase reaction

Substituting, $\qquad -141.8 \times 10^3 \text{ J} = -2.303(8.314 \text{ J/K})(298.15 \text{ K})\log K_p$
$$\log K_p = 24.84$$
$$K_p = \mathbf{6.9 \times 10^{24}}$$

Balanced equation: $2Cl_2(g) + 2H_2O(g) \rightleftarrows 4HCl(g) + O_2(g)$ $\quad \Delta H° = +115$ kJ/mol rxn

$$K_p = 4.6 \times 10^{-14} \quad \text{at } 25°C$$

At 400°C (673 K): Substituting into the van't Hoff equation, we have

$$\ln \frac{K_{p\ T_2}}{K_{p\ T_1}} = \frac{\Delta H°(T_2 - T_1)}{RT_1 T_2}$$

$$\ln \frac{K_{p\ 673K}}{4.6 \times 10^{-14}} = \frac{(115 \times 10^3 \text{ J})(673 \text{ K} - 298 \text{ K})}{(8.314 \text{ J/K})(298 \text{ K})(673 \text{ K})} = 25.9$$

$$\frac{K_{p\ 673K}}{4.6 \times 10^{-14}} = e^{25.9} = 1.8 \times 10^{11}$$

$$K_{p\ 673K} = \mathbf{8.3 \times 10^{-3}}$$

$$K_c = K_p(RT)^{-\Delta n} = (8.3 \times 10^{-3})(0.0821 \times 673)^{-(5-4)} = \mathbf{1.5 \times 10^{-4}}$$

At 800°C (1073 K): Substituting into the van't Hoff equation, we have

$$\ln \frac{K_{p\ 1073K}}{4.6 \times 10^{-14}} = \frac{(115 \times 10^3 \text{ J})(1073 \text{ K} - 298 \text{ K})}{(8.314 \text{ J/K})(298 \text{ K})(1073 \text{ K})} = 33.5$$

$$K_{p\ 1073K} = \mathbf{16}$$

$$K_c = K_p(RT)^{-\Delta n} = (16)(0.0821 \times 1073)^{-(5-4)} = \mathbf{0.18}$$

Balanced equation: $CO(g) + H_2O(g) \rightleftarrows CO_2(g) + H_2(g)$

(a) Plan: (1) Calculate $\Delta H°_{rxn}$ and $\Delta S°_{rxn}$ from data in Appendix K and substitute into the Gibbs-Helmholtz equation to determine $\Delta G°_{rxn}$.

(2) Calculate K_P, using $\Delta G° = -2.303RT \log K_p$.

(1) $\Delta H°_{rxn} = [\Delta H°_{f\ CO_2(g)} + \Delta H°_{f\ H_2(g)}] - [\Delta H°_{f\ CO(g)} + \Delta H°_{f\ H_2O(g)}]$

$\quad = [(1 \text{ mol})(-393.5 \text{ kJ/mol}) + (1 \text{ mol})(0 \text{ kJ/mol})]$

$\qquad\qquad - [(1 \text{ mol})(-110.5 \text{ kJ/mol}) + (1 \text{ mol})(-241.8 \text{ kJ/mol})]$

$\quad = -41.2 \text{ kJ/mol rxn}$

$\Delta S°_{rxn} = [S°_{CO_2(g)} + S°_{H_2(g)}] - [S°_{CO(g)} + S°_{H_2O(g)}]$

$\quad = [(1 \text{ mol})(213.6 \text{ J/mol·K}) + (1 \text{ mol})(130.6 \text{ J/mol·K})]$

$\qquad\qquad - [(1 \text{ mol})(197.6 \text{ J/mol·K}) + (1 \text{ mol})(188.7 \text{ J/mol·K})]$

$\quad = -42.1 \text{ J/K}$

$\Delta G° = \Delta H° - T\Delta S° = -41.2 \times 10^3 \text{ J} - (298 \text{ K})(-42.1 \text{ J/K}) = -2.87 \times 10^4 \text{ J/mol rxn}$

(2) Substituting into $\Delta G° = -2.303RT \log K_p$, we have

$$-2.87 \times 10^4 \text{ J} = -2.303(8.314 \text{ J/K})(298 \text{ K})\log K_p$$

$$\log K_p = 5.03$$

$$K_p = \mathbf{1.1 \times 10^5}$$

(b) If we assume that $\Delta H°$ and $\Delta S°$ for the reaction are independent of temperature, we can determine $\Delta G°$ at 200°C (473 K) by substituting into the Gibbs-Helmholtz equation.

$$\Delta G° = \Delta H° - T\Delta S° = -41.2 \times 10^3 \text{ J} - (473 \text{ K})(-42.1 \text{ J/K}) = -2.13 \times 10^4 \text{ J/mol rxn}$$

Then,

$$\Delta G° = -2.303RT \log K_p$$
$$-2.13 \times 10^4 \text{ J} = -2.303(8.314 \text{ J/K})(473 \text{ K})\log K_p$$
$$\log K_p = 2.35$$
$$K_p = 2.2 \times 10^2$$

(c) $\Delta G°_{rxn} = [\Delta G°_f \, {}_{CO_2(g)} + \Delta G°_f \, {}_{H_2(g)}] - [\Delta G°_f \, {}_{CO(g)} + \Delta G°_f \, {}_{H_2O(g)}]$
  $= [(1 \text{ mol})(-394.4 \text{ kJ/mol}) + (1 \text{ mol})(0 \text{ kJ/mol})] - [(1 \text{ mol})(-137.2 \text{ kJ/mol}) + (1 \text{ mol})(-228.6 \text{ kJ/mol})]$
  $= $ **-28.6 kJ/mol rxn or -2.86 × 10⁴ J/mol rxn**

Substituting into $\Delta G° = -2.303RT \log K_p$, we have

$$-2.86 \times 10^4 \text{ J} = -2.303(8.314 \text{ J/K})(298 \text{ K})\log K_p$$
$$\log K_p = 5.01$$
$$K_p = 1.0 \times 10^5$$

17-82. *Refer to Sections 17-11 and 17-12, and Example 17-17.*

(a) $\Delta G°_{rxn} = -2.303RT \log K_p$
  $= -2.303(8.314 \text{ J/K})(298.15 \text{ K})(\log 4.3 \times 10^6)$
  $= $ **-3.79 × 10⁴ J/mol rxn or -37.9 kJ/mol rxn at 25°C**

(b) Using the van't Hoff equation,

$$\ln \frac{K_{p \, T_2}}{K_{p \, T_1}} = \frac{\Delta H°(T_2 - T_1)}{RT_1 T_2}$$

$$\ln \frac{K_{p \, 1073 K}}{4.3 \times 10^6} = \frac{(-78.58 \times 10^3 \text{ J})(1073 \text{ K} - 298 \text{ K})}{(8.314 \text{ J/K})(298 \text{ K})(1073 \text{ K})} = -22.9$$

$$\frac{K_{p \, 1073 K}}{4.3 \times 10^6} = e^{-22.9} = 1.13 \times 10^{-10}$$

$$K_{p \, 1073 K} = 4.9 \times 10^{-4}$$

(c) $\Delta G°_{rxn} = -2.303(8.314 \text{ J/K})(1073 \text{ K})(\log 4.9 \times 10^{-4}) = $ **+6.80 × 10⁴ J/mol rxn**
  or +68.0 kJ/mol rxn at 800°C

(d) The forward reaction at 800°C is nonspontaneous (ΔG is positive) whereas at 25°C the forward reaction is spontaneous (ΔG is negative).

(e) The reaction mixture is heated to speed up the rate at which equilibrium is reached, not to shift the equilibrium toward more product. Heating actually decreases the amount of product present at equilibrium since the reaction is exothermic (ΔH is negative).

(f) The basic purpose of a catalyst is to speed up the reaction of interest. Recall from Chapter 16 that the presence of a catalyst increases the rates of both forward and reverse reactions to the same extent without affecting equilibrium. However, the catalyst increases the rate at which the reaction goes to equilibrium. Hopefully, the yield of desired product is adequate at equilibrium.

17-84. *Refer to Sections 17-9 and 17-4.*

Balanced equation: $2CO(g) \rightleftharpoons C(\text{graphite}) + CO_2(g)$ $K_p = 1.11 \times 10^{21}$ at 25°C

Plan: (1) Calculate K_c from K_p.
  (2) Calculate the reaction quotient, Q_c, and compare the value with K_c.

(1) $K_c = K_p(RT)^{-\Delta n_{gas}} = (1.11 \times 10^{21})(0.0821 \text{ L·atm/mol·K} \times 298 \text{ K})^{-(1-2)} = \mathbf{2.72 \times 10^{22}}$

(2) $Q_c = \dfrac{[CO_2]}{[CO]^2} = \dfrac{(2 \text{ mol/1 L})}{(3 \text{ mol/1 L})^2} = 0.2$ (to 1 significant figure)

Since $Q_c << K_c$, the reaction will favor the products. Because K_c is so large, the reaction will go to near completion. The concentration of CO will decrease to nearly zero, the concentration of CO_2 will increase to nearly 3.5 M and nearly 1.5 moles of C(graphite) will be formed.

17-86. *Refer to Sections 17-2 , 17-3 and 17-9, Example 17-2, and Exercise 17-16 Solution.*

Balanced equation: $H_2(g) + Br_2(g) \rightleftarrows 2HBr(g)$ $\qquad K_c = \dfrac{[HBr]^2}{[H_2][Br_2]} = 7.9 \times 10^{11}$ $\quad$ at 500 K

(a) $K_p = \dfrac{(P_{HBr})^2}{(P_{H_2})(P_{Br_2})} = K_c(RT)^{\Delta n_{gas}} = (7.9 \times 10^{11})(0.0821 \text{ L·atm/mol·K} \times 500 \text{ K})^{(2-2)} = \mathbf{7.9 \times 10^{11}}$

(b) $1/2\ H_2(g) + 1/2\ Br_2(g) \rightleftarrows HBr(g)$ $\qquad K_p' = \dfrac{(P_{HBr})}{(P_{H_2})^{1/2}(P_{Br_2})^{1/2}} = \sqrt{K_p} = \mathbf{8.9 \times 10^5}$

(c) $2HBr(g) \rightleftarrows H_2(g) + Br_2(g)$ $\qquad K_p'' = \dfrac{(P_{H_2})(P_{Br_2})}{(P_{HBr})^2} = \dfrac{1}{K_p} = \mathbf{1.3 \times 10^{-12}}$

(d) $4HBr(g) \rightleftarrows 2H_2(g) + 2Br_2(g)$ $\qquad K_p''' = \dfrac{(P_{H_2})^2(P_{Br_2})^2}{(P_{HBr})^4} = \dfrac{1}{K_p{}^2} = \mathbf{1.6 \times 10^{-24}}$

17-88. *Refer to Sections 17-5 and 17-9.*

Balanced equation: $CO(g) + 3H_2(g) \rightleftarrows CH_4(g) + H_2O(g)$

(1) $K_c = \dfrac{[CH_4][H_2O]}{[CO][H_2]^3} = \dfrac{\left(\dfrac{1.21 \times 10^{-4} \text{ mol}}{0.100 \text{ L}}\right)\left(\dfrac{5.63 \times 10^{-8} \text{ mol}}{0.100 \text{ L}}\right)}{\left(\dfrac{1.21 \times 10^{-4} \text{ mol}}{0.100 \text{ L}}\right)\left(\dfrac{2.47 \times 10^{-4} \text{ mol}}{0.100 \text{ L}}\right)^3} = 37.4$

(2) $K_p = K_c(RT)^{\Delta n_{gas}} = (37.4)(0.0821 \text{ L·atm/mol·K} \times 1133 \text{ K})^{(2-4)} = \mathbf{4.32 \times 10^{-3}}$

17-90. *Refer to Section 17-11 and Appendix K.*

Balanced equation: $2SO_2(g) + O_2(g) \rightleftarrows 2SO_3(g)$ $\qquad K_p = 6.98 \times 10^{24}$ at 298 K

(1) $\Delta G^\circ_{rxn} = -2.303RT \log K_p = -2.303(8.314 \text{ J/K})(298 \text{ K})\log (6.98 \times 10^{24}) = -1.42 \times 10^5 \text{ J}$ or **-142 kJ**

(2) Given: $\Delta G^\circ_f\ SO_2(g) = -300.194 \text{ kJ/mol}$

$\Delta G^\circ_{rxn} = [2\Delta G^\circ_f\ SO_3(g)] - [2\Delta G^\circ_f\ SO_2(g) + \Delta G^\circ_f\ O_2(g)]$

$-142 \text{ kJ} = [2\Delta G^\circ_f\ SO_3(g)] - [(2 \text{ mol})(-300.194 \text{ kJ/mol}) + (1 \text{ mol})(0 \text{ kJ/mol})]$

Solving, $\Delta G^\circ_f\ SO_3(g) = \mathbf{-371 \text{ kJ/mol}}$ $\quad$ (In Appendix K, $\Delta G^\circ_f\ SO_3(g) = -371.1 \text{ kJ/mol}$)

242

Plan: (1) Treat the vaporization process as a chemical reaction, then establish the relationship between K_{eq} and P_{H_2O}.

       (2) Use the van't Hoff equation to evaluate P_{H_2O} at 50°C.

(1) For vaporization: $H_2O(\ell) \rightleftarrows H_2O(g)$ $K_{eq} = P_{H_2O}$

At temperature $T_1 = 100°C = 373$ K $K_1 = P_{H_2O}$ at 373 K = 1 atm
At temperature $T_2 = 50°C = 323$ K $K_2 = P_{H_2O}$ at 323 K = unknown

(2) van't Hoff equation:

$$\log \frac{K_2}{K_1} = \frac{\Delta H°(T_2 - T_1)}{2.303 R T_1 T_2} \quad \text{where } \Delta H° = 40.66 \text{ kJ/mol}$$

$$\log \frac{K_2}{1 \text{ atm}} = \frac{(40660 \text{ J})(623 \text{ K} - 373 \text{ K})}{2.303(8.314 \text{ J/K})(323 \text{ K})(373 \text{ K})}$$

$$\log \frac{K_2}{1 \text{ atm}} = -0.881$$

$$K_2 = P_{H_2O} \text{ at 323 K} = \textbf{0.13 atm}$$

Note: In Appendix E, the vapor pressure of water, P_{H_2O}, at 323 K = 0.122 atm. The answers agree very well.

18 Ionic Equilibria I: Acids and Bases

18-2. *Refer to Sections 18-1 and 4-2, and Examples 18-1 and 18-2.*

(a) $? \, M \, NaCl = \dfrac{mol \, NaCl}{L \, soln} = \dfrac{17.52 \, g/58.44 \, g/mol}{0.125 \, L} = \textbf{2.40} \, \textbf{\textit{M}} \, \textbf{NaCl}$

(b) $? \, M \, H_2SO_4 = \dfrac{mol \, H_2SO_4}{L \, soln} = \dfrac{55.0 \, g/98.1 \, g/mol}{0.575 \, L} = \textbf{0.975} \, \textbf{\textit{M}} \, \textbf{H}_2\textbf{SO}_4$

(c) $? \, M \, C_6H_5OH = \dfrac{mol \, C_6H_5OH}{L \, soln} = \dfrac{0.135 \, g/94.1 \, g/mol}{1.5 \, L} = \textbf{9.6} \times \textbf{10}^{-4} \, \textbf{\textit{M}} \, \textbf{C}_6\textbf{H}_5\textbf{OH}$

18-4. *Refer to Sections 18-1 and 4-2, and Example 18-1.*

These compounds are strong electrolytes.

(a) $0.25 \, M \, HBr(aq) \rightarrow 0.25 \, M \, H^+(aq) + 0.25 \, M \, Br^-(aq)$ $[H^+] = [Br^-] = \textbf{0.25} \, \textbf{\textit{M}}$

(b) $0.055 \, M \, KOH(aq) \rightarrow 0.055 \, M \, K^+(aq) + 0.055 \, M \, OH^-(aq)$ $[K^+] = [OH^-] = \textbf{0.055} \, \textbf{\textit{M}}$

(c) $0.0020 \, M \, CaCl_2(aq) \rightarrow 0.0020 \, Ca^{2+}(aq) + 0.0040 \, M \, Cl^-(aq)$ $[Ca^{2+}] = \textbf{0.0020} \, \textbf{\textit{M}}; \, [Cl^-] = \textbf{0.0040} \, \textbf{\textit{M}}$

18-6. *Refer to Section 18-1 and Example 18-2.*

These compounds are strong electrolytes.

(a) $? \, M \, KOH = \dfrac{(2.5 \, g/56.1 \, g/mol)}{1.50 \, L} = 0.030 \, M$

	KOH	$\rightarrow$	K$^+$	+	OH$^-$
initial	0.030 *M*		0 *M*		0 *M*
change	- 0.030 *M*		+ 0.030 *M*		+ 0.030 *M*
final	0 *M*		0.030 *M*		0.030 *M*

Therefore, $[K^+] = \textbf{0.030} \, \textbf{\textit{M}}$ and $[OH^-] = \textbf{0.030} \, \textbf{\textit{M}}$

(b) $? \, M \, Ba(OH)_2 = \dfrac{(0.72 \, g/171 \, g/mol)}{0.250 \, L} = 0.017 \, M$ Therefore, $[Ba^{2+}] = \textbf{0.017} \, \textbf{\textit{M}}$ and $[OH^-] = \textbf{0.034} \, \textbf{\textit{M}}$

(c) $? \, M \, Ca(NO_3)_2 = \dfrac{(2.64 \, g/164 \, g/mol)}{0.100 \, L} = 0.161 \, M$ Therefore, $[Ca^{2+}] = \textbf{0.161} \, \textbf{\textit{M}}$ and $[NO_3^-] = \textbf{0.322} \, \textbf{\textit{M}}$

18-8. *Refer to Section 18-2.*

(a) $H_2O(\ell) + H_2O(\ell) \rightleftarrows H_3O^+(aq) + OH^-(aq)$

(b) $K_c = K_w = [H_3O^+][OH^-]$

(c) The equilibrium constant, K_w, is known as the ion product for water.

(d) In pure water, $[H_3O^+] = [OH^-]$. At 25°C, $[OH^-] = [H_3O^+] = 1.0 \times 10^{-7} \, M$.

(e) In acidic solutions, $[H_3O^+] > [OH^-]$. In basic solutions, $[OH^-] > [H_3O^+]$.

18-10. Refer to Section 18-2.

(a) A 0.10 M solution of NaOH does have 2 sources of OH^- ion: (1) OH^- from the complete dissociation of NaOH and (2) OH^- from the ionization of water. Since Source 1 dominates Source 2, the concentration of OH^- produced by the ionization of water is therefore neglected.

(b) From Source 1, $NaOH(aq) \rightarrow Na^+(aq) + OH^-(aq)$, $[OH^-] = 0.10\ M$. To calculate $[OH^-]$ from Source 2, consider the ionization of water in a 0.10 M solution of NaOH.

Let x = moles per liter of H_3O^+ and OH^- produced by the ionization of water.

	$2H_2O(\ell)$	$\rightleftarrows$	$H_3O^+(aq)$	+	$OH^-(aq)$
initial			0 M		0.10 M
change			+ x M		+ x M
at equilibrium			x M		(0.10 + x) M

$$K_w = [H_3O^+][OH^-] = x(0.10 + x) = 1.0 \times 10^{-14}$$

We know that x << 0.10. Let us make the approximation that $0.10 + x \approx 0.10$.

Then $x \times 0.10 = 1.0 \times 10^{-14}$ and $x = 1.0 \times 10^{-13}$. So, our approximation that x << 0.10 is a good one.

Therefore, $[OH^-]$ from 0.10 M NaOH = 0.10 M

$[OH^-]$ from the ionization of water = $1.0 \times 10^{-13}\ M$, and can be neglected.

18-12. Refer to Section 18-2, Example 18-3, and Exercises 18-4, 18-5 and 18-6 Solutions.

In pure water, $[H_3O^+]$ at 25°C = $1.0 \times 10^{-7}\ M$.

To determine $[H_3O^+]$ in basic solution, use the K_w expression: $K_w = [H_3O^+][OH^-] = 1.0 \times 10^{-14}$.

Solution	$[OH^-]$	$[H_3O^+] = K_w/[OH^-]$
0.055 M KOH	0.055 M	$1.8 \times 10^{-13}\ M$
0.025 M Sr(OH)$_2$	0.050 M	$2.0 \times 10^{-13}\ M$
0.017 M Ba(OH)$_2$	0.034 M	$2.9 \times 10^{-13}\ M$
pure water	$1.0 \times 10^{-7}\ M$	$1.0 \times 10^{-7}\ M$

The $[H_3O^+]$ in basic solutions is much lower than that in pure water.

18-14. Refer to Section 18-2 and Example 18-3.

$$[H^+] = \frac{K_w}{[OH^-]} = \frac{1.00 \times 10^{-14}}{6.32 \times 10^{-6}} = 1.58 \times 10^{-9}\ M \text{ at } 25°C$$

18-16. Refer to Section 18-3 and Appendix A.2.

When working with base 10 logarithms, the number of significant digits in the mantissa (the decimal part of the logarithm) gives the number of significant figures in the antilogarithm.

(a) log 0.000052 = **-4.28**

(b) log 5.7 = **0.76**

(c) $\log (5.8 \times 10^{-12})$ = **-11.24**

(d) $\log (4.9 \times 10^{-7})$ = **-6.31**

18-18. *Refer to Section 18-3, Example 18-4 and Exercise 18-12 Solution.*

Solution	$[H_3O^+]$	pH = $-\log[H_3O^+]$
0.055 M KOH	$1.8 \times 10^{-13}\ M$	12.74
0.025 M Sr(OH)$_2$	$2.0 \times 10^{-13}\ M$	12.70
0.017 M Ba(OH)$_2$	$2.9 \times 10^{-13}\ M$	12.54
pure water	$1.0 \times 10^{-7}\ M$	7.00

18-20. *Refer to Sections 18-2 and 18-3, and Example 18-4.*

(a) $[H^+] = [HCl] = 0.500\ M$

$pH = -\log[H^+] = -\log(0.500) = \mathbf{0.301}$

(b) $[H^+] = [HNO_3] = 0.030\ M$

$pH = -\log[H^+] = -\log(0.030) = \mathbf{1.52}$

(c) $[H^+] = [HClO_4] = \dfrac{0.75 \text{ g HClO}_4}{1 \text{ L}} \times \dfrac{1 \text{ mol HClO}_4}{100.45 \text{ g HClO}_4} = 7.5 \times 10^{-3}\ M$

$pH = -\log(7.5 \times 10^{-3}\ M) = \mathbf{2.12}$

18-22. *Refer to Sections 18-2 and 18-3, and Example 18-6.*

$[OH^-] = [NaOH] = 2.5 \times 10^{-4}\ M$ $\qquad\qquad$ since NaOH is a strong base

Method 1: $\quad [H^+] = \dfrac{K_w}{[OH^-]} = \dfrac{1.00 \times 10^{-14}}{2.5 \times 10^{-4}} = 4.0 \times 10^{-11}$ at 25°C

$\qquad\qquad pH = -\log[H^+] = -\log(4.0 \times 10^{-11}) = \mathbf{10.40}$

Method 2: $\quad pOH = -\log[OH^-] = 3.60;\quad pH = 14.00 - pOH = 14.00 - 3.60 = \mathbf{10.40}$

18-24. *Refer to Section 18-3 and Example 18-5.*

We know: $pH = -\log[H^+] = 3.52$. Therefore, $[H^+] = $ antilog $(-3.52) = 3.0 \times 10^{-4}\ M$

Since HNO$_3$ is a strong acid, $[HNO_3] = [H^+] = \mathbf{3.0 \times 10^{-4}\ M}$

18-26. *Refer to Section 18-3 and Example 18-6.*

	Solution	$[H_3O^+]$	$[OH^-]$	pH	pOH
(a)	0.055 M NaOH	$1.8 \times 10^{-13}\ M$	0.055 M	12.74	1.26
(b)	0.055 M HCl	0.055 M	$1.8 \times 10^{-13}\ M$	1.26	12.74
(c)	0.055 M Ca(OH)$_2$	$9.1 \times 10^{-14}\ M$	0.110 M	13.04	0.96

Balanced equation for the ionization of a weak acid: $HA(aq) + H_2O(\ell) \rightleftarrows H_3O^+(aq) + A^-(aq)$

In Chapter 17 (Section 17-10), we stated that for heterogeneous equilibria, terms for pure liquids and pure solids do not appear in K expressions because their activity is taken as 1. In the discussion in Section 18-4, these terms are temporarily included.

$$K_c = \frac{[H_3O^+][A^-]}{[HA][H_2O]} \qquad\qquad K_a = K_c[H_2O] = \frac{[H_3O^+][A^-]}{[HA]}$$

The symbol for the equilibrium constant, $K_c[H_2O]$, is K_a. It is called the acid ionization constant.

18-30.   *Refer to Sections 18-2 and 18-3, and Table 18-2.*

(a) K_w at $37°C = 2.38 \times 10^{-14} = [H_3O^+][OH^-]$.   For pure water, $[H^+] = [OH^-]$

   $[H_3O^+] = \sqrt{K_w} = \sqrt{2.38 \times 10^{-14}} = 1.54 \times 10^{-7}$ at $37°C$

   $pH = -\log [H_3O^+] = -\log (1.54 \times 10^{-7}) = \textbf{6.812 at 37°C}$

(b) Pure water at $37°C$ is **neutral** since $[H_3O^+] = [OH^-]$. The pH of a neutral system is 7.0 only at $25°C$.

18-32.   *Refer to Section 18-4 and Exercise 18-28 Solution.*

Reaction of proton-accepting weak base, B, with water: $B(aq) + H_2O(\ell) \rightleftarrows BH^+(aq) + OH^-(aq)$

$$K_c = \frac{[BH^+][OH^-]}{[B][H_2O]}$$

$$K_b = K_c[H_2O] = \frac{[BH^+][OH^-]}{[B]}$$

K_b, the ionization constant for bases, is used for calculations involving the equilibria of weak bases.

18-34.   *Refer to Section 18-4 and Example 18-8.*

Balanced equation: $HX + H_2O \rightleftarrows H_3O^+ + X^-$

Since HX is 1.07% ionized, $[HX]_{reacted} = 0.083\ M \times 0.0107 = 8.9 \times 10^{-4}\ M$

	HX	+	H$_2$O(ℓ)	$\rightleftarrows$	H$_3$O$^+$	+	X$^-$
initial	0.083 M				$\approx$ 0 M		0 M
change	- 8.9 $\times$ 10^{-4} M				+ 8.9 $\times$ 10^{-4} M		+ 8.9 $\times$ 10^{-4} M
at equilibrium	0.082 M				8.9 $\times$ 10^{-4} M		8.9 $\times$ 10^{-4} M

$pH = -\log (8.9 \times 10^{-4}) = \textbf{3.05}$

$$K_a = \frac{[H_3O^+][X^-]}{[HX]} = \frac{(8.9 \times 10^{-4})^2}{0.083} = \textbf{9.6} \times \textbf{10}^{-6}$$

18-36. *Refer to Section 18-4 and Example 18-9.*

Balanced equation: $C_3H_7COOH + H_2O \rightleftarrows H_3O^+ + C_3H_7COO^-$

Let x = mol/L of C_3H_7COOH that reacts. Then
        x = mol/L of H_3O^+ produced = mol/L $C_3H_7COO^-$ produced.

Since pH = 3.21, $[H_3O^+]$ = x = antilog (-3.21) = $6.2 \times 10^{-4} M$

	C_3H_7COOH	+	$H_2O(\ell)$	$\rightleftarrows$	H_3O^+	+	$C_3H_7COO^-$
initial	0.025 M				$\approx 0 M$		0 M
change	- x M				+ x M		+ x M
at equilibrium	(0.025 - x) M				x M		x M

$$K_a = \frac{[H_3O^+][C_3H_7COO^-]}{[C_3H_7COOH]} = \frac{(6.2 \times 10^{-4})^2}{0.025 - 6.2 \times 10^{-4}} = \mathbf{1.6 \times 10^{-5}}$$

18-38. *Refer to Section 18-4, Example 18-10 and Appendix F.*

Balanced equation: $C_6H_5COOH + H_2O \rightleftarrows H_3O^+ + C_6H_5COO^-$ $K_a = 6.3 \times 10^{-5}$

Let x = mol/L of C_6H_5COOH that reacts. Then
        x = mol/L of H_3O^+ produced = mol/L $C_6H_5COO^-$ produced.

	C_6H_5COOH	+	$H_2O(\ell)$	$\rightleftarrows$	H_3O^+	+	$C_6H_5COO^-$
initial	0.35 M				$\approx 0 M$		0 M
change	- x M				+ x M		+ x M
at equilibrium	(0.35 - x) M				x M		x M

$$K_a = \frac{[H_3O^+][C_6H_5COO^-]}{[C_6H_5COOH]} = \frac{x^2}{0.35 - x} = 6.3 \times 10^{-5}$$

Assume that 0.35 - x $\approx$ 0.35. Then $x^2/0.35 = 6.3 \times 10^{-5}$ and x = 4.7×10^{-3}. The simplifying assumption is justified since 4.7×10^{-3} is much less than 5% of 0.35 (= 0.02).

Therefore at equilibrium: $[C_6H_5COOH] = \mathbf{0.35\ M}$
           $[H_3O^+] = [C_6H_5COO^-] = \mathbf{4.7 \times 10^{-3}\ M}$
           $[OH^-] = K_w/[H_3O^+] = \mathbf{2.1 \times 10^{-12}\ M}$

18-40. *Refer to Section 18-4, Examples 18-10 and 18-11, and Appendix F.*

Balanced equation: $HCOOH + H_2O \rightleftarrows H_3O^+ + HCOO^-$ $K_a = 1.8 \times 10^{-4}$

Let x = [HCOOH] that ionizes. Then
        x = $[H_3O^+]$ produced = $[HCOO^-]$ produced.

	HCOOH	+	H_2O	$\rightleftarrows$	H_3O^+	+	$HCOO^-$
initial	0.0500 M				$\approx 0 M$		0 M
change	- x M				+ x M		+ x M
at equilibrium	(0.0500 - x) M				x M		x M

$$K_a = \frac{[H_3O^+][HCOO^-]}{[HCOOH]} = \frac{x^2}{0.0500 - x} = 1.8 \times 10^{-4}$$

Assuming $0.0500 - x \approx 0.0500$, then $x^2/0.0500 = 1.8 \times 10^{-4}$ and $x = 3.0 \times 10^{-3}$. However, the simplifying assumption may not be justified since 3.0×10^{-3} is 6% of 0.0500.

Let us check ourselves by solving the original quadratic equation: $x^2 + (1.8 \times 10^{-4})x - 9.0 \times 10^{-6} = 0$

$$x = \frac{-(1.8 \times 10^{-4}) \pm \sqrt{(1.8 \times 10^{-4})^2 - 4(1)(-9.0 \times 10^{-6})}}{2(1)} = \frac{-1.8 \times 10^{-4} \pm 6.0 \times 10^{-3}}{2}$$

$$= 2.9 \times 10^{-3} \text{ or } -6.2 \times 10^{-3} \text{ (discard)}$$

The two solutions for x are slightly different. Therefore, in this case, if an approximate answer is desired, the solution using the simplifying assumption is adequate. However, the correct answer is obtained by solving the quadratic equation.

$$\% \text{ ionization} = \frac{[HCOOH]_{\text{ionized}}}{[HCOOH]_{\text{initial}}} \times 100\% = \frac{2.9 \times 10^{-3}\ M}{0.0500\ M} \times 100\% = \textbf{5.8\%}$$

18-42. Refer to Section 18-4, Example 18-12 and Appendix F.

For benzoic acid $\qquad$ $pK_a = -\log K_a = -\log (6.3 \times 10^{-5}) = \textbf{4.20}$

For hydrocyanic acid $\qquad$ $pK_a = -\log K_a = -\log (4.0 \times 10^{-10}) = \textbf{9.40}$

18-44. Refer to Section 18-4, Example 18-13 and Table 18-6.

(a) Recall that for a series of weak bases, as K_b increases, $[OH^-]$ increases, pOH decreases and pH increases.

    i. highest pH - dimethylamine $\qquad\qquad$ ii. lowest pH - pyridine

    iii. highest pOH - pyridine $\qquad\qquad\quad$ iv. lowest pOH - dimethylamine

(b) Consider the dissociation of a weak base: $B + H_2O \rightleftarrows BH^+ + OH^-$. As K_b increases, $[BH^+]$ increases.

    i. highest $[BH^+]$ - dimethylamine $\qquad$ ii. lowest $[BH^+]$ - pyridine

---

18-46. Refer to Section 18-4, Table 18-6 and Appendix G.

Balanced equation: $C_5H_5N(aq) + H_2O(\ell) \rightleftarrows C_5H_5NH^+(aq) + OH^-(aq)$

Since a 0.00500 M C_5H_5N solution is 0.053% ionized, $[C_5H_5N]_{\text{reacted}} = 0.00500\ M \times 0.00053 = 2.7 \times 10^{-6}\ M$

	C_5H_5N	+	$H_2O(\ell)$	$\rightleftarrows$	$C_5H_5NH^+$	+	OH^-
initial	0.00500 M				$\approx 0\ M$		$\approx 0\ M$
change	$-2.7 \times 10^{-6}\ M$				$+2.7 \times 10^{-6}\ M$		$+2.7 \times 10^{-6}\ M$
at equilibrium	0.00500 M				$2.7 \times 10^{-6}\ M$		$2.7 \times 10^{-6}\ M$

$K_b = \dfrac{[C_5H_5NH^+][OH^-]}{[C_5H_5N]} = \dfrac{(2.7 \times 10^{-6})^2}{0.00500} = 1.5 \times 10^{-9}$ $\qquad\qquad$ $pK_b = -\log (1.5 \times 10^{-9}) = \textbf{8.82}$

18-48. Refer to Section 18-4, Table 18-6 and Appendix G.

Balanced equation: $CH_3NH_2(aq) + H_2O(\ell) \rightleftarrows CH_3NH_3^+(aq) + OH^-(aq)$

$K_b = \dfrac{[CH_3NH_3^+][OH^-]}{[CH_3NH_2]} = \dfrac{(2.0 \times 10^{-3})^2}{0.0080} = \textbf{5.0} \times \textbf{10}^{-4}$ $\qquad$ This agrees with the K_b value given in Appendix G and Table 18-6.

18-50. *Refer to Section 18-4 , Example 18-14 and Appendix G.*

(a) Balanced equation: $NH_3 + H_2O \rightleftarrows NH_4^+ + OH^-$ $K_b = 1.8 \times 10^{-5}$

Let $x = [NH_3]$ that ionizes. Then
  $x = [NH_4^+]$ produced $= [OH^-]$ produced.

	NH_3	$+$	H_2O	$\rightleftarrows$	NH_4^+	$+$	OH^-
initial	0.10 M				$\approx 0\ M$		0 M
change	- x M				+ x M		+ x M
at equilibrium	(0.10 - x) M				x M		x M

$K_b = \dfrac{[NH_4^+][OH^-]}{[NH_3]} = 1.8 \times 10^{-5} = \dfrac{x^2}{0.10 - x} \approx \dfrac{x^2}{0.10}$ Solving, $x = 1.3 \times 10^{-3}$

Since 1.3×10^{-3} is less than 5% of 0.10, the approximation is justified.

Therefore, $[OH^-] = \mathbf{1.3 \times 10^{-3}\ M}$

$\text{\% ionization} = \dfrac{[NH_3]_{\text{ionized}}}{[NH_3]_{\text{initial}}} \times 100\% = \dfrac{1.3 \times 10^{-3}\ M}{0.10\ M} \times 100\% = \mathbf{1.3\%}$

$pH = -\log \dfrac{K_w}{[OH^-]} = -\log \left[\dfrac{1.0 \times 10^{-14}}{1.3 \times 10^{-3}} \right] = \mathbf{11.11}$

(b) Balanced equation: $CH_3NH_2 + H_2O \rightleftarrows CH_3NH_3^+ + OH^-$ $K_b = 5.0 \times 10^{-4}$

Let $x = [CH_3NH_2]$ that ionizes. Then
  $x = [CH_3NH_3^+]$ produced $= [OH^-]$ produced.

	CH_3NH_2	$+$	H_2O	$\rightleftarrows$	$CH_3NH_3^+$	$+$	OH^-
initial	0.10 M				$\approx 0\ M$		0 M
change	- x M				+ x M		+ x M
at equilibrium	(0.10 - x) M				x M		x M

$K_b = \dfrac{[CH_3NH_3^+][OH^-]}{[CH_3NH_2]} = 5.0 \times 10^{-4} = \dfrac{x^2}{0.10 - x} \approx \dfrac{x^2}{0.10}$ Solving, $x = 7.1 \times 10^{-3}$

Since 7.1×10^{-3} is greater than 5% of 0.10 ($= 5.0 \times 10^{-3}$), we cannot use the simplifying assumption; the correct answer is obtained by solving the original quadratic equation: $x^2 + (5.0 \times 10^{-4})x - 5.0 \times 10^{-5} = 0$

$x = \dfrac{-(5.0 \times 10^{-4}) \pm \sqrt{(5.0 \times 10^{-4})^2 - 4(1)(-5.0 \times 10^{-5})}}{2(1)} = \dfrac{-5.0 \times 10^{-4} \pm 1.42 \times 10^{-3}}{2}$

$= 6.8 \times 10^{-3}$ or -7.4×10^{-3} (discard)

Therefore, $[OH^-] = \mathbf{6.8 \times 10^{-3}\ M}$

$\text{\% ionization} = \dfrac{[CH_3NH_2]_{\text{ionized}}}{[CH_3NH_2]_{\text{initial}}} \times 100\% = \dfrac{6.8 \times 10^{-3}\ M}{0.10\ M} \times 100\% = \mathbf{6.8\%}$

$pH = -\log \dfrac{K_w}{[OH^-]} = -\log \left[\dfrac{1.0 \times 10^{-14}}{6.8 \times 10^{-3}} \right] = \mathbf{11.83}$

Note: Even if the simplifying assumption may not always be applicable, it is a good idea to always use it in the beginning for several reasons. (1) It is quick and easy to use, and (2) it usually works. (3) Even when it does not yield the correct answer, it *usually* gives a reasonable estimate of the true value. For example, in this exercise, using the simplifying assumption gave the answer, $x = 7.1 \times 10^{-3}$, which was fairly close to the correct answer, $x = 6.8 \times 10^{-3}$.

(1) For triethylamine, $(C_2H_5)_3N$

Balanced equation: $(C_2H_5)_3N + H_2O \rightleftarrows (C_2H_5)_3NH^+ + OH^-$ $\qquad K_b = 5.2 \times 10^{-4}$

Let $x = [(C_2H_5)_3N]$ that ionizes. Then $x = [(C_2H_5)_3NH^+]$ produced $= [OH^-]$ produced.

$$K_b = \frac{[(C_2H_5)_3NH^+][OH^-]}{[(C_2H_5)_3N]} = 5.2 \times 10^{-4} = \frac{x^2}{0.015 - x} \approx \frac{x^2}{0.015} \qquad \text{Solving, } x = 2.8 \times 10^{-3}$$

In this case, 2.8×10^{-3} is greater than 5% of 0.015 ($= 7.5 \times 10^{-4}$). A simplifying assumption cannot be made and the original quadratic equation must be solved: $x^2 + (5.2 \times 10^{-4})x - 7.8 \times 10^{-6} = 0$

$$x = \frac{-(5.2 \times 10^{-4}) \pm \sqrt{(5.2 \times 10^{-4})^2 - 4(1)(-7.8 \times 10^{-6})}}{2(1)} = \frac{-5.2 \times 10^{-4} \pm 5.6 \times 10^{-3}}{2}$$

$$= 2.5 \times 10^{-3} \text{ or } -3.1 \times 10^{-3} \text{ (discard)}$$

Therefore, $[OH^-] = x = \mathbf{2.5 \times 10^{-3}} \, \boldsymbol{M}$

(2) For trimethylamine, $(CH_3)_3N$

Balanced equation: $(CH_3)_3N + H_2O \rightleftarrows (CH_3)_3NH^+ + OH^-$ $\qquad K_b = 7.4 \times 10^{-5}$

Let $x = [(CH_3)_3N]$ that ionizes. Then $x = [(CH_3)_3NH^+]$ produced $= [OH^-]$ produced.

$$K_b = \frac{[(CH_3)_3NH^+][OH^-]}{[(CH_3)_3N]} = 7.4 \times 10^{-5} = \frac{x^2}{0.015 - x} \approx \frac{x^2}{0.015} \qquad \text{Solving, } x = 1.1 \times 10^{-3}$$

In this case, 1.1×10^{-3} is greater than 5% of 0.015 ($= 7.5 \times 10^{-4}$). A simplifying assumption cannot be made and the original quadratic equation must be solved: $x^2 + (7.4 \times 10^{-5})x - 1.1 \times 10^{-6} = 0$

$$x = \frac{-(7.4 \times 10^{-5}) \pm \sqrt{(7.4 \times 10^{-5})^2 - 4(1)(-1.1 \times 10^{-6})}}{2(1)} = \frac{-7.4 \times 10^{-5} \pm 2.1 \times 10^{-3}}{2}$$

$$= 1.0 \times 10^{-3} \text{ or } -1.1 \times 10^{-3} \text{ (discard)}$$

Therefore, $[OH^-] = x = \mathbf{1.0 \times 10^{-3}} \, \boldsymbol{M}$

Yes, $[OH^-]$ in triethylamine(aq) is greater than $[OH^-]$ in trimethylamine(aq) for the same concentration.

(a) Acid-base indicators are organic compounds which behave as weak acids or bases and exhibit different colors in solutions with different acidities.

(b) The essential characteristic of an acid-base indicator is that the conjugate acid-base pair must exhibit different colors. Consider the weak acid indicator, HIn. In solution, HIn dissociates slightly as follows:

$$HIn + H_2O \rightleftarrows H_3O^+ + In^-$$
$$\text{acid} \qquad\qquad\qquad \text{conjugate base}$$

HIn dominates in more acidic solutions with one characteristic color; In^- dominates in more basic solutions with another color.

(c) The color of an acid-base indicator in an aqueous solution depends upon the ratio, $[In^-]/[HIn]$, which in turn depends upon $[H^+]$ and the K_a value of the indicator. A general rule of thumb: If $[In^-]/[HIn] \leq 0.1$, then the indicator will show its true acid color. If $[In^-]/[HIn] \geq 10$, then the indicator will show its true base color.

18-56. *Refer to Section 18-5.*

Balanced equation: $HIn + H_2O \rightleftarrows H_3O^+ + In^-$

At pH 8.2, $[HIn] = [In^-]$; $[H_3O^+] = $ antilog $(-pH) = 6 \times 10^{-9} M$

Substituting into the K_a expression, we have: $K_a = \dfrac{[H_3O^+][In^-]}{[HIn]} = 6 \times 10^{-9}$ since $[In^-]$ and $[HIn]$ are equal and cancel.

$$pK_a = 8.2$$

18-58. *Refer to Section 18-6 and Appendix F.*

This is an example of the common ion effect; the common ion in this case is H_3O^+.

Balanced equations: $HCl + H_2O \rightarrow H_3O^+ + Cl^-$ (to completion)

$HCOOH + H_2O \rightleftarrows H_3O^+ + HCOO^-$ (reversible) $K_a = 1.8 \times 10^{-4}$

Since HCl ionizes completely, $[H_3O^+]$ from $HCl = [HCl]_{initial} = 0.40 M$

Let   x = [HCOOH] that ionizes. Then
x = $[H_3O^+]$ produced from the ionization of HCOOH = $[HCOO^-]$ produced from the ionization of HCOOH

	HCOOH	+	H$_2$O	$\rightleftarrows$	H$_3$O$^+$	+	HCOO$^-$
initial	1.00 M				0.40 M		0 M
change	- x M				+ x M		+ x M
at equilibrium	(1.00 - x) M				(0.40 + x) M		x M

$$K_a = \frac{[H_3O^+][HCOO^-]}{[HCOOH]} = \frac{(0.40 + x)(x)}{(1.00 - x)} = 1.8 \times 10^{-4} \approx \frac{0.40x}{1.00}$$

Solving, $[HCOO^-] = x = \mathbf{4.5 \times 10^{-4}\ M}$

18-60. *Refer to Section 18-6, Example 18-16 and Appendix F.*

(a) Balanced equations: $KF \rightarrow K^+ + F^-$ (to completion)

$HF + H_2O \rightleftarrows H_3O^+ + F^-$ (reversible) $K_a = 7.2 \times 10^{-4}$

Since KF dissociates completely, $[F^-]$ from the salt $= [KF]_{initial} = 0.25 M$

Let   x = [HF] that ionizes. Then
x = $[H_3O^+]$ produced from HF
x = $[F^-]$ produced from HF.

	HF	+	H$_2$O	$\rightleftarrows$	H$_3$O$^+$	+	F$^-$
initial	0.15 M				$\approx$ 0 M		0.25 M
change	- x M				+ x M		+ x M
at equilibrium	(0.15 - x) M				x M		(0.25 + x) M

$$K_a = \frac{[H_3O^+][F^-]}{[HF]} = \frac{(x)(0.25 + x)}{(0.15 - x)} = 7.2 \times 10^{-4} \approx \frac{x(0.25)}{(0.15)}$$ Solving, $x = 4.3 \times 10^{-4}$

Therefore, $[H_3O^+] = 4.3 \times 10^{-4}\ M$; pH = **3.36**

Alternatively, this problem can be solved using the Henderson-Hasselbalch equation:

$$pH = pK_a + \log \frac{[salt]}{[acid]}$$ Substituting, pH $= 3.14 + \log \dfrac{(0.25)}{(0.15)} = \mathbf{3.36}$

(b) Note that the Henderson-Hasselbalch equation cannot be used in this case without alteration since it is really valid only for solutions containing a weak monoprotic acid and a soluble, ionic salt of the weak acid with a *univalent* cation. The cation, Ba^{2+}, in $Ba(CH_3COO)_2$, is divalent.

Since $Ba(CH_3COO)_2$ dissociates totally, $[CH_3COO^-]$ from the salt $= 2 \times [Ba(CH_3COO)_2]_{initial} = 0.050\ M$

Let $x = [CH_3COOH]$ that ionizes. Then
$x = [H_3O^+]$ produced from CH_3COOH
$x = [CH_3COO^-]$ produced from CH_3COOH.

	CH_3COOH	$+$	H_2O	$\rightleftarrows$	H_3O^+	$+$	CH_3COO^-
initial	0.050 M				$\approx 0\ M$		0.050 M
change	- x M				+ x M		+ x M
at equilibrium	(0.050 - x) M				x M		(0.050 + x) M

$$K_a = \frac{[H_3O^+][CH_3COO^-]}{[CH_3COOH]} = \frac{(x)(0.050 + x)}{(0.050 - x)} = 1.8 \times 10^{-5} \approx \frac{x(0.050)}{(0.050)} \qquad \text{Solving, } x = 1.8 \times 10^{-5}$$

Therefore, $[H_3O^+] = 1.8 \times 10^{-5}\ M$; pH $= $ **4.74**

18-62. *Refer to Section 18-6, Example 18-17 and Appendix G.*

Balanced equations: $NH_4NO_3 \rightarrow NH_4^+ + NO_3^-$ (to completion)

$NH_3 + H_2O \rightleftarrows NH_4^+ + OH^-$ (reversible) $K_b = 1.8 \times 10^{-5}$

(a) Since NH_4NO_3 is a soluble salt, $[NH_4^+]$ from the salt $= [NH_4NO_3]_{initial} = 0.20\ M$

Let $x = [NH_3]$ that ionizes. Then
$x = [NH_4^+]$ produced from NH_3
$x = [OH^-]$ produced from NH_3.

	NH_3	$+$	H_2O	$\rightleftarrows$	NH_4^+	$+$	OH^-
initial	0.30 M				0.20 M		$\approx 0\ M$
change	- x M				+ x M		+ x M
at equilibrium	(0.30 - x) M				(0.20 + x) M		x M

$$K_b = \frac{[NH_4^+][OH^-]}{[NH_3]} = \frac{(0.20 + x)(x)}{(0.30 - x)} = 1.8 \times 10^{-5} \approx \frac{(0.20)(x)}{(0.30)} \qquad \text{Solving, } x = 2.7 \times 10^{-5}$$

Therefore, $[OH^-] = $ **2.7 × 10⁻⁵** M;   pOH $= 4.57$;   pH $= $ **9.43**

Alternatively, using the Henderson-Hasselbalch equation,

$$pOH = pK_b + \log \frac{[salt]}{[base]} = -\log (1.8 \times 10^{-5}) + \log \frac{(0.20)}{(0.30)} = 4.57; \quad pH = \textbf{9.43}$$

(b) Since $(NH_4)_2SO_4$ is a soluble salt, $[NH_4^+]$ from the salt $= 2 \times [(NH_4)_2SO_4]_{initial} = 0.40\ M$

Using the Henderson-Hasselbalch equation,

$$pOH = pK_b + \log \frac{[salt]}{[base]} = -\log (1.8 \times 10^{-5}) + \log \frac{(0.40)}{(0.15)} = 5.17$$

$[OH^-] = $ antilogarithm $(-5.17) = $ **6.8 × 10⁻⁶** M

pH $= 14 - pOH = $ **8.83**

(a) Balanced equations: $NaHCO_3 \xrightarrow{H_2O} Na^+ + HCO_3^-$ (to completion)

$\qquad\qquad\qquad\qquad\quad H_2CO_3 + H_2O \rightleftarrows H_3O^+ + HCO_3^-$ (reversible)

When a small amount of base is added to the buffer: $H_2CO_3 + OH^- \rightarrow HCO_3^- + H_2O$

When a small amount of acid is added to the buffer: $HCO_3^- + H_3O^+ \rightarrow H_2CO_3 + H_2O$

(b) Balanced equations: $NaH_2PO_2 \xrightarrow{H_2O} Na^+ + H_2PO_4^-$ (to completion)

$\qquad\qquad\qquad\qquad\quad Na_2HPO_4 \xrightarrow{H_2O} 2Na^+ + HPO_4^{2-}$ (to completion)

$\qquad\qquad\qquad\qquad\quad H_2PO_4^- + H_2O \rightleftarrows H_3O^+ + HPO_4^{2-}$ (reversible)

When a small amount of base is added to the buffer: $H_2PO_4^- + OH^- \rightarrow HPO_4^{2-} + H_2O$

When a small amount of acid is added to the buffer: $HPO_4^{2-} + H_3O^+ \rightarrow H_2PO_4^- + H_2O$

18-66. *Refer to Section 18-6.*

Balanced equations: $NaHCOO \rightarrow Na^+ + HCOO^-$ (to completion)

$\qquad\qquad\qquad\qquad HCOOH + H_2O \rightleftarrows H_3O^+ + HCOO^-$ (reversible)

When the soluble salt sodium formate (NaHCOO) is added to a formic acid solution, the salt undergoes complete dissociation in water to produce the common ion, $HCOO^-$. This causes the original equilibrium involving the weak acid to shift to the left. As a result, the fraction of HCOOH molecules that undergo ionization in aqueous solution will be **less**.

18-68. *Refer to Sections 18-1 and 18-3.*

Balanced equation: $HClO_4 + H_2O \rightarrow H_3O^+ + ClO_4^-$

$[H_3O^+] = [HClO_4] = 0.10\ M$; pH $= -\log(0.10) = $ **1.00**

Perchloric acid, $HClO_4$, is a strong acid and dissociates completely into its ions, even in the presence of a supplier of common ion, $KClO_4$. A solution of 0.10 M $HClO_4$ and 0.10 M $KClO_4$ is *not* a buffer. There is no species present that could react with any added acid.

18-70. *Refer to Sections 18-6 and 18-7, Example 18-18 and Appendix G.*

From Section 18-6, we learned that if the concentrations of the weak acid or base and its salt are $\approx 0.05\ M$ or greater, and the salt contains a univalent cation, then

for a weak acid buffer: $\quad [H_3O^+] = \dfrac{[\text{acid}]}{[\text{salt}]} \times K_a = \dfrac{\text{mol acid}}{\text{mol salt}} \times K_a$

for a weak base buffer: $\quad [OH^-] = \dfrac{[\text{base}]}{[\text{salt}]} \times K_b = \dfrac{\text{mol base}}{\text{mol salt}} \times K_b$

Original NH_3/NH_4^+ buffer: (K_b for $NH_3 = 1.8 \times 10^{-5}$)

$[NH_4Cl] = \dfrac{12.78\ \text{g } NH_4Cl/L}{53.49\ \text{g/mol}} = 0.2389\ M$

$[OH^-] = \dfrac{[NH_3]}{[NH_4Cl]} \times K_b = \dfrac{0.400\ M}{0.2389\ M} \times (1.8 \times 10^{-5}) = 3.0 \times 10^{-5}\ M$;

$[H_3O^+] = \dfrac{K_w}{[OH^-]} = 3.3 \times 10^{-10}\ M$

pH $= 9.48$

New NH_3/NH_4^+ buffer: When 0.155 mol per liter of HCl is added to the original buffer presented in (a), it reacts with the base component of the buffer, NH_3, to form more of the acid component, NH_4^+ (the conjugate acid of NH_3). A new buffer solution is created with a slightly more acidic pH. In this type of problem, always perform the acid-base limiting reactant problem first, then the equilibrium calculation.

	HCl	+	NH_3	$\rightarrow$	NH_4Cl
initial	0.155 mol		0.400 mol		0.239 mol
change	- 0.155 mol		- 0.155 mol		+ 0.155 mol
after reaction	0 mol		0.245 mol		0.394 mol

$$[OH^-] = \frac{\text{mol } NH_3}{\text{mol } NH_4Cl} \times K_b = \frac{0.245 \text{ mol}}{0.394 \text{ mol}} \times (1.8 \times 10^{-5}) = 1.1 \times 10^{-5}$$

$$[H_3O^+] = \frac{K_w}{[OH^-]} = 9.1 \times 10^{-10}$$

new pH = 9.04

The change in pH = final pH - initial pH = 9.04 - 9.48 = -0.44; **the pH decreases by 0.43 units.**

18-72. *Refer to Sections 18-6, 18-7 and 18-8, Example 18-18, Exercise 18-70 Solution, and Appendix G.*

(a) Original NH_3/NH_4^+ buffer: (K_b for $NH_3 = 1.8 \times 10^{-5}$)

$$[OH^-] = \frac{[NH_3]}{[NH_4Cl]} \times K_b = \frac{1.00 \, M}{0.80 \, M} \times (1.8 \times 10^{-5}) = 2.2 \times 10^{-5} \, M$$

$$[H_3O^+] = \frac{K_w}{[OH^-]} = 4.5 \times 10^{-10} \, M$$

pH = **9.34**

(b) New NH_3/NH_4^+ buffer: When 0.10 mol per liter of HCl is added to the original buffer presented in (a), it reacts with the base component of the buffer, NH_3, to form more of the acid component, NH_4^+ (the conjugate acid of NH_3). A new buffer solution is created with a slightly more acidic pH. In this type of problem, always perform the acid-base limiting reactant problem first, then the equilibrium calculation.

(1)	HCl	+	NH_3	$\rightarrow$	NH_4Cl
initial	0.10 mol		1.00 mol		0.80 mol
change	- 0.10 mol		- 0.10 mol		+ 0.10 mol
after reaction	0 mol		0.90 mol		0.90 mol

$$\text{(2) } [OH^-] = \frac{\text{mol } NH_3}{\text{mol } NH_4Cl} \times K_b = \frac{0.90 \text{ mol}}{0.90 \text{ mol}} \times (1.8 \times 10^{-5}) = 1.8 \times 10^{-5}$$

$$[H_3O^+] = \frac{K_w}{[OH^-]} = 5.6 \times 10^{-10}$$

pH = **9.26**

(c) This is a simple strong acid/strong base neutralization problem.

Plan: (1) Find the concentration and the number of moles of NaOH from the pH of the solution.
(2) Perform the limiting reactant testing for the acid-base reaction.
(3) Determine the pH of the final solution.

(1) Since pH = 9.34, pOH = 14.00 - 9.34 = 4.66; $[OH^-] = 2.19 \times 10^{-5} \, M$

Therefore, 1.00 L would contain 2.19×10^{-5} mol NaOH

(2)

	NaOH	+	HCl	→	NaCl
initial	2.19×10^{-5} mol		0.10 mol		0 mol
change	$- 2.19 \times 10^{-5}$ mol		$- 2.19 \times 10^{-5}$ mol		$+ 2.19 \times 10^{-5}$ mol
after reaction	0 mol		0.10 mol		2.19×10^{-5} mol

(3) The number of moles of HCl is essentially unaffected by the presence of 2.19×10^{-5} moles of NaOH. Therefore, $[H_3O^+] = [HCl] = 0.10$ mol/L; pH = **1.00**

18-74. *Refer to Section 18-6 and Exercise 18-72 Solution.*

Balanced equations:　$NaCH_2BrCOO \rightarrow Na^+ + CH_2BrCOO^-$　　　(to completion)
　　　　　　　　　$CH_2BrCOOH + H_2O \rightleftarrows H_3O^+ + CH_2BrCOO^-$　　(reversible)　　$K_a = 2.0 \times 10^{-3}$

Since pH = 3.10, $[H_3O^+] = 7.9 \times 10^{-4}\ M$

Let　　　x = $[CH_2BrCOOH]$.　　Then
　　0.20 - x = $[NaCH_2BrCOO]$.

For a weak acid buffer:　　　　　　　　　$[H_3O^+] = \dfrac{[CH_2BrCOOH]}{[NaCH_2BrCOO]} \times K_a$

$$7.9 \times 10^{-4} = \left[\frac{x}{0.20 - x}\right](2.0 \times 10^{-3})$$
$$1.6 \times 10^{-4} - (7.9 \times 10^{-4})x = (2.0 \times 10^{-3})x$$
$$1.6 \times 10^{-4} = (2.8 \times 10^{-3})x$$
$$x = 5.7 \times 10^{-2}$$

Therefore,　　　$[CH_2BrCOOH] = x = \mathbf{5.7 \times 10^{-2}\ M}$
　　　　　　　$[NaCH_2BrCOO] = 0.20 - x = \mathbf{0.14\ M}$

18-76. *Refer to Section 18-6 and Exercise 58 Solution.*

Balanced equation:　$C_2H_5NH_2 + H_2O \rightleftarrows C_2H_5NH_3^+ + OH^-$　　　　　$K_b = 4.7 \times 10^{-4}$

This is an example of the common ion effect;　the common ion in this case is OH^-;　$[OH^-]_{initial} = 0.0010\ M$

Let　　　x = $[C_2H_5NH_2]$ that ionizes.　Then
　　　　x = $[OH^-]$ produced from $C_2H_5NH_2$ and
　　　　x = $[C_2H_5NH_3^+]$ produced from $C_2H_5NH_2$.

	$C_2H_5NH_2$	+	H_2O	$\rightleftarrows$	$C_2H_5NH_3^+$	+	OH^-
initial	0.015 M				0 M		0.0010 M
change	- x M				+ x M		+ x M
at equilibrium	(0.015 - x) M				x M		(0.0010 + x) M

$$K_b = \frac{[C_2H_5NH_3^+][OH^-]}{[C_2H_5NH_2]} = \frac{(x)(0.0010 + x)}{(0.015 - x)} = 4.7 \times 10^{-4} \approx \frac{x(0.0010)}{0.015} \qquad \text{Solving, } x = 7.1 \times 10^{-3}$$

However, x has the same order of magnitude as 0.0010, so the simplifying assumption does not hold.　We must solve the original quadratic equation:　$x^2 + 0.0015x - 7.1 \times 10^{-6} = 0$

$$x = \frac{-0.0015 \pm \sqrt{(0.0015)^2 - 4(1)(-7.1 \times 10^{-6})}}{2(1)} = \frac{-0.0015 \pm 0.0055}{2} = 0.0020 \text{ or } -0.0035 \text{ (discard)}$$

Therefore,　$[C_2H_5NH_3^+] = x = \mathbf{0.0020\ M}$

18-78. *Refer to Sections 18-6 and 18-8, and Example 18-20.*

Balanced equations: $NaCH_3CH_2COO + H_2O \rightarrow Na^+ + CH_3CH_2COO^-$ (to completion)

$CH_3CH_2COOH + H_2O \rightleftarrows H_3O^+ + CH_3CH_2COO^-$ (reversible) $K_a = 1.3 \times 10^{-5}$

In this buffer system: $[H_3O^+] = 5.0 \times 10^{-6} M$ since pH = 5.30

$[NaCH_3CH_2COO] = 0.50 M$

$[acid] = [CH_3CH_2COOH] = \dfrac{[H_3O^+][salt]}{K_a} = \dfrac{(5.0 \times 10^{-6})(0.50)}{(1.3 \times 10^{-5})} = \mathbf{0.19\ M}$ since $K_a = \dfrac{[H_3O^+][salt]}{[acid]}$

18-80. *Refer to Sections 18-6 and 18-8, Appendix F, and General Algebraic Principles.*

Plan: (1) Perform the acid-base neutralization limiting reactant problem.
   (2) Determine the volumes of acetic acid and sodium hydroxide that must be mixed without adding additional water by substituting into the modified K_a expression.

(1) Let V_A = volume (in liters) of 0.1500 M CH_3COOH $V_A \times 0.1500\ M$ = initial moles of CH_3COOH
   V_B = volume (in liters) of 0.1000 M NaOH $V_B \times 0.1000\ M$ = initial moles of NaOH

In order to produce a buffer solution, NaOH must be consumed and is therefore the limiting reactant in the acid-base neutralization reaction.

	CH_3COOH	+	NaOH	$\rightarrow$	$NaCH_3COO$	+	H_2O
initial	0.1500 M × V_A mol		0.1000 M × V_B mol		0 mol		
change	- 0.1000 M × V_B mol		- 0.1000 M × V_B mol		+ 0.1000 M × V_B mol		
after rxn	(0.1500 V_A - 0.1000 V_B) mol		0 mol		0.1000 M × V_B mol		

(2) For a weak acid buffer: pH = 4.50; $[H^+] = 3.2 \times 10^{-5}$

$$[H^+] = K_a \times \dfrac{mol\ CH_3COOH}{mol\ NaCH_3COO}$$

Substituting,

$$3.2 \times 10^{-5} = (1.8 \times 10^{-5}) \times \dfrac{0.1500\ V_A - 0.1000\ V_B}{0.1000\ M \times V_B}$$

$$1.8 = \dfrac{0.1500\ V_A}{0.1000\ V_B} - \dfrac{0.1000\ V_B}{0.1000\ V_B}$$

$$2.8 = \dfrac{0.1500\ V_A}{0.1000\ V_B}$$

$$1.9 = \dfrac{V_A}{V_B}$$

$$V_A = 1.9\ V_B$$

Since $V_A + V_B = 1.000$ L
Substituting, $1.9\ V_B + V_B = 1.000$ L
          $2.9\ V_B = 1.000$ L
          V_B = volume of NaOH = **0.34 L**
          V_A = volume of CH_3COOH = 1.000 L - 0.34 L = **0.66 L**

Balanced equations: $Ca(CH_3COO)_2 \rightarrow Ca^{2+} + 2CH_3COO^-$ (to completion)

$CH_3COOH + H_2O \rightleftarrows H_3O^+ + CH_3COO^-$ (reversible) $K_a = 1.8 \times 10^{-5}$

Recall: for dilution, $M_1V_1 = M_2V_2$. In this instance, the acetic acid solution and the calcium acetate solution are diluting each other.

$$[CH_3COOH]_{initial} = M_2 = \frac{M_1V_1}{V_2} = \frac{1.25\ M \times 500\ mL}{500\ mL + 500\ mL} = 0.625\ M$$

$$[Ca(CH_3COO)_2]_{initial} = M_2 = \frac{M_1V_1}{V_2} = \frac{0.500\ M \times 500\ mL}{500\ mL + 500\ mL} = 0.250\ M$$

Therefore,

$$[CH_3COO^-]_{initial} = 2 \times [Ca(CH_3COO)_2]_{initial} = 2 \times 0.250\ M = 0.500\ M$$

Let $x = [CH_3COOH]$ that ionizes. Then
   $x = [H_3O^+]$ produced from $CH_3COOH = [CH_3COO^-]$ produced from CH_3COOH.

	CH_3COOH	+	H_2O	$\rightleftarrows$	H_3O^+	+	CH_3COO^-
initial	0.625 M				$\approx 0\ M$		0.500 M
change	- x M				+ x M		+ x M
at equilibrium	(0.625 - x) M				x M		(0.500 + x) M

$$K_a = \frac{[H_3O^+][CH_3COO^-]}{[CH_3COOH]} = \frac{(x)(0.500 + x)}{(0.625 - x)} = 1.8 \times 10^{-5} \approx \frac{x(0.500)}{(0.625)} \qquad \text{Solving, } x = 2.3 \times 10^{-5}$$

The simplifying assumption is justified and we have

(a) $[CH_3COOH] = \mathbf{0.625\ M}$

(b) $[Ca^{2+}] = [Ca(CH_3COO)_2]_{initial} = \mathbf{0.250\ M}$

(c) $[CH_3COO^-] = \mathbf{0.500\ M}$

(d) $[H^+] = \mathbf{2.3 \times 10^{-5}\ M}$

(e) $pH = -log\ (2.3 \times 10^{-5}) = \mathbf{4.65}$

Balanced equations: $NaCH_2ClCOO \rightarrow Na^+ + CH_2ClCOO^-$ (to completion)

$CH_2ClCOOH + H_2O \rightleftarrows H_3O^+ + CH_2ClCOO^-$ (reversible) $K_a = 1.4 \times 10^{-3}$

$[CH_2ClCOO^-]_{initial} = [NaCH_2ClCOO] = 0.015\ M$

Since pH = 3.00, $[H_3O^+] = 1.0 \times 10^{-3}\ M$
Therefore, $1.0 \times 10^{-3}\ M = [CH_3ClCOOH]_{ionized} = [H_3O^+]_{produced\ from\ CH_3ClCOOH}$
$= [CH_2ClCOO^-]_{produced\ from\ CH_3ClCOOH}.$

Let $x = [CH_2ClCOOH]_{initial}$

	$CH_2ClCOOH$	+	H_2O	$\rightleftarrows$	H_3O^+	+	CH_2ClCOO^-
initial	x M				$\approx 0\ M$		0.015 M
change	- $1.0 \times 10^{-3}\ M$				+ $1.0 \times 10^{-3}\ M$		+ $1.0 \times 10^{-3}\ M$
at equilibrium	(x - 1.0×10^{-3}) M				$1.0 \times 10^{-3}\ M$		0.016 M

$$K_a = \frac{[H_3O^+][CH_2ClCOO^-]}{[CH_2ClCOOH]} = \frac{(1.0 \times 10^{-3})(0.016)}{(x - 1.0 \times 10^{-3})} = 1.4 \times 10^{-3} \qquad \text{Solving, } x = 1.1 \times 10^{-2}$$

Therefore, $[CH_2ClCOOH] = \mathbf{1.1 \times 10^{-2}} \, \boldsymbol{M}$

18-86. *Refer to Section 18-9, Example 18-21, Table 18-10 and Appendix F.*

Balanced equations: $\quad H_3AsO_4 + H_2O \rightleftarrows H_3O^+ + H_2AsO_4^- \qquad K_1 = 2.5 \times 10^{-4}$
$\qquad\qquad\qquad\qquad\quad H_2AsO_4^- + H_2O \rightleftarrows H_3O^+ + HAsO_4^{2-} \qquad K_2 = 5.6 \times 10^{-8}$
$\qquad\qquad\qquad\qquad\quad HAsO_4^{2-} + H_2O \rightleftarrows H_3O^+ + AsO_4^{3-} \qquad K_3 = 3.0 \times 10^{-13}$

First Step:

Let $x = [H_3AsO_4]_{ionized}$. $\qquad$ Then $[H_3AsO_4] = (0.200 - x) \, M$
$\qquad\qquad\qquad\qquad\qquad\qquad\qquad\qquad [H_3O^+] = [H_2AsO_4^-] = x \, M$

$$K_1 = \frac{[H_3O^+][H_2AsO_4^-]}{[H_3AsO_4]} = \frac{x^2}{(0.200 - x)} = 2.5 \times 10^{-4} \approx \frac{x^2}{0.200} \qquad \text{Solving, } x = 7.1 \times 10^{-3}$$

Since 7.1×10^{-3} is only 3% of 0.200, the simplifying assumption is valid.

Therefore, $\quad [H_3O^+] = [H_2AsO_4^-] = x = 7.1 \times 10^{-3} \, M$
$\qquad\qquad\quad [H_3AsO_4] = 0.200 - x = 0.193 \, M$

Second Step:

Let $y = [H_2AsO_4^-]_{ionized}$. $\qquad$ Then $[H_2AsO_4^-] = (7.1 \times 10^{-3} - y) \, M$
$\qquad\qquad\qquad\qquad\qquad\qquad\qquad\qquad\quad [H_3O^+] = (7.1 \times 10^{-3} + y) \, M$
$\qquad\qquad\qquad\qquad\qquad\qquad\qquad\qquad\quad [HAsO_4^{2-}] = y \, M$

$$K_2 = \frac{[H_3O^+][HAsO_4^{2-}]}{[H_2AsO_4^-]} = \frac{(7.1 \times 10^{-3} + y)(y)}{(7.1 \times 10^{-3} - y)} = 5.6 \times 10^{-8} \approx \frac{(7.1 \times 10^{-3})y}{(7.1 \times 10^{-3})}$$

Solving, $y = 5.6 \times 10^{-8}$

Therefore, because the simplifying assumptions are valid, $\quad [H_3O^+] = [H_2AsO_4^-] = 7.1 \times 10^{-3} \, M$
$\qquad\qquad\qquad\qquad\qquad\qquad\qquad\qquad\qquad\qquad\qquad\qquad\qquad\qquad\quad [HAsO_4^{2-}] = 5.6 \times 10^{-8} \, M$

Third Step:

Let $z = [HAsO_4^{2-}]_{ionized}$. $\qquad$ Then $[HAsO_4^{2-}] = (5.6 \times 10^{-8} - z) \, M$
$\qquad\qquad\qquad\qquad\qquad\qquad\qquad\qquad\quad [H_3O^+] = (7.1 \times 10^{-3} + z) \, M$
$\qquad\qquad\qquad\qquad\qquad\qquad\qquad\qquad\quad [AsO_4^{3-}] = z \, M$

$$K_3 = \frac{[H_3O^+][AsO_4^{3-}]}{[HAsO_4^-]} = \frac{(7.1 \times 10^{-3} + z)(z)}{(5.6 \times 10^{-8} - z)} = 3.0 \times 10^{-13} = \frac{(7.1 \times 10^{-3})z}{(5.6 \times 10^{-8})} \qquad \text{Solving, } z = 2.4 \times 10^{-18}$$

Therefore, $[AsO_4^{3-}] = 2.4 \times 10^{-18} \, M$ because the simplifying assumptions are valid.

0.200 *M* H$_3$AsO$_4$ Solution		0.100 *M* H$_3$PO$_4$ Solution	
Species	Concentration (*M*)	Species	Concentration (*M*)
H_3AsO_4	0.193	H_3PO_4	0.076
H_3O^+	0.0071	H_3O^+	0.024
$H_2AsO_4^-$	0.0071	$H_2PO_4^-$	0.024
$HAsO_4^{2-}$	5.6×10^{-8}	HPO_4^{2-}	6.2×10^{-8}
OH^-	1.4×10^{-12}	OH^-	4.2×10^{-13}
AsO_4^{3-}	2.4×10^{-18}	PO_4^{3-}	9.3×10^{-19}

Balanced equations: $H_2SeO_4 + H_2O \rightleftarrows H_3O^+ + HSeO_4^-$ $\qquad$ K_1 = very large

$\qquad\qquad\qquad\quad$ $HSeO_4^- + H_2O \rightleftarrows H_3O^+ + SeO_4^{2-}$ $\qquad$ $K_2 = 1.2 \times 10^{-2}$

First Step:

$\quad$ Since K_1 is very large, H_2SeO_4 is a strong electrolyte and totally dissociates into H_3O^+ and $HSeO_4^-$. Therefore,

$\qquad\qquad$ $[H_2SeO_4] = 0\ M$ $\qquad\qquad$ $[HSeO_4^-] = 0.10\ M$ $\qquad\qquad$ $[H_3O^+] = 0.10\ M$

Second Step:

$\quad$ Let x = $[HSeO_4^-]_{ionized}$. $\qquad$ Then $[HSeO_4^-] = (0.10 - x)\ M$

$\qquad\qquad\qquad\qquad\qquad\qquad\qquad\quad$ $[H_3O^+] = (0.10 + x)\ M$

$\qquad\qquad\qquad\qquad\qquad\qquad\qquad\quad$ $[SeO_4^{2-}] = x\ M$

$$K_2 = \frac{[H_3O^+][SeO_4^{2-}]}{[HSeO_4^-]} = \frac{(0.10 + x)(x)}{(0.10 - x)} = 1.2 \times 10^{-2} \approx \frac{(0.10)x}{(0.10)} \qquad \text{Solving, x} = 1.2 \times 10^{-2}$$

In this case, 1.2×10^{-2} is greater than 5% of 0.10 (= 5.0×10^{-3}). A simplifying assumption cannot be made and the original quadratic equation must be solved: $x^2 + 0.11x - 1.2 \times 10^{-3} = 0$

$$x = \frac{-0.11 \pm \sqrt{(0.11)^2 - 4(1)(-1.2 \times 10^{-3})}}{2(1)} = \frac{-0.11 \pm 0.13}{2} = 0.01 \text{ or } -0.24 \text{ (discard)}$$

Therefore, $\quad$ $[H_3O^+] = 0.10 + x = \mathbf{0.11\ M}$ $\qquad\qquad\qquad$ $[HSeO_4^-] = 0.10 - x = \mathbf{0.09\ M}$

$\qquad\qquad\quad$ $[OH^-] = K_w/[H_3O^+] = \mathbf{9.1 \times 10^{-14}\ M}$ $\qquad\qquad$ $[SeO_4^{2-}] = x = \mathbf{0.01\ M}$

Balanced equations: $(COOH)_2 + H_2O \rightleftarrows H_3O^+ + COOCOOH^-$ $\qquad$ $K_1 = 5.9 \times 10^{-2}$

$\qquad\qquad\qquad\qquad$ $COOCOOH^- + H_2O \rightleftarrows H_3O^+ + (COO)_2^{2-}$ $\qquad$ $K_2 = 6.4 \times 10^{-5}$

First Step:

$\quad$ Let x = $[(COOH)_2]_{ionized}$. $\qquad$ Then $[(COOH)_2] = (0.20 - x)\ M$

$\qquad\qquad\qquad\qquad\qquad\qquad\qquad\quad$ $[H_3O^+] = [COOCOOH^-] = x\ M$

$$K_1 = \frac{[H_3O^+][COOCOOH^-]}{[(COOH)_2]} = \frac{x^2}{(0.20 - x)} = 5.9 \times 10^{-2} \approx \frac{x^2}{0.20} \qquad \text{Solving, x} = 0.11$$

Since 0.11 is more than 50% of 0.20, the simplifying assumptions do not hold and we must solve the original quadratic equation: $x^2 + (5.9 \times 10^{-2})x - 1.2 \times 10^{-2} = 0$

$$x = \frac{-5.9 \times 10^{-2} \pm \sqrt{(5.9 \times 10^{-2})^2 - 4(1)(-1.2 \times 10^{-2})}}{2(1)} = \frac{-5.9 \times 10^{-2} \pm 0.23}{2} = 0.085 \text{ or } -0.14 \text{(discard)}$$

Therefore, $\quad$ $[(COOH)_2] = 0.20 - x = 0.12\ M$

$\qquad\qquad\quad$ $[H_3O^+] = [COOCOOH^-] = x = 0.085\ M$

Second Step:

$\quad$ Let y = $[COOCOOH^-]_{ionized}$. $\qquad$ Then $[COOCOOH^-] = (0.085 - y)\ M$

$\qquad\qquad\qquad\qquad\qquad\qquad\qquad\qquad\quad$ $[H_3O^+] = (0.085 + y)\ M$

$\qquad\qquad\qquad\qquad\qquad\qquad\qquad\qquad\quad$ $[(COO)_2^{2-}] = y\ M$

$$K_2 = \frac{[H_3O^+][(COO)_2^{2-}]}{[COOCOOH^-]} = \frac{(0.085 + y)y}{(0.085 - y)} = 6.4 \times 10^{-5} \approx \frac{0.085y}{0.085} \qquad \text{Solving, y} = 6.4 \times 10^{-5}$$

Therefore, since the simplifying assumptions are valid,

(a) $[H_3O^+] = 0.085\ M$; pH = **1.07** $\qquad\qquad\qquad$ (b) $[(COO)_2^{2-}] = \mathbf{6.4 \times 10^{-5}\ M}$

Balanced equations: $HClO + H_2O \rightleftarrows H_3O^+ + ClO^-$ (reversible) $K_a = 3.5 \times 10^{-8}$

$HIO + H_2O \rightleftarrows H_3O^+ + IO^-$ (reversible) $K_a = 2.3 \times 10^{-11}$

In the aqueous solution, two equilibria involving H_3O^+ are operating simultaneously. We could solve two equations (the K_a expressions) for two unknowns ($[HClO]_{ionized}$ and $[HIO]_{ionized}$). However, after carefully examining the two-acid system, we can deduce that

(1) the acids are weak with small K_a values, so the simplifying assumptions will hold, and
(2) the K_a for HIO is about 1000 times smaller than that for HClO, so we can ignore the contribution of H_3O^+ ions from the ionization of HIO.

Plan: Calculate the pH from the ionization of only the weak acid, HClO.

Let $x = [HClO]_{ionized}$.   Then, $x = [H_3O^+]_{produced} = [ClO^-]_{produced}$.

$$K_a = \frac{[H_3O^+][ClO^-]}{[HClO]} = 3.5 \times 10^{-8} = \frac{x^2}{0.12 - x} \approx \frac{x^2}{0.12} \qquad \text{Solving, } x = 6.5 \times 10^{-5}$$

Therefore, $[H_3O^+] = x = 6.5 \times 10^{-5}\,M$;   pH = **4.19**

Balanced equation: $C_2H_4OCOOH + H_2O \rightleftarrows H_3O^+ + C_2H_4OCOO^-$ $K_a = 8.4 \times 10^{-4}$

Let $x = [C_2H_4OCOOH]$ that ionizes. Then $x = [H_3O^+]$ produced $= [C_2H_4OCOO^-]$ produced.

$$K_a = \frac{[H_3O^+][C_2H_4OCOO^-]}{[C_2H_4OCOOH]} = \frac{x^2}{0.100 - x} = 8.4 \times 10^{-4} \approx \frac{x^2}{0.100} \qquad \text{Solving, } x = 9.2 \times 10^{-3}$$

However, 9.2×10^{-3} is more than 5% of 0.100. A simplifying assumption *cannot* be made; we must solve the original quadratic equation: $x^2 + (8.4 \times 10^{-4})x - 8.4 \times 10^{-5} = 0$

$$x = \frac{-(8.4 \times 10^{-4}) \pm \sqrt{(8.4 \times 10^{-4})^2 - 4(1)(-8.4 \times 10^{-5})}}{2(1)} = \frac{-8.4 \times 10^{-4} \pm 1.8 \times 10^{-2}}{2}$$

$$= 8.6 \times 10^{-3} \text{ or } -9.4 \times 10^{-3} \text{ (discard)}$$

Therefore, $[H_3O^+] = 8.6 \times 10^{-3}\,M$; pH = **2.07**

Balanced equation: $C_5H_7O_4COOH + H_2O \rightleftarrows H_3O^+ + C_5H_7O_4COO^-$ $K_a = 7.9 \times 10^{-5}$

Let $x = [C_5H_7O_4COOH]$ that ionizes. Then $x = [H_3O^+]$ produced $= [C_5H_7O_4COO^-]$ produced.

$$K_a = \frac{[H_3O^+][C_5H_7O_4COO^-]}{[C_5H_7O_4COOH]} = \frac{x^2}{0.100 - x} = 7.9 \times 10^{-5} \approx \frac{x^2}{0.100} \qquad \text{Solving, } x = 2.8 \times 10^{-3}$$

The simplifying assumption is justified since 2.8×10^{-3} is less than 5% of 0.100 $(= 5 \times 10^{-3})$.

Therefore, $[H_3O^+] = 2.8 \times 10^{-3}\,M$; pH = **2.55**

19 Ionic Equilibria II: Hydrolysis

19-2. *Refer to the Introduction to Chapter 19.*

Dilute aqueous solutions of weak acids, such as HF and HNO_2, contain relatively few ions because they are weak electrolytes and dissociate only slightly to form ions.

19-4. *Refer to the Introduction to Chapter 19.*

(a) Solvolysis is the reaction of a substance with the solvent in which it is dissolved. Common solvents used include $H_2O(\ell)$, $NH_3(\ell)$, $H_2SO_4(\ell)$ and $CH_3COOH(\ell)$. There are many others. For example, glacial acetic acid, $CH_3COOH(\ell)$, is commonly used in non-aqueous titrations with weak acids:

$$C_6H_5NH_3^+ + CH_3COOH \rightleftarrows CH_3COOH_2^+ + C_6H_5NH_2$$
$$\text{solute} \qquad \text{solvent} \qquad\qquad\qquad \text{aniline}$$

(b) Hydrolysis is the reaction of a substance with the solvent, water, or its ions, OH^- and H_3O^+, e.g., the hydrolysis of the weak acid, CH_3COOH:

$$CH_3COOH + H_2O \rightleftarrows H_3O^+ + CH_3COO^-$$
$$\text{solute} \qquad \text{solvent}$$

19-6. *Refer to the Introduction to Chapter 19 and Section 19-5.*

Consider a cation which does not react appreciably with water. The relative acid strength of such a cation is low compared to water. Dissolution of these cations will have essentially no effect on the pH of the solution.

19-8. *Refer to Sections 19-1, 19-2 and 19-3.*

The pH of aqueous salt solutions depend on whether or not the ions produced by the dissociation of the salt will hydrolyze (react with water). If the cation of the salt hydrolyzes more than the anion, the solution is acidic. If the anion hydrolyzes more than the cation, the solution is basic. If the cation and the anion hydrolyze to the same extent or if neither hydrolyzes appreciably, the resulting solution is neutral.

19-10. *Refer to Section 19-2.*

The solution of a salt derived from a strong soluble base and weak acid is basic because the anion of a weak acid reacts with water (hydrolysis) to form hydroxide ions. Consider the soluble salt NaClO prepared by reacting NaOH, a strong soluble base, and HClO, a weak acid. The salt dissociates completely in water and the conjugate base of the weak acid, ClO^-, hydrolyzes, producing OH^- ions.

$$NaClO(s) \xrightarrow[100\%]{H_2O} Na^+(aq) + ClO^-(aq)$$

$$ClO^-(aq) + H_2O(\ell) \rightleftarrows HClO(aq) + OH^-(aq)$$

19-12. *Refer to Section 19-2, Example 19-1 and Appendix F.*

Balanced equation: $N_3^-(aq) + H_2O(\ell) \rightleftarrows HN_3(aq) + OH^-(aq)$ (K_a for hydrazoic acid, $HN_3 = 1.9 \times 10^{-5}$)

$$K = K_b = \frac{[HN_3][OH^-]}{[N_3^-]} = \frac{K_w}{K_{a(HN_3)}} = \frac{1.0 \times 10^{-14}}{1.9 \times 10^{-5}} = \mathbf{5.3 \times 10^{-10}}$$

19-14. *Refer to Section 19-2, Example 19-1 and Appendix F.*

(a) for NO_2^-, $K_b = \dfrac{K_w}{K_{a(HNO_2)}} = \dfrac{1.0 \times 10^{-14}}{4.5 \times 10^{-4}} = \mathbf{2.2 \times 10^{-11}}$

(b) for ClO^-, $K_b = \dfrac{K_w}{K_{a(HClO)}} = \dfrac{1.0 \times 10^{-14}}{3.5 \times 10^{-8}} = \mathbf{2.9 \times 10^{-7}}$

(c) for $HCOO^-$, $K_b = \dfrac{K_w}{K_{a(HCOOH)}} = \dfrac{1.0 \times 10^{-14}}{1.8 \times 10^{-4}} = \mathbf{5.6 \times 10^{-11}}$

The mathematical relationship between K_a, the ionization constant for a weak acid, and K_b, the base hydrolysis constant for the anion of the weak acid is $K_w = K_a \times K_b$. The weaker the acid, the smaller is its K_a, the more its anion will hydrolyze, and the larger is K_b for the anion.

19-16. *Refer to Section 19-2, Example 19-2, and Exercises 19-14 and 19-15 Solutions.*

(a) Balanced equations: $NaNO_2 \rightarrow Na^+ + NO_2^-$ (to completion)

  $NO_2^- + H_2O \rightleftarrows HNO_2 + OH^-$ (reversible)

Let $x = [NO_2^-]_{hydrolyzed}$ Then, $0.15 - x = [NO_2^-]$; $x = [HNO_2] = [OH^-]$

$K_b = \dfrac{[HNO_2][OH^-]}{[NO_2^-]} = \dfrac{x^2}{0.15 - x} = 2.2 \times 10^{-11} \approx \dfrac{x^2}{0.15}$ Solving, $x = 1.8 \times 10^{-6}$

% hydrolysis $= \dfrac{[NO_2^-]_{hydrolyzed}}{[NO_2^-]_{initial}} \times 100\% = \dfrac{1.8 \times 10^{-6}\ M}{0.15\ M} \times 100\% = \mathbf{1.2 \times 10^{-3}\ \%}$

(b) Balanced equations: $NaClO \rightarrow Na^+ + ClO^-$ (to completion)

  $ClO^- + H_2O \rightleftarrows HClO + OH^-$ (reversible)

Let $x = [ClO^-]_{hydrolyzed}$ Then, $0.15 - x = [ClO^-]$; $x = [HClO] = [OH^-]$

$K_b = \dfrac{[HClO][OH^-]}{[ClO^-]} = \dfrac{x^2}{0.15 - x} = 2.9 \times 10^{-7} \approx \dfrac{x^2}{0.15}$ Solving, $x = 2.1 \times 10^{-4}$

% hydrolysis $= \dfrac{[ClO^-]_{hydrolyzed}}{[ClO^-]_{initial}} \times 100\% = \dfrac{2.1 \times 10^{-4}\ M}{0.15\ M} \times 100\% = \mathbf{0.14\%}$

(c) Balanced equations: $NaHCOO \rightarrow Na^+ + HCOO^-$ (to completion)

  $HCOO^- + H_2O \rightleftarrows HCOOH + OH^-$ (reversible)

Let $x = [HCOO^-]_{hydrolyzed}$ Then, $0.15 - x = [HCOO^-]$; $x = [HCOOH] = [OH^-]$

$K_b = \dfrac{[HCOOH][OH^-]}{[HCOO^-]} = \dfrac{x^2}{0.15 - x} = 5.6 \times 10^{-11} \approx \dfrac{x^2}{0.15}$ Solving, $x = 2.9 \times 10^{-6}$

% hydrolysis $= \dfrac{[HCOO^-]_{hydrolyzed}}{[HCOO^-]_{initial}} \times 100\% = \dfrac{2.9 \times 10^{-6}\ M}{0.15\ M} \times 100\% = \mathbf{1.9 \times 10^{-3}\ \%}$

19-18. *Refer to Section 19-2, Example 19-2 and Exercise 19-12 Solution.*

Balanced equations: $KF \rightarrow K^+ + F^-$ (to completion)

$\qquad\qquad\qquad\quad F^- + H_2O \rightleftarrows HF + OH^-$ (reversible)

Let $x = [F^-]_{hydrolyzed}$ Then, $0.15 - x = [F^-]$; $x = [HF] = [OH^-]$

$$K_b = \frac{[HF][OH^-]}{[F^-]} = \frac{x^2}{0.15 - x} = 1.4 \times 10^{-11} \approx \frac{x^2}{0.15} \qquad \text{Solving, } x = 1.5 \times 10^{-6}$$

$[OH^-] = x = 1.5 \times 10^{-6}\,M$; $pOH = 5.84$; $pH = 14 - pOH = \mathbf{8.16}$

19-20. *Refer to Section 19-3.*

NH_4Cl	ammonium chloride	$CH_3NH_3NO_3$	methylammonium nitrate
$C_5H_5NHClO_4$	pyridinium perchlorate	$[(CH_3)_3NH]_2SO_4$	trimethylammonium sulfate

---

19-22. *Refer to Section 19-3, Example 19-3 and Appendix G.*

(a) Balanced equations: $NH_4NO_3 \rightarrow NH_4^+ + NO_3^-$ (to completion)

$\qquad\qquad\qquad\qquad\quad NH_4^+ + H_2O \rightleftarrows NH_3 + H_3O^+$ (reversible)

Let $x = [NH_4^+]_{hydrolyzed}$ Then, $0.15 - x = [NH_4^+]$; $x = [NH_3] = [H_3O^+]$

$$K_a = \frac{K_w}{K_{b(NH_3)}} = \frac{1.0 \times 10^{-14}}{1.8 \times 10^{-5}} = 5.6 \times 10^{-10} = \frac{[NH_3][H_3O^+]}{[NH_4^+]} = \frac{x^2}{0.15 - x} \approx \frac{x^2}{0.15}$$

Solving, $x = 9.2 \times 10^{-6}$ Therefore, $[H_3O^+] = 9.2 \times 10^{-6}\,M$; $pH = \mathbf{5.04}$

(b) Balanced equations: $CH_3NH_3NO_3 \rightarrow CH_3NH_3^+ + NO_3^-$ (to completion)

$\qquad\qquad\qquad\qquad\quad CH_3NH_3^+ + H_2O \rightleftarrows CH_3NH_2 + H_3O^+$ (reversible)

Let $x = [CH_3NH_3^+]_{hydrolyzed}$ Then, $0.15 - x = [CH_3NH_3^+]$; $x = [CH_3NH_2] = [H_3O^+]$

$$K_a = \frac{K_w}{K_{b(CH_3NH_2)}} = \frac{1.0 \times 10^{-14}}{5.0 \times 10^{-4}} = 2.0 \times 10^{-11} = \frac{[CH_3NH_2][H_3O^+]}{[CH_3NH_3^+]} = \frac{x^2}{0.15 - x} \approx \frac{x^2}{0.15}$$

Solving, $x = 1.7 \times 10^{-6}$ Therefore, $[H_3O^+] = 1.7 \times 10^{-6}\,M$; $pH = \mathbf{5.77}$

(c) Balanced equations: $C_6H_5NH_3NO_3 \rightarrow C_6H_5NH_3^+ + NO_3^-$ (to completion)

$\qquad\qquad\qquad\qquad\quad C_6H_5NH_3^+ + H_2O \rightleftarrows C_6H_5NH_2 + H_3O^+$ (reversible)

Let $x = [C_6H_5NH_3^+]_{hydrolyzed}$ Then, $0.15 - x = [C_6H_5NH_3^+]$; $x = [C_6H_5NH_2] = [H_3O^+]$

$$K_a = \frac{K_w}{K_{b(C_6H_5NH_2)}} = \frac{1.0 \times 10^{-14}}{4.2 \times 10^{-10}} = 2.4 \times 10^{-5} = \frac{[C_6H_5NH_2][H_3O^+]}{[C_6H_5NH_3^+]} = \frac{x^2}{0.15 - x} \approx \frac{x^2}{0.15}$$

Solving, $x = 1.9 \times 10^{-3}$ Therefore, $[H_3O^+] = 1.9 \times 10^{-3}\,M$; $pH = \mathbf{2.72}$

19-24. *Refer to Section 19-3.*

(a) NH_4Br and NH_4NO_3 are both salts derived from monoprotic strong acids and the weak base, NH_3. Since the concentration of NH_4^+ is the same in each solution, the pH values will be identical.

(b) NH_4ClO_4 and NH_4Cl are both salts derived from monoprotic strong acids and the weak base, NH_3. Since the concentration of NH_4^+ is the same in each solution, the pH values will be identical.

19-26. *Refer to Section 19-4 and Appendices F and G.*

(a) A salt of a weak acid and a weak base for which $K_a = K_b$ gives a neutral solution, e.g., ammonium acetate, NH_4CH_3COO. $K_{a(CH_3COOH)} = K_{b(NH_3)} = 1.8 \times 10^{-5}$.

(b) A salt of a weak acid and weak base for which $K_a > K_b$ gives an acidic solution, e.g., pyridinium fluoride, C_5H_5NHF. $K_{a(HF)} = 7.2 \times 10^{-4}$; $K_{b(C_5H_5N)} = 1.5 \times 10^{-9}$.

(c) A salt of a weak acid and weak base for which $K_b > K_a$ gives a basic solution, e.g., methylammonium cyanide, CH_3NH_3CN. $K_{a(HCN)} = 4.0 \times 10^{-10}$; $K_{b(CH_3NH_2)} = 5.0 \times 10^{-4}$.

19-28. *Refer to Section 19-5 and Table 19-2.*

The cations that will reaction with water to form H^+ (or H_3O^+) ions are

(b) $[Be(OH_2)_4]^{2+} + H_2O \rightleftarrows [Be(OH)(OH_2)_3]^+ + H_3O^+$

(c) $[Al(OH_2)_6]^{3+} + H_2O \rightleftarrows [Al(OH)(OH_2)_5]^{2+} + H_3O^+$

(d) $[Fe(OH_2)_6]^{3+} + H_2O \rightleftarrows [Fe(OH)(OH_2)_5]^{2+} + H_3O^+$

(e) $[Cu(OH_2)_6]^{3+} + H_2O \rightleftarrows [Cu(OH)(OH_2)_5]^{2+} + H_3O^+$

19-30. *Refer to Section 19-5, Example 19-4 and Table 19-2.*

(a) Balanced equations: $\quad Al(NO_3)_3 \rightarrow Al^{3+} + 3NO_3^-$ $\qquad$ (to completion)

$\qquad\qquad\qquad\qquad\quad Al^{3+} + 2H_2O \rightleftarrows Al(OH)^{2+} + H_3O^+$ $\qquad$ (reversible) $\qquad K_a = 1.2 \times 10^{-5}$

Let x = $[Al^{3+}]_{hydrolyzed}$ $\qquad$ Then, 0.15 - x = $[Al^{3+}]$; x = $[Al(OH)^{2+}]$ = $[H_3O^+]$

$$K_a = \frac{[Al(OH)^{2+}][H_3O^+]}{[Al^{3+}]} = \frac{x^2}{0.15 - x} = 1.2 \times 10^{-5} \approx \frac{x^2}{0.15} \qquad \text{Solving, x} = 1.3 \times 10^{-3}$$

Therefore, $[H_3O^+] = 1.3 \times 10^{-3} M$; pH = **2.89**

$$\% \text{ hydrolysis} = \frac{[Al^{3+}]_{hydrolyzed}}{[Al^{3+}]_{initial}} \times 100\% = \frac{1.3 \times 10^{-3} M}{0.15 M} \times 100\% = \mathbf{0.87\%}$$

(b) Balanced equations: $\quad Co(ClO_4)_2 \rightarrow Co^{2+} + 2ClO_4^-$ $\qquad$ (to completion)

$\qquad\qquad\qquad\qquad\quad Co^{2+} + 2H_2O \rightleftarrows Co(OH)^+ + H_3O^+$ $\qquad$ (reversible) $\qquad K_a = 5.0 \times 10^{-10}$

Let x = $[Co^{2+}]_{hydrolyzed}$ $\qquad$ Then, 0.075 - x = $[Co^{2+}]$; x = $[Co(OH)^+]$ = $[H_3O^+]$

$$K_a = \frac{[Co(OH)^+][H_3O^+]}{[Co^{2+}]} = \frac{x^2}{0.075 - x} = 5.0 \times 10^{-10} \approx \frac{x^2}{0.075} \qquad \text{Solving, x} = 6.1 \times 10^{-6}$$

Therefore, $[H_3O^+] = 6.1 \times 10^{-6} M$; pH = **5.21**

$$\% \text{ hydrolysis} = \frac{[Co^{2+}]_{hydrolyzed}}{[Co^{2+}]_{initial}} \times 100\% = \frac{6.1 \times 10^{-6} M}{0.075 M} \times 100\% = \mathbf{8.2 \times 10^{-3} \%}$$

(c) Balanced equations: $\quad MgCl_2 \rightarrow Mg^{2+} + 2Cl^-$ $\qquad$ (to completion)

$\qquad\qquad\qquad\qquad\quad Mg^{2+} + 2H_2O \rightleftarrows Mg(OH)^+ + H_3O^+$ $\qquad$ (reversible) $\qquad K_a = 3.0 \times 10^{-12}$

Let x = $[Mg^{2+}]_{hydrolyzed}$ $\qquad$ Then, 0.15 - x = $[Mg^{2+}]$; x = $[Mg(OH)^+]$ = $[H_3O^+]$

$$K_a = \frac{[Mg(OH)^+][H_3O^+]}{[Mg^{2+}]} = \frac{x^2}{0.15 - x} = 3.0 \times 10^{-12} \approx \frac{x^2}{0.15} \qquad \text{Solving, x} = 6.7 \times 10^{-7}$$

Therefore, $[H_3O^+] = 6.7 \times 10^{-7} M$; $pH = \mathbf{6.17}$ (ignoring the H_3O^+ produced by the ionization of water)

$$\% \text{ hydrolysis} = \frac{[Mg^{2+}]_{hydrolyzed}}{[Mg^{2+}]_{initial}} \times 100\% = \frac{6.7 \times 10^{-7} M}{0.15 M} \times 100\% = \mathbf{4.5 \times 10^{-4} \%}$$

Note: To calculate the actual $[H_3O^+]$, let $x = [OH^-] = [H_3O^+]$ produced by the ionization of water.

Therefore $[H_3O^+]_{total} = [H_3O^+]$ produced by hydrolysis + $[H_3O^+]$ produced by the ionization of water
$$= 6.7 \times 10^{-7} + x$$

We know that $K_w = 1.0 \times 10^{-14} = [H_3O^+][OH^-] = (6.7 \times 10^{-7} + x)(x) = (6.7 \times 10^{-7})x + x^2$

Solving the quadratic equation: $x^2 + (6.7 \times 10^{-7})x - 1.0 \times 10^{-14} = 0$, we have $x = 1.5 \times 10^{-8}$

Therefore, $[H_3O^+]_{total} = 6.7 \times 10^{-7} + x = 6.9 \times 10^{-7} M$; $pH = \mathbf{6.16}$

19-32. Refer to Section 19-6 and Figure 19-3.

Example of a titration of a strong acid with a strong base:

(a) When no base is added, the pH of the solution is determined by the initial strong acid concentration.

(b) At the point halfway to the equivalence point, only half of the base required to titrate all of the acid has been added. The strong base is the limiting reactant and the pH of the resulting solution is less than 7. It is calculated from the concentration of the remaining acid.

(c) At the equivalence point, only water and the salt of the strong acid and strong base are present. Since neither the cation nor the anion of the salt hydrolyzes appreciably, the pH is 7.

(d) Past the equivalence point, the strong acid is the limiting reactant and the pH of the solution is greater than 7. It is determined from the concentration of the excess strong base.

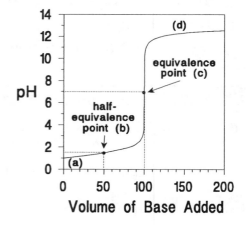

The graph compares well with Figure 19-3a.

19-34. Refer to Section 19-6, Exercise 19-32 Solution and Appendix A.2.

Balanced equation: $HNO_3 + NaOH \rightarrow NaNO_3 + H_2O$

A 25.0 mL sample of $0.125 M$ HNO_3 is titrated with $0.100 M$ NaOH.

(a) Initially: $[H^+] = [HNO_3] = 0.125 M$; $pH = \mathbf{0.903}$

For the rest of the exercise, the plan is straightforward: for the neutralization reaction between HNO_3 and NaOH, perform the limiting reactant problem. The pH is determined from the concentration of excess HNO_3 or NaOH.

(b) Addition of 5.0 mL of $0.100 M$ NaOH:

? mmol $HNO_3 = 0.125 M \times 25.0$ mL $= 3.13$ mmol HNO_3
? mmol NaOH $= 0.100 M \times 5.0$ mL $= 0.50$ mmol NaOH

Before the equivalence point, NaOH is the limiting reactant. The pH is determined from the concentration of excess HNO_3 remaining. The salt produced, $NaNO_3$, is the salt of a strong acid and a strong base. It will not affect the pH of the solution.

	HNO$_3$	+	NaOH	→	NaNO$_3$	+	H$_2$O
initial	3.13 mmol		0.50 mmol		0 mmol		
change	- 0.50 mmol		- 0.50 mmol		+ 0.50 mmol		
after reaction	2.63 mmol		0 mmol		0.50 mmol		

$$[H^+] = \frac{\text{mmol excess HNO}_3}{\text{total volume (mL)}} = \frac{2.63 \text{ mmol}}{(25.0 \text{ mL} + 5.0 \text{ mL})} = 0.088 \ M; \ pH = \mathbf{1.06}$$

(c) Addition of 12.5 mL of 0.100 M NaOH:

? mmol HNO$_3$ = 0.125 M × 25.0 mL = 3.13 mmol HNO$_3$
? mmol NaOH = 0.100 M × 12.5 mL = 1.25 mmol NaOH

	HNO$_3$	+	NaOH	→	NaNO$_3$	+	H$_2$O
initial	3.13 mmol		1.25 mmol		0 mmol		
change	- 1.25 mmol		- 1.25 mmol		+ 1.25 mmol		
after reaction	1.88 mmol		0 mmol		1.25 mmol		

$$[H^+] = \frac{\text{mmol excess HNO}_3}{\text{total volume (mL)}} = \frac{1.88 \text{ mmol}}{(25.0 \text{ mL} + 12.5 \text{ mL})} = 0.0501 \ M; \ pH = \mathbf{1.300}$$

(d) Addition of 25.0 mL of 0.100 M NaOH:

? mmol HNO$_3$ = 0.125 M × 25.0 mL = 3.13 mmol HNO$_3$
? mmol NaOH = 0.100 M × 25.0 mL = 2.50 mmol NaOH

	HNO$_3$	+	NaOH	→	NaNO$_3$	+	H$_2$O
initial	3.13 mmol		2.50 mmol		0 mmol		
change	- 2.50 mmol		- 2.50 mmol		+ 2.50 mmol		
after reaction	0.63 mmol		0 mmol		2.50 mmol		

$$[H^+] = \frac{\text{mmol excess HNO}_3}{\text{total volume (mL)}} = \frac{0.63 \text{ mmol}}{(25.0 \text{ mL} + 25.0 \text{ mL})} = 0.0126 \ M; \ pH = \mathbf{1.900}$$

(e) Addition of 31.25 mL of 0.100 M NaOH:

? mmol HNO$_3$ = 0.125 M × 25.0 mL = 3.13 mmol HNO$_3$
? mmol NaOH = 0.100 M × 31.25 mL = 3.13 mmol NaOH

Since the mmoles of HNO$_3$ equal the mmoles of NaOH, the titration is at the equivalence point; there is no excess acid or base in the solution, only salt and water. The ions in any salt derived from a strong acid and a strong soluble base do not hydrolyze, so the pH of water is not affected. Therefore, **pH = 7**.

	HNO$_3$	+	NaOH	→	NaNO$_3$	+	H$_2$O
initial	3.13 mmol		3.13 mmol		0 mmol		
change	- 3.13 mmol		- 3.13 mmol		+ 3.13 mmol		
after reaction	0 mmol		0 mmol		3.13 mmol		

(f) Addition of 37.5 mL of 0.100 M NaOH:

? mmol HNO$_3$ = 0.125 M × 25.0 mL = 3.13 mmol HNO$_3$
? mmol NaOH = 0.100 M × 37.5 mL = 3.75 mmol NaOH

After the equivalence point, HNO$_3$ is the limiting reactant. The pH is determined from the concentration of excess NaOH.

	HNO$_3$	+	NaOH	→	NaNO$_3$	+	H$_2$O
initial	3.13 mmol		3.75 mmol		0 mmol		
change	- 3.13 mmol		- 3.13 mmol		+ 3.13 mmol		
after reaction	0 mmol		0.62 mmol		3.13 mmol		

$$[OH^-] = \frac{\text{mmol excess NaOH}}{\text{total volume (mL)}} = \frac{0.62 \text{ mmol}}{(25.0 \text{ mL} + 37.5 \text{ mL})} = 0.0099 \ M; \ \text{pOH} = 2.00; \ \text{pH} = \mathbf{12.00}$$

19-36. *Refer to Section 19-7, Table 19-3 and Figure 19-4.*

The calculations for determining the pH at every point in the titration of 1 liter of 0.0200 M CH$_3$COOH with solid NaOH, assuming no volume change, can be divided into 4 types.

(1) Initially, the pH is determined by the concentration of the weak acid, CH$_3$COOH.

(2) Before the equivalence point, the pH is determined by the buffer solution consisting of the unreacted CH$_3$COOH and NaCH$_3$COO produced by the reaction. Each calculation is a limiting reactant problem using the original concentration of CH$_3$COOH. For example, at point (c) in the following table:

	CH$_3$COOH	+	NaOH	→	NaCH$_3$COO	+	H$_2$O
initial	20.0 mmol		8.00 mmol		0 mmol		
change	- 8.00 mmol		- 8.00 mmol		+ 8.00 mmol		
after reaction	12.0 mmol		0 mmol		8.00 mmol		

After the reaction, we have a 1 liter buffer solution consisting of 12.0 mmol CH$_3$COOH and 8.00 mmol NaCH$_3$COO.

$$[H_3O^+] = K_a \times \frac{\text{mmol CH}_3\text{COOH}}{\text{mmol NaCH}_3\text{COO}} = (1.8 \times 10^{-5}) \frac{12.0 \text{ mmol}}{8.00 \text{ mmol}} = 2.7 \times 10^{-5} \ M; \ \text{pH} = 4.57$$

Halfway to the equivalence point (i.e., when half of the required amount of base needed to reach the equivalence point is added), pH = pK_a. At point (d):

	CH$_3$COOH	+	NaOH	→	NaCH$_3$COO	+	H$_2$O
initial	20.0 mmol		10.00 mmol		0 mmol		
change	- 10.00 mmol		- 10.00 mmol		+ 10.00 mmol		
after reaction	10.0 mmol		0 mmol		10.00 mmol		

$$[H_3O^+] = K_a \times \frac{\text{mmol CH}_3\text{COOH}}{\text{mmol NaCH}_3\text{COO}} = (1.8 \times 10^{-5}) \frac{10.0 \text{ mmol}}{10.00 \text{ mmol}} = 1.8 \times 10^{-5} \ M; \ \text{pH} = \text{p}K_a = 4.74$$

(3) At the equivalence point, there is no excess acid or base. The concentration of NaCH$_3$COO determines the pH of the system (Refer to Exercise 19-18 Solution). At point (h):

	CH$_3$COOH	+	NaOH	→	NaCH$_3$COO	+	H$_2$O
initial	20.0 mmol		20.0 mmol		0 mmol		
change	- 20.0 mmol		- 20.0 mmol		+ 20.0 mmol		
after reaction	0 mmol		0 mmol		20.0 mmol		

$$[NaCH_3COO] = \frac{0.0200 \text{ mol}}{1.00 \text{ L}} = 0.0100 \ M$$

If aqueous NaOH had been added,
[NaCH$_3$COO] = 0.0200 mol/total volume (L)

Then the anion of the salt hydrolyzes to produce a basic solution: CH$_3$COO$^-$ + H$_2$O $\rightleftarrows$ CH$_3$COOH + OH$^-$

Let $x = [CH_3COO^-]_{hydrolyzed}$ Then, $0.0200 - x = [CH_3COO^-]$; $x = [CH_3COOH] = [OH^-]$

$$K_b = \frac{K_w}{K_{a(CH_3COOH)}} = 5.6 \times 10^{-10} = \frac{[CH_3COOH][OH^-]}{[CH_3COO^-]} = \frac{x^2}{0.0200 - x} \approx \frac{x^2}{0.0200} \quad \text{Solving, } x = 3.3 \times 10^{-6}$$

$[OH^-] = x = 3.3 \times 10^{-6}\,M$; pOH = 5.48; pH = 8.52

(4) After the equivalence point, the pH is determined directly from the concentration of *excess* NaOH since CH_3COOH is now the limiting reactant. In the presence of the strong base, the effect of the weak base, CH_3COO^-, derived from the salt is negligible. For example, at point (j):

	CH_3COOH	+	NaOH	→	$NaCH_3COO$	+	H_2O
initial	20.0 mmol		24.0 mmol		0 mmol		
change	- 20.0 mmol		- 20.0 mmol		+ 20.0 mmol		
after reaction	0 mmol		4.0 mmol		20.0 mmol		

$$[OH^-] = [NaOH]_{excess} = \frac{0.0040\ mol}{1.00\ L} = 4.0 \times 10^{-3}\,M; \quad pOH = 2.40; \quad pH = 11.60$$

Note: If aqueous NaOH had been added, $[NaOH]_{excess} = 0.0040$ mol/total volume (L)

Data Table:

	Mol NaOH Added	Type of Solution	$[H_3O^+]$ (M)	$[OH^-]$ (M)	pH	pOH
(a)	none	weak acid	6.0×10^{-4}	1.7×10^{-11}	3.22	10.78
(b)	0.00400	buffer	7.2×10^{-5}	1.4×10^{-10}	4.14	9.86
(c)	0.00800	buffer	2.7×10^{-5}	3.7×10^{-10}	4.57	9.43
(d)	0.01000	buffer	1.8×10^{-5}	5.6×10^{-10}	4.74 (= pK_a)	9.26
	(halfway to the equivalence point)					
(e)	0.01400	buffer	7.7×10^{-6}	1.3×10^{-9}	5.11	8.89
(f)	0.01800	buffer	2.0×10^{-6}	5.0×10^{-9}	5.70	8.30
(g)	0.01900	buffer	9×10^{-7}	1.1×10^{-8}	6.0	7.98
(h)	0.0200	salt	3.0×10^{-9}	3.3×10^{-6}	8.52	5.48
	(at the equivalence point)					
(i)	0.0210	strong base	1.0×10^{-11}	1.0×10^{-3}	11.00	3.0
(j)	0.0240	strong base	2.5×10^{-12}	4.0×10^{-3}	11.60	2.40
(k)	0.0300	strong base	1.0×10^{-12}	1.0×10^{-2}	12.00	2.00

Titration Curve: CH_3COOH vs. NaOH

An appropriate indicator would change color in the pH range, 7 - 10. From Table 19-3, **phenolphthalein** is the best indicator for this titration.

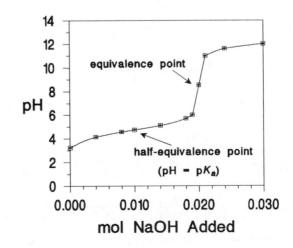

269

Balanced equation: $CH_3COOH + KOH \rightarrow KCH_3COO + H_2O$

The resultant solution at the equivalence point of any acid-base reaction contains only salt and water. The pH is determined from the concentration of the salt. Even when the volume of solution is given, it is not necessary to use that information. We can calculate the concentration of a salt derived from a monoprotic acid and a base with one OH group by:

$$[salt] = \frac{M_A M_B}{M_A + M_B} \qquad \text{where} \quad \begin{array}{l} M_A = \text{molarity of the acid} \\ M_B = \text{molarity of the base} \end{array}$$

Derivation:

Let V_A = volume (in L) of acid with molarity M_A

$\quad\ \ V_B$ = volume (in L) of base with molarity M_B

At the equivalence point, mol acid $(M_A V_A)$ = mol base $(M_B V_B)$ = mol salt produced

$$[salt] = \frac{\text{mol acid}}{\text{total volume}} = \frac{M_A V_A}{V_A + V_B} = \frac{M_A}{\left[1 + \dfrac{V_B}{V_A}\right]} = \frac{M_A}{\left[1 + \dfrac{M_A}{M_B}\right]} = \frac{M_A M_B}{M_A + M_B}$$

Note: $\dfrac{V_B}{V_A} = \dfrac{M_A}{M_B}$, since $M_A V_A = M_B V_B$ for a monoprotic acid reacting with a base with one OH group.

(a) $M_A = 1.000\ M$, $M_B = 0.150\ M$

(1) $[KCH_3COO] = \dfrac{M_A M_B}{M_A + M_B} = \dfrac{(1.000)(0.150)}{1.000 + 0.150} = 0.130\ M$

(2) The anion of the soluble salt hydrolyzes to form a basic solution:

$$CH_3COO^- + H_2O \rightleftharpoons CH_3COOH + OH^-$$

$$K_b = \frac{K_w}{K_{a(CH_3COOH)}} = \frac{1.0 \times 10^{-14}}{1.8 \times 10^{-5}} = 5.6 \times 10^{-10} = \frac{[CH_3OOH][OH^-]}{[CH_3COO^-]} = \frac{x^2}{0.130 - x} \approx \frac{x^2}{0.130}$$

Solving, $x = 8.5 \times 10^{-6}$; $[OH^-] = 8.5 \times 10^{-6}\ M$; pOH = 5.07; pH = **8.93**

(b) $M_A = 0.100\ M$, $M_B = 0.150\ M$

(1) $[KCH_3COO] = \dfrac{M_A M_B}{M_A + M_B} = \dfrac{(0.100)(0.150)}{0.100 + 0.150} = 0.0600\ M$

(2) $K_b = \dfrac{K_w}{K_{a(CH_3COOH)}} = \dfrac{1.0 \times 10^{-14}}{1.8 \times 10^{-5}} = 5.6 \times 10^{-10} = \dfrac{[CH_3OOH][OH^-]}{[CH_3COO^-]} = \dfrac{x^2}{0.0600 - x} \approx \dfrac{x^2}{0.0600}$

Solving, $x = 5.8 \times 10^{-6}$; $[OH^-] = 5.8 \times 10^{-6}\ M$; pOH = 5.23; pH = **8.76**

(c) $M_A = 0.0100\ M$, $M_B = 0.150\ M$

(1) $[KCH_3COO] = \dfrac{M_A M_B}{M_A + M_B} = \dfrac{(0.0100)(0.150)}{0.0100 + 0.150} = 9.38 \times 10^{-3}\ M$

(2) $K_b = \dfrac{[CH_3OOH][OH^-]}{[CH_3COO^-]} = \dfrac{x^2}{9.38 \times 10^{-3} - x} \approx \dfrac{x^2}{9.38 \times 10^{-3}}$

Solving, $x = 2.3 \times 10^{-6}$; $[OH^-] = 2.3 \times 10^{-6}\ M$; pOH = 5.64; pH = **8.36**

19-40. Refer to Sections 19-1, 19-2, 19-3, 19-4 and 19-5.

(a) $(NH_4)HSO_4$ acidic (salt of weak base and a fairly strong acid)   (Section 19-3)

(b) $(NH_4)_2SO_4$ acidic (salt of a weak base and strong acid)   (Section 19-3)

(c) KCl neutral (salt of a strong base and strong acid)   (Section 19-1)

(d) $LiBrO$ basic (salt of a strong base and weak acid)   (Section 19-2)

(e) $Al(NO_3)_3$ acidic (salt of a small, highly charged cation)   (Section 19-5)

19-42. Refer to Section 19-5.

If a cation reacts appreciably with water, its acid strength must be greater than that of water. The pH of the solution will be less than 7.

19-44. Refer to Section 19-4, Figure 19-6, Table 19-3 and Exercise 19-36 Solution.

The calculations for determining the pH in the titration of 1 liter of 0.0100 M HNO_3 with gaseous NH_3 can be divided into 4 types:

(1) Initially, the pH is determined by the concentration of the strong acid, HNO_3.

(2) Before the equivalence point, the pH is essentially determined by the concentration of *excess* strong acid, HNO_3, since NH_3 is the limiting reagent. In the presence of the strong acid, the effect of the weak acid, NH_4^+, derived from the salt is negligible. For example, consider the limiting reactant problem for point (c) in the following table:

	HNO_3	+	NH_3	→	NH_4NO_3	+	H_2O
initial	10.00 mmol		4.00 mmol		0 mmol		
change	- 4.00 mmol		- 4.00 mmol		+ 4.00 mmol		
after reaction	6.00 mmol		0 mmol		4.00 mmol		

$$[H^+] = [HNO_3]_{excess} = \frac{0.00600 \text{ mol}}{1.00 \text{ L}} = 6.0 \times 10^{-3} \, M; \quad pH = 2.22$$

(3) At the equivalence point, there is no excess acid or base. The pH of the salt solution, NH_4NO_3, is <7 since we have a solution of the salt of a weak base and a strong acid. The concentration of NH_4NO_3 determines the pH of the system. At point (g):

	HNO_3	+	NH_3	→	NH_4NO_3	+	H_2O
initial	10.0 mmol		10.0 mmol		0 mmol		
change	- 10.0 mmol		- 10.0 mmol		+ 10.0 mmol		
after reaction	0 mmol		0 mmol		10.0 mmol		

$$[NH_4NO_3] = \frac{0.0100 \text{ mol}}{1.00 \text{ L}} = 0.0100 \, M$$

Then the cation of the salt hydrolyzes to produce an acidic solution: $NH_4^+ + H_2O \rightleftarrows NH_3 + H_3O^+$

Let x = $[NH_4^+]_{hydrolyzed}$ Then, 0.0100 - x = $[NH_4^+]$; x = $[NH_3]$ = $[H_3O^+]$

$$K_a = \frac{K_w}{K_{b(NH_3)}} = 5.6 \times 10^{-10} = \frac{[NH_3][H_3O^+]}{[NH_4^+]} = \frac{x^2}{0.0100 - x} \approx \frac{x^2}{0.0100} \qquad \text{Solving, x} = 2.4 \times 10^{-6}$$

$$[H_3O^+] = x = 2.4 \times 10^{-6} \, M; \quad pH = 5.62$$

(4) After the equivalence point, the pH is determined by the buffer solution consisting of excess NH_3 and NH_4^+ produced by the neutralization reaction. For example, at point (i):

	HNO_3	$+$	NH_3	$\rightarrow$	NH_4NO_3
initial	10.00 mmol		13.00 mmol		0 mmol
change	- 10.00 mmol		- 10.00 mmol		+ 10.00 mmol
after reaction	0 mmol		3.00 mmol		10.00 mmol

After the reaction, we have a 1 liter solution of NH_3 and the soluble salt, NH_4NO_3. OH^- ion is produced by:

$$NH_3 + H_2O \rightleftharpoons NH_4^+ + OH^- \qquad\qquad K_b = 1.8 \times 10^{-5}$$

$$[OH^-] = K_b \times \frac{\text{mol } NH_3}{\text{mol } NH_4NO_3} = (1.8 \times 10^{-5})\frac{3.0 \text{ mmol}}{10.0 \text{ mmol}} = 5.4 \times 10^{-6}\, M$$

Therefore, pH = 8.73

Data Table:

	Mol NH_3 Added	Type of Solution	$[H_3O^+]$ (M)	$[OH^-]$ (M)	pH	pOH
(a)	none	strong acid	1.00×10^{-2}	1.00×10^{-12}	2.00	14.00
(b)	0.00100	strong acid	9.0×10^{-9}	1.1×10^{-12}	2.05	11.95
(c)	0.00400	strong acid	6.0×10^{-3}	1.7×10^{-12}	2.22	11.78
(d)	0.00500	strong acid	5.0×10^{-3}	2.0×10^{-12}	2.30	11.70
	(halfway to the equivalence point - has no real significance in this case)					
(e)	0.00900	strong acid	1.0×10^{-3}	1.0×10^{-11}	3.00	11.00
(f)	0.00950	strong acid	5×10^{-4}	2×10^{-11}	3.3	10.7
(g)	0.0100	salt	2.4×10^{-6}	4.2×10^{-9}	5.62	8.38
	(at the equivalence point)					
(h)	0.0105	buffer	1.1×10^{-8}	9×10^{-7}	8.0	6.0
(i)	0.0130	buffer	1.8×10^{-9}	5.4×10^{-6}	8.73	5.27

Titration Curve: HNO_3 vs. NH_3

The major difference between this titration curve of the HNO_3/NH_3 reaction and the other titrations is that the system is buffered after the equivalence point, not before the equivalence point. A satisfactory indicator for this titration is methyl red.

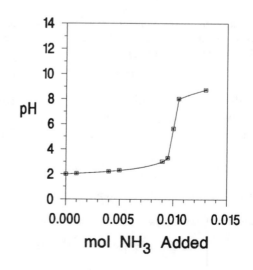

19-46. *Refer to Sections 19-2 and 19-7, Table 19-3, and Exercise 19-38 Solution.*

Balanced equation: $CH_3COOH + NaOH \rightarrow NaCH_3COO + H_2O$

The solution at the equivalence point of any acid-base reaction contains only salt and water. The pH is determined from the concentration of the salt. When the volumes of the acid and base are not given, we can calculate the concentration of a salt derived from a monoprotic acid and a base with one OH group by:

$$[salt] = \frac{M_A M_B}{M_A + M_B} \qquad \text{where} \quad \begin{array}{l} M_A = \text{molarity of the acid} \\ M_B = \text{molarity of the base} \end{array}$$

Note: See Exercise 19-38 Solution for the derivation.

Plan: (1) Calculate the concentration of $NaCH_3COO$.
(2) Determine the pH.

(1) $[NaCH_3COO] = \dfrac{M_A M_B}{M_A + M_B} = \dfrac{(0.020\ M)(0.015\ M)}{(0.020\ M + 0.015\ M)} = 0.0086\ M$

(2) The anion of the soluble salt hydrolyzes to form a basic solution: $CH_3COO^- + H_2O \rightleftarrows CH_3COOH + OH^-$

Let $x = [CH_3COO^-]_{hydrolyzed}$ Then, $0.0086 - x = [CH_3COO^-]$; $x = [CH_3COOH] = [OH^-]$

$K_b = \dfrac{K_w}{K_{a(CH_3COOH)}} = \dfrac{1.0 \times 10^{-14}}{1.8 \times 10^{-5}} = 5.6 \times 10^{-10} = \dfrac{[CH_3COOH][OH^-]}{[CH_3COO^-]} = \dfrac{x^2}{0.0086 - x} \approx \dfrac{x^2}{0.0086}$

Solving, $x = 2.2 \times 10^{-6}$; $[OH^-] = 2.2 \times 10^{-6}\ M$; pOH = 5.66; pH = **8.34**

An appropriate indicator for this titration would be **phenolphthalein**.

19-48. *Refer to Sections 19-1, 19-2 and 19-3.*

Balanced equations:

(a) $Na_2CO_3 \rightarrow 2Na^+ + CO_3^{2-}$
$CO_3^{2-} + H_2O \rightleftarrows HCO_3^- + OH^-$ $K_{b(1)} = 2.1 \times 10^{-4}$
$HCO_3^- + H_2O \rightleftarrows H_2CO_3 + OH^-$ $K_{b(2)} = 2.4 \times 10^{-8}$

(b) $Na_2SO_4 \rightarrow 2Na^+ + SO_4^{2-}$
$SO_4^{2-} + H_2O \rightleftarrows HSO_4^- + OH^-$ $K_{b(1)} = 8.3 \times 10^{-13}$
$HSO_4^- + H_2O \rightleftarrows H_2SO_4 + OH^-$ $K_{b(2)} = $ very small

(c) $(NH_4)_2SO_4 \rightarrow 2NH_4^+ + SO_4^{2-}$
$NH_4^+ + H_2O \rightleftarrows NH_3 + H_3O^+$ $K_a = 5.6 \times 10^{-10}$
$SO_4^{2-} + H_2O \rightleftarrows HSO_4^- + OH^-$ $K_{b(1)} = 8.3 \times 10^{-13}$
$HSO_4^- + H_2O \rightleftarrows H_2SO_4 + OH^-$ $K_{b(2)} = $ very small

(d) $Na_3PO_4 \rightarrow 3Na^+ + PO_4^{3-}$
$PO_4^{3-} + H_2O \rightleftarrows HPO_4^{2-} + OH^-$ $K_{b(1)} = 2.8 \times 10^{-2}$
$HPO_4^{2-} + H_2O \rightleftarrows H_2PO_4^- + OH^-$ $K_{b(2)} = 1.6 \times 10^{-7}$
$H_2PO_4^- + H_2O \rightleftarrows H_3PO_4 + OH^-$ $K_{b(3)} = 1.3 \times 10^{-12}$

$(NH_4)_2SO_4$ definitely could not be used in cleaning materials since it produces an acidic solution, not a basic solution. Also, Na_2SO_4 cannot be used either since SO_4^{2-} is an extremely weak base (has a very small K_b).

For phenol (C_6H_5OH), $K_a = 1.3 \times 10^{-10}$

Balanced equation: $C_6H_5OH + H_2O \rightarrow H_3O^+ + C_6H_5O^-$

Let x $= [C_6H_5OH]_{ionized}$.   Then $[C_6H_5OH] = (0.0100 - x)\ M$
                                                  $[H_3O^+] = [C_6H_5O^-] = x\ M$

$$K_a = \frac{[H_3O^+][C_6H_5O^-]}{[C_6H_5OH]} = \frac{x^2}{0.0100 - x} \approx \frac{x^2}{0.0100} = 1.3 \times 10^{-10}$$

Solving,   x $= [H_3O^+] = 1.1 \times 10^{-6}\ M$
                  pH $= $ **5.96**

20 Ionic Equilibria III: The Solubility Product Principle

20-2.	**Refer to Section 20-1.**

The solubility product principle states that the solubility product expression for a slightly soluble compound is the product of the concentrations of its constituent ions, each raised to the power that corresponds to the number of ions in one formula unit of the compound. The quantity, K_{sp}, is constant at constant temperature for a saturated solution of the compound, when the system is at equilibrium. The significance of the solubility product is that it can be used to calculate the concentrations of the ions in solutions for such slightly soluble compounds.

20-4.	**Refer to Section 20-1.**

The molar solubility of a compound is the number of moles of the compound that dissolve to produce one liter of saturated solution.

20-6.	**Refer to Section 20-1.**

(a) $Co_3(AsO_4)_2(s) \rightleftarrows 3Co^{2+}(aq) + 2AsO_4^{3-}(aq)$ $\qquad$ $K_{sp} = [Co^{2+}]^3[AsO_4^{3-}]^2$

(b) $Hg_2I_2(s) \rightleftarrows Hg_2^{2+}(aq) + 2I^-(aq)$ $\qquad$ $K_{sp} = [Hg_2^{2+}][I^-]^2$

(b) $HgI_2(s) \rightleftarrows Hg^{2+}(aq) + 2I^-(aq)$ $\qquad$ $K_{sp} = [Hg^{2+}][I^-]^2$

(d) $Ag_2CO_3(s) \rightleftarrows 2Ag^+(aq) + CO_3^{2-}(aq)$ $\qquad$ $K_{sp} = [Ag^+]^2[CO_3^{2-}]$

20-8.	**Refer to Section 20-2, Examples 20-1 and 20-2, and Appendix H.**

Plan: (1) Calculate the molar solubility of the slightly soluble salt, which is the number of moles of the salt that will dissolve in 1 liter of solution.
(2) Determine the concentrations of the ions in solution.
(3) Substitute the ion concentrations into the K_{sp} expression to calculate K_{sp}.

(a) Balanced equation: $SrCrO_4(s) \rightleftarrows Sr^{2+}(aq) + CrO_4^{2-}(aq)$ $\qquad$ $K_{sp} = [Sr^{2+}][CrO_4^{2-}]$

(1) molar solubility (mol $SrCrO_4$/L) $= \dfrac{1.2 \text{ mg } SrCrO_4}{1 \text{ mL}} \times \dfrac{1000 \text{ mL}}{1 \text{ L}} \times \dfrac{1 \text{ g}}{1000 \text{ mg}} \times \dfrac{1 \text{ mol } SrCrO_4}{204 \text{ g } SrCrO_4}$
$\qquad = 5.9 \times 10^{-3}$ mol $SrCrO_4$/L (dissolved)

(2) $[Sr^{2+}] = [CrO_4^{2-}] = $ molar solubility $= 5.9 \times 10^{-3}$ M

(3) $K_{sp} = [Sr^{2+}][CrO_4^{2-}] = (5.9 \times 10^{-3})^2 = \mathbf{3.5 \times 10^{-5}}$ $\qquad$ (3.6×10^{-5} from Appendix H)

(b) Balanced equation: $BiI_3(s) \rightleftarrows Bi^{3+}(aq) + 3I^-(aq)$ $\qquad$ $K_{sp} = [Bi^{3+}][I^-]^3$

(1) molar solubility $= \dfrac{7.7 \times 10^{-3} \text{ g } BiI_3}{1 \text{ L}} \times \dfrac{1 \text{ mol } BiI_3}{590 \text{ g } BiI_3} = 1.3 \times 10^{-5}$ mol BiI_3/L (dissolved)

(2) $[Bi^{3+}] = $ molar solubility $= 1.3 \times 10^{-5}$ M
$\quad [I^-] = 3 \times$ molar solubility $= 3.9 \times 10^{-5}$ M

(3) $K_{sp} = [Bi^{3+}][I^-]^3 = (1.3 \times 10^{-5})(3.9 \times 10^{-5})^3 = \mathbf{7.7 \times 10^{-19}}$ $\qquad$ (8.1×10^{-19} from Appendix H)

(c) Balanced equation: $Fe(OH)_2(s) \rightleftarrows Fe^{2+}(aq) + 2OH^-(aq)$ $K_{sp} = [Fe^{2+}][OH^-]^2$

(1) molar solubility $= \dfrac{1.1 \times 10^{-3} \text{ g Fe(OH)}_2}{1 \text{ L}} \times \dfrac{1 \text{ mol Fe(OH)}_2}{89.9 \text{ g Fe(OH)}_2} = 1.2 \times 10^{-5}$ mol Fe(OH)$_2$/L (dissolved)

(2) $[Fe^{2+}]$ = molar solubility = $1.2 \times 10^{-5} M$

$[OH^-]$ = 2 × molar solubility = $2.4 \times 10^{-5} M$

(3) $K_{sp} = [Fe^{2+}][OH^-]^2 = (1.2 \times 10^{-5})(2.4 \times 10^{-5})^2 = \mathbf{6.9 \times 10^{-15}}$ (7.9×10^{-15} from Appendix H)

(d) Balanced equation: $SnI_2(s) \rightleftarrows Sn^{2+}(aq) + 2I^-(aq)$ $K_{sp} = [Sn^{2+}][I^-]^2$

(1) molar solubility $= \dfrac{10.9 \text{ g SnI}_2}{1 \text{ L}} \times \dfrac{1 \text{ mol SnI}_2}{372.5 \text{ g SnI}_2} = 2.93 \times 10^{-2}$ mol SnI$_2$/L (dissolved)

(2) $[Sn^{2+}]$ = molar solubility = $2.93 \times 10^{-2} M$

$[CN^-]$ = 2 × molar solubility = $5.86 \times 10^{-2} M$

(3) $K_{sp} = [Sn^{2+}][I^-]^2 = (2.93 \times 10^{-2})(5.86 \times 10^{-2})^2 = \mathbf{1.00 \times 10^{-4}}$ (1.0×10^{-4} from Appendix H)

20-10. *Refer to Section 20-2, Table 20-1 and Exercise 20-8 Solution.*

Compound	Molar Solubility (M)	K_{sp} (calculated)
SrCrO$_4$	5.9×10^{-3}	$[Sr^{2+}][CrO_4^{2-}] = 3.5 \times 10^{-5}$
BiI$_3$	1.3×10^{-5}	$[Bi^{3+}][I^-]^3 = 7.7 \times 10^{-19}$
Fe(OH)$_2$	1.2×10^{-5}	$[Fe^{2+}][OH^-]^2 = 6.9 \times 10^{-15}$
SnI$_2$	2.93×10^{-2}	$[Sn^{2+}][I^-]^2 = 1.00 \times 10^{-4}$

(a) SnI$_2$ has the highest molar solubility.

(b) Fe(OH)$_2$ has the lowest molar solubility.

(c) SnI$_2$ has the largest K_{sp}.

(d) BiI$_3$ has the smallest K_{sp}.

20-12. *Refer to Section 20-3, Example 20-3 and Appendix H.*

(a) Balanced equation: $CuI(s) \rightleftarrows Cu^+(aq) + I^-(aq)$ $K_{sp} = 5.1 \times 10^{-12}$

Let x = molar solubility of CuI. Then, $[Cu^+] = [I^-] = x$

$K_{sp} = [Cu^+][I^-] = x^2 = 5.1 \times 10^{-12}$ Solving, x $= 2.3 \times 10^{-6}$

Therefore, molar solubility = $\mathbf{2.3 \times 10^{-6}}$ **mol CuI/L** (dissolved)

$[Cu^+] = \mathbf{2.3 \times 10^{-6} M}$

$[I^-] = \mathbf{2.3 \times 10^{-6} M}$

solubility (g/L) $= \dfrac{2.3 \times 10^{-6} \text{ mol CuI}}{1 \text{ L}} \times \dfrac{190.4 \text{ g CuI}}{1 \text{ mol CuI}} = \mathbf{4.4 \times 10^{-4} \text{ g/L}}$

(b) Balanced equation: $Ba_3(PO_4)_2(s) \rightleftarrows 3Ba^{2+}(aq) + 2PO_4^{3-}(aq)$ $K_{sp} = 1.3 \times 10^{-29}$

Let x = molar solubility of Ba$_3$(PO$_4$)$_2$. Then, $[Ba^{2+}] = 3x$ and $[PO_4^{3-}] = 2x$

$K_{sp} = [Ba^{2+}]^3[PO_4^{3-}]^2 = (3x)^3(2x)^2 = 108x^5 = 1.3 \times 10^{-29}$ Solving, x $= 6.5 \times 10^{-7}$

Therefore, molar solubility = $\mathbf{6.5 \times 10^{-7}}$ **mol Ba$_3$(PO$_4$)$_2$/L** (dissolved)

$[Ba^{2+}] = 3x = \mathbf{2.0 \times 10^{-6} M}$

$[PO_4^{3-}] = 2x = \mathbf{1.3 \times 10^{-6} M}$

solubility (g/L) $= \dfrac{6.5 \times 10^{-7} \text{ mol Ba}_3(PO_4)_2}{1 \text{ L}} \times \dfrac{602 \text{ g Ba}_3(PO_4)_2}{1 \text{ mol Ba}_3(PO_4)_2} = \mathbf{3.9 \times 10^{-4} \text{ g/L}}$

(c) Balanced equation: $PbF_2(s) \rightleftarrows Pb^{2+}(aq) + 2F^-(aq)$ $\qquad K_{sp} = 3.7 \times 10^{-8}$

Let x = molar solubility of PbF_2. Then, $[Pb^{2+}] = x$ and $[F^-] = 2x$

$K_{sp} = [Pb^{2+}][F^-]^2 = (x)(2x)^2 = 4x^3 = 3.7 \times 10^{-8}$ $\qquad$ Solving, x = 2.1×10^{-3}

Therefore, molar solubility = 2.1×10^{-3} mol PbF_2/L (dissolved)

$\qquad [Pb^{2+}] = x = 2.1 \times 10^{-3}$ M

$\qquad [F^-] = 2x = 4.2 \times 10^{-3}$ M

$\qquad$ solubility (g/L) = $\dfrac{2.1 \times 10^{-3} \text{ mol } PbF_2}{1 \text{ L}} \times \dfrac{245 \text{ g } PbF_2}{1 \text{ mol } PbF_2} = \textbf{0.51 g/L}$

(d) Balanced equation: $Pb_3(PO_4)_2(s) \rightleftarrows 3Pb^{2+}(aq) + 2PO_4^{3-}(aq)$ $\qquad K_{sp} = 3.0 \times 10^{-44}$

Let x = molar solubility of $Pb_3(PO_4)_2$ Then, $[Pb^{2+}] = 3x$ and $[PO_4^{3-}] = 2x$

$K_{sp} = [Pb^{2+}]^3[PO_4^{3-}]^2 = (3x)^3(2x)^2 = 108x^5 = 3.0 \times 10^{-44}$ $\qquad$ Solving, x = 7.7×10^{-10}

Therefore, molar solubility = 7.7×10^{-10} mol $Pb_3(PO_4)_2$/L (dissolved)

$\qquad [Pb^{2+}] = 3x = 2.3 \times 10^{-9}$ M

$\qquad [PO_4^{3-}] = 2x = 1.5 \times 10^{-9}$ M

$\qquad$ solubility (g/L) = $\dfrac{7.7 \times 10^{-10} \text{ mol } Pb_3(PO_4)_2}{1 \text{ L}} \times \dfrac{811.54 \text{ g } Pb_3(PO_4)_2}{1 \text{ mol } Pb_3(PO_4)_2} = \textbf{6.2} \times \textbf{10}^{-7}$ **g/L**

20-14. *Refer to Section 20-3, Table 20-1 and Exercise 20-12 Solution.*

Compound	K_{sp}	Molar Solubility (M)	Solubility (g/L)
CuI	$[Cu^+][I^-] = 5.1 \times 10^{-12}$	2.3×10^{-6}	4.4×10^{-4}
$Ba_3(PO_4)_2$	$[Ba^{2+}]^3[PO_4^{3-}]^2 = 1.3 \times 10^{-29}$	6.5×10^{-7}	3.9×10^{-4}
PbF_2	$[Pb^{2+}][F^-]^2 = 3.7 \times 10^{-8}$	2.1×10^{-3}	0.51
$Pb_3(PO_4)_2$	$[Pb^{2+}]^3[PO_4^{3-}]^2 = 3.0 \times 10^{-44}$	7.7×10^{-10}	6.2×10^{-7}

(a) PbF_2 has the highest molar solubility.

(b) $Pb_3(PO_4)_2$ has the lowest molar solubility.

(c) PbF_2 has the highest solubility (g/L).

(d) $Pb_3(PO_4)_2$ has the lowest solubility (g/L).

20-16. *Refer to Section 20-3 and Example 20-4.*

This is a common ion effect problem similar to some buffer problems.

Balanced equations: $K_2SO_4(aq) \rightarrow 2K^+(aq) + SO_4^{2-}(aq)$ $\qquad$ (to completion)

$\qquad\qquad\qquad\qquad Ag_2SO_4(s) \rightleftarrows 2Ag^+(aq) + SO_4^{2-}(aq)$ $\qquad$ (reversible) $\qquad K_{sp} = 1.7 \times 10^{-5}$

Let $\quad$ x = molar solubility of Ag_2SO_4 in 0.15 M K_2SO_4. Then,

$\qquad [Ag^+] = 2x$ (from Ag_2SO_4)

$\qquad [SO_4^{2-}] = x$ (from Ag_2SO_4) + 0.15 M (from K_2SO_4)

$K_{sp} = [Ag^+]^2[SO_4^{2-}] = (2x)^2(x + 0.15) = 1.7 \times 10^{-5} \approx (2x)^2(0.15)$ $\qquad$ Solving, x = 5.3×10^{-3}

Therefore, molar solubility = 5.3×10^{-3} mol Ag_2SO_4/L 0.15 M K_2SO_4 soln

(a) Balanced equations: $NaOH(aq) \rightarrow Na^+(aq) + OH^-(aq)$ (to completion)

$Mg(OH)_2(s) \rightleftarrows Mg^{2+}(aq) + 2OH^-(aq)$ (reversible) $K_{sp} = 1.5 \times 10^{-11}$

Let   x = molar solubility of $Mg(OH)_2$ in 0.015 M NaOH. Then,

$[Mg^{2+}]$ = x (from $Mg(OH)_2$)

$[OH^-]$ = 2x (from $Mg(OH)_2$) + 0.015 M (from NaOH)

$K_{sp} = [Mg^{2+}][OH^-]^2 = (x)(2x + 0.015)^2 = 1.5 \times 10^{-11} \approx (x)(0.015)^2$ Solving, x = 6.7×10^{-8}

Therefore, molar solubility = **6.7×10^{-8} mol $Mg(OH)_2$/L 0.015 M NaOH soln**

(b) Balanced equations: $MgCl_2(aq) \rightarrow Mg^{2+}(aq) + 2Cl^-(aq)$ (to completion)

$Mg(OH)_2(s) \rightleftarrows Mg^{2+}(aq) + 2OH^-(aq)$ (reversible) $K_{sp} = 1.5 \times 10^{-11}$

Let   x = molar solubility of $Mg(OH)_2$ in 0.015 M $MgCl_2$. Then,

$[Mg^{2+}]$ = x (from $Mg(OH)_2$) + 0.015 M (from $MgCl_2$)

$[OH^-]$ = 2x (from $Mg(OH)_2$)

$K_{sp} = [Mg^{2+}][OH^-]^2 = (x + 0.015)(2x)^2 = 1.5 \times 10^{-11} \approx (0.015)(2x)^2$ Solving, x = 1.6×10^{-5}

Therefore, molar solubility = **1.6×10^{-5} mol $Mg(OH)_2$/L 0.015 M $MgCl_2$ soln**

20-20.   *Refer to Section 20-3, Example 20-4 and Appendix H.*

Plan: Determine separately the molar solubility of $BaCrO_4$ and Ag_2CrO_4 in 0.20 M K_2CrO_4. The salt with the higher molar solubility is more soluble in the K_2CrO_4 solution.

(1) Molar solubility of $BaCrO_4$ in 0.20 M K_2CrO_4:

Balanced equations: $K_2CrO_4(aq) \rightarrow 2K^+(aq) + CrO_4^{2-}(aq)$ (to completion)

$BaCrO_4(s) \rightleftarrows Ba^{2+}(aq) + CrO_4^{2-}(aq)$ (reversible) $K_{sp} = 2.0 \times 10^{-10}$

Let   x = molar solubility of $BaCrO_4$ in 0.20 M K_2CrO_4. Then,

$[Ba^{2+}]$ = x (from $BaCrO_4$)

$[CrO_4^{2-}]$ = x (from $BaCrO_4$) + 0.20 M (from K_2CrO_4)

$K_{sp} = [Ba^{2+}][CrO_4^{2-}] = (x)(x + 0.20) = 2.0 \times 10^{-10} \approx (x)(0.20)$ Solving, x = 1.0×10^{-9}

Therefore, molar solubility = 1.0×10^{-9} mol $BaCrO_4$/L 0.20 M K_2CrO_4

(2) Molar solubility of Ag_2CrO_4 in 0.20 M K_2CrO_4:

Balanced equations: $K_2CrO_4(aq) \rightarrow 2K^+(aq) + CrO_4^{2-}(aq)$ (to completion)

$Ag_2CrO_4(s) \rightleftarrows 2Ag^+(aq) + CrO_4^{2-}(aq)$ (reversible) $K_{sp} = 9.0 \times 10^{-12}$

Let   y = molar solubility of Ag_2CrO_4 in 0.20 M K_2CrO_4. Then,

$[Ag^+]$ = 2y (from Ag_2CrO_4)

$[CrO_4^{2-}]$ = y (from Ag_2CrO_4) + 0.20 M (from K_2CrO_4)

$K_{sp} = [Ag^+]^2[CrO_4^{2-}] = (2y)^2(y + 0.20) = 2.0 \times 10^{-10} \approx (2y)^2(0.20)$ Solving, y = 3.4×10^{-6}

Therefore, molar solubility = **3.4×10^{-6} mol Ag_2CrO_4/L 0.20 M K_2CrO_4**

Since the molar solubility of Ag_2CrO_4 in 0.20 M K_2CrO_4 is greater than that of $BaCrO_4$, **Ag_2CrO_4** is more soluble on a per mole basis. (Note: Ag_2CrO_4 is more soluble on a per gram basis also.)

20-22. *Refer to Section 20-3, Example 20-5 and Appendix H.*

Balanced equations: $Pb(NO_3)_2(aq) + 2NaCl(aq) \rightarrow PbCl_2(s) + 2NaNO_3(aq)$ (Will precipitation occur?)
$PbCl_2(s) \rightleftarrows Pb^{2+}(aq) + 2Cl^-(aq)$ (reversible) $K_{sp} = 1.7 \times 10^{-5}$

Plan: (1) Calculate the concentration of Pb^{2+} ions and Cl^- ions at the instant of mixing before combination occurs.
(2) Determine the reaction quotient, Q_{sp}. If $Q_{sp} > K_{sp}$, then precipitation will occur.

(1) $[Pb^{2+}] = [Pb(NO_3)_2] = \dfrac{5.0 \text{ g } Pb(NO_3)_2/331 \text{ g/mol}}{1.00 \text{ L soln}} = 0.015 \ M \ Pb^{2+}$

$[Cl^-] = [NaCl] = 0.010 \ M \ Cl^-$

(2) $Q_{sp} = [Pb^{2+}][Cl^-]^2 = (0.015)(0.010)^2 = 1.5 \times 10^{-6}$
Since $Q_{sp} < K_{sp}$, precipitation will **not** occur.

20-24. *Refer to Section 20-3, Example 20-5 and Appendix H.*

Balanced equations: $CuCl_2(aq) + 2NaOH(aq) \rightarrow Cu(OH)_2(s) + 2NaCl(aq)$ (Will precipitation occur?)
$Cu(OH)_2(s) \rightleftarrows Cu^{2+}(aq) + 2OH^-(aq)$ (reversible) $K_{sp} = 1.6 \times 10^{-19}$

Plan: (1) Calculate the concentrations of Cu^{2+} ions and Cl^- ions at the instant of mixing before reaction occurs using $M_1V_1 = M_2V_2$.
(2) Determine the reaction quotient, Q_{sp}. If $Q_{sp} > K_{sp}$, a precipitate will form.

(1) $[Cu^{2+}] = [CuCl_2] = \dfrac{M_1V_1}{V_2} = \dfrac{0.010 \ M \times 1.00 \text{ L}}{(1.00 \text{ L} + 0.010 \text{ L})} = 9.9 \times 10^{-3} \ M$

$[OH^-] = [NaOH] = \dfrac{M_1V_1}{V_2} = \dfrac{0.010 \ M \times 0.010 \text{ L}}{(1.00 \text{ L} + 0.010 \text{ L})} = 9.9 \times 10^{-5} \ M$

(2) $Q_{sp} = [Cu^{2+}][OH^-]^2 = (9.9 \times 10^{-3})(9.9 \times 10^{-5})^2 = 9.7 \times 10^{-11}$
Since $Q_{sp} > K_{sp}$, a precipitate **will** form.

20-26. *Refer to Section 20-3, Example 20-6 and Appendix H.*

Balanced equations: (i) $2KOH(aq) + Zn(NO_3)_2(aq) \rightarrow Zn(OH)_2(s) + 2KNO_3(aq)$
$Zn(OH)_2(s) \rightleftarrows Zn^{2+}(aq) + 2OH^-(aq)$ $K_{sp} = 4.5 \times 10^{-17}$

(ii) $K_2CO_3(aq) + Zn(NO_3)_2(aq) \rightarrow ZnCO_3(s) + 2KNO_3(aq)$
$ZnCO_3(s) \rightleftarrows Zn^{2+}(aq) + CO_3^{2-}(aq)$ $K_{sp} = 1.5 \times 10^{-11}$

(iii) $K_2S(aq) + Zn(NO_3)_2(aq) \rightarrow ZnS(s) + 2KNO_3(aq)$
$ZnS(s) \rightleftarrows Zn^{2+}(aq) + S^{2-}(aq)$ $K_{sp} = 1.1 \times 10^{-21}$

Plan: Substitute the value of K_{sp} and the given ion concentration into each K_{sp} equilibrium expression and solve for the other ion concentration. An ion concentration just greater than the calculated one will initiate precipitation.

(a) (i) $K_{sp} = [Zn^{2+}][OH^-]^2$ (ii) $K_{sp} = [Zn^{2+}][CO_3^{2-}]$ (iii) $K_{sp} = [Zn^{2+}][S^{2-}]$
$4.5 \times 10^{-17} = [Zn^{2+}](0.0015)^2$ $1.5 \times 10^{-11} = [Zn^{2+}](0.0015)$ $1.1 \times 10^{-21} = [Zn^{2+}](0.0015)$
$[Zn^{2+}] = \mathbf{2.0 \times 10^{-11}} \ M$ $[Zn^{2+}] = \mathbf{1.0 \times 10^{-8}} \ M$ $[Zn^{2+}] = \mathbf{7.3 \times 10^{-19}} \ M$

(b) (i) $K_{sp} = [Zn^{2+}][OH^-]^2$ (ii) $K_{sp} = [Zn^{2+}][CO_3^{2-}]$ (iii) $K_{sp} = [Zn^{2+}][S^{2-}]$
$4.5 \times 10^{-17} = (0.0015)[OH^-]^2$ $1.5 \times 10^{-11} = (0.0015)[CO_3^{2-}]$ $1.1 \times 10^{-21} = (0.0015)[S^{2-}]$
$[OH^-] = \mathbf{1.7 \times 10^{-7}} \ M$ $[CO_3^{2-}] = \mathbf{1.0 \times 10^{-8}} \ M$ $[S^{2-}] = \mathbf{7.3 \times 10^{-19}} \ M$

Balanced equations: $Pb^{2+}(aq) + 2NaI(aq) \rightarrow PbI_2(s) + 2Na^+(aq)$

$PbI_2(s) \rightleftarrows Pb^{2+}(aq) + 2I^-(aq)$ (reversible) $K_{sp} = 8.7 \times 10^{-9}$

Plan: (1) Do the limiting reactant problem to determine which reactant is in excess.

(2) If Pb^{2+} is the limiting reactant, calculate its concentration in the presence of the excess amount of NaI.

(3) Calculate %Pb^{2+} ions remaining in solution.

(1) ? mol Pb^{2+} = 0.0100 $M \times$ 1.00 L = 0.0100 mol Pb^{2+}

? mol NaI = 0.103 mol NaI

	$Pb^{2+}(aq)$	+	$2NaI(aq)$	$\rightarrow$	$PbI_2(s)$	+	$2Na^+(aq)$
initial	0.0100 mol		0.103 mol		0 mol		0 mol
change	- 0.0100 mol		- 0.020 mol		+ 0.0100 mol		+ 0.0200 mol
after reaction	0 mol		0.083 mol		0.0100 mol		0.0200 mol

Pb^{2+} is the limiting reactant; NaI is in excess.

(2) In the resulting solution, $[I^-]$ = [NaI] = 0.083 mol/1.00 L = 0.083 M

Let   x = molar solubility of PbI_2 in 0.083 M NaI. Then,

$[Pb^{2+}]$ = x (from PbI_2)

$[I^-]$ = 2x (from PbI_2) + 0.083 M (from NaI)

$K_{sp} = [Pb^{2+}][I^-]^2 = (x)(2x + 0.083)^2 = 8.7 \times 10^{-9} \approx (x)(0.083)^2$ Solving, x = 1.3×10^{-6}

Therefore, $[Pb^{2+}]$ = x = $1.3 \times 10^{-6} M$

(3) Therefore, % Pb^{2+} in solution = $\dfrac{1.3 \times 10^{-6} M}{0.0100 M} \times 100\%$ = **0.013%**

Fractional precipitation refers to a separation process whereby some ions are removed from solution by precipitation, leaving other ions with similar properties in solution.

Balanced equations: $Cu^+(aq) + NaCl(aq) \rightarrow CuCl(s) + Na^+(aq)$

$CuCl(s) \rightleftarrows Cu^+(aq) + Cl^-(aq)$ (reversible) $K_{sp} = 1.9 \times 10^{-7}$

$Ag^+(aq) + NaCl(aq) \rightarrow AgCl(s) + Na^+(aq)$

$AgCl(s) \rightleftarrows Ag^+(aq) + Cl^-(aq)$ (reversible) $K_{sp} = 1.8 \times 10^{-10}$

$Au^+(aq) + NaCl(aq) \rightarrow AuCl(s) + Na^+(aq)$

$AuCl(s) \rightleftarrows Au^+(aq) + Cl^-(aq)$ (reversible) $K_{sp} = 2.0 \times 10^{-13}$

(a) Because the compounds to be precipitated are all of the same molecular type, i.e., their cation to anion ratios are the same, the one with the smallest K_{sp} is the least soluble and will precipitate first.   Hence, **AuCl** (K_{sp} = 2.0×10^{-13}) will precipitate first, then AgCl (K_{sp} = 1.8×10^{-10}), and finally CuCl (K_{sp} = 1.9×10^{-7}).

(b) AgCl will begin to precipitate when $Q_{sp(AgCl)} = K_{sp(AgCl)} = [Ag^+][Cl^-]$. At this point,

$$[Cl^-] = \frac{K_{sp(AgCl)}}{[Ag^+]} = \frac{1.8 \times 10^{-10}}{0.10} = 1.8 \times 10^{-9}\,M$$

At this concentration of $[Cl^-]$, the $[Au^+]$ still in solution is governed by the K_{sp} expression for AuCl: $K_{sp(AuCl)} = [Au^+][Cl^-]$.

$$[Au^+] = \frac{K_{sp(AuCl)}}{[Cl^-]} = \frac{2.0 \times 10^{-13}}{1.8 \times 10^{-9}} = 1.1 \times 10^{-4}\,M$$

Therefore, the percentage of Au^+ that is still in solution when $[Cl^-] = 1.8 \times 10^{-9}\,M$ is

$$\%\ Au^+\ \text{in solution} = \frac{[Au^+]}{[Au^+]_{initial}} \times 100\% = \frac{1.1 \times 10^{-4}\,M}{0.10\,M} \times 100\% = 0.11\%$$

$$\%\ Au^+\ \text{precipitated out} = 100.00\% - 0.11\% = \mathbf{99.89\%}$$

(c) CuCl will begin to precipitate when $Q_{sp(CuCl)} = K_{sp(CuCl)} = [Cu^+][Cl^-]$

$$[Cl^-] = \frac{K_{sp(AuCl)}}{[Cu^+]} = \frac{1.9 \times 10^{-7}}{0.10} = 1.9 \times 10^{-6}\,M$$

At this concentration of Cl^-, $[Au^+]$ and $[Ag^+]$ in solution are governed by their K_{sp} expressions:

$$[Au^+] = \frac{K_{sp(AuCl)}}{[Cl^-]} = \frac{2.0 \times 10^{-13}}{1.9 \times 10^{-6}} = 1.1 \times 10^{-7}\,M$$

$$[Ag^+] = \frac{K_{sp(AgCl)}}{[Cl^-]} = \frac{1.8 \times 10^{-10}}{1.9 \times 10^{-6}} = 9.5 \times 10^{-5}\,M$$

20-34. *Refer to Section 20-4, Example 20-8 and 20-9, and Appendix H.*

Balanced equations: $K_2SO_4(aq) + Pb(NO_3)_2(aq) \rightarrow PbSO_4(s) + 2KNO_3(aq)$

$PbSO_4(s) \rightleftarrows Pb^{2+}(aq) + SO_4^{2-}(aq)$ $\qquad\qquad K_{sp} = 1.8 \times 10^{-8}$

$K_2CrO_4(aq) + Pb(NO_3)_2(aq) \rightarrow PbCrO_4(s) + 2KNO_3(aq)$

$PbCrO_4(s) \rightleftarrows Pb^{2+}(aq) + CrO_4^{2-}(aq)$ $\qquad\qquad K_{sp} = 1.8 \times 10^{-14}$

(a) Both $PbSO_4$ and $PbCrO_4$ are of the same molecular type, i.e., their cation to anion ratio is 1:1. **$PbCrO_4$** has the smaller K_{sp}, so it is less soluble and will precipitate first.

(b) $PbCrO_4$ will begin to precipitate when $Q_{sp(PbCrO_4)} = K_{sp(PbCrO_4)} = [Pb^{2+}][CrO_4^{2-}]$. At this point,

$$[Pb^{2+}] = \frac{K_{sp(PbCrO_4)}}{[CrO_4^{2-}]} = \frac{1.8 \times 10^{-14}}{0.050} = \mathbf{3.6 \times 10^{-13}\,M}$$

(c) $PbSO_4$ will begin to precipitate when $Q_{sp(PbSO_4)} = K_{sp(PbSO_4)} = [Pb^{2+}][SO_4^{2-}]$. At this point,

$$[Pb^{2+}] = \frac{K_{sp(PbSO_4)}}{[SO_4^{2-}]} = \frac{1.8 \times 10^{-8}}{0.050} = \mathbf{3.6 \times 10^{-7}\,M}$$

(d) When $PbSO_4$ begins to precipitate in (c), the concentration of SO_4^{2-} in solution is still the original concentration, **0.050 M**. The CrO_4^{2-} concentration can be calculated by substituting the concentration of Pb^{2+} obtained in (c) into the K_{sp} expression for $PbCrO_4$:

$$[CrO_4^{2-}] = \frac{K_{sp(PbCrO_4)}}{[Pb^{2+}]} = \frac{1.8 \times 10^{-14}}{3.6 \times 10^{-7}} = \mathbf{5.0 \times 10^{-8}\,M}$$

20-36. Refer to Section 20-5, Example 20-10 , and Appendices G and H.

Plan: Calculate the concentrations of Mg^{2+} and OH^- ions and determine Q_{sp}. If $Q_{sp} > K_{sp}$ for $Mg(OH)_2$, then precipitation will occur.

Two equilibrium must be considered:

$$Mg(OH)_2(s) \rightleftharpoons Mg^{2+}(aq) + 2OH^-(aq) \qquad K_{sp} = 1.5 \times 10^{-11}$$
$$NH_3(aq) + H_2O(\ell) \rightleftharpoons NH_4^+(aq) + OH^-(aq) \qquad K_b = 1.8 \times 10^{-5}$$

We recognize that the given solution is a buffer. The NH_3/NH_4^+ equilibrium determines $[OH^-]$. Recall from Chapter 18:

$$[OH^-] = K_b \times \frac{[base]}{[salt]} = (1.8 \times 10^{-5}) \times \frac{0.075\ M}{3.5\ M} = 3.9 \times 10^{-7}\ M$$

At this low $[OH^-]$, we must include the effect of the ionization of water.

Let $x = [OH^-]$ and $[H^+]$ produced by the ionization of water

Therefore, $K_w = [H^+][OH^-] = (x)(3.9 \times 10^{-7} + x) = 1.0 \times 10^{-14}$

Solving the quadratic equation, $x^2 + (3.9 \times 10^{-7})x - 1.0 \times 10^{-14} = 0$,

$$x = \frac{-3.9 \times 10^{-7} \pm \sqrt{(3.9 \times 10^{-7})^2 - 4(1)(-1.0 \times 10^{-14})}}{2(1)} = \frac{-3.9 \times 10^{-7} \pm 4.4 \times 10^{-7}}{2}$$

$$= 2 \times 10^{-8} \text{ or } 8.3 \times 10^{-7} \text{ (discard)}$$

Therefore, $[OH^-] = 3.9 \times 10^{-7}\ M + 0.2 \times 10^{-7}\ M = 4.1 \times 10^{-7}\ M$
$[Mg^{2+}] = [Mg(NO_3)_2] = 0.080\ M$ since $Mg(NO_3)_2$ is a soluble salt

For $Mg(OH)_2$, $Q_{sp} = [Mg^{2+}][OH^-]^2 = (0.080)(4.1 \times 10^{-7})^2 = 1.3 \times 10^{-14}$,
$Q_{sp} < K_{sp}$, $Mg(OH)_2$ will **not** precipitate.

Since $[OH^-] = 4.1 \times 10^{-7}\ M$, pOH = 6.39; pH = **7.61**

20-38. Refer to Section 20-5.

Balanced equations: $CaF_2(s) \rightleftharpoons Ca^{2+}(aq) + 2F^-(aq)$ $\qquad K_{sp} = 3.9 \times 10^{-11}$
$HF(aq) + H_2O(\ell) \rightleftharpoons H_3O^+(aq) + F^-(aq)$ $\qquad K_a = 7.2 \times 10^{-4}$

When equilibria are present in the same solution, all relevant equilibrium expressions must be satisfied:

$$K_{sp} = [Ca^{2+}][F^-]^2 \quad \text{and} \quad K_a = \frac{[H_3O^+][F^-]}{[HF]}$$

In this case, the solid CaF_2 is allowed to dissolve in a solution where $[H_3O^+]$ is buffered at $0.00500\ M$ and $[HF] = 0.10\ M$, which is a source of common ion, F^-.

Let x = molar solubility of CaF_2. Then, x = mol/L of Ca^{2+} produced by dissolution of CaF_2
$2x$ = mol/L of F^- produced by dissolution of CaF_2
y = mol/L of F^- produced by dissociation of HF

Therefore, at equilibrium:

$CaF_2(s)$	$\rightleftharpoons$	$Ca^{2+}(aq)$	+	$2F^-(aq)$
		$x\ M$		$(2x + y)\ M$

$HF(aq)$	+	$H_2O(\ell)$	$\rightleftharpoons$	$H_3O^+(aq)$	+	$F^-(aq)$
$(0.10 - y)\ M$				$0.0050\ M$		$(2x + y)\ M$

Plan: (1) Use the K_a equilibrium expression to find the value of y in terms of x. Note: the H_3O^+ concentration is constant because it is buffered at $0.0050\ M$.
(2) Substitute for y in the K_{sp} expression to calculate x, the molar solubility.

(1) $K_a = \dfrac{[H_3O^+][F^-]}{[HF]} = \dfrac{(0.0050)(2x + y)}{(0.10 - y)} = 7.2 \times 10^{-4}$

Therefore,

$$(0.0050)(2x + y) = (7.2 \times 10^{-4})(0.10 - y)$$
$$0.010x + 0.0050\,y = 7.2 \times 10^{-5} - 7.2 \times 10^{-4}\,y$$
$$0.010x + 0.0057y = 7.2 \times 10^{-5}$$
$$y = \dfrac{7.2 \times 10^{-5} - 0.010x}{0.0057}$$
$$y = 0.0126 - 1.75x$$

(2) $K_{sp} = [Ca^{2+}][F^-]^2 = (x)(2x + y)^2 = 3.9 \times 10^{-11}$

Substituting for y,

$$3.9 \times 10^{-11} = (x)(2x + (0.0126 - 1.75x))^2$$
$$= (x)(0.0126)^2$$
$$= x(1.59 \times 10^{-4})$$
$$x = 2.5 \times 10^{-7}$$

Therefore, molar solubility of CaF_2 in this system is $\mathbf{2.5 \times 10^{-7}}$ **mol** $\mathbf{CaF_2/L}$

20-40. *Refer to Section 20-5, Example 20-10, and Appendices G and H.*

Balanced equations: $Mn(OH)_2(s) \rightleftarrows Mn^{2+}(aq) + 2OH^-(aq)$ $K_{sp} = 4.6 \times 10^{-14}$
   $NH_3(aq) + H_2O(\ell) \rightleftarrows NH_4^+(aq) + OH^-(aq)$ $K_b = 1.8 \times 10^{-5}$

Plan: Calculate the concentrations of Mn^{2+} and OH^- ions and determine Q_{sp} for $Mn(OH)_2$. If $Q_{sp} > K_{sp}$, then precipitation will occur.

$[Mn^{2+}] = [Mn(NO_3)_2] = 2.0 \times 10^{-5}\,M$
$[OH^-]$ is determined from the ionization of NH_3

Let x = $[NH_3]$ that ionizes.   Then, $1.0 \times 10^{-3} - x = [NH_3]$; x = $[OH^-] = [NH_4^+]$

$K_b = \dfrac{[NH_3][OH^-]}{[NH_3]} = \dfrac{x^2}{1.0 \times 10^{-3} - x} = 1.8 \times 10^{-5} \approx \dfrac{x^2}{1.0 \times 10^{-3}}$ Solving, x = 1.3×10^{-4}

Since the value for x is greater than 5% of 1.0×10^{-3}, the simplifying assumption may not hold. When we solve the original quadratic equation, $x^2 + (1.8 \times 10^{-5})x - 1.8 \times 10^{-8} = 0$, x = 1.3×10^{-4}. In this case, the simplifying assumption was adequate to 2 significant figures. Therefore, $[OH^-] = 1.3 \times 10^{-4}\,M$.

$Q_{sp} = [Mn^{2+}][OH^-]^2 = (2.0 \times 10^{-5})(1.3 \times 10^{-4})^2 = 3.4 \times 10^{-13}$

Thus, $Q_{sp} > K_{sp}$ by a factor of 7. Therefore **a precipitate will form** but will not be seen.

20-42. *Refer to Section 20-5 and Appendices F and H.*

Balanced equations: $MnS(s) \rightleftarrows Mn^{2+}(aq) + S^{2-}(aq)$ $K_{sp} = 5.1 \times 10^{-15}$

   $H_2S(aq) + H_2O(\ell) \rightleftarrows H_3O^+(aq) + HS^-(aq)$ $K_1 = 1.0 \times 10^{-7}$
   $HS^-(aq) + H_2O(\ell) \rightleftarrows H_3O^+(aq) + S^{2-}(aq)$ $K_2 = 1.3 \times 10^{-13}$

   or $H_2S(aq) + 2H_2O(\ell) \rightleftarrows 2H_3O^+(aq) + S^{2-}(aq)$ $K = K_1K_2 = 1.3 \times 10^{-20}$

Plan: (1) Determine the concentration of S^{2-} necessary to initiate the precipitation of MnS, using the K_{sp} expression, $K_{sp} = [Mn^{2+}][S^{2-}]$.
   (2) Calculate the concentration of H_3O^+ and pH necessary to produce the S^{2-} concentration calculated in (1), using the combined K expression of H_2S, $K = K_1K_2$.

(1) $[Mn^{2+}] = [MnCl_2] = 0.0100\,M$

$[S^{2-}] = \dfrac{K_{sp}}{[Mn^{2+}]} = \dfrac{5.1 \times 10^{-15}}{0.0100} = 5.1 \times 10^{-13}\,M$

(2) $[H_2S] = [H_2S]_{initial} = 0.100\ M$ since K_1 and K_2 are so small

$$K = K_1K_2 = \frac{[H_3O^+]^2[S^{2-}]}{[H_2S]} \qquad \text{Solving, } [H_3O^+] = \sqrt{\frac{K[H_2S]}{[S^{2-}]}} = \sqrt{\frac{(1.3 \times 10^{-20})(0.100)}{(5.1 \times 10^{-13})}} = 5.0 \times 10^{-5}\ M$$

pH = **4.30**

20-44. *Refer to Section 20-3, Example 20-3 and Appendix H.*

Balanced equation: $Mn(OH)_2(s) \rightleftharpoons Mn^{2+}(aq) + 2OH^-(aq)$ $K_{sp} = 4.6 \times 10^{-14}$

(a) Plan: Calculate the molar solubility of $Mn(OH)_2$, then determine $[OH^-]$ and pH.

     Let   x = molar solubility of $Mn(OH)_2$. Then
            x = moles/L of Mn^{2+}
         2x = moles/L of OH^-

$K_{sp} = [Mn^{2+}][OH^-]^2 = (x)(2x)^2 = 4x^3 = 4.6 \times 10^{-14}$ Solving, x $= 2.3 \times 10^{-5}$
molar solubility $= 2.3 \times 10^{-5}$ mol $Mn(OH)_2$/L (dissolved)
$[OH^-] = 2x = 4.5 \times 10^{-5}\ M$;   pOH = 4.35;   pH = **9.65**

Note: Values for molar solubility and $[OH^-]$ were rounded to 2 significant figures after calculating them.

(b) solubility (g/100 mL) $= \dfrac{2.3 \times 10^{-5}\ \text{mol } Mn(OH)_2}{1\ L} \times \dfrac{89.0\ g}{1\ mol} \times \dfrac{0.1\ L}{100\ mL} = \mathbf{2.0 \times 10^{-4}}$ **g/100 mL**

Note: Remember, do not divide by 100 mL; you want 100 mL to remain in the denominator.

20-46. *Refer to Section 20-6.*

A slightly soluble compound will dissolve when the concentration of its ions in solution are reduced to such a level that $Q_{sp} < K_{sp}$. The following hydroxides and carbonates dissolve in strong acid, such as nitric acid.

(a) $Cu(OH)_2(s) \rightleftharpoons Cu^{2+}(aq) + 2OH^-(aq)$
$$\underline{2H^+(aq) + 2OH^-(aq) \rightarrow 2H_2O(\ell)}$$
$$Cu(OH)_2(s) + 2H^+(aq) \rightarrow Cu^{2+}(aq) + 2H_2O(\ell)$$

The H^+ from the acid reacts with OH^- lowering the concentration of OH^- by forming H_2O, a weak electrolyte in an acid/base neutralization reaction. Whenever $[OH^-]$ is low enough such that $[Cu^{2+}][OH^-]^2 < K_{sp}$, $Cu(OH)_2(s)$ will dissolve.

(b) $Al(OH)_3(s) \rightleftharpoons Al^{3+}(aq) + 3OH^-(aq)$
$$\underline{3H^+(aq) + 3OH^-(aq) \rightarrow 3H_2O(\ell)}$$
$$Al(OH)_3(s) + 3H^+(aq) \rightarrow Al^{3+}(aq) + 3H_2O(\ell)$$

The H^+ from the acid reacts with OH^- and thus lowers the concentration of OH^- in an acid/base neutralization reaction. Whenever $[OH^-]$ is low enough such that $[Al^{3+}][OH^-]^3 < K_{sp}$, $Al(OH)_3(s)$ will dissolve.

(c) $MnCO_3(s) \rightleftharpoons Mn^{2+}(aq) + CO_3^{2-}(aq)$
$$\underline{2H^+(aq) + CO_3^{2-}(aq) \rightarrow CO_2(g) + H_2O(\ell)}$$
$$MnCO_3(s) + 2H^+(aq) \rightarrow Mn^{2+}(aq) + CO_2(g) + H_2O(\ell)$$

The H^+ from the acid removes CO_3^{2-} from solution in a reaction which forms $CO_2(g)$ and $H_2O(\ell)$. Whenever $[CO_3^{2-}]$ is low enough such that $[Mn^{2+}][CO_3^{2-}] < K_{sp}$, $MnCO_3(s)$ will dissolve.

(d) $(PbOH)_2CO_3(s) \rightleftharpoons 2Pb^{2+}(aq) + 2OH^-(aq) + CO_3^{2-}(aq)$
$$\underline{4H^+(aq) + 2OH^-(aq) + CO_3^{2-}(aq) \rightarrow 3H_2O(\ell) + CO_2(g)}$$
$$(PbOH)_2CO_3(s) + 4H^+(aq) \rightarrow 2Pb^{2+}(aq) + 3H_2O(\ell) + CO_2(g)$$

The H^+ from the acid removes both OH^- and CO_3^{2-} ions from solution by forming $H_2O(\ell)$ and $CO_2(g)$. When $[OH^-]$ and $[CO_3^{2-}]$ are low enough such that $[Pb^{2+}][OH^-]^2[CO_3^{2-}] < K_{sp}$, $(PbOH)_2CO_3(s)$ will dissolve.

20-48. *Refer to Section 20-6.*

Nonoxidizing acids dissolve some insoluble sulfides, including MnS and CuS. The H^+ ions react with S^{2-} ions to form gaseous H_2S, which bubbles out of the solutions. The Q_{sp} of the sulfide becomes less than the corresponding K_{sp} value and the metal sulfide dissolves.

(a)
$$MnS(s) \rightleftarrows Mn^{2+}(aq) + S^{2-}(aq)$$
$$\underline{2H^+(aq) + S^{2-}(aq) \rightarrow H_2S(g)}$$
$$MnS(s) + 2H^+(aq) \rightarrow Mn^{2+}(aq) + H_2S(g)$$

(b)
$$CuS(s) \rightleftarrows Cu^{2+}(aq) + S^{2-}(aq)$$
$$\underline{2H^+(aq) + S^{2-}(aq) \rightarrow H_2S(g)}$$
$$CuS(s) + 2H^+(aq) \rightarrow Cu^{2+}(aq) + H_2S(g)$$

20-50. *Refer to Sections 20-5 and 20-6, and Appendix F.*

Balanced equation: $MnS(s) \rightleftarrows Mn^{2+}(aq) + S^{2-}(aq)$

To make $MnS(s)$ more soluble, the above equilibrium must be shifted to the right. Applying LeChatelier's Principle, any process which will reduce either $[Mn^{2+}]$ or $[S^{2-}]$ will do this. In the presence of 0.10 M HCl (a strong acid), competing equilibria will lower $[S^{2-}]$ by producing the weak acids, HS^- and H_2S:

$$S^{2-} + H^+ \rightleftarrows HS^- \qquad K_{1'} = \frac{1}{K_{2\,H_2S}} = \frac{1}{1.3 \times 10^{-13}} = 7.7 \times 10^{12}$$

$$HS^- + H^+ \rightleftarrows H_2S \qquad K_{2'} = \frac{1}{K_{1\,H_2S}} = \frac{1}{1.0 \times 10^{-7}} = 1.0 \times 10^{7}$$

The equilibrium constants are very large, so the above equilibria are shifted far to the right, greatly reducing $[S^{2-}]$. In addition, $H_2S(g)$ will bubble out of solution when its solubility is exceeded, thus removing more S^{2-} from the system. On the other hand, the soluble salt, $Mn(NO_3)_2$, would **not** become more soluble in the presence of 0.10 M HCl. The H^+ ions will not remove NO_3^- ions from solution. The Cl^- ions do not remove Mn^{2+} from solution.

20-52. *Refer to Section 20-3, Example 20-5 and Appendix H.*

Balanced equations: $BaCl_2(aq) + 2NaF(aq) \rightarrow BaF_2(s) + 2NaCl(aq)$ (Will precipitation occur?)
$\qquad\qquad\qquad\quad BaF_2(s) \rightleftarrows Ba^{2+}(aq) + 2F^-(aq)$ (reversible) $K_{sp} = 1.7 \times 10^{-6}$

Plan: (a) Calculate the concentration of Ba^{2+} ions and F^- ions at the instant of mixing before any reaction occurs, using $M_1V_1 = M_2V_2$.
        (b) Determine the reaction quotient, Q_{sp}. If $Q_{sp} > K_{sp}$, then precipitation will occur.

(a) $[Ba^{2+}] = [BaCl_2] = \dfrac{M_1V_1}{V_2} = \dfrac{0.0030\ M \times 0.025\ L}{(0.025\ L + 0.050\ L)} = 1.0 \times 10^{-3}\ M$

$[F^-] = [NaF] = \dfrac{M_1V_1}{V_2} = \dfrac{0.050\ M \times 0.050\ L}{(0.025\ L + 0.050\ L)} = 3.3 \times 10^{-2}\ M$

(b) $Q_{sp} = [Ba^{2+}][F^-]^2 = (1.0 \times 10^{-3})(3.3 \times 10^{-2})^2 = 1.1 \times 10^{-6}$

Since $Q_{sp} < K_{sp}$, precipitation will **not** occur.

20-54. *Refer to Section 20-4, Examples 20-8 and 20-9, and Exercise 20-32 and 20-34 Solutions.*

Balanced equations: $AgBr(s) \rightleftarrows Ag^+(aq) + Br^-(aq)$ (reversible) $K_{sp} = 3.3 \times 10^{-13}$
$\qquad\qquad\qquad\quad AgI(s) \rightleftarrows Ag^+(aq) + I^-(aq)$ (reversible) $K_{sp} = 1.5 \times 10^{-16}$

(a) Because the compounds to be precipitated are all of the same molecular type, i.e., their cation to anion ratios are the same, the one with the smallest K_{sp} is the least soluble and will precipitate first. Hence, **AgI** ($K_{sp} = 1.5 \times 10^{-16}$) will precipitate first, then AgBr ($K_{sp} = 3.3 \times 10^{-13}$).

(b) AgBr will begin to precipitate when $Q_{sp(AgBr)} = K_{sp(AgBr)} = [Ag^+][Br^-]$. At this point,

$$[Ag^+] = \frac{K_{sp(AgBr)}}{[Br^-]} = \frac{3.3 \times 10^{-13}}{0.015} = 2.2 \times 10^{-11} \, M$$

At this concentration of $[Ag^+]$, the $[I^-]$ still in solution is governed by the K_{sp} expression for AgI: $K_{sp(AgI)} = [Ag^+][I^-]$.

$$[I^-] = \frac{K_{sp(AgI)}}{[Ag^+]} = \frac{1.5 \times 10^{-16}}{2.2 \times 10^{-11}} = 6.8 \times 10^{-6} \, M$$

Therefore, the percentage of I^- that is still in solution when AgBr begins to precipitate is

$$\% \, I^- \text{ in solution} = \frac{[I^-]}{[I^-]_{initial}} \times 100\% = \frac{6.8 \times 10^{-6} \, M}{0.015 \, M} \times 100\% = 0.045\%$$

and $\% \, I^-$ removed from solution $= 100.000\% - 0.045\% = \textbf{99.955\%}$

20-56.   *Refer to Section 20-6 and Appendix H.*

Balanced equation: $CaF_2(s) \rightleftarrows Ca^{2+}(aq) + 2F^-(aq)$ $\qquad\qquad\qquad$ $K_{sp} = 3.9 \times 10^{-11}$

Plan: (1) Calculate $[F^-]$.
$\qquad$ (2) Substitute the value into the K_{sp} expression and solve for $[Ca^{2+}]$.

(1) $[F^-] = \dfrac{1 \text{ mg F}^-}{1 \text{ L soln}} \times \dfrac{1 \text{ g F}^-}{1000 \text{ mg F}^-} \times \dfrac{1 \text{ mol F}^-}{19.0 \text{ g F}^-} = 5.3 \times 10^{-5} \, M$ (good to 1 significant figure; will round later)

(2) $K_{sp} = [Ca^{2+}][F^-]^2$

$$[Ca^{2+}] = \frac{[K_{sp}]}{[F^-]^2} = \frac{3.9 \times 10^{-11}}{(5.3 \times 10^{-5})^2} = 1.4 \times 10^{-2} \, M$$

? amount of Ca^{2+} (g/L) $= \dfrac{0.014 \text{ mol Ca}^{2+}}{1 \text{ L soln}} \times \dfrac{40.1 \text{ g Ca}^{2+}}{1 \text{ mol Ca}^{2+}} = \textbf{0.6 g/L}$ (1 significant figure)

20-58.   *Refer to Section 20-3, Example 20-3 and Appendix H.*

Balanced equation: $MgCO_3(s) \rightleftarrows Mg^{2+}(aq) + CO_3^{2-}(aq)$ $\qquad\qquad$ $K_{sp} = 4.0 \times 10^{-5}$

Plan: (1) Calculate the molar solubility of $MgCO_3$.
$\qquad$ (2) Determine the mass of $MgCO_3$ that would dissolve in 15 L of water to produce a saturated solution.
$\qquad$ (3) Determine the percent loss of $MgCO_3$.

(1) Let x = molar solubility of $MgCO_3$. Then, $[Mg^{2+}] = [CO_3^{2-}] = x$

$\quad K_{sp} = [Mg^{2+}][CO_3^{2-}] = x^2 = 4.0 \times 10^{-5}$ $\qquad\qquad$ Solving, x $= 6.3 \times 10^{-3}$

$\quad$ Therefore,   molar solubility $= 6.3 \times 10^{-3}$ mol $MgCO_3$/L (dissolved)

(2) ? g $MgCO_3$ dissolve in 15 L water $= \dfrac{6.3 \times 10^{-3} \text{ mol MgCO}_3}{1 \text{ L}} \times \dfrac{84.3 \text{ g MgCO}_3}{1 \text{ mol MgCO}_3} \times 15 \text{ L} = 8.0 \text{ g}$

(3) $\%$ loss of $MgCO_3 = \dfrac{MgCO_3 \text{ lost}}{\text{initial } MgCO_3} \times 100 = \dfrac{8.0 \text{ g}}{28 \text{ g}} \times 100 = \textbf{29\%}$

21 Electrochemistry

21-2. *Refer to Section 4-7 and Example 4-9.*

In a redox reaction,

(a) oxidizing agents are the species that   (1) gain or appear to gain electrons,
                                                       (2) are reduced, and
                                                       (3) oxidize other substances.

(b) Reducing agents are the species that   (1) lose or appear to lose electrons,
                                                       (2) are oxidized and
                                                       (3) reduce other substances.

$$\overset{+3\ -2}{Fe_2O_3(s)} + \overset{+2\ -2}{3CO(g)} \rightarrow \overset{0}{2Fe(s)} + \overset{+4\ -2}{3CO_2(g)}$$

Consider the reaction: $Fe_2O_3(s) + 3CO(g) \rightarrow 2Fe(s) + 3CO_2(g)$

Fe_2O_3 is the oxidizing agent because it contains Fe, which is being reduced from an oxidation state of $+3$ to 0.
CO is the reducing agent because it contains C, which is being oxidized from an oxidation state of $+2$ to $+4$.

21-4. *Refer to Sections 4-7.*

(a) oxidation: $3(FeS + 4H_2O \rightarrow SO_4^{2-} + Fe^{2+} + 8H^+ + 8e^-)$
reduction: $8(3e^- + NO_3^- + 4H^+ \rightarrow NO + 2H_2O)$
balanced equation: $3FeS + 8NO_3^- + 8H^+ \rightarrow 8NO + 3SO_4^{2-} + 3Fe^{2+} + 2H_2O$

(b) oxidation: $3(Sn^{2+} \rightarrow Sn^{4+} + 2e^-)$
reduction: $6e^- + Cr_2O_7^{2-} + 14H^+ \rightarrow 2Cr^{3+} + 7H_2O$
balanced equation: $3Sn^{2+} + Cr_2O_7^{2-} + 14H^+ \rightarrow 3Sn^{4+} + 2Cr^{3+} + 7H_2O$

(c) oxidation: $S^{2-} + 8OH^- \rightarrow SO_4^{2-} + 4H_2O + 8e^-$
reduction: $4(2e^- + Cl_2 \rightarrow 2Cl^-)$
balanced equation: $S^{2-} + 4Cl_2 + 8OH^- \rightarrow SO_4^{2-} + 8Cl^- + 4H_2O$

21-6. *Refer to Sections 21-2, 21-3 and 21-9.*

The cathode is defined as the electrode at which reduction occurs, i.e., where electrons are consumed, regardless of whether the electrochemical cell is an electrolytic or voltaic cell. In both electrolytic and voltaic cells, the electrons flow through the wire from the anode, where electrons are produced, to the cathode, where electrons are consumed. In an electrolytic cell, the dc source forces the electrons to travel nonspontaneously through the wire. Thus, the electrons flow from the positive electrode (the anode) to the negative electrode (the cathode). However, in a voltaic cell, the electrons flow spontaneously, *away* from the negative electrode (the anode) and toward the positive electrode (the cathode).

(a) The statement, "The positive electrode in any electrochemical cell is the one toward which the electrons flow through the wire," is **false**. It holds for voltaic cells, but not for electrolytic cells.

(b) The statement, "The cathode in any electrochemical cell is the negative electrode," is also **false**. It holds for any electrolytic cell, but not for a voltaic cell.

(a) Magnesium metal is too reactive in water to be obtained by the electrolysis of $MgCl_2(aq)$. In other words, $H_2O(\ell)$ is more easily reduced to $OH^-(aq)$ and $H_2(g)$ than is $Mg^{2+}(aq)$ to $Mg(s)$. In electrochemical reactions, the species that is most easily reduced (or oxidized) will be reduced (or oxidized) first.

(b) Sodium ions do not appear in the overall cell reaction for the electrolysis of $NaCl(aq)$ because Na^+ ions are spectator ions and do not react. Since H_2O is more easily reduced than Na^+ ions, the reduction reaction involves H_2O:

reduction at cathode:	$2e^- + 2H_2O(\ell) \rightarrow H_2(g) + 2OH^-(aq)$
oxidation at anode:	$2Cl^-(aq) \rightarrow Cl_2(g) + 2e^-$
overall cell reaction:	$2Cl^- + 2H_2O(\ell) \rightarrow H_2(g) + Cl_2(g) + 2OH^-(aq)$

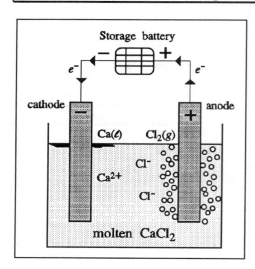

Electrolysis of molten calcium chloride:

oxidation:	$2Cl^-(molten) \rightarrow Cl_2(g) + 2e^-$
reduction:	$Ca^{2+}(molten) + 2e^- \rightarrow Ca(\ell)$
overall cell reaction:	$CaCl_2(\ell) \rightarrow Ca(\ell) + Cl_2(g)$

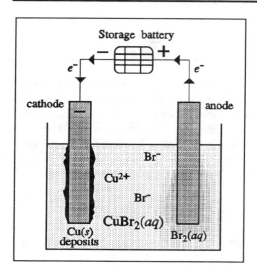

Electrolysis of aqueous copper(II) bromide:

oxidation:	$2Br^-(aq) \rightarrow Br_2(aq) + 2e^-$
reduction:	$Cu^{2+}(aq) + 2e^- \rightarrow Cu(s)$
overall cell reaction:	$Cu^{2+}(aq) + 2Br^-(aq) \rightarrow Cu(s) + Br_2(aq)$

21-14. *Refer to Section 21-6 and the Key Terms for Chapter 21.*

(a) A coulomb (C) is the amount of electrical charge that passes a given point when one ampere of current flows for one second.

(b) Electrical current is the motion of electrons or ions through a conducting medium.

(c) An ampere (A) is the practical unit of electrical current equal to the transfer of 1 coulomb per second. So, $1 \text{ A} = 1 \text{ C/s}$.

(d) A faraday of electricity corresponds to the charge on 6.022×10^{23} electrons, or 96,487 coulombs. It is the amount of electricity that reduces 1 equivalent weight of a substance at the cathode and oxidizes 1 equivalent weight of a substance at the anode.

21-16. *Refer to Section 21-6.*

$$? \text{ coulombs/electron} = \frac{1 \text{ faraday}}{6.022 \times 10^{23} \ e^-} \times \frac{96487 \text{ coulombs}}{1 \text{ faraday}} = \mathbf{1.602 \times 10^{-19} \ C/}e^-$$

21-18. *Refer to Section 21-6.*

(i) Recall that 1 faraday of electricity is equivalent to 1 mole of electrons passing through a system. Consider the general balanced half-reaction:

$$M^{n+} + ne^- \rightarrow M$$

In accordance with stoichiometry, the production of 1 mole of M requires n moles of electrons, hence n faradays of electricity.

Balanced Half-Reaction	No. of Faradays/1 mol Free Metal
(a) $Co^{3+}(aq) + 3e^- \rightarrow Co(s)$	3
(b) $Hg^{2+}(aq) + 2e^- \rightarrow Hg(\ell)$	2
(c) $Hg_2^{2+}(aq) + 2e^- \rightarrow 2Hg(\ell)$	1

(ii) The amount of charge required to deposit 1.00 g of each of the metals according to the reactions above:

(a) $? \text{ coulombs} = 1.00 \text{ g Co} \times \dfrac{1 \text{ mol Co}}{58.93 \text{ g Co}} \times \dfrac{3 \text{ mol } e^-}{1 \text{ mol Co}} \times \dfrac{96500 \text{ C}}{1 \text{ mol } e^-} = \mathbf{4.91 \times 10^3 \ C}$

(b) $? \text{ coulombs} = 1.00 \text{ g Hg} \times \dfrac{1 \text{ mol Hg}}{200.6 \text{ g Hg}} \times \dfrac{2 \text{ mol } e^-}{1 \text{ mol Hg}} \times \dfrac{96500 \text{ C}}{1 \text{ mol } e^-} = \mathbf{962 \ C}$

(c) $? \text{ coulombs} = 1.00 \text{ g Hg} \times \dfrac{1 \text{ mol Hg}}{200.6 \text{ g Hg}} \times \dfrac{2 \text{ mol } e^-}{2 \text{ mol Hg}} \times \dfrac{96500 \text{ C}}{1 \text{ mol } e^-} = \mathbf{481 \ C}$

21-20. *Refer to Section 21-6 and Example 21-2.*

Plan: (1) Determine the half-reaction involving Cl_2.
(2) Calculate the moles of Cl_2 produced at the experimental conditions at 88% efficiency.
(3) Calculate the volume of Cl_2 produced using the ideal gas law, $PV = nRT$.

(1) Balanced half-reaction: $2Cl^- \rightarrow Cl_2 + 2e^-$

(2) $? \text{ mol } Cl_2 = 5.0 \text{ hr} \times \dfrac{3600 \text{ s}}{1 \text{ hr}} \times \dfrac{1.5 \text{ C}}{1 \text{ s}} \times \dfrac{1 \text{ mol } e^-}{96500 \text{ C}} \times \dfrac{1 \text{ mol } Cl_2}{2 \text{ mol } e^-} \times \dfrac{88}{100} = 0.12 \text{ mol } Cl_2$

(3) $V = \dfrac{nRT}{P} = \dfrac{(0.12 \text{ mol})(0.0821 \text{ L·atm/mol·K})(5°C + 273°)}{(735/760 \text{ atm})} = \mathbf{2.9 \ L \ Cl_2}$

289

Balanced half-reaction: $Ag^+ + e^- \rightarrow Ag$

? coulombs $= 0.775$ mg Ag $\times \dfrac{1 \text{ g Ag}}{1000 \text{ mg Ag}} \times \dfrac{1 \text{ mol Ag}}{107.9 \text{ g Ag}} \times \dfrac{1 \text{ mol } e^-}{1 \text{ mol Ag}} \times \dfrac{96500 \text{ C}}{1 \text{ mol } e^-} = \textbf{0.693 C}$

Balanced half-reaction: $Ag^+ + e^- \rightarrow Ag$

? g Ag $= 20.0$ min $\times \dfrac{60 \text{ s}}{1 \text{ min}} \times \dfrac{2.50 \text{ C}}{1 \text{ s}} \times \dfrac{1 \text{ mol } e^-}{96500 \text{ C}} \times \dfrac{1 \text{ mol Ag}}{1 \text{ mol } e^-} \times \dfrac{107.9 \text{ g Ag}}{1 \text{ mol Ag}} = \textbf{3.35 g Ag}$

Balanced half-reactions: anode $2I^- \rightarrow I_2 + 2e^-$

cathode $2H_2O + 2e^- \rightarrow H_2 + 2OH^-$

(a) The number of faradays passing through the cell is equivalent to the number of moles of electrons passing through the cell.

? faradays $= 47.7 \times 10^{-3}$ mol $I_2 \times \dfrac{2 \text{ mol } e^-}{1 \text{ mol } I_2} \times \dfrac{1 \text{ faraday}}{1 \text{ mol } e^-} = \textbf{0.0954 faradays}$

(b) ? coulombs $= 0.0954$ faradays $\times \dfrac{96500 \text{ C}}{1 \text{ faraday}} = \textbf{9.21} \times \textbf{10}^3 \textbf{ C}$

(c) ? $L_{STP} H_2 = 9.21 \times 10^3$ C $\times \dfrac{1 \text{ mol } e^-}{96500 \text{ C}} \times \dfrac{1 \text{ mol } H_2}{2 \text{ mol } e^-} \times \dfrac{22.4 \text{ } L_{STP} H_2}{1 \text{ mol } H_2} = \textbf{1.07 } L_{STP} \textbf{ H}_2$

(d) Plan: (1) Determine the moles of OH^- formed.

(2) Calculate $[OH^-]$, pOH and pH.

(1) ? mol $OH^- = 9.21 \times 10^3$ C $\times \dfrac{1 \text{ mol } e^-}{96500 \text{ C}} \times \dfrac{2 \text{ mol } OH^-}{2 \text{ mol } e^-} = 0.0954$ mol OH^-

(2) $[OH^-] = \dfrac{0.0954 \text{ mol } OH^-}{0.500 \text{ L}} = 0.191 \text{ } M \text{ } OH^-$; pOH $= 0.719$; pH $= \textbf{13.281}$

Balanced half-reaction: $Cu^{2+} + 2e^- \rightarrow Cu$

(1) Plan: M, L $CuCl_2$ soln $\overset{(i)}{\Longrightarrow}$ mol Cu^{2+} reacted $\overset{(ii)}{\Longrightarrow}$ mol e^- reacted $\overset{(iii)}{\Longrightarrow}$ time required

(i) Original moles of $Cu^{2+} = (0.333 \text{ } M)(0.250 \text{ L}) = 8.33 \times 10^{-2}$ mol

Final moles of $Cu^{2+} = (0.167 \text{ } M)(0.250 \text{ L}) = 4.18 \times 10^{-2}$ mol

? mol Cu^{2+} reacted $= 8.33 \times 10^{-2}$ mol - 4.18×10^{-2} mol $= 4.15 \times 10^{-2}$ mol Cu^{2+}

(ii) ? mol e^- reacted $= 2 \times$ mol Cu^{2+} reacted $= 2 \times 4.15 \times 10^{-2}$ mol $= 8.30 \times 10^{-2}$ mol e^-

(iii) ? time required (s) $= 8.30 \times 10^{-2}$ mol $e^- \times \dfrac{96500 \text{ C}}{1 \text{ mol } e^-} \times \dfrac{1 \text{ amp-s}}{1 \text{ C}} \times \dfrac{1}{0.75 \text{ amp}} = 1.07 \times 10^4 \text{ s} \equiv \textbf{2.97 hr}$

(2) ? mass of Cu $= 4.15 \times 10^{-2}$ mol Cu $\times \dfrac{63.55 \text{ g Cu}}{1 \text{ mol Cu}} = \textbf{2.64 g Cu}$

21-30. *Refer to Section 21-6 and Example 21-1.*

Balanced half-reaction: $M^{2+} + 2e^- \rightarrow M$

Plan: (1) Determine the number of moles of metal that can be plated out.
(2) Calculate the atomic weight of the metal and identify.

(1) $? \text{ mol metal} = 14{,}475 \text{ C} \times \dfrac{1 \text{ mol } e^-}{96500 \text{ C}} \times \dfrac{1 \text{ mol metal}}{2 \text{ mol } e^-} = 0.0750 \text{ mol metal}$

(2) $\text{atomic weight (g/mol)} = \dfrac{15.54 \text{ g}}{0.0750 \text{ mol}} = 207.2 \text{ g/mol}$ Therefore, the metal is **lead**.

21-32. *Refer to Section 21-6 and Example 21-1.*

Balanced half-reactions: $Cd \rightarrow Cd^{2+} + 2e^-$ $Ag^+ + e^- \rightarrow Ag$ $Fe^{2+} \rightarrow Fe^{3+} + e^-$

(a) $? \text{ faradays} = 1.00 \text{ g Cd} \times \dfrac{1 \text{ mol Cd}}{112.4 \text{ g Cd}} \times \dfrac{2 \text{ mol } e^-}{1 \text{ mol Cd}} \times \dfrac{1 \text{ faraday}}{1 \text{ mol } e^-} = \mathbf{0.0178 \text{ faradays}}$

(b) $? \text{ g Ag} = 0.0178 \text{ faradays} \times \dfrac{1 \text{ mol } e^-}{1 \text{ faraday}} \times \dfrac{1 \text{ mol Ag}}{1 \text{ mol } e^-} \times \dfrac{107.9 \text{ g Ag}}{1 \text{ mol Ag}} = \mathbf{1.92 \text{ g Ag}}$

(c) $? \text{ g Fe(NO}_3)_3 = 0.0178 \text{ faraday} \times \dfrac{1 \text{ mol } e^-}{1 \text{ faraday}} \times \dfrac{1 \text{ mol Fe}^{3+}}{1 \text{ mol } e^-} \times \dfrac{1 \text{ mol Fe(NO}_3)_3}{1 \text{ mol Fe}^{3+}} \times \dfrac{241.9 \text{ g Fe(NO}_3)_3}{1 \text{ mol Fe(NO}_3)_3}$

$\qquad = \mathbf{4.31 \text{ g Fe(NO}_3)_3}$

21-34. *Refer to the Introduction to Voltaic or Galvanic Cells, Section 21-9 and Figure 21-7.*

(a) In a voltaic cell, the solutions in the two half-cells must be kept separate in order to produce usable electrical energy since electricity is only produced when electron transfer is forced to occur through the external circuit. If the two half-cells were mixed, electron transfer would happen directly in the solution and could not be exploited to give electricity.

(b) A salt bridge in a voltaic or galvanic cell has three functions: it allows electrical contact between the two solutions; it prevents mixing of the electrode solutions; and it maintains electrical neutrality in each half-cell.

21-36. *Refer to Sections 21-9, 21-10 and 21-11, and Figure 21-7.*

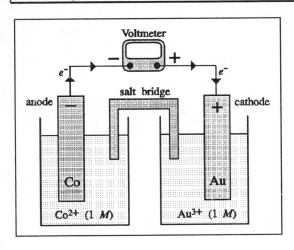

Voltaic cell:

oxidation at anode:	$3(Co \rightarrow Co^{2+} + 2e^-)$
reduction at cathode:	$2(Au^{3+} + 3e^- \rightarrow Au)$
overall cell reaction:	$3Co + 2Au^{3+} \rightarrow 3Co^{2+} + 2Au$

Balanced equation: $Al(s) + 3Ag^+(aq) \rightarrow Al^{3+}(aq) + 3Ag(s)$

(a) reduction half-reaction: $Ag^+(aq) + e^- \rightarrow Ag(s)$

(b) oxidation half-reaction: $Al(s) \rightarrow Al^{3+}(aq) + 3e^-$

(c) Al is the anode.

(d) Ag is the cathode.

(e) Refer to cell diagram at right.

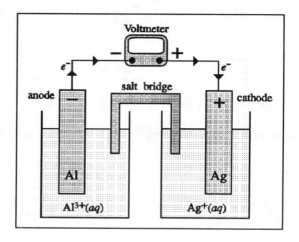

No electricity is produced when $Cu(s)$ is placed into $AgNO_3(aq)$ even though a spontaneous redox reaction occurs:

$$Cu(s) + 2AgNO_3(aq) \rightarrow Cu(NO_3)_2(aq) + 2Ag(s)$$

The electron transfer occurs within the solution; it is not forced to occur through an external circuit where it would produce useful electrical energy.

If the sign of the standard reduction potential, $E°$, of a half-reaction is positive, the half-reaction is the cathodic (reduction) reaction when connected to the standard hydrogen electrode (SHE). Half-reactions with more positive $E°$ values have greater tendencies to occur in the forward direction. Hence, the magnitude of a half-cell potential measures the spontaneity of the forward reaction. If the $E°$ of a half-reaction is negative, the half-reaction is the anodic (oxidation) reaction when connected to the SHE. Half-reactions with more negative $E°$ values have greater tendencies to occur in the reverse direction.

(a) The substance that is the stronger oxidizing agent is the more easily reduced and has the more positive reduction potential. Therefore, in order of increasing strength,

$$K^+ (-2.9 \text{ V}) < Na^+ (-2.7 \text{ V}) < Fe^{2+} (-0.4 \text{ V}) < Cu^{2+} (0.3 \text{ V}) < Cu^+ (0.5 \text{ V}) < Ag^+ (0.8 \text{ V}) < F_2 (2.9 \text{ V})$$

(b) Under standard state conditions, both F_2 and Ag^+ can oxidize Cu, since their standard reduction potentials are more positive than those for Cu^+ and Cu^{2+}. Also, under appropriate conditions, Cu^+ can oxidize Cu to Cu^{2+}.

The activity of a metal is based on how easily it oxidizes to positively-charged ions. Therefore, a more active metal loses electrons more readily, is more easily oxidized and is a better reducing agent. The strength of a reducing agent increases as its standard reduction potential becomes more negative.

most active   Eu (-3.4 V) > Ra (-2.9 V) > Rh (0.80 V)   least active

Only Eu is more active than Li (-3.0 V). Eu and Ra are more active than H (0.00 V). All three are more active than Au (1.5 V).

(a) cell diagram:

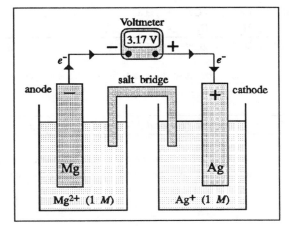

		$E°$
oxidation at anode:	$Mg \rightarrow Mg^{2+} + 2e^-$	+2.37 V
reduction at cathode:	$2(Ag^+ + e^- \rightarrow Ag)$	+0.7994 V
cell reaction:	$Mg + 2Ag^+ \rightarrow Mg^{2+} + 2Ag$	$E°_{cell} = $ **+3.17 V**

(b) cell diagram:

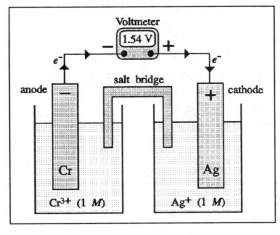

		$E°$
oxidation at anode:	$Cr \rightarrow Cr^{3+} + 3e^-$	+0.74 V
reduction at cathode:	$3(Ag^+ + e^- \rightarrow Ag)$	+0.7994 V
cell reaction:	$Cr + 3Ag^+ \rightarrow Cr^{3+} + 3Ag$	$E°_{cell} = $ **+1.54 V**

21-50. *Refer to Sections 21-16 and 21-17, Examples 21-4 and 21-5, and Appendix J.*

Plan: Calculate $E°_{cell}$ for each reaction as written. If $E°_{cell}$ is positive, the reaction is spontaneous and will go as written. If $E°_{cell}$ is negative, the reaction is nonspontaneous and will not go as written; the reverse reaction is spontaneous.

			$E°$
(a) reduction:		$2(Fe^{3+} + e^- \rightarrow Fe^{2+})$	+0.771 V
oxidation:		$Sn^{2+} \rightarrow Sn^{4+} + 2e^-$	-0.15 V
cell reaction:		$2Fe^{3+} + Sn^{2+} \rightarrow 2Fe^{2+} + Sn^{4+}$	$E°_{cell} = +0.62$ V

Yes, Fe^{3+} will oxidize Sn^{2+} to Sn^{4+} because the reaction is spontaneous ($E°_{cell} > 0$).

		$E°$
(b) reduction:	$Cr_2O_7^{2-} + 14H^+ + 6e^- \rightarrow 2Cr^{3+} + 7H_2O$	+1.33 V
oxidation:	$3(2F^- \rightarrow F_2 + 2e^-)$	-2.87 V
cell reaction:	$Cr_2O_7^{2-} + 14H^+ + 6F^- \rightarrow 2Cr^{3+} + 3F_2 + 7H_2O$	$E°_{cell} = -1.54$ V

No, $Cr_2O_7^{2-}$ ions cannot oxidize F^- ions to F_2 because the reaction is not spontaneous ($E°_{cell} < 0$).

21-52. *Refer to Sections 21-16 and 21-17, Examples 21-4 and 21-5, and Appendix J.*

Plan: Calculate $E°_{cell}$ for each reaction as written. If $E°_{cell}$ is positive, the reaction is spontaneous and will go as written. If $E°_{cell}$ is negative, the reaction is nonspontaneous and will not go as written; the reverse reaction is spontaneous.

		$E°$
(a) reduction:	$5(Cr_2O_7^{2-} + 14H^+ + 6e^- \rightarrow 2Cr^{3+} + 7H_2O)$	+1.33 V
oxidation:	$6(Mn^{2+} + 4H_2O \rightarrow MnO_4^- + 8H^+ + 5e^-)$	-1.51 V
cell rxn:	$5Cr_2O_7^{2-} + 22H^+ + 6Mn^{2+} \rightarrow 10\,Cr^{3+} + 11H_2O + 6MnO_4^-$	$E°_{cell} = -0.18$ V

No, $Cr_2O_7^{2-}$ ions cannot oxidize Mn^{2+} ions to MnO_4^- ions since the reaction is non-spontaneous ($E°_{cell} < 0$).

		$E°$
(b) reduction (1):	$SO_4^{2-} + 4H^+ + 2e^- \rightarrow H_2SO_3 + H_2O$	+0.17 V
reduction (2):	$SO_4^{2-} + 4H^+ + 2e^- \rightarrow SO_2 + 2H_2O$	+0.20 V
oxidation:	$H_3AsO_3 + H_2O \rightarrow H_3AsO_4 + 2H^+ + 2e^-$	-0.58 V

No matter which sulfate reduction reaction is used, $E°_{cell} (= E°_{cathode} + E°_{anode}) < 0$. Therefore, **no**, SO_4^{2-} ions cannot oxidize H_3AsO_3 to H_3AsO_4.

21-54. *Refer to Section 21-16, Example 21-4 and Exercise 21-36 Solution.*

Refer to the cell diagram in Exercise 21-36 Solution.

		$E°$
reduction:	$2(Au^{3+} + 3e^- \rightarrow Au)$	+1.50 V
oxidation:	$3(Co \rightarrow Co^{2+} + 2e^-)$	+0.28 V
cell reaction:	$2Au^{3+} + 3Co \rightarrow 2Au + 3Co^{2+}$	$E°_{cell} = +1.78$ V

21-56. *Refer to Sections 21-10 and 21-16, and Appendix J.*

(a) Consider the voltaic cell: $Zn/Zn^{2+}||Sn^{2+}/Sn$

    (i)   cell reaction: $Zn + Sn^{2+} \rightarrow Zn^{2+} + Sn$

    (ii) oxidation half-reaction at anode: $Zn \rightarrow Zn^{2+} + 2e^-$ $E° = +0.763$ V
         reduction half-reaction at cathode: $Sn^{2+} + 2e^- \rightarrow Sn$ $E° = -0.14$ V

    (iii) $E°_{cell} = E°_{anode} + E°_{cathode} = +0.763$ V $+ (-0.14$ V$) = \mathbf{+0.62}$ **V**

    (iv) Yes, the standard reaction occurs as written since $E°_{cell} > 0$.

(b) Consider the voltaic cell: $Ag/Ag^+||Cd^{2+}/Cd$

    (i)   cell reaction: $2Ag + Cd^{2+} \rightarrow 2Ag^+ + Cd$

    (ii) oxidation half-reaction at anode: $Ag \rightarrow Ag^+ + e^-$ $E° = -0.7994$ V
         reduction half-reaction at cathode: $Cd^{2+} + 2e^- \rightarrow Cd$ $E° = -0.403$ V

    (iii) $E°_{cell} = E°_{cathode} + E°_{anode} = (-0.7994$ V$) + (-0.403$ V$) = \mathbf{-1.202}$ **V**

    (iv) No, the standard reaction will not occur as written since $E°_{cell} < 0$; the reverse reaction will occur.

21-58. *Refer to Sections 21-16 and 21-17, and Appendix J.*

Plan: Calculate $E°_{cell}$ for each reaction as written. If $E°_{cell}$ is positive, the reaction is spontaneous.

		$E°$
(a) reduction:	$2(Cu^{2+} + 2e^- \rightarrow Cu)$	$+0.337$ V
oxidation:	$Si + 6OH^- \rightarrow SiO_3^{2-} + 3H_2O + 4e^-$	$+1.70$ V
cell reaction:	$2Cu^{2+} + Si + 6OH^- \rightarrow 2Cu + SiO_3^{2-} + 3H_2O$	$E°_{cell} = +2.04$ V

Yes, the reaction is spontaneous; $E°_{cell} > 0$.

		$E°$
(b) reduction:	$Ag_2CrO_4 + 2e^- \rightarrow 2Ag + CrO_4^{2-}$	$+0.446$ V
oxidation:	$Zn + 4CN^- \rightarrow Zn(CN)_4^{2-} + 2e^-$	$+1.26$ V
cell reaction:	$Ag_2CrO_4 + Zn + 4CN^- \rightarrow 2Ag + CrO_4^{2-} + Zn(CN)_4^{2-}$	$E°_{cell} = +1.71$ V

Yes, the reaction is spontaneous; $E°_{cell} > 0$.

		$E°$
(c) reduction:	$MnO_2 + 4H^+ + 2e^- \rightarrow Mn^{2+} + 2H_2O$	$+1.23$ V
oxidation:	$Sr \rightarrow Sr^{2+} + 2e^-$	$+2.89$ V
cell reaction:	$MnO_2 + 4H^+ + Sr \rightarrow Mn^{2+} + 2H_2O + Sr^{2+}$	$E°_{cell} = +4.12$ V

Yes, the reaction is spontaneous; $E°_{cell} > 0$.

		$E°$
(d) reduction:	$ZnS + 2e^- \rightarrow Zn + S^{2-}$	-1.44 V
oxidation:	$Cl_2 + 2H_2O \rightarrow 2HClO + 2H^+ + 2e^-$	-1.63 V
cell reaction:	$ZnS + Cl_2 + 2H_2O \rightarrow Zn + 2HClO + 2H^+ + S^{2-}$	$E°_{cell} = -3.07$ V
or	$ZnS + Cl_2 + 2H_2O \rightarrow Zn + 2HClO + H_2S$	since H_2S is a weak acid

No, the reaction is non-spontaneous; $E°_{cell} < 0$.

21-60. *Refer to Sections 21-16 and 21-17, and Appendix J.*

The substance that is the stronger reducing agent is the more easily oxidized. The reduced form of a species is a stronger reducing agent when the half-reaction has a more negative standard reduction potential. The stronger reducing agents are given below.

(a) H_2 (0.000 V) > Ag (0.7994 V)

(b) Sn (-0.14 V) > Pb (-0.126 V)

(c) Hg (0.855 V) > Au (1.68 V)

(d) Cl^- in basic soln (0.62 V or 0.89 V) > Cl^- in acidic soln (1.36 V)

(e) H_2S (0.14 V) > HCl, i.e., Cl^- in acidic soln (1.36 V) .

(f) Fe^{2+} (+0.771 V) > Br^- (1.08 V)

21-62. *Refer to Sections 21-21 and 15-15.*

The half-reactions can be added together so that the desired half-reaction is obtained:

reduction half-reaction:	$Yb^{3+} + 3e^- \rightarrow Yb$	(1)
oxidation half-reaction:	$Yb \rightarrow Yb^{2+} + 2e^-$	(2)
net half-reaction:	$Yb^{3+} + e^- \rightarrow Yb^{2+}$	(3)

Since ΔG is a state function, we can write:

$$\Delta G^\circ_{rxn\ (3)} = \Delta G^\circ_{rxn\ (1)} + \Delta G^\circ_{rxn\ (2)}$$
$$\Delta G^\circ_{Yb^{3+}/Yb^{2+}} = \Delta G^\circ_{Yb^{3+}/Yb} + \Delta G^\circ_{Yb/Yb^{2+}}$$
$$-nFE^\circ_{Yb^{3+}/Yb^{2+}} = -nFE^\circ_{Yb^{3+}/Yb} + (-nFE^\circ_{Yb/Yb^{2+}})$$
$$-(1)E^\circ_{Yb^{3+}/Yb^{2+}} = -(3)(-2.267\ V) - (2)(2.797\ V)$$
$$E^\circ_{Yb^{3+}/Yb^{2+}} = \textbf{-1.207 V}$$

21-64. *Refer to Section 21-18.*

The tarnish on silver, Ag_2S, can be removed by boiling the silverware in slightly salty water (to improve the water's conductivity) in an aluminum pan. The reaction is an oxidation-reduction reaction that occurs spontaneously. The Ag in Ag_2S is reduced back to silver, while the Al in the pan is oxidized to Al^{3+}. This is then a version of a voltaic cell.

reduction reaction (at cathode - the silverware): $3[Ag_2S(s) + 2H^+(aq) + 2e^- \rightarrow 2Ag(s) + H_2S(aq)]$
oxidation reaction (at anode - the aluminum pan): $2[Al(s) \rightarrow Al^{3+}(aq) + 3e^-]$
overall reaction: $3Ag_2S(s) + 6H^+ + 2Al \rightarrow 6Ag(s) + 3H_2S(aq) + 2Al^{3+}(aq)$

21-66. *Refer to Section 21-20.*

The Nernst equation is used to calculate electrode potentials or cell potentials when the concentrations and partial pressures are other than standard state values. The Nernst equation using base 10 logarithms is given by:

$$E = E^\circ - \frac{2.303RT}{nF} \log Q$$

or

$$E = E^\circ - \frac{RT}{nF} \ln Q$$

where

E = potential at nonstandard conditions (V)
E° = standard potential (V)
R = gas constant, 8.314 J/mol·K
T = absolute temperature (K); $T = °C + 273.15°$
F = Faraday's constant, 96500 J/V·mol e^-
n = number of moles of e^- transferred
Q = reaction quotient

Substituting at 25°C, using base 10 logarithms:

$$E = E° - \frac{(2.303)(8.314)(298.15)}{n(96500)} \log Q$$

$$= E° - \frac{0.0592}{n} \log Q$$

using natural logarithms:

$$E = E° - \frac{(8.314)(298.15)}{n(96500)} \ln Q$$

$$= E° - \frac{0.0257}{n} \ln Q$$

21-68. *Refer to Section 21-20 and Appendix J.*

Balanced reduction half-reaction: $Zn^{2+} + 2e^- \rightarrow Zn$ $\qquad\qquad E° = -0.763$ V

For the standard half-cell, $[Zn^{2+}] = 1\ M$. Substituting these data into the Nernst equation, we have

$$E = E° - \frac{0.0592}{n} \log \frac{1}{[Zn^{2+}]} = -0.763\ V - \frac{0.0592}{2} \log \frac{1}{1} = -0.763\ V$$

Therefore, the Nernst equation predicts that the voltage of a standard half-cell equals $E°$.

21-70. *Refer to Section 21-20, Example 21-8 and Appendix J.*

Balanced half-reaction: $2H^+(aq) + 2e^- \rightarrow H_2(g)$ $\qquad\qquad E° = 0.000$ V

Given: $[H^+] = [HClO_4] = 2.00 \times 10^{-4}\ M$ $\qquad P_{H_2} = 3.00$ atm

$$E = E° - \frac{0.0592}{n} \log \frac{P_{H_2}}{[H^+]^2} = 0.000\ V - \frac{0.0592}{2} \log \frac{3.00}{(2.00 \times 10^{-4})^2} = 0.000\ V - 0.233\ V = -\mathbf{0.233\ V}$$

21-72. *Refer to Section 21-20, Example 21-10 and Appendix J.*

			$E°$
(a) oxidation half-reaction:	$Zn(s) \rightarrow Zn^{2+}(aq) + 2e^-$		$+0.763$ V
reduction half-reaction:	$Cl_2(g) + 2e^- \rightarrow 2Cl^-(aq)$		$+1.360$ V
cell reaction:	$Zn(s) + Cl_2(g) \rightarrow Zn^{2+}(aq) + 2Cl^-(aq)$		$E°_{cell} = +\mathbf{2.123\ V}$

(b) $[Zn^{2+}] = [ZnCl_2] = 0.15\ M$: assume $[Cl^-]$ in other half cell $= 2 \times [ZnCl_2] = 0.30\ M$; $P_{Cl_2} = 1.0$ atm

$$E_{cell} = E°_{cell} - \frac{0.0592}{n} \log \frac{[Zn^{2+}][Cl^-]^2}{P_{Cl_2}} = +2.123\ V - \frac{0.0592}{2} \log \frac{(0.15)(0.30)^2}{(1.0)} = +\mathbf{2.178\ V}$$

21-74. *Refer to Section 21-20, Example 21-11 and Appendix J.*

Balanced half-reaction: $F_2(g) + 2e^- \rightarrow 2F^-(aq)$ $\qquad\qquad E° = +2.87$ V

Applying the Nernst equation:

$$E = E° - \frac{0.0592}{n} \log \frac{[F^-]^2}{P_{F_2}}$$

Substituting,

$$+2.75\ V = +2.87\ V - \frac{0.0592}{2} \log \frac{(0.40)^2}{P_{F_2}}$$

$$0.12\ V = \frac{0.0592}{2} \log \frac{0.16}{P_{F_2}}$$

$$4.1 = \log \frac{0.16}{P_{F_2}}$$

Taking the antilogarithm of both sides, $\qquad 1.3 \times 10^4 = \dfrac{0.16}{P_{F_2}}$

Therefore, $\qquad\qquad\qquad\qquad\qquad P_{F_2} = \mathbf{1.2 \times 10^{-5}\ atm}$

Balanced half-reactions: (1) $2H^+(aq) + 2e^- \rightarrow H_2(g)$ $E^\circ = 0.0000$ V

                                  (2) $Ag^+(aq) + e^- \rightarrow Ag(s)$ $E^\circ = 0.7994$ V

(a) Balanced equation: $H_2(g) + 2Ag^+(aq) \rightarrow 2H^+(aq) + 2Ag(s)$ $E^\circ = 0.7994$ V

$$E = E^\circ - \frac{0.0592}{n} \log \frac{[H^+]^2}{P_{H_2}[Ag^+]^2} = 0.7994 \text{ V} - \frac{0.0592}{2} \log \frac{(1.00 \times 10^{-3})^2}{(10.0)(4.96 \times 10^{-3})^2} = \mathbf{0.870 \text{ V}}$$

(b) Balanced equation: $H_2(1.00 \text{ atm}) + 2H^+(pH = 3.47) \rightarrow 2H^+(pH = 5.97) + H_2(1.00 \text{ atm})$

    For pH $= 5.97$, $[H^+] = 1.07 \times 10^{-6}$ M

    For pH $= 3.47$, $[H^+] = 3.39 \times 10^{-4}$ M

$$E = E^\circ - \frac{0.0592}{n} \log \frac{[H^+]^2 P_{H_2}}{P_{H_2}[H^+]^2} = 0.0000 \text{ V} - \frac{0.0592}{2} \log \frac{(1.07 \times 10^{-6})^2(1.00)}{(1.00)(3.39 \times 10^{-4})^2} = \mathbf{0.148 \text{ V}}$$

(c) Balanced equation: $H_2(0.0361 \text{ atm}) + 2H^+(0.0100 \text{ } M) \rightarrow 2H^+(0.0100 \text{ } M) + H_2(5.98 \times 10^{-4} \text{ atm})$

$$E = E^\circ - \frac{0.0592}{n} \log \frac{[H^+]^2 P_{H_2}}{P_{H_2}[H^+]^2} = 0.0000 \text{ V} - \frac{0.0592}{2} \log \frac{(0.0100)^2(5.98 \times 10^{-4})}{(0.0361)(0.0100)^2} = \mathbf{0.0527 \text{ V}}$$

Balanced half-reaction: $Cu^{2+} + 2e^- \rightarrow Cu$ $E^\circ = 0.337$ V

For a concentration cell: $E_{cell} = E^\circ_{cell} - \dfrac{0.0592}{n} \log \dfrac{[\text{dilute solution}]}{[\text{concentrated solution}]}$

Substituting, $0.045 \text{ V} = 0 \text{ V} - \dfrac{0.0592}{2} \log \dfrac{[Cu^{2+}]_B}{0.75}$

$$\log \frac{[Cu^{2+}]_B}{0.75} = -1.52$$

Taking the antilogarithm, $\dfrac{[Cu^{2+}]_B}{0.75} = 0.030$

Therefore, $[Cu^{2+}]_B = \mathbf{0.023 \text{ } M}$

(a) Plan: (1) Determine E°_{cell}.

               (2) Use the Nernst equation to find the ratio of Zn^{2+} to Ni^{2+}.

		E°
(1) reduction half-reaction:	$Ni^{2+} + 2e^- \rightarrow Ni$	-0.25 V
oxidation half-reaction:	$Zn \rightarrow Zn^{2+} + 2e^-$	$+0.763$ V
cell reaction:	$Ni^{2+} + Zn \rightarrow Ni + Zn^{2+}$	$E^\circ_{cell} = +0.513$ V

(2) Using the Nernst equation: $E_{cell} = E^\circ_{cell} - \dfrac{0.0592}{n} \log \dfrac{[Zn^{2+}]}{[Ni^{2+}]}$

Substituting, $0 = 0.513 \text{ V} - \dfrac{0.0592}{2} \log \dfrac{[Zn^{2+}]}{[Ni^{2+}]}$

$$\log \frac{[Zn^{2+}]}{[Ni^{2+}]} = 17.33 \text{ or } 17.3 \text{ (3 significant figures)}$$

$$\frac{[Zn^{2+}]}{[Ni^{2+}]} = \mathbf{2 \times 10^{17}} \text{ (1 significant figures)}$$

(b) Since the cell starts at standard conditions, $[Ni^{2+}]_{initial} = [Zn^{2+}]_{initial} = 1.00\ M$. Also, for every 1 mole of Zn^{2+} produced, there is 1 mole of Ni^{2+} lost. Therefore,

Let $x = mol/L$ of Ni^{2+} that reacted. Then, $x = mol/L$ of Zn^{2+} that were produced.

	Ni^{2+}	$+$	Zn	$\rightarrow$	Ni	$+$	Zn^{2+}
initial	1.00 M		-		-		1.00 M
change	- x M						+ x M
after reaction	(1.00 - x) M						(1.00 + x) M

Therefore, $[Zn^{2+}] + [Ni^{2+}] = (1.00 + x)\ M + (1.00 - x)\ M = 2.00\ M$

We know from (a) that $[Zn^{2+}] = (2 \times 10^{17})[Ni^{2+}]$. Substituting for $[Zn^{2+}]$ and solving for $[Ni^{2+}]$,

$$2.00\ M = (2 \times 10^{17})[Ni^{2+}] + [Ni^{2+}]$$
$$= [Ni^{2+}](2 \times 10^{17} + 1) \approx (2 \times 10^{17})[Ni^{2+}]$$
$$[Ni^{2+}] = 1 \times 10^{-17}\ M$$
$$[Zn^{2+}] = 2.00\ M - [Ni^{2+}] = \mathbf{2.00\ M}\ \text{(to 3 significant figures)}$$

21-82. *Refer to Section 21-21.*

Because $\Delta G° = -nFE°_{cell}$ and $\Delta G° = -RT \ln K$, the signs and magnitudes of $E°_{cell}$, $\Delta G°$ and K are related as shown in the following table for different types of reactions under standard state conditions.

Forward Reaction	$E°_{cell}$	$\Delta G°$	K
spontaneous	+	-	>1
at equilibrium	0	0	1
non-spontaneous	-	+	<1

From the above equations, it is seen that the value of K is related to the value of $\Delta G°$ and $E°$ of the cell, but not ΔG and E of the cell. $E°$, $\Delta G°$ and K are indicators of the thermodynamic tendency of an oxidation-reduction reaction to occur under standard conditions.

On the other hand, E and ΔG are related to the value of Q and are indicators of the spontaneity of a reaction under any given conditions. The reaction proceeds until $Q = K$ at which point $\Delta G = 0$ and $E_{cell} = 0$. Then:

$$\log K = \frac{nFE°_{cell}}{2.303RT}$$

21-84. *Refer to Section 21-21 and Examples 21-12, 21-13 and 21-14.*

			$E°$
(a) reduction:	$Sn^{4+} + 2e^- \rightarrow Sn^{2+}$		+0.15 V
oxidation:	$2(Fe^{2+} \rightarrow Fe^{3+} + e^-)$		-0.771 V
cell reaction:	$Sn^{4+} + 2Fe^{2+} \rightarrow Sn^{2+} + 2Fe^{3+}$		$E°_{cell} = \mathbf{-0.62\ V}$

The reaction is not spontaneous as written under standard conditions since $E°_{cell} < 0$.

$$\Delta G° = -nFE°_{cell} = -(2\ \text{mol}\ e^-)(96500\ \text{J/V·mol}\ e^-)(-0.62\ V) = \mathbf{1.2 \times 10^5\ J/mol\ rxn}\ \text{or}\ \mathbf{120\ kJ/mol\ rxn}$$

$$\text{at } 25°C,\ E°_{cell} = \frac{2.303RT \log K}{nF} = \frac{(2.303)(8.314\ \text{J/mol·K})(298.15\ K)}{n(96500\ \text{J/V·mol}\ e^-)} \log K = \frac{0.0592}{n} \log K$$

$$\text{Therefore, } \log K = \frac{nE°_{cell}}{0.0592} = \frac{(2)(-0.62)}{0.0592} = -21 \qquad \text{Solving, } K = 10^{-21}$$

(b)	reduction:	$Cu^+ + e^- \rightarrow Cu$	$E°$ +0.521 V
	oxidation:	$Cu^+ \rightarrow Cu^{2+} + e^-$	-0.153 V
	cell reaction:	$2Cu^+ \rightarrow Cu + Cu^{2+}$	$E°_{cell} = +0.368$ V

The reaction is spontaneous as written under standard conditions since $E°_{cell} > 0$.

$$\Delta G° = -nFE°_{cell} = -(1 \text{ mol } e^-)(96500 \text{ J/V·mol } e^-)(0.368 \text{ V}) = -3.55 \times 10^4 \text{ J/mol rxn or } -35.5 \text{ kJ/mol rxn}$$

at 25°C, $\log K = \dfrac{nE°_{cell}}{0.0592} = \dfrac{(1)(0.368)}{0.0592} = 6.22$ \qquad Solving, $K = 10^{6.22}$ or 1.6×10^6

(c)	reduction:	$2(MnO_4^- + 2H_2O + 3e^- \rightarrow MnO_2 + 4OH^-)$	$E°$ +0.588 V
	oxidation:	$3(Zn + 2OH^- \rightarrow Zn(OH)_2 + 2e^-)$	+1.245 V
	cell reaction:	$2MnO_4^- + 3Zn + 4H_2O \rightarrow 2MnO_2 + 3Zn(OH)_2 + 2OH^-$	$E°_{cell} = +1.833$ V

The reaction is spontaneous as written under standard conditions since $E°_{cell} > 0$.

$$\Delta G° = -nFE°_{cell} = -(6 \text{ mol } e^-)(96500 \text{ J/V·mol } e^-)(1.833 \text{ V}) = -1.061 \times 10^6 \text{ J/mol rxn or } -1061 \text{ kJ/mol rxn}$$

at 25°C, $\log K = \dfrac{nE°_{cell}}{0.0592} = \dfrac{(6)(1.833)}{0.0592} = 186$ \qquad Solving, $K = 10^{186}$

21-86. *Refer to Section 21-21, Examples 21-13 and 21-14, and Appendix J.*

		$E°$
reduction half-reaction:	$2H_2O + 2e^- \rightarrow 2OH^- + H_2$	-0.8277 V
oxidation half-reaction:	$2(K \rightarrow K^+ + e^-)$	+2.925 V
cell reaction:	$2K + 2H_2O \rightarrow 2K^+ + 2OH^- + H_2$	$E°_{cell} = +2.097$ V

Note: Since the $E°$ value has 4 significant figures, we must use values for the constants (F, R and T) that also have at least 4 significant figures.

at 25°C, $\log K = \dfrac{nFE°_{cell}}{2.303RT} = \dfrac{(2 \text{ mol})(96487 \text{ J/V·mol})(+2.097 \text{ V})}{(2.303)(8.314 \text{ J/mol·K})(298.15 \text{ K})} = 70.89$ \quad Solving, $K = 10^{70.89}$ or 7.7×10^{70}

21-88. *Refer to Section 21-21 and Examples 21-13 and 21-14.*

		$E°$
reduction half-reaction:	$PbSO_4 + 2e^- \rightarrow Pb + SO_4^{2-}$	-0.356 V
oxidation half-reaction:	$Pb + 2I^- \rightarrow PbI_2 + 2e^-)$	+0.365 V
cell reaction:	$PbSO_4 + 2I^- \rightarrow PbI_2 + SO_4^{2-}$	$E°_{cell} = +0.009$ V

$\ln K = \dfrac{nFE°_{cell}}{RT} = \dfrac{(2 \text{ mol})(96500 \text{ J/V·mol})(+0.009 \text{ V})}{(8.314 \text{ J/mol·K})(298 \text{ K})} = 0.7$ \qquad Solving, $K = 2$ (to 1 significant figure)

21-90. *Refer to Section 21-21 and Example 21-12.*

Balanced half-reaction: \qquad $\frac{1}{2} H_2O_2 + H^+ + e^- \rightarrow H_2O$ \hfill $E° = +1.77$ V

$$\Delta G° = -nFE° = -(1 \text{ mol})(96500 \text{ J/V·mol})(1.77 \text{ V}) = -1.71 \times 10^5 \text{ J/mol rxn or } -171 \text{ kJ/mol rxn}$$

(a) The dry cell (Leclanchè cell) is shown in Figure 21-16b. The container is made of zinc, which also acts as one of the electrodes. The other electrode is a carbon rod in the center of the cell. The cell is filled with a moist mixture of NH_4Cl, MnO_2, $ZnCl_2$ and a porous inert filler. The cell is separated from the zinc container by a porous paper. Dry cells are sealed to keep moisture from evaporating. As the cell operates, the Zn electrode is the anode and is oxidized to Zn^{2+} ions. The ammonium ion is reduced to give NH_3 and H_2 at the carbon cathode. The ammonia produced combines with Zn^{2+} ion and forms a soluble compound containing the complex ion, $Zn(NH_3)_4^{2+}$; H_2 is removed by being oxidized by MnO_2. This type of battery cannot be recharged.

(b) The lead storage battery is shown in Figure 21-17. It consists of a group of lead plates bearing compressed spongy lead alternating with a group of lead plates bearing lead(IV) oxide, PbO_2. The electrodes are immersed in a solution of about 40% sulfuric acid. When the cell discharges, the spongy lead is oxidized to give Pb^{2+} ions which then combine with sulfate ions to form insoluble $PbSO_4$, coating the anode. Electrons produced at the anode by oxidation of spongy lead travel through the external circuit to the cathode and reduce lead(IV) to lead(II) in the presence of H^+. The cathode also becomes coated with insoluble lead sulfate. The lead storage battery can be recharged by reversal of all reactions.

(c) The hydrogen-oxygen fuel cell is shown in Figure 21-18. Hydrogen (the fuel) is supplied to the anode compartment. Oxygen is fed into the cathode compartment. Oxygen is reduced at the cathode to OH^- ions. The OH^- ions migrate through the electrolyte, an aqueous solution of a base, to the anode, where H_2 is oxidized to H_2O. The net reaction of the cell is the same as the burning of hydrogen in oxygen to form water, but combustion does not occur. Rather, most of the chemical energy, produced from the destruction of H-H and O-O bonds and the formation of O-H bonds, is converted directly into electrical energy.

(a) When attempting to recharge an Leclanchè cell (a dry cell), the electrodes are reversed; the zinc container which is the anode under normal operation becomes the cathode. The reaction expected is the reduction of Zn^{2+} to zinc metal:

$$Zn^{2+} + 2e^- \rightarrow Zn \qquad\qquad E° = -0.763 \text{ V}$$

(b) Recharging the battery means reversing the actual cell reaction to yield:

$$H_2 + 2NH_3 + Zn^{2+} \rightarrow Zn + 2NH_4^+ \qquad E_{cell} = -1.6 \text{ V}$$

This is essentially an impossible task because each of the original products have been permanently removed from the system.

(1) NH_3 and Zn^{2+} have reacted together to give a very stable zinc-ammonia complex:

$$Zn^{2+} + 4NH_3 \rightarrow Zn(NH_3)_4^{2+} \qquad\qquad K = \frac{1}{K_d} = 2.9 \times 10^9 \text{ (Appendix I)}$$

The zinc complex is more difficult than the free Zn^{2+} to reduce as deduced from the more negative standard reduction potential:

$$Zn(NH_3)_4^{2+} + 2e^- \rightarrow Zn + 4NH_3. \qquad E° = -1.04 \text{ V}$$

(2) H_2 has reacted with MnO_2 to give the solid, $MnO(OH)$: $H_2 + 2MnO_2 \rightarrow 2MnO(OH)$.

Since there are essentially none of the original products, the recharging cannot occur.

A fuel cell is different from a dry cell or storage cell because:

(1) the reactant, the fuel (usually H_2) and oxygen are fed into the cell continuously and the products are constantly removed. Hence, the fuel cell creates chemical energy, but does not store it. It can operate indefinitely as long as fuel is available.

(2) The electrodes are made of an inert material such as platinum and do not react during the electrochemical process.

(3) Many fuel cells are non-polluting; , e.g., the H_2/O_2 fuel cell whose only product is H_2O.

21-98. *Refer to Sections 21-16, 21-20 and 21-6.*

Consider: $Zn(s)/Zn^{2+}(aq)||Fe^{3+}(aq)/Fe(s)$

(a) oxidation half reaction (at anode): $3(Zn(s) \rightarrow Zn^{2+}(aq) + 2e^-)$
     reduction half-reaction (at cathode): $2(Fe^{3+}(aq) + 3e^- \rightarrow Fe(s))$
     overall cell reaction: $3Zn(s) + 2Fe^{3+}(aq) \rightarrow 3Zn^{2+}(aq) + 2Fe(s)$

(b) $E^\circ_{cell} = E^\circ_{anode} + E^\circ_{cathode} = (+0.763 \text{ V}) + (-0.036 \text{ V}) = \textbf{+0.727 V}$

(c) $E = E^\circ - \dfrac{0.0592}{n} \log \dfrac{[Zn^{2+}]^3}{[Fe^{2+}]^2} = +0.727 \text{ V} - \dfrac{0.0592}{6} \log \dfrac{(1.00 \times 10^{-3})^3}{(10.0)^2} = +0.727 \text{ V} - (-0.109 \text{ V}) = \textbf{+0.836 V}$

(d) The minimum mass that the zinc electrode can have is the mass of Zn lost when 150 mA passes through the cell for 15 minutes.

$$? \text{ g Zn} = 15 \text{ min} \times \frac{60 \text{ s}}{1 \text{ min}} \times \frac{0.150 \text{ C}}{1 \text{ s}} \times \frac{1 \text{ mol } e^-}{96500 \text{ C}} \times \frac{1 \text{ mol Zn}}{2 \text{ mol } e^-} \times \frac{65.39 \text{ g}}{1 \text{ mol Zn}} = \textbf{0.0457 g Zn}$$

21-100. *Refer to Section 21-6.*

Balanced half-reaction: $Cu^{2+} + 2e^- \rightarrow Cu$

$$? \text{ coulombs} = 0.025 \text{ L soln} \times \frac{0.175 \text{ mol Cu}}{1 \text{ L soln}} \times \frac{2 \text{ mol } e^-}{1 \text{ mol Cu}} \times \frac{96500 \text{ C}}{1 \text{ mol } e^-} = \textbf{840 C}$$

21-102. *Refer to Section 21-6.*

Balanced equations: $UO_2(s) + 4HF(g) \rightarrow UF_4(s) + 2H_2O(\ell)$
                     $UF_4(s) + 2Mg(s) \rightarrow U(s) + 2MgF_2(s)$

(a) ox. no. U in $UO_2(s)$: +4

(b) ox. no. U in $UF_4(s)$: +4

(c) ox. no. U in $U(s)$: 0

(d) reducing agent: $Mg(s)$

(e) substance reduced (oxidizing agent): $UF_4(s)$

(f) U is being reduced from +4 oxidation number in $UF_4(s)$ to 0 in $U(s)$. Therefore, 4 moles of electrons are required to reduce 1 mole of $UF_4(s)$.

$$? \text{ coulombs/s} = \frac{0.500 \text{ g UF}_4}{1 \text{ min}} \times \frac{1 \text{ min}}{60 \text{ s}} \times \frac{1 \text{ mol UF}_4}{314 \text{ g UF}_4} \times \frac{4 \text{ mol } e^-}{1 \text{ mol UF}_4} \times \frac{96500 \text{ C}}{1 \text{ mol } e^-} = \textbf{10.2 C/s or 10.2 A}$$

(g) Plan: (1) Determine the number of moles of HF(g).
  (2) Calculate the volume of HF(g) using the ideal gas law, $PV = nRT$.

(1) $? \text{ mol HF} = 0.500 \text{ g U} \times \dfrac{1 \text{ mol U}}{238 \text{ g U}} \times \dfrac{1 \text{ mol UF}_4}{1 \text{ mol U}} \times \dfrac{4 \text{ mol HF}}{1 \text{ mol UF}_4} = 8.40 \times 10^{-3} \text{ mol HF}$

(2) $V = \dfrac{nRT}{P} = \dfrac{(8.40 \times 10^{-3} \text{ mol})(0.0821 \text{ L·atm/mol·K})(298 \text{ K})}{(10.0 \text{ atm})} = \mathbf{0.0206 \text{ L HF}(g)}$

(h) Plan: Determine the mass of U that can be prepared from 0.500 g Mg and compare.

$? \text{ g U} = 0.500 \text{ g Mg} \times \dfrac{1 \text{ mol Mg}}{24.30 \text{ g Mg}} \times \dfrac{1 \text{ mol U}}{2 \text{ mol Mg}} \times \dfrac{238.0 \text{ g U}}{1 \text{ mol U}} = 2.45 \text{ g U}$

Yes, 0.500 g Mg is more than enough to prepare 0.50 g U. In fact, 0.500 g Mg can ideally produce 2.45 g U.

21-104. *Refer to Section 21-8 and Figures 21-5 and 21-6.*

(a) Electroplating is a process that plates metal onto a cathodic surface by electrolysis.

(b) A simple silver electroplating apparatus for a jeweler consists of a dc generator (a battery) with the negative lead attached to the piece of jewelry (cathode) and the positive lead attached to a piece of silver metal (anode). The jewelry and the silver metal are both immersed in a beaker containing an aqueous solution of a silver salt such as AgNO$_3$. During electroplating, the Ag metal at the anode will be oxidized to Ag$^+$ ions, and the Ag$^+$ ions in solution will be reduced to Ag metal and plated onto the jewelry at the cathode.

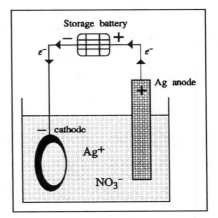

(c) Highly purified silver as the anode is not necessary in an electroplating operation. As the electrolytic cell operates, Ag and other metals from an impure Ag anode oxidize to form metal cations in solution. However, only Ag$^+$ ions are reduced to Ag metal at the cathode because of its ease of reduction and higher concentration. Such a preference can be enhanced by controlling the operating voltage just above the threshold required to electroplate silver. Hence, the extra cost of highly purified silver would make its purchase an unwise decision.

21-106. *Refer to Sections 21-16 and 21-17.*

(a) Given:

			$E°$
(1) H$_2$O/H$_2$,OH$^-$	2H$_2$O + 2e^- $\rightarrow$ H$_2$ + 2OH$^-$		-0.828 V
(2) H$^+$/H$_2$	2H$^+$ + 2e^- $\rightarrow$ H$_2$		0.000 V
(3) O$_2$,H$^+$/H$_2$O	O$_2$ + 4H$^+$ + 4e^- $\rightarrow$ 2H$_2$O		$+1.229$ V
(4) O$_2$,H$_2$O/OH$^-$	O$_2$ + 2H$_2$O + 4e^- $\rightarrow$ 4OH$^-$		$+0.401$ V

Combining half-reactions (1) and (3) would give the greatest voltage:

$E°_{\text{cell}} = E°_{\text{cathode}} + E°_{\text{anode}} = +1.229 \text{ V} + 0.828 \text{ V} = 2.057 \text{ V}$

(b) reduction at cathode: O$_2$ + 4H$^+$ + 4e^- $\rightarrow$ 2H$_2$O
 oxidation at anode: 2(H$_2$ + 2OH$^-$ $\rightarrow$ 2H$_2$O + 2e^-)
 ───
 cell reaction: O$_2$ + 2H$_2$ + 4H$^+$ + 4OH$^-$ $\rightarrow$ 6H$_2$O
   or O$_2$ + 2H$_2$ + 4H$_2$O $\rightarrow$ 6H$_2$O
   or O$_2$ + 2H$_2$ $\rightarrow$ 2H$_2$O

21-108. *Refer to Section 21-21 and Appendix J.*

(a) Plan: The K_{sp} value for $AgBr(s)$ is the equilibrium constant for: $AgBr(s) \rightleftarrows Ag^+(aq) + Br^-(aq)$. It can be estimated from data in Appendix J. Choose the appropriate oxidation and reduction half-reactions that produce the above reaction and calculate E°_{cell} and K_{sp} at 25°C.

		E°
reduction half-reaction:	$AgBr(s) + e^- \rightarrow Ag(s) + Br^-(aq)$	+0.10 V
oxidation half-reaction:	$Ag(s) \rightarrow Ag^+(aq) + e^-$	-0.7994 V
cell reaction:	$AgBr(s) \rightarrow Ag^+(aq) + Br^-(aq)$	E°_{cell} = -0.70 V

$$\log K = \log K_{sp} = \frac{nFE^\circ_{cell}}{2.303RT} = \frac{(1\text{ mol})(96500\text{ J/V·mol})(-0.70\text{ V})}{(2.303)(8.314\text{ J/mol·K})(298\text{ K})} = -12$$

Solving, K_{sp} = **10^{-12}**

(From Appendix H, K_{sp} for AgBr = 3.3×10^{-13})

(b) $\Delta G^\circ = -nFE^\circ_{cell}$ = -(1 mol)(96500 1/V·mol)(-0.70 V) = **+68,000 J/mol rxn** or **+68 kJ/mol rxn**

21-110. *Refer to Section 21-21.*

Balanced equations: (i) $\frac{1}{3}Al^{3+} + 3e^- \rightarrow \frac{1}{3}Al$ $\qquad \Delta G^\circ$ = 160.4 kJ/mol rxn

(ii) $Al^{3+} + 3e^- \rightarrow Al$ $\qquad \Delta G^\circ$ = 481.2 kJ/mol rxn

(i) $$E^\circ = -\frac{\Delta G^\circ}{nF} = \frac{(+160400\text{ J})}{(1\text{ mol }e^-)(96487\text{ J/V·mol }e^-)} = \mathbf{-1.662\text{ V}}$$

(ii) $$E^\circ = -\frac{\Delta G^\circ}{nF} = \frac{(+481200\text{ J})}{(3\text{ mol }e^-)(96487\text{ J/V·mol }e^-)} = \mathbf{-1.662\text{ V}}$$

22 Metals I: Metallurgy

22-2. *Refer to Section 22-1.*

Metallurgy is the commercial extraction of metals from their ores and the preparation of metals for use. It includes

(1) mining the ore,
(2) pretreatment of the ore,
(3) reduction of the ore to the free metal,
(4) refining or purifying the metal, and
(5) alloying, if needed.

22-4. *Refer to Section 22-1, Table 22-1 and Figure 22-1.*

Anion Name	Formula	Example	Mineral Name
oxide	O^{2-}	Fe_2O_3	hematite
sulfide	S^{2-}	Cu_2S	chalcocite
chloride	Cl^-	$NaCl$	halite (rock salt)
carbonate	CO_3^{2-}	$CaCO_3$	limestone
sulfate	SO_4^{2-}	$BaSO_4$	barite
silicate	Si_xO_y	$Al_2(Si_2O_8)(OH)_4$	kaolinite

The silicates are the most widespread minerals. However, extraction of metals from silicates is very difficult.

22-6. *Refer to Section 22-2 and Figure 22-2.*

High density sulfide ores can be separated from the less dense gangue after pulverization by several methods. One way involves a cyclone separator in which the lighter impurities are blown away.

22-8. *Refer to Section 22-2.*

The flotation method of separating a crushed ore from the gangue is a physical separation method used with ores, e.g., sulfides, carbonates or silicates, which either are not "wet" by water or can be made water repellent by treatment. Their surfaces are covered by layers of oil or other flotation agents. A stream of air is blown through a swirled suspension of such an ore in a mixture of water and oil; bubbles form on the oil surfaces of the mineral particles, causing them to rise to the surface of the suspension. The bubbles are prevented from breaking and escaping by a layer of oil and emulsifying agent. A frothy ore concentrate forms on the surface. No chemical changes are involved.

22-10. *Refer to Section 22-3 and Table 22-2.*

Aluminum and the metals of Groups IA and IIA are metals which are easily oxidized to ions that are difficult to reduce. So, we predict that electrolysis would be required to obtain the free metals from the molten, anhydrous salts, KCl, Al_2O_3 and $MgSO_4$.

For the electrolysis of a brine solution:

oxidation half-reaction:	$2Cl^- \rightarrow Cl_2 + 2e^-$
reduction half-reaction:	$2e^- + 2H_2O \rightarrow 2OH^- + H_2$
balanced net ionic equation:	$2Cl^- + 2H_2O \rightarrow 2OH^- + H_2 + Cl_2$
formula unit equation:	$2NaCl + 2H_2O \rightarrow 2NaOH + H_2 + Cl_2$ (Na$^+$ is the spectator ion)

If 1 mole of electrons, i.e., 1 faraday, passes through the cell at 100% efficiency, then 1 mole of NaOH, 1/2 mole of H_2 and 1/2 mole of Cl_2 are produced.

Therefore,   ? g NaOH produced = 1 mol = **39.997 g NaOH**

? g H_2 produced = 0.5 mol = **1.008 g H_2**

? g Cl_2 produced = 0.5 mol = **35.45 g Cl_2**

22-14. *Refer to Table 22-3.*

(a) $2Al_2O_3$ (cryolite solution) $\xrightarrow{\text{electrolysis}}$ $4Al(\ell) + 3O_2(g)$ involves reduction of Al^{3+} to elemental Al

(b) $PbSO_4(s) + PbS(s) \rightarrow 2Pb(\ell) + 2SO_2(g)$ involves reduction of Pb^{2+} to elemental Pb

(c) $2TaCl_5(g) + 5Mg(\ell) \rightarrow 2Ta(s) + 5MgCl_2(\ell)$ involves reduction of Ta^{5+} to elemental Ta

---

22-16. *Refer to Section 22-7.*

The basic oxygen furnace is used to purify pig iron, which is the iron obtained from the blast furnace process. It is impure and contains carbon, among other substances, but it can be converted to steel by burning out most of the carbon with oxygen in a basic oxygen furnace. The method involves blowing oxygen through the molten iron at high temperatures. The carbon is converted to carbon monoxide and finally to carbon dioxide.

22-18. *Refer to Sections 22-7 and 22-9.*

(a) The procedure for obtaining Fe from Fe_2O_3 or Fe_3O_4 is as follows:

(1) The oxides are reduced in blast furnaces by CO. First, coke (C), limestone ($CaCO_3$) and the crushed ore (Fe_2O_3 or Fe_3O_4 in very hard SiO_2 rock) are loaded into the top of the blast furnace.

(2) Most of the oxides are reduced to molten iron by CO, although some are reduced by coke directly. Carbon dioxide, a reaction product, reacts with excess coke to provide more CO to reduce the next charge of iron ore.

(3) The obtained product contains C as an impurity and is called pig iron. It can be remelted and cooled into cast iron. Alternatively, if some C is removed and other metals, such as Mn, Cr, Ni, W, Mo and V, are added to increase the tensile strength, the mixture is known as steel.

(b) The procedure for obtaining Au from very low grade ores by the cyanide process is as follows:

(1) The ore containing native Au is mixed with a solution of NaCN and converted to an aqueous slurry.

(2) Air is bubbled through the agitated slurry to oxidize the gold metal to a water soluble complex ion, $[Au(CN)_2]^-$.

(3) Free gold can then be regenerated by reduction of $[Au(CN)_2]^-$ with zinc or by electrolytic reduction.

Fe-containing minerals:

Fe_2O_3	hematite	oxidation number of Fe = +3
Fe_3O_4	magnetite	oxidation number of Fe = +8/3
FeS_2	iron pyrite	oxidation number of Fe = +2
$CuFeS_2$	chalcopyrite	oxidation number of Fe = +2

22-22. *Refer to Section 21-8 and Figure 21-5.*

Impure metallic Cu obtained from the chemical reduction of Cu_2S and CuS can be refined with the following arrangement.

(1) Thin sheets of very pure Cu are made cathodes by connecting them to the negative terminal of a d.c. generator. Impure chunks of copper connected to the positive terminal function as anodes. The electrodes are immersed in a solution of $CuSO_4$ and H_2SO_4.

(2) When the cell operates, Cu from impure anodes is oxidized and goes into solution as Cu^{2+} ions; Cu^{2+} ions from the solution are reduced and plate out as metallic Cu on the pure Cu cathode.

22-24. *Refer to Sections 22-7, 15-8 and 15-15.*

Balanced equation: $FeO(s) + CO(g) \rightarrow Fe(s) + CO_2(g)$

(a) $\Delta H°_{800} = [\Delta H°_f\ _{Fe(s)} + \Delta H°_f\ _{CO_2(g)}] - [\Delta H°_f\ _{FeO(s)} + \Delta H°_f\ _{CO(g)}]$

$= [(0\ kJ) + (-394\ kJ)] - [(-268\ kJ) + (-111\ kJ)]$

$= $ **-15 kJ/mol rxn**

Yes, this is a favorable enthalpy change, because the value is negative, implying an exothermic reaction.

(b) $\Delta G°_{800} = [\Delta G°_f\ _{Fe(s)} + \Delta G°_f\ _{CO_2(g)}] - [\Delta G°_f\ _{FeO(s)} + \Delta G°_f\ _{CO(g)}]$

$= [(0\ kJ) + (-396\ kJ)] - [(-219\ kJ) + -182\ kJ)]$

$= $ **+5 kJ/mol rxn**

No, this is not a spontaneous reaction since $\Delta G > 0$.

(c) Recall the Gibbs-Helmholtz equation: $\Delta G = \Delta H - T\Delta S$

Therefore, $\Delta S°_{800} = \dfrac{\Delta H°_{800} - \Delta G°_{800}}{T} = \dfrac{(-15\ kJ) - (+5\ kJ)}{800\ K} = $ **-0.025 kJ/(mol rxn)·K** or **-25 J/(mol rxn)·K**

22-26. *Refer to Sections 22-8 and 4-7.*

(a) Balanced equation: $\overset{+1\ -2}{2Cu_2S(\ell)} + \overset{0}{3O_2(g)} \rightarrow \overset{+1\ -2}{2Cu_2O(\ell)} + \overset{+4\ -2}{2SO_2(g)}$

oxidizing agent: O_2 (O is being reduced)
reducing agent: Cu_2S (S is being oxidized)

In the balanced equation, the total increase in oxidation number equals the total decease in oxidation number. This is true in this example:

increase in oxidation number = $|[2 \times$ ox. no. S in $SO_2] - [2 \times$ ox. no. S in $Cu_2S]|$

$= |[2 \times (+4)] - [2 \times (-2)]|$

$= 12$

decrease in oxidation number = $|[2 \times$ ox no. O in $Cu_2O + 4 \times$ ox. no. O in $SO_2] - [6 \times$ ox. no. O in $O_2]|$

$= |[2 \times (-2) + 4 \times (-2)] - [6 \times 0]|$

$= 12$

(b) Balanced equation: $\overset{+1\ -2}{2Cu_2O(\ell)} + \overset{+1\ -2}{Cu_2S(\ell)} \rightarrow \overset{0}{6Cu(\ell)} + \overset{+4\ -2}{SO_2(g)}$

oxidizing agent: Cu_2O (Cu is being reduced)
                   Cu_2S (Cu is being reduced)

reducing agent: Cu_2S (S is being oxidized)

The total increase in oxidation number equals the total decrease in oxidation number, so the reaction is balanced.

increase in oxidation number $= |[1 \times$ ox. no. S in $SO_2] - [1 \times$ ox. no. S in $Cu_2S]|$
$= |[1 \times (+4)] - [1 \times (-2)]|$
$= 6$

decrease in oxidation number $= |[6 \times$ ox. no. Cu in free Cu]
$- [4 \times$ ox. no. Cu in $Cu_2O + 2 \times$ ox. no. Cu in $Cu_2S]|$
$= |[6 \times 0] - [4 \times (+1) + 2 \times (+1)]|$
$= 6$

22-28. *Refer to Section 21-6 and Example 21-1.*

Balanced half-reaction: $Cu^{2+} + 2e^- \rightarrow Cu$

? g Cu $= 5.00$ hr $\times \dfrac{3600\text{ s}}{1\text{ hr}} \times \dfrac{3.00\text{ C}}{1\text{ s}} \times \dfrac{1\text{ mol }e^-}{96500\text{ C}} \times \dfrac{1\text{ mol Cu}}{2\text{ mol }e^-} \times \dfrac{63.55\text{ g Cu}}{1\text{ mol Cu}} = $ **17.8 g Cu**

22-30. *Refer to Sections 22-6 and 21-6, and Exercise 14a.*

Plan: (1) Calculate the time necessary to convert all of the aluminum in 45 pounds of Al_2O_3 to aluminum metal.
        (2) Calculate the number of moles of oxygen in 45 pounds of Al_2O_3 and use the ideal gas law, $PV = nRT$, to determine the volume of O_2 gas produced.

(1) Balanced reduction half-reaction: $Al^{3+} + 3e^- \rightarrow Al$

? time (s) $= 45.0$ lb $Al_2O_3 \times \dfrac{454\text{ g }Al_2O_3}{1.00\text{ lb }Al_2O_3} \times \dfrac{1\text{ mol }Al_2O_3}{102\text{ g }Al_2O_3} \times \dfrac{2\text{ mol Al}}{1\text{ mol }Al_2O_3} \times \dfrac{3\text{ mol }e^-}{1\text{ mol Al}} \times \dfrac{96500\text{ C}}{1\text{ mol }e^-} \times \dfrac{1\text{ s}}{0.900\text{ C}}$
$= \mathbf{1.29 \times 10^8}$ **s** or **4.09 years**

(2) Let us assume that all of the oxygen in Al_2O_3 is converted to $O_2(g)$.

? mol $O_2 = 45.0$ lb $Al_2O_3 \times \dfrac{454\text{ g }Al_2O_3}{1.00\text{ lb }Al_2O_3} \times \dfrac{1\text{ mol }Al_2O_3}{102\text{ g }Al_2O_3} \times \dfrac{3\text{ mol O}}{1\text{ mol }Al_2O_3} \times \dfrac{1\text{ mol }O_2}{2\text{ mol O}} = 3.00 \times 10^2$ mol O_2

$V = \dfrac{nRT}{P} = \dfrac{(3.00 \times 10^2\text{ mol})(0.0821\text{ L·atm/mol·K})(125°C + 273°)}{(785/760)\text{ atm}} = \mathbf{9.49 \times 10^3\text{ L } O_2}$

22-32. *Refer to Sections 22-3, 22-6 and 22-7, Table 22-3 and Appendix K.*

The data from Appendix K gives:

Compound	ΔH_f°		Extractive Metallurgy Method
$HgS(s)$	-58.2	kJ/mol	heating HgS
$Fe_2O_3(s)$	-824.2	kJ/mol	chemical reduction by CO
$Al_2O_3(s)$	-1676	kJ/mol	electrolysis of molten Al_2O_3

As the heats of formation of minerals become more exothermic, i.e., more negative, their thermodynamic stability increases. And so the difficulty by which free metals can be extracted from the minerals also increases. In other words, the more active is the metal, the easier it is to form compounds and the more difficult it is to retrieve the metal from its compounds. This relationship is obvious in the methods by which the metals are removed from their mineral matrix as shown in the third column of the above table: heating is a less severe metallurgic process, whereas electrolysis is a more severe method.

22-34. *Refer to Section 3-6.*

$$? \text{ tons seawater} = 1.0 \text{ ton Mg} \times \frac{100 \text{ tons seawater}}{0.13 \text{ ton Mg}} \times \frac{100}{82} = \mathbf{9.4 \times 10^2 \text{ tons seawater}}$$

23 Metals II: Properties and Reactions

23-2. *Refer to the Introduction to Chapter 23 and Section 6-1.*

The representative elements have valence electrons in s or s and p orbitals in the outermost occupied energy level, whereas the d-transition metals must have a partially filled set of d orbitals.

23-4. *Refer to Chapter 6.*

Metals are located at the left side of the periodic table and therefore, in comparison with nonmetals, have (a) fewer outer shell electrons, (b) lower electronegativities, (c) more negative standard reduction potentials and (d) less endothermic ionization energies.

23-6. *Refer to Sections 23-1 and 23-4, and Tables 23-1 and 23-3.*

(a) Alkali metals are larger than alkaline earth metals in the same period due to the increased effective nuclear charge of the alkaline earth metals.

(b) Alkaline earth metals have higher densities since they are both heavier and smaller than alkali metals of the same period.

(c) Alkali metals have lower first ionization energies than alkaline earth metals of the same period due to both the increased effective nuclear charge and decreased size of the alkaline earth metals.

(d) Alkali metals have much higher second ionization energies than alkaline earth metals of the same period. This is because removal of a second electron from an alkali metal ion involves destroying the very stable noble gas electronic configuration of the ion whereas removal of a second electron from an alkaline earth metal ion involves creating a stable noble gas configuration.

23-8. *Refer to Section 23-4.*

physical properties: Alkaline earth metals are silvery-white, malleable, ductile metals, somewhat harder than alkali metals, and are excellent electrical and thermal conductors.

chemical properties: Alkaline earth metals are easily oxidized and thus are strong reducing agents. They are not as reactive as IA metals, but are too reactive to occur as free elements in nature. Alkaline earths are characterized by the loss of 2 electrons per metal atom and form basic metal oxides (except BeO) which react with water to produce hydroxides.

23-10. *Refer to Sections 5-17 and 7-2, and Appendix B.*

(a) Mg [Ne] ↑↓ $\overline{}$ 3s   (b) Mg^{2+} [Ne]   (c) Na [Ne] ↑ $\overline{}$ 3s   (d) Na$^+$ [Ne]

(e) Sn [Kr] ↑↓ ↑↓ ↑↓ ↑↓ ↑↓ $\overline{}$ 4d   ↑↓ $\overline{}$ 5s   ↑ ↑ __ $\overline{}$ 5p

(f) Sn^{2+} [Kr] ↑↓ ↑↓ ↑↓ ↑↓ ↑↓ $\overline{}$ 4d   ↑↓ $\overline{}$ 5s

(g) Sn^{4+} [Kr] ↑↓ ↑↓ ↑↓ ↑↓ ↑↓ $\overline{}$ 4d

23-12. *Refer to Sections 23-1 and 23-4.*

The alkali metals (Group IA) and the alkaline earth metals (Group IIA) are not found free in nature because they are so easily oxidized. Their primary sources are seawater, brines of their soluble salts and deposits of sea salt. The metals are obtained from the electrolysis of their molten salts.

23-14. *Refer to Sections 23-1, 23-2, 23-4 and 23-5, and Tables 23-1 and 23-3.*

The metals in Group IA and Group IIA have standard reduction potentials that are more negative than that for H_2. Therefore, the alkali metals and alkaline earth metals are stronger reducing agents than H_2. They should be above H_2 in the activity series. Consequently, they all react with acids by reducing acidic H^+ ions to H_2, while they are oxidized to +1 (Group IA) or +2 (Group IIA) ions. All but Be reduce H_2O to H_2 gas.

23-16. *Refer to Section 23-6.*

(a) Calcium metal is used (1) as a reducing agent in the metallurgy of U, Th and other metals, (2) as a scavenger to remove dissolved impurities in molten metals and residual gases in vacuum tubes, and (3) as a component in many alloys. Slaked lime, $Ca(OH)_2$, is a cheap base used in industry and is also a major component of mortar and lime plaster. Careful heating of gypsum, $CaSO_4 \cdot 2H_2O$, produces plaster of Paris, $2CaSO_4 \cdot H_2O$.

(b) Magnesium metal is used (1) in photographic flash accessories, fireworks and incendiary bombs, (2) as a component in alloys for structural purposes, and (3) as a reagent in organic syntheses. Magnesia, MgO, is an excellent heat insulator used in furnaces, ovens and crucibles. Milk of magnesia, an aqueous suspension of $Mg(OH)_2$, is a stomach antacid and laxative. Anhydrous $MgSO_4$ and $Mg(ClO_4)_2$ are used as drying agents.

23-18. *Refer to Section 23-2 and Table 23-2.*

Let M = alkali metal, X = halogen

(a) $2M + 2H_2O \rightarrow 2MOH + H_2$ (b) $12M + P_4 \rightarrow 4M_3P$ (c) $2M + X_2 \rightarrow 2MX$

23-20. *Refer to Section 23-5 and Table 23-4.*

Let M = alkaline earth metal

(a) $M + 2H_2O \rightarrow M(OH)_2 + H_2$ (b) $6M + P_4 \rightarrow 2M_3P_2$ (c) $M + Cl_2 \rightarrow MCl_2$

23-22. *Refer to Section 23-2 and the Key Terms for Chapter 23.*

Diagonal similarities refer to chemical similarities of Period 2 elements of a certain group to Period 3 elements, one group to the right. This effect is particularly evident toward the left side of the periodic table. One example is the pair, B and Si, which are both metalloids with similar properties. Another example is the pair, Li and Mg. They have similar ionic charge densities and electronegativities; their compounds are similar in many ways:

(1) Li is the only IA metal that combines with N_2 to form a nitride, Li_3N. Mg readily forms the nitride, Mg_3N_2.
(2) Li and Mg both form carbides.
(3) The solubilities of Li compounds are similar to those of Mg compounds.
(4) Li and Mg form normal oxides, Li_2O and MgO, when oxidized in air at 1 atm pressure, while the other members of Group IA form peroxides and superoxides.

Hydration energy is the energy released when a mole of ions in the gaseous phase forms a mole of ions in the aqueous phase. The higher the charge to size ratio of a cation, the stronger is its interaction with polar water molecules and the more exothermic is its hydration energy. Therefore, hydration energies of the alkaline earth metals become less exothermic from top to bottom within a group because the size of the ions increases, whereas the charge of the ions remains +2.

Standard reduction potentials of the alkaline earth metals are, in general, very negative, indicating that alkaline earth metals are easily oxidized and hence are good reducing agents. Progressing down Group IIA, the atoms are larger, the outer electrons are more easily lost, the metals become better reducing agents and standard reduction potentials become more negative.

Note: The metal hydroxide product is in the solid phase because only stoichiometric amounts of water are added.

(a) Balanced equation: $Li(s) + H_2O(\ell) \rightarrow LiOH(s) + 1/2\ H_2(g)$

$\Delta H^\circ_{rxn} = [\Delta H^\circ_{f\ LiOH(s)} + 1/2\ \Delta H^\circ_{f\ H_2(g)}] - [\Delta H^\circ_{f\ Li(s)} + \Delta H^\circ_{f\ H_2O(\ell)}]$

$= [(1\ mol)(-487.23\ kJ/mol) + (1/2\ mol)(0\ kJ/mol)] - [(1\ mol)(0\ kJ/mol) + (1\ mol)(-285.8\ kJ/mol)]$

$= $ **-201.4 kJ/mol rxn**

(b) Balanced equation: $K(s) + H_2O(\ell) \rightarrow KOH(s) + 1/2\ H_2(g)$

$\Delta H^\circ_{rxn} = [\Delta H^\circ_{f\ KOH(s)} + 1/2\ \Delta H^\circ_{f\ H_2(g)}] - [\Delta H^\circ_{f\ K(s)} + \Delta H^\circ_{f\ H_2O(\ell)}]$

$= [(1\ mol)(-424.7\ kJ/mol) + (1/2\ mol)(0\ kJ/mol)] - [(1\ mol)(0\ kJ/mol) + (1\ mol)(-285.8\ kJ/mol)]$

$= $ **-138.9 kJ/mol rxn**

(c) Balanced equation: $Ca(s) + 2H_2O(\ell) \rightarrow Ca(OH)_2(s) + H_2(g)$

$\Delta H^\circ_{rxn} = [\Delta H^\circ_{f\ Ca(OH)_2(s)} + \Delta H^\circ_{f\ H_2(g)}] - [\Delta H^\circ_{f\ Ca(s)} + 2\Delta H^\circ_{f\ H_2O(\ell)}]$

$= [(1\ mol)(-986.6\ kJ/mol) + (1\ mol)(0\ kJ/mol)] - [(1\ mol)(0\ kJ/mol) + (2\ mol)(-285.8\ kJ/mol)]$

$= $ **-415.0 kJ/mol rxn**

In explanation, refer to the data table:

Cation	E° (V)	ΔH°_{rxn} from above (kJ)
Li^+	-3.045	-201.4
K^+	-2.925	-138.9
Ca^{2+}	-2.87	-415.0

From the E° values, we see that the relative strengths of reducing agents are : Li > K > Ca. It is expected that for the reactions of these metals with water, ΔH°_{rxn} for Li would be more negative than ΔH°_{rxn} for K, which is in turn more negative than ΔH°_{rxn} for Ca. This trend is true only for Li and K. ΔH°_{rxn} for Ca is much more negative than predicted since it is a IIA metal (Li and K are IA metals) and reacts with twice as much water to produce twice as much H_2 gas.

23-30. *Refer to Section 23-8.*

The following are properties of most transition elements:

(1) All are metals.
(2) Most are harder, more brittle and have higher melting points and boiling points and higher heats of vaporization than nontransition metals.
(3) Their ions and compounds are usually colored.
(4) They form many complex ions.
(5) With few exceptions, they exhibit multiple oxidation states.
(6) Many of the metals and their compounds are paramagnetic.
(7) Many of the metals and their compounds are effective catalysts.

23-32. *Refer to Section 23-9, Table 23-7 and Appendix B.*

(a) V [Ar] $3d^3 4s^2$

(b) Fe [Ar] $3d^6 4s^2$

(c) Cu [Ar] $3d^{10} 4s^1$

(d) Zn [Ar] $3d^{10} 4s^2$

(e) Fe^{3+} [Ar] $3d^5$

(f) Ni^{2+} [Ar] $3d^8$

(g) Ag [Kr] $4d^{10} 5s^1$

(h) Ag^+ [Kr] $4d^{10}$

23-34. *Refer to Section 23-10.*

Group VIIIB consists of three columns of three metals each, which have no counterparts among the representative elements. Each horizontal row is called a triad and is named after the best-known metal of the row; they are called the iron (Fe, Co, Ni), palladium (Ru, Rh, Pd) and platinum (Os, Ir, Pt) triads.

23-36. *Refer to Section 23-11 and Table 23-10.*

The acidity and the covalent nature of transition metal oxides generally increases with increasing oxidation state of the metals. This is shown by the oxides of chromium.

Cr Oxide	Ox. No. of Cr	Character
CrO	+2	basic
Cr_2O_3	+3	amphoteric
CrO_3	+6	weakly acidic/acidic

23-38. *Refer to Section 23-11.*

Chromium(VI) oxide, CrO_3, is the acid anhydride of chromic acid, H_2CrO_4, and dichromic acid, $H_2Cr_2O_7$. Recall that there is no change in oxidation state when an acid anhydride is converted to the corresponding acid and so the oxidation state of Cr is +6 in both acids.

$$CrO_3 + H_2O \rightarrow H_2CrO_4 \qquad\qquad 2CrO_3 + H_2O \rightarrow H_2Cr_2O_7$$

23-40. *Refer to Section 23-2, Appendix K and Exercise 23-39 Solution.*

Balanced equation: $Rb(s) + H_2O(\ell) \rightarrow RbOH(aq) + 1/2\ H_2(g)$

$$\Delta H^\circ_{rxn} = [\Delta H^\circ_{f\ RbOH(aq)} + 1/2\ \Delta H^\circ_{f\ H_2(g)}] - [\Delta H^\circ_{f\ Rb(s)} + \Delta H^\circ_{f\ H_2O(\ell)}]$$
$$= [(1\ mol)(-481.16\ kJ/mol) + (1/2\ mol)(0\ kJ/mol)] - [(1\ mol)(0\ kJ/mol) + (1\ mol)(-285.8\ kJ/mol)]$$
$$= \mathbf{-195.4\ kJ/mol\ Rb(s)}$$

$$\Delta S^\circ_{rxn} = [S^\circ_{f\ RbOH(aq)} + 1/2\ S^\circ_{f\ H_2(g)}] - [S^\circ_{f\ Rb(s)} + S^\circ_{f\ H_2O(\ell)}]$$
$$= [(1\ mol)(110.75\ J/mol\cdot K) + (1/2\ mol)(130.60\ J/mol\cdot K)]$$
$$- [(1\ mol)(76.78\ J/mol\cdot K) + (1\ mol)(69.91\ J/mol\cdot K)]$$
$$= \mathbf{29.4\ J/K\ per\ 1\ mol\ Rb(s)}$$

$$\Delta G^\circ_{rxn} = [\Delta G^\circ_{f\ RbOH(aq)} + 1/2\ \Delta G^\circ_{f\ H_2(g)}] - [\Delta G^\circ_{f\ Rb(s)} + \Delta H^\circ_{f\ H_2O(\ell)}]$$
$$= [(1\ mol)(-441.24\ kJ/mol) + (1/2\ mol)(0\ kJ/mol)] - [(1\ mol)(0\ kJ/mol) + (1\ mol)(-237.2\ kJ/mol)]$$
$$= \mathbf{-204.0\ kJ/mol\ Rb(s)}$$

In Exercise 23-39, the ΔG_{rxn} was calculated for the following reaction: $Na(s) + H_2O(\ell) \rightarrow NaOH(aq) + 1/2\ H_2(g)$

$$\Delta G^\circ_{rxn} = [\Delta G^\circ_{f\ NaOH(aq)} + 1/2\ \Delta G^\circ_{f\ H_2(g)}] - [\Delta G^\circ_{f\ Na(s)} + \Delta G^\circ_{f\ H_2O(\ell)}]$$
$$= [(1\ mol)(-419.2\ kJ/mol) + (1/2\ mol)(0\ kJ/mol)] - [(1\ mol)(0\ kJ/mol) + (1\ mol)(-237.2\ kJ/mol)]$$
$$= -182.0\ kJ/mol\ Na(s)$$

Therefore, the reaction between Rb(*s*) and water is more spontaneous than the reaction between Na(*s*) and water, since the ΔG° for the reaction between Rb(*s*) and water is more negative.

28-42. *Refer to Section 3-4.*

Balanced equation: $3Co_3O_4 + 8Al \rightarrow 9Co + 4Al_2O_3$

Plan:
(1) Calculate the theoretical yield of Co metal.
(2) Calculate the mass of Co_3O_4 required to produce the theoretical yield of Co.

(1) $$\%\ yield = \frac{actual\ yield}{theoretical\ yield} \times 100\%$$

Substituting,

$$69.3\% = \frac{175\ g}{theoretical\ yield} \times 100\%$$

Solving, theoretical yield = 253 g Co

(2) $$?\ g\ Co_3O_4 = 253\ g\ Co \times \frac{1\ mol\ Co}{58.9\ g\ Co} \times \frac{3\ mol\ Co_3O_4}{9\ mol\ Co} \times \frac{241\ g\ Co_3O_4}{1\ mol\ Co_3O_4} = \mathbf{345\ g\ Co_3O_4}$$

24 Some Nonmetals

24-2.   Refer to Section 24-1.

The inert nature of the noble gases and their low abundances in the atmosphere were two factors causing their late discovery.

24-4.   Refer to Section 24-1.

(a) In order of increasing size: He $<$ Ne $<$ Ar $<$ Kr $<$ Xe $<$ Rn.

(b) In order of increasing melting points and boiling points: He $<$ Ne $<$ Ar $<$ Kr $<$ Xe $<$ Rn. This order is because the only forces of attraction between the monatomic noble gas atoms are London dispersion forces, which increase with increasing atomic size.

(c) See (b).

(d) In order of increasing density: He $<$ Ne $<$ Ar $<$ Kr $<$ Xe $<$ Rn. This order is because the masses of noble gases increase going down the group, but the molar volumes of all these gases are the same at constant temperature and pressure.

(e) In order of increasing first ionization energy: Rn $<$ Xe $<$ Kr $<$ Ar $<$ Ne $<$ He. From top to bottom within a group, the outer electrons are further from the nucleus, and thus the first ionization energy, the energy required to remove the outermost electron from the attractive force of the nucleus, decreases.

24-6.   Refer to Section 24-2.

The accidental preparation of $O_2^+PtF_6^-$ by the reaction of O_2 with PtF_6 led Bartlett to reason that xenon should also be oxidized by PtF_6, since the first ionization energy of molecular oxygen is actually slightly larger than that of xenon. He obtained a red crystalline solid initially believed to be $Xe^+PtF_6^-$, but now known to be more complex. At present, Xe and Kr are the only noble gases known to form compounds, mostly combining with F and O. Our textbook discusses the compounds of Xe.

24-8.   Refer to Section 24-2.

Balanced equation: $XeF_4(s) + F_2(g) \rightarrow XeF_6(s)$

$$? \text{ g } XeF_6 = 3.62 \text{ g } XeF_4 \times \frac{1 \text{ mol } XeF_4}{207 \text{ g } XeF_4} \times \frac{1 \text{ mol } XeF_6}{1 \text{ mol } XeF_4} \times \frac{245 \text{ g } XeF_6}{1 \text{ mol } XeF_6} = \textbf{4.28 g } \textbf{XeF}_6$$

24-10.   Refer to Section 24-3 and Table 24-4.

X_2 $\boxed{\text{:}\overset{..}{X}\overset{..}{X}\text{:}}$ $S = N - A = [2 \times 8 \text{ (for X)}] - [2 \times 7 \text{ (for X)}] = 16 - 14 = 2$

A halogen molecule contains one nonpolar covalent single bond, as shown by its Lewis structure.

As we descend the VIIA family from F_2 to I_2, the size of a halogen atom increases and so the bond length increases. The strength of the X-X bond varies; it increases from F_2 to Cl_2, then decreases from Cl_2 to Br_2 to I_2.

315

24-12. *Refer to Section 24-4 and Table 24-4.*

(a) In order of increasing atomic radii: F < Cl < Br < I < At

(b) In order of increasing ionic radii: F^- < Cl^- < Br^- < I^- < At^-

(c) In order of increasing electronegativity: At < I < Br < Cl < F

(d) In order of increasing melting points and boiling points: F_2 < Cl_2 < Br_2 < I_2 < At_2

In nature, the halogens exist as nonpolar diatomic molecules. London dispersion forces are the only forces of attraction acting between the molecules. These forces increase with increasing molecular size.

(e) See (d).

(f) In order of increasing standard reduction potentials: At_2 < I_2 < Br_2 < Cl_2 < F_2

F_2 has the most positive standard reduction potential and therefore is the strongest of all common oxidizing agents. Oxidizing strengths of the diatomic halogen molecules decrease down Group VIIA.

24-14. *Refer to Section 24-4.*

Christie's preparation of F_2 is not a direct chemical oxidation, but rather it involves the formation of unstable MnF_4, which spontaneously decomposes into MnF_3 and F_2.

24-16. *Refer to Section 24-5.*

(a) $2NaI + Cl_2 \rightarrow 2NaCl + I_2$

(b) $NaCl + Br_2 \rightarrow$ no reaction

(c) $2NaI + Br_2 \rightarrow 2NaBr + I_2$

(d) $2NaBr + Cl_2 \rightarrow 2NaCl + Br_2$

(e) $NaF + I_2 \rightarrow$ no reaction

24-18. *Refer to Section 24-5.*

F_2 is a strong oxidizing agent, whereas F^- ions are difficult to oxidize. Highly oxidized cations will not be easily reduced in the presence of F^- ions.

$$2Fe + 3F_2 \rightarrow 2FeF_3 \text{ (only)} \qquad Cu + F_2 \rightarrow CuF_2$$

I_2 is only a mild oxidizing agent, but I^- ions are fairly easily oxidized. I^- ions are able to reduce cations in high oxidation states and stabilize low oxidation states.

$$Fe + I_2 \rightarrow FeI_2 \text{ (only)} \qquad 2Cu + I_2 \rightarrow 2CuI \text{ (only)}$$

24-20. *Refer to Section 24-6.*

Hydrogen bromide, $HBr(g)$, is a colorless gas which dissolves in water to give hydrobromic acid, $HBr(aq)$. The latter is a strong acid which completely dissociates in aqueous solutions giving $H_3O^+(aq)$ and $Br^-(aq)$.

24-22. *Refer to Section 24-6.*

Hydrofluoric acid is used to etch glass by reacting with the silicates in glass to produce a very volatile and thermodynamically stable compound, silicon tetrafluoride, SiF_4. For example,

$$CaSiO_3(s) + 6HF(aq) \rightarrow CaF_2(s) + SiF_4(g) + 3H_2O(\ell)$$

24-24. *Refer to Sections 24-13 and 4-10, and Table 24-9.*

(a)	$KBrO_3$	potassium bromate
(b)	$KBrO$	potassium hypobromite
(c)	$NaClO_4$	sodium perchlorate
(d)	$NaClO_2$	sodium chlorite
(e)	$HBrO$	hypobromous acid
(f)	$HBrO_3$	bromic acid
(g)	HIO_3	iodic acid
(h)	$HClO_4$	perchloric acid

24-26. *Refer to Sections 24-6 and 24-7.*

(a) $X_2 + H_2O \rightarrow HX + HOX$ (X = Cl, Br, I)

(b) $X_2 + 2NaOH \rightarrow NaX + NaOX + H_2O$ (X = Cl, Br, I)

(c) $Ba(ClO_2)_2 + H_2SO_4 \rightarrow BaSO_4 + 2HClO_2$

(d) $NaClO_4 + H_2SO_4 \rightarrow NaHSO_4 + HClO_4$ (explosive)

24-28. *Refer to Sections 24-7 and 10-5.*

The strongest acid is

(a) HClO since for most ternary acids containing different elements in the same oxidation state from the same group in the periodic table, acid strengths increase with increasing electronegativity of the central element.

(b) $HClO_4$ since the acid strengths of most ternary acids containing the same central element increase with increasing oxidation state of the central element and with increasing numbers of oxygen atoms.

(c) $HClO_4$ since we know $HClO_4$ is one of our 7 strong acids and the others are not.

24-30. *Refer to Section 24-8 and Table 24-6.*

oxide, O^{2-}	[Ne]	or	$1s^2\ 2s^2\ 2p^6$
sulfide, S^{2-}	[Ar]	or	$1s^2\ 2s^2\ 2p^6\ 3s^2\ 3p^6$
selenide, Se^{2-}	[Kr]	or	$1s^2\ 2s^2\ 2p^6\ 3s^2\ 3p^6\ 3d^{10}\ 4s^2\ 4p^6$

24-32. *Refer to Section 24-8.*

Every Group VIA element has six valence electrons in the highest energy level, and is therefore two electrons away from achieving an octet of electrons. This is why all of them exhibit an oxidation state of -2. In addition, Group VIA elements below oxygen, since they have empty *d* orbitals available for containing electron pairs, can share their six valence electrons to various degrees with other elements to give different positive oxidation states up to +6. An oxidation state of -3 is impossible because it would require placing an electron into the next higher energy *d* or *s* orbitals. An oxidation state of +7 is impossible because the Group VIA elements only possess six valence electrons to share or to lose.

(a) H_2S

$$H:\overset{..}{\underset{..}{S}}:H$$

The Lewis dot formula predicts 4 regions of high electron density around the central S atom, a tetrahedral electronic geometry and a bent (angular) molecular geometry. The S atom has sp^3 hybridization. The valence electrons available for bonding from H are in the $1s$ orbitals. The three-dimensional structure is shown below (Section 8-9).

(b) SF_6

The Lewis dot formula predicts 6 regions of high electron density around the central S atom and an octahedral electronic and molecular geometry. The S atom has sp^3d^2 hybridization. The valence electron available for bonding from each F is in a $2p$ orbital and the three dimensional structure is shown below (Section 8-12).

(c) SF_4

The Lewis dot formula predicts 5 regions of high electron density around the central S atom, a trigonal bipyramidal electronic geometry and a see-saw molecular geometry. The S atom has sp^3d hybridization. The valence electron available for bonding from each F is in a $2p$ orbital. The three-dimensional structure is shown below (Section 8-11).

(d) SO_2

The Lewis dot formulas for the two resonance structures (one is shown) predict 3 regions of high electron density around the central S atom, a trigonal planar electronic geometry and an angular molecular geometry. The S atom has sp^2 hybridization. The valence electrons available for bonding from the O atoms are in $2p$ orbitals. A delocalized π bond is formed between the remaining p orbitals of S and O. The three dimensional structure is shown below (Table 8-3).

(e) SO_3

The Lewis dot formulas for the three resonance structures (one is shown) predict 3 regions of high electron density around the central S atom and trigonal planar electronic and molecular geometries. The S atom has sp^2 hybridization. The valence electrons available from the O atoms are in $2p$ orbitals. A delocalized π bond is formed between the remaining p orbitals of S and O. The three-dimensional structure is shown below (Section 8-6).

(a)

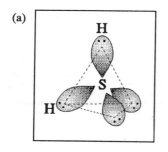

(b)

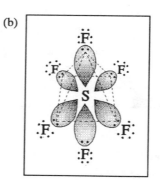

(c)

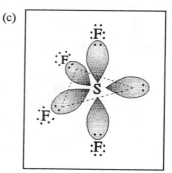

(d)

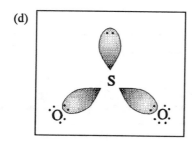

(e)

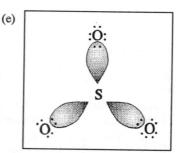

Note: Pi bonding is not shown in (d) and (e).

The oxidation state of S in these molecules is: (a) H_2S, -2; (b) SF_6, +6; (c) SF_4, +4; (d) SO_2, +4; (e) SO_3, +6.

(a) $E + 3F_2(\text{excess}) \rightarrow EF_6$ (E = S, Se, Te)   (c) $E + O_2 \rightarrow EO_2$ (E = S, Se, Te)

(b) $O_2 + 2H_2 \rightarrow 2H_2O$
    $E + H_2 \rightarrow H_2E$ (E = S, Se, Te)

24-38. *Refer to Section 24-10.*

Group VIA hydrides dissociate in two stages. Their acid ionization constants, K_1 and K_2, are shown below:

		H_2S	H_2Se	H_2Te
$H_2E \rightleftharpoons H^+ + HE^-$	K_1:	1.0×10^{-7}	1.9×10^{-4}	2.3×10^{-3}
$HE^- \rightleftharpoons H^+ + E^{2-}$	K_2:	1.3×10^{-13}	$\approx 10^{-11}$	$\approx 1.6 \times 10^{-11}$

Acid strength increases upon descending the group: $H_2O < H_2S < H_2Se < H_2Te$. This results from the corresponding decrease in the average E-H bond energy.

24-40. *Refer to Sections 24-11 and 24-12.*

(a) $NaOH + H_2SO_4 \rightarrow NaHSO_4 + H_2O$ (e) $NaOH + H_2SeO_4 \rightarrow NaHSeO_4 + H_2O$

(b) $2NaOH + H_2SO_4 \rightarrow Na_2SO_4 + 2H_2O$ (f) $2NaOH + H_2SeO_4 \rightarrow Na_2SeO_4 + 2H_2O$

(c) $NaOH + H_2SO_3 \rightarrow NaHSO_3 + H_2O$ (g) $NaOH + TeO_2 \rightarrow NaHTeO_3$

(d) $2NaOH + H_2SO_3 \rightarrow Na_2SO_3 + 2H_2O$ (h) $2NaOH + TeO_2 \rightarrow Na_2TeO_3 + H_2O$

24-42. *Refer to Sections 24-11 and 3-5.*

Balanced equations: (1) $4FeS_2 + 11 O_2 \rightarrow 2Fe_2O_3 + 8SO_2$
                    (2) $2SO_2 + O_2 \rightarrow 2SO_3$
                    (3) $SO_3 + H_2SO_4 \rightarrow H_2S_2O_7$
                    (4) $H_2S_2O_7 + H_2O \rightarrow 2H_2SO_4$

$$? \text{ kg } H_2SO_4 = 1.00 \text{ kg FeS}_2 \times \frac{1000 \text{ g FeS}_2}{1 \text{ kg FeS}_2} \times \frac{1 \text{ mol FeS}_2}{120.0 \text{ g FeS}_2} \times \frac{8 \text{ mol SO}_2}{4 \text{ mol FeS}_2} \times \frac{2 \text{ mol SO}_3}{2 \text{ mol SO}_2} \times \frac{1 \text{ mol H}_2S_2O_7}{1 \text{ mol SO}_3}$$

$$\times \frac{2 \text{ mol H}_2SO_4}{1 \text{ mol H}_2S_2O_7} \times \frac{98.1 \text{ g H}_2SO_4}{1 \text{ mol H}_2SO_4} \times \frac{1 \text{ kg H}_2SO_4}{1000 \text{ g H}_2SO_4}$$

$$= 3.27 \text{ kg } H_2SO_4 \text{ total}$$

However, half of the H_2SO_4 was previously added in Step (3).

Therefore, the net mass of H_2SO_4 produced = 3.27 kg/2 = **1.64 kg H_2SO_4**

24-44. *Refer to Sections 25-11, 17-2 and 17-5.*

Balanced equation: $2SO_2(g) + O_2(g) \rightarrow 2SO_3(g)$

Let $x = [O_2]_{\text{reacted}}$. Then,
    $1.00 - 2x = [SO_2]$
      $5.00 - x = [O_2]$
          $2x = [SO_3]$

	$2SO_2$	$+$	O_2	$\rightleftarrows$	$2SO_3$
initial	1.00 M		5.00 M		0 M
change	$- 2x\ M$		$- x\ M$		$+ 2x\ M$
at equilibrium	$(1.00 - 2x)\ M$		$(5.00 - x)\ M$		$2x\ M$

However, $[SO_3] = 2x = 78.3\%$ of $[SO_2]_{initial} = 0.783 \times 1.00\ M = 0.783\ M$

Therefore, $K_c = \dfrac{[SO_3]^2}{[SO_2]^2[O_2]} = \dfrac{(2x)^2}{(1.00 - 2x)^2(5.00 - x)} = \dfrac{(0.783)^2}{(1.00 - 0.783)^2(5.00 - 0.783/2)} = \mathbf{2.8}$

24-46. *Refer to the Introduction to Section 24-13, Table 24-7 and Appendix B.*

N	[He] $2s^2\,2p^3$	P	[Ne] $3s^2\,3p^3$	As	[Ar] $3d^{10}\,4s^2\,4p^3$

Sb   [Kr] $4d^{10}\,5s^2\,5p^3$ Bi   [Xe] $4f^{14}\,5d^{10}\,6s^2\,6p^3$

N^{3-} [Ne] or $1s^2\,2s^2\,2p^6$ P^{3-} [Ar] or $1s^2\,2s^2\,2p^6\,3s^2\,3p^6$

24-48. *Refer to Section 24-13.*

The nitrogen cycle is the complex series of reactions by which nitrogen is slowly but continually recycled in the atmosphere, lithosphere (earth) and hydrosphere (water). Atmospheric nitrogen is made accessible to us and other life-forms in mainly two ways.

(1) A class of plants, called legumes, have bacteria which extract N_2 directly, converting it to NH_3. This nitrogen fixation process, catalyzed by an enzyme produced by the bacteria, is highly efficient at usual temperatures and pressures.

(2) N_2 and O_2 react in the atmosphere near lightning, forming NO and NO_2, which dissolve in rainwater and fall to earth. These nitrogen compounds are absorbed and incorporated into plants forming amino acids and proteins. The plants are eaten by animals or die and decay, releasing their nitrogen to the environment. The animals, in turn, excrete waste and/or die, releasing their nitrogen to the environment.

24-50. *Refer to Sections 24-14 and 17-6, and Example 17-7.*

The Haber process is the economically important industrial process for making ammonia, NH_3, from atmospheric N_2, according to:

$$N_2(g) + 3H_2(g) \rightleftarrows 2NH_3(g) \qquad \Delta H° = -92 \text{ kJ/mol}$$

(1) effect of temperature: The reaction is exothermic ($\Delta H < 0$), so one might expect that to increase the amount of NH_3, one would need to lower the temperature. This action would increase the *relative* amount of NH_3 present, however, the reaction rates are lowered as well. So, Haber investigated other ways to increase the yield.

(2) effect of pressure: In Chapter 17, we learned that increasing the pressure favors the reaction that produces the smaller number of moles of gas (forward in this case). This reaction is run under pressures ranging from 200 to 1000 atmospheres to increase the yield of NH_3.

(3) effect of catalyst: The addition of finely divided iron and small amounts of selected oxides speeds up both the forward and reverse reactions. This allow NH_3 to be produced not only faster but at a lower temperature, which increases the yield of NH_3 and extends the life of the equipment.

Oxidation number of N:

(a) N_2 0   (b) NO   +2   (c) N_2O_4 +4   (d) HNO_3 +5   (e) HNO_2 +3

24-54. *Refer to Chapter 8 and the Sections as stated.*

(a) NH_2Br

| H:N:Br: |
| H |

The Lewis dot formula predicts 4 regions of high electron density around the central N atom, a tetrahedral electronic geometry and a pyramidal molecular geometry. The N atom has sp^3 hybridization (Sections 8-8 and 24-14). The three-dimensional structure is shown below.

(b) HN_3

| H:N::N::N: |

Around the outer two N atoms, the Lewis dot formula predicts 3 regions of high electron density, a trigonal planar electronic geometry, a bent molecular geometry and sp^2 hybridization. The Lewis dot formula also predicts 2 regions of high electron density around the central N atom, a linear electronic and molecular geometry and sp hybridization for the central N atom (Sections 8-5, 8-13 and 24-14). The three-dimensional structure is shown below.

(c) N_2O_2

| :O::N:N::O: |

The Lewis dot formula predicts 3 regions of high electron density, trigonal planar electronic geometry and angular molecular geometry around each N atom. The N atoms have sp^2 hybridization (Sections 8-13 and 24-15). The three-dimensional structure is shown below.

(d) NO_2^+

| [:O::N::O:]⁺ |

The Lewis dot formula predicts 2 regions of high electron density, a linear electronic and ionic geometry around the N atom and sp hybridization for the N atom (Section 24-15). The three-dimensional structure is shown on the next page.

NO_3^-

| [:O::N:O:]⁻ |
| :O: |

The Lewis dot formulas for the three resonance structures (one is shown) predicts 3 regions of high electron density around the central N atom and a trigonal planar electronic and ionic geometry. The N atom has sp^2 hybridization (Section 24-16). The three-dimensional structure is shown on the next page.

(e) HNO_3

| H:O:N::O: |
| :O: |

The Lewis dot formulas for the two resonance structures (one is shown) predicts 3 regions of high electron density for the N atom, and a trigonal planar electronic and molecular geometry about the N atom. The N atom has sp^2 hybridization (Section 24-16). The three-dimensional structure is shown on the next page.

(f) NO_2^-

| [:O::N:O:]⁻ |

The Lewis dot formulas for the two resonance structures (one is shown) predicts 3 regions of high electron density for the central N atom, and a trigonal planar electronic geometry and a bent ionic geometry. The N atom has sp^2 hybridization (Section 24-16). The three-dimensional structure is shown on the next page.

(a) NH_2Br

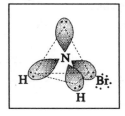

(b) HN_3

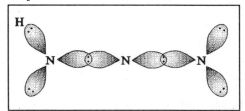

(c) N_2O_2

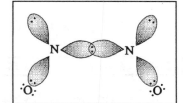

(d) NO_2^+

NO_3^-

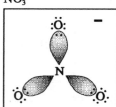

(e) HNO_3

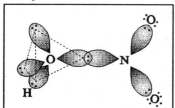

(f) NO_2^-

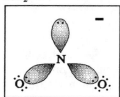

Note: Pi bonding is not shown.

24-56. *Refer to the Sections as stated.*

(a) $2KN_3(s) \overset{\Delta}{\to} 2K(\ell) + 3N_2(g)$ (Section 24-13)

(b) $NH_3(g) + HCl(g) \to NH_4Cl(s)$ (Section 24-14)

(c) $NH_3(aq) + HCl(aq) \to NH_4Cl(aq)$ (Section 24-14)

(d) $2NH_4NO_3(s) \overset{\Delta}{\to} 2N_2(g) + 4H_2O(g) + O_2(g)$ (Sections 24-14 and 24-16)

(e) $4NH_3(g) + 5O_2(g) \overset{\Delta}{\underset{Pt}{\rightleftharpoons}} 4NO(g) + 6H_2O(g)$ (Section 24-16)

(f) $2N_2O(g) \overset{\Delta}{\to} 2N_2(g) + O_2(g)$ (Section 24-15)

(g) $3NO_2(g) + H_2O(\ell) \to 2HNO_3(aq) + NO(g)$ (Section 24-16)

24-58. *Refer to Section 24-14.*

$Co^{2+}(aq) + 6NH_3(aq) \to [Co(NH_3)_6]^{2+}(aq)$ $BF_3(g) + :NH_3(g) \to F_3B:NH_3(s)$

24-60. *Refer to Exercise 8-38 Solution and the Sections as stated.*

	Molecule	Structure	Polarity of Molecule	
(a)	NH_3	pyramidal	polar	(Section 8-8)
(b)	NH_2Cl	distorted pyramidal	polar	(Section 8-8)
(c)	NO	linear (heteronuclear diatomic)	polar	(Section 24-15)
(d)	NH_2OH	unsymmetric	polar	(Section 24-14)
(e)	HNO_3	unsymmetric	polar	(Section 24-16)

24-62. *Refer to Section 24-15.*

Nitrogen oxide, NO, is very reactive because each NO molecule contains an unpaired electron. Note: Species with unpaired electrons, resulting from the cleavage of chemical bonds, are called radicals.

24-64. *Refer to Sections 24-15, 17-6 and 17-7.*

NO_2, a brown gas, is very reactive because it contains 1 unpaired electron and easily dimerizes to form the colorless gas, N_2O_4, in the following equilibrium reaction:

$$2NO_2(g) \rightleftarrows N_2O_4(g) + \text{heat} \qquad\qquad \Delta H° = -57.2 \text{ kJ/mol rxn}$$

At room temperature, there is sufficient NO_2 present in the equilibrium mixture to give it a brown color. When the system is cooled, the equilibrium shifts to the right, brown NO_2 gas is converted to colorless N_2O_4 gas and the mixture loses color.

24-66. *Refer to Section 6-8.*

When an acid anhydride reacts with water, the corresponding acid is formed. There is no change in oxidation number of the elements.

(a) $N_2O_5(s) + H_2O(\ell) \rightarrow 2HNO_3(\ell)$

(b) $N_2O_3(g) + H_2O(\ell) \rightarrow 2HNO_2(\ell)$

(c) $P_4O_{10}(s) + 6H_2O(\ell) \rightarrow 4H_3PO_4(\ell)$

(d) $P_4O_6(s) + 6H_2O(\ell) \rightarrow 4H_3PO_3(\ell)$

24-68. *Refer to Section 24-16.*

The function of sodium nitrite, $NaNO_2$, as a food additive is two-fold:

(1) it inhibits the oxidation of blood, preventing the discoloring of red meat, and
(2) it prevents the growth of botulism bacteria.

There is now some controversy regarding this food additive because nitrites are suspected of combining with amines under the acidic conditions of the stomach to produce carcinogenic nitrosoamines.

24-70. *Refer to Section 13-16 and Example 13-8.*

Consider the Ar face-centered cubic structure with a unit cell edge represented as a, shown here. For Ar, $a = 5.43$ Å

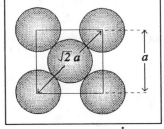

The hypotenuse, d, of a isosceles right angle triangle is determined by simple geometry:

$$d^2 = a^2 + a^2$$
$$d = \sqrt{a^2 + a^2} = \sqrt{2a^2} = \sqrt{2}a = \sqrt{2} \times 5.43 \text{ Å} = 7.68 \text{ Å}$$

By inspection, the hypotenuse is equal to 4 Ar radii. Therefore, the apparent radius of Ar $= \dfrac{7.68 \text{ Å}}{4} = 1.92$ Å

24-72. *Refer to Sections 24-2.*

Balanced equation: $XeF_4(s) + F_2(g) \rightarrow XeF_6(s)$

$$\begin{aligned}
\Delta H°_{rxn} &= [\Delta H°_f \text{ }_{XeF_6(s)}] - [\Delta H°_f \text{ }_{XeF_4(s)} + \Delta H°_f \text{ }_{F_2(g)}] \\
&= [(1 \text{ mol})(-402 \text{ kJ/mol})] - [(1 \text{ mol})(-261.5 \text{ kJ/mol}) + (1 \text{ mol})(0 \text{ kJ/mol})] \\
&= \textbf{-140 kJ/mol rxn}
\end{aligned}$$

Balanced equations: (1) $4N(g) \rightarrow 2$:N$\equiv$N:(g) $\qquad\qquad \Delta H_{rxn\,1}$

(2) $4N(g) \rightarrow$:N$\underset{\underset{\ddot{N}}{}}{\overset{\overset{\ddot{N}}{}}{\diamondsuit}}$N: (g) $\qquad\qquad \Delta H_{rxn\,2}$

(3) $4P(g) \rightarrow 2$:P$\equiv$P:(g) $\qquad\qquad \Delta H_{rxn\,3}$

(4) $4P(g) \rightarrow$:P$\underset{\underset{\ddot{P}}{}}{\overset{\overset{\ddot{P}}{}}{\diamondsuit}}$P: (g) $\qquad\qquad \Delta H_{rxn\,4}$

(1) $\Delta H_{rxn\,1} = \Sigma$ B.E.$_{reactants} - \Sigma$ B.E.$_{products} = 0 - (2\text{ mol})(\text{B.E.}_{N\equiv N}) = -(2\text{ mol})(946\text{ kJ/mol}) = -1892\text{ kJ}$

$\Delta H_{rxn\,2} = \Sigma$ B.E.$_{reactants} - \Sigma$ B.E.$_{products} = 0 - (6\text{ mol})(\text{B.E.}_{N\text{-}N}) = -(6\text{ mol})(159\text{ kJ/mol}) = -954\text{ kJ}$

Reaction 1 is more exothermic than Reaction 2 ($\Delta H_{rxn\,1}$ is more negative than $\Delta H_{rxn\,2}$).
Therefore, the formation of N_2 molecules by Reaction 1 is the predicted result.

(2) $\Delta H_{rxn\,3} = \Sigma$ B.E.$_{reactants} - \Sigma$ B.E.$_{products} = 0 - (2\text{ mol})(\text{B.E.}_{P\equiv P}) = -(2\text{ mol})(485\text{ kJ/mol}) = -970\text{ kJ}$

$\Delta H_{rxn\,4} = \Sigma$ B.E.$_{reactants} - \Sigma$ B.E.$_{products} = 0 - (6\text{ mol})(\text{B.E.}_{P\text{-}P}) = -(6\text{ mol})(213\text{ kJ/mol}) = -1280\text{ kJ}$

Reaction 4 is more exothermic than Reaction 3 ($\Delta H_{rxn\,4}$ is more negative than $\Delta H_{rxn\,3}$).
Therefore, the formation of P_4 molecules by Reaction 4 is the predicted result.

? volume of earth's crust = (volume of the earth) - (volume of the earth minus the crust)

$$= \tfrac{4}{3}\pi(r_{earth})^3 - \tfrac{4}{3}\pi(r_{earth\ minus\ crust})^3$$

$$= \frac{4}{3}(3.14157)\left[6400\text{ km} \times \frac{1000\text{ m}}{1\text{ km}} \times \frac{100\text{ cm}}{1\text{ m}}\right]^3$$

$$-\frac{4}{3}(3.14157)\left[(6400\text{ km} - 50\text{ km}) \times \frac{1000\text{ m}}{1\text{ km}} \times \frac{100\text{ cm}}{1\text{ m}}\right]^3$$

$$= 1.098 \times 10^{27}\text{ cm}^3 - 1.073 \times 10^{27}\text{ cm}^3$$

$$= 2.5 \times 10^{25}\text{ cm}^3$$

? mass Si $= 2.5 \times 10^{25}\text{ cm}^3\text{ crust} \times \dfrac{3.5\text{ g crust}}{1\text{ cm}^3\text{ crust}} \times \dfrac{25.7\text{ g Si}}{100\text{ g crust}} = \mathbf{2.2 \times 10^{25}\text{ g Si}}$

25 Coordination Compounds

25-2. *Refer to Section 25-1 and Table 25-1.*

$NiSO_4 \cdot 6H_2O$ $\equiv$ $[Ni(OH_2)_6]SO_4$

$Cu(NO_3)_2 \cdot 4NH_3$ $\equiv$ $[Cu(NH_3)_4](NO_3)_2$

$Ni(NO_3)_2 \cdot 6NH_3$ $\equiv$ $[Ni(NH_3)_6](NO_3)_2$

25-4. *Refer to Sections 25-3 and 25-5, and Table 25-6.*

The coordination number of a metal atom or ion in a complex is the number of donor atoms to which it is coordinated. It is not necessarily equal to the number of ligands. The most common values of coordination numbers are 2, 4, 5 and 6.

25-6. *Refer to Section 25-1 and Table 25-1.*

Formula	Coordination Sphere	Charge on the Complex
$[Pt(NH_3)_2Cl_4]$	$[Pt(NH_3)_2Cl_4]^0$	zero
$[Pt(NH_3)_3Cl_3]Cl$	$[Pt(NH_3)_3Cl_3]^+$	$+1$
$[Pt(NH_3)_4Cl_2]Cl_2$	$[Pt(NH_3)_4Cl_2]^{2+}$	$+2$
$[Pt(NH_3)_5Cl]Cl_3$	$[Pt(NH_3)_5Cl]^{3+}$	$+3$
$[Pt(NH_3)_6]Cl_4$	$[Pt(NH_3)_6]^{4+}$	$+4$

25-8. *Refer to Sections 25-3 and 25-4.*

Complex	Ligand(s)		Coordination Number	Ox. No. of Central Metal	
(a) $[Co(NH_3)_2(NO_2)_4]^-$	NH_3	ammine	6	Co	$+3$
	NO_2^-	nitro			
(b) $[Cr(NH_3)_5Cl]Cl_2$	NH_3	ammine	6	Cr	$+3$
	Cl^-	chloro			
(c) $K_4[Fe(CN)_6]$	CN^-	cyano	6	Fe	$+2$
(d) $[Pd(NH_3)_4]^{2+}$	NH_3	ammine	4	Pd	$+2$

25-10. *Refer to Section 25-3 and Table 25-4.*

Polydentate ligands such as ethylenediamine (en) and ethylenediaminetetraacetato (edta), cause ring formation to occur in a complex. This phenomenon is called chelation, resulting from a polydentate ligand bonding to the central metal atom or ion through two or more donor atoms at the same time. For example, consider the $[Co(en)_3]^{3+}$ ion, in which three bidentate ethylenediamine ligands bond to the Co^{3+} ion creating 3 rings in the complex. See figure on next page.

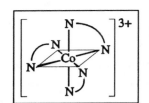 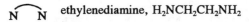 ethylenediamine, $H_2NCH_2CH_2NH_2$

25-12. *Refer to Section 25-2.*

The metal hydroxides that dissolve in an excess of aqueous NH_3 to form ammine complexes are derived from the twelve metals of the cobalt, nickel, copper and zinc families. The only cation of these metals that behaves differently is Hg_2^{2+}. Therefore, when excess NH_3 is added:

(a) $Zn(OH)_2(s)$ will dissolve   (b) $Cr(OH)_3(s)$ will not dissolve   (c) $Fe(OH)_2(s)$ will not dissolve

(d) $Ni(OH)_2(s)$ will dissolve   (e) $Cd(OH)_2(s)$ will dissolve

---

25-14. *Refer to Section 25-2 and Table 25-2.*

In general terms, we may represent the reaction in which a metal cation reacts in aqueous NH_3 to form an insoluble metal hydroxide by the following reaction:

$$M^{n+} + nNH_3 + nH_2O \rightarrow M(OH)_n(s) + nNH_4^+.$$

(a) $Cu^{2+} + 2NH_3 + 2H_2O \rightarrow Cu(OH)_2(s) + 2NH_4^+$

(b) $Zn^{2+} + 2NH_3 + 2H_2O \rightarrow Zn(OH)_2(s) + 2NH_4^+$

(c) $Fe^{3+} + 3NH_3 + 3H_2O \rightarrow Fe(OH)_3(s) + 3NH_4^+$

25-16. *Refer to Sections 25-3 and 25-4.*

(a) $[Ni(CO)_4]$ tetracarbonylnickel(0)

(b) $Na_2[Co(OH_2)_2(OH)_4]$ sodium diaquatetrahydroxocobaltate(II)

(c) $[Ag(NH_3)_2]Br$ diamminesilver(I) bromide

(d) $[Cr(en)_3](NO_3)_3$ tris(ethylenediamine)chromium(III) nitrate

(e) $[Pt(NH_3)_4(NO_2)_2]F_2$ tetraamminedinitroplatinum(IV) fluoride

(f) $K_2[Cu(CN)_4]$ potassium tetracyanocuprate(II)

25-18. *Refer to Sections 25-3 and 25-4.*

(a) diamminedichlorozinc $[Zn(NH_3)_2Cl_2]$

(b) tin(IV) hexacyanoferrate(II) $Sn[Fe(CN)_6]$

(c) tetracyanoplatinate(II) ion $[Pt(CN)_4]^{2-}$

(d) potassium hexacyanochromate(III) $K_3[Cr(CN)_6]$

326

(e) tetraammineplatinum(II) ion $[Pt(NH_3)_4]^{2+}$

(f) hexaamminenickel(II) bromide $[Ni(NH_3)_6]Br_2$

(g) tetraamminecopper(II) pentacyanohydroxoferrate(III) $[Cu(NH_3)_4]_3[Fe(CN)_5(OH)]_2$

25-20. *Refer to Sections 25-3 and 25-4.*

(a) $[Ag(NH_3)_2]^+$ diamminesilver(I) ion
    $[Pt(NH_3)_4]^{2+}$ tetraamineplatinum(II) ion
    $[Cr(OH_2)_6]^{3+}$ hexaaquachromium(III) ion

(b) $[Ni(en)_3]^{2+}$ tris(ethylenediamine)nickel(II) ion
    $[Co(en)_3]^{3+}$ tris(ethylenediamine)cobalt(III) ion
    $[Cr(en)_3]^{3+}$ tris(ethylenediamine)chromium(III) ion

(c) $[Co(en)_2(NO_2)_2]^+$ bis(ethylenediamine)dinitrocobalt(III) ion
    $[CoBr_2(en)_2]^+$ dibromobis(ethylenediamine)cobalt(III) ion
    $[Ni(en)_2(NO)_2]^{2+}$ bis(ethylenediamine)dinitrosylnickel(II) ion

(d) $[FeCl(dien)(en)]^{2+}$ chloro(diethylenetriamine)(ethylenediamine)iron(III) ion
    $[Cr(OH_2)(dien)(ox)]^+$ aqua(diethylenetriamine)(oxalato)chromium(III) ion
    $[RuCl(dien)(en)]^{2+}$ chloro(diethylenetriamine)(ethylenediamine)ruthenium(III) ion

(e) $[Co(OH_2)_3(dien)]^{3+}$ triaqua(diethylenetriamine)cobalt(III) ion
    $[Cr(NH_3)_3(dien)]^{3+}$ triammine(diethylenetriamine)chromium(III) ion
    $[Fe(NH_3)_3(dien)]^{2+}$ triammine(diethylenetriamine)iron(II) ion

---

25-22. *Refer to Section 25-7.*

(a) MA_2B_4, an octahedral complex, can exist as two geometrical isomers: *cis* and *trans*.

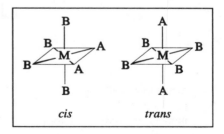

cis   *trans*

(b) The octahedral complex, MA_3B_3, can also exist in two geometrical forms. The facial isomer (*fac*) has three identical ligands at the corners of a trigonal face. The meridional isomer (*mer*) has three identical ligands at three corners of a square plane.

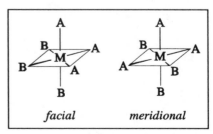

facial   *meridional*

None of these geometric isomers are optical isomers since they all are superimposable on their mirror images, i.e., the compound and its mirror image are actually the same compound.

(a) $[Cr(NH_3)_4(OH_2)_2]^{3+}$

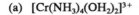

tetraammine-*cis*-diaquachromium(III) ion

tetraammine-*trans*-diaquachromium(III) ion

Structures I and II are mirror images that are superimposable if the vertical axis is rotated 180°; therefore, they are simply different representations of the same compound. Structures III and IV are also the same compound. Hence both the *cis* and *trans* geometrical isomers of $[Cr(NH_3)_4(OH_2)_2]^{3+}$ have no optical isomers, so the total number of isomers is 2.

(b) $[Cr(NH_3)_3(OH_2)_3]^{3+}$

fac-triamminetriaquachromium(III) ion

mer-triamminetriaquachromium(III) ion

Structures I and II are identical. Hence, the facial (*fac*) geometrical isomer has no optical isomer. Structures III and IV are also identical. So, the meridional (*mer*) geometrical isomer also has no optical isomer. The total number of isomers is 2 for $[Cr(NH_3)_3(OH_2)_3]^{3+}$.

(c) $[Cr(en)_3]^{3+}$

tris(ethylenediammine)chromium(III) ion

Structures I and II are mirror images of each other that are not superimposable. Therefore, $[Cr(en)_3]^{3+}$ has 2 optical isomers.

(d) $[Pt(en)_2Cl_2]Cl_2$

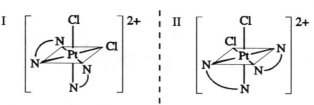

I, II — cis-dichlorobis(ethylenediamine)platinum(IV) ion

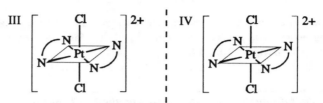

III, IV — trans-dichlorobis(ethylenediamine)platinum(IV) ion

Structures I and II are mirror images that are not superimposable; hence the *cis* geometrical isomer consists of a pair of optical isomers. Structures III and IV are identical; the *trans* geometrical isomer has no optical isomer. Therefore, $[Pt(en)_2Cl_2]Cl_2$ has a total of 3 isomers.

(e) $[Cr(NH_3)_2Br_2Cl_2]^-$

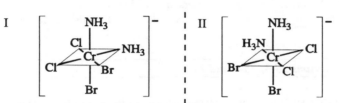

I, II — cis-diammine-cis-dibromo-cis-dichlorochromate(III) ion

III, IV — trans-diammine-cis-dibromo-cis-dichlorochromate(III) ion

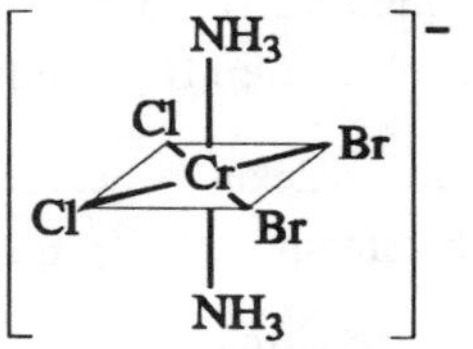

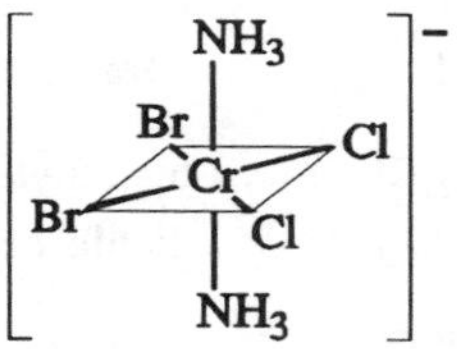

V, VI — cis-diammine-trans-dibromo-cis-dichlorochromate(III) ion

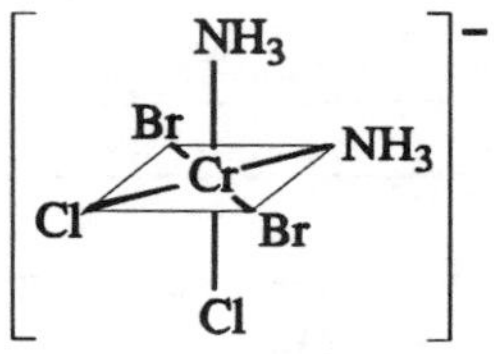

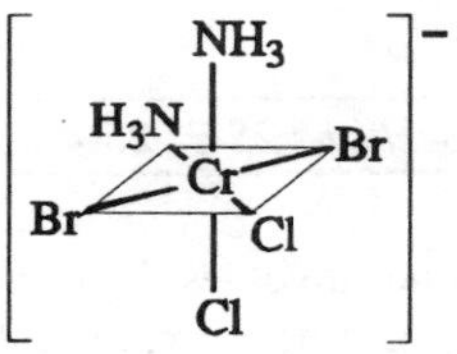

VII, VIII — cis-diammine-cis-dibromo-trans-dichlorochromate(III) ion

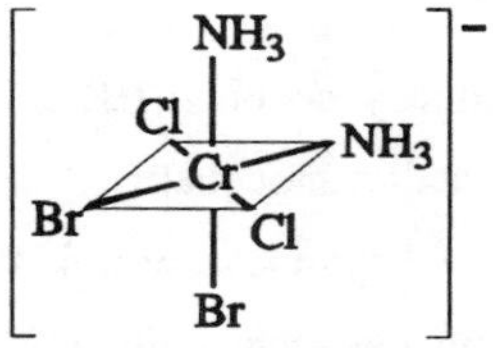

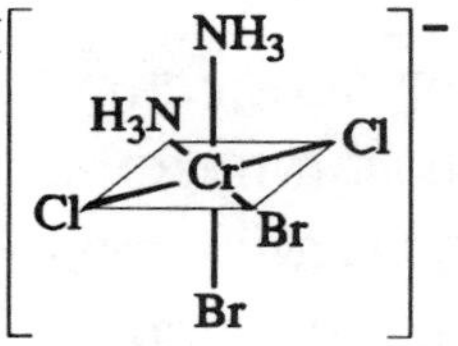

329

(e) $[Cr(NH_3)_2Br_2Cl_2]^-$

(continued)

IX

$$\begin{bmatrix} & NH_3 & \\ Cl & | & Br \\ & Cr & \\ Br & | & Cl \\ & NH_3 & \end{bmatrix}^-$$

X

$$\begin{bmatrix} & NH_3 & \\ Br & | & Cl \\ & Cr & \\ Cl & | & Br \\ & NH_3 & \end{bmatrix}^-$$

trans-diammine-*trans*-dibromo-*trans*-dichlorochromate(III) ion

Structures I and II are mirror images that are not identical and are optical isomers. The mirror image pairs of the remaining geometrical isomers are identical and therefore, they have no optical isomers. The total number of isomers for $[Cr(NH_3)_2Br_2Cl_2]^-$ is 6.

25-26. *Refer to the Introduction to Section 25-6 and the Key Terms for Chapter 25.*

Isomers are substances that have the same number and kinds of atoms, but arranged differently. Structural isomers, as applied to coordination compounds, are isomers whose differences involve having more than a single coordination sphere or different donor atoms on the same ligand. They contain different atom-to-atom bonding sequences. Stereoisomers, on the other hand, are isomers that differ only in the way that atoms are oriented in space, and therefore involve only one coordination sphere and the same ligands and donor atoms.

25-28. *Refer to Sections 25-3, 25-4 and 25-6.*

An ionization isomer results from the exchange of ions inside and outside the coordination sphere.

(a) $[Cr(NH_3)_4I_2]Br$ tetraamminediiodochromium(III) bromide

     $[Cr(NH_3)_4BrI]I$ tetraamminebromoiodochromium(III) iodide

(b) $[Ni(en)_2(NO_2)_2]Cl_2$ bis(ethylenediamine)dinitronickel(IV) chloride

     $[NiCl_2(en)_2](NO_2)_2$ dichlorobis(ethylenediamine)nickel(IV) nitrite

(c) $[Fe(NH_3)_5CN]SO_4$ pentaamminecyanoiron(III) sulfate

     $[Fe(NH_3)_5SO_4]CN$ pentaamminesulfatoiron(III) cyanide

---

25-30. *Refer to Sections 25-3, 25-4 and 25-6.*

A coordination isomer involves the exchange of ligands between a complex cation and a complex anion of the same compound, forming another complex cation and complex anion.

(a) There are 5 possible coordination isomers of $[Co(NH_3)_6][Cr(CN)_6]$:

$[Co(NH_3)_5(CN)][Cr(NH_3)(CN)_5]$ pentaamminecyanocobalt(III) amminepentacyanochromate(III)

$[Co(NH_3)_4(CN)_2][Cr(NH_3)_2(CN)_4]$ tetraamminedicyanocobalt(III) diamminetetracyanochromate(III)

$[Cr(NH_3)_4(CN)_2][Co(NH_3)_2(CN)_4]$ tetraamminedicyanochromium(III) diamminetetracyanocobaltate(III)

$[Cr(NH_3)_5(CN)][Co(NH_3)(CN)_5]$ pentaamminecyanochromium(III) amminepentacyanocobaltate(III)

$[Cr(NH_3)_6][Co(CN)_6]$ hexaamminechromium(III) hexacyanocobaltate(III)

(b) $[Cu(en)_2][Ni(en)(CN)_4]$ bis(ethylenediamine)copper(II) tetracyano(ethylenediamine)nickelate(II)

	Ionic Geometry	Hybridization of Central Metal Ion
(a) $[Ag(NH_3)_2]^+$	linear	sp
(b) $[Fe(en)_3]^{3+}$	octahedral	d^2sp^3 or sp^3d^2
(c) $[Co(NH_3)_6]^{2+}$	octahedral	d^2sp^3 or sp^3d^2
(d) $[Co(NH_3)_6]^{3+}$	octahedral	d^2sp^3 or sp^3d^2
(e) $[Pt(NH_3)_4]^{2+}$	square planar	dsp^2 or sp^2d
(f) $[PtCl_4]^{2-}$	square planar	dsp^2 or sp^2d

(a)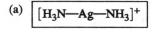

(b)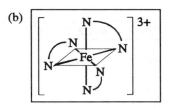

(c)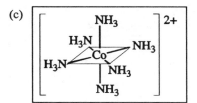

(d)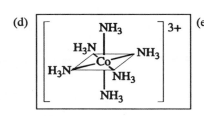

(e)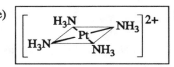

(f)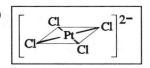

According to Crystal Field Theory, different ligands repel the metal electrons to different extents, thus removing the degeneracy of the d orbitals on the metal. The energy separation between the d orbitals, Δ_{oct}, depends on the crystal field strength of the ligands. Weak field ligands form high spin complexes (outer orbital complexes in Valence Bond Theory) and strong field ligands form low spin complexes (inner orbital complexes in Valence Bond Theory). The spectrochemical series is the arrangement of ligands in order of increasing ligand field strength:

$$I^- < Br^- < Cl^- < F^- < OH^- < H_2O < (COO)_2^{2-} < NH_3 < en < NO_2^- < CN^-$$
$$\text{weak field ligands} \qquad\qquad\qquad\qquad \text{strong field ligands}$$

As shown in Table 25-8, low spin configurations exist only for octahedral complexes having metal ions with d^4 - d^7 configurations. For d^1 - d^3 and d^8 - d^{10} ions, only one possibility exists, which is designated as high spin.

Using the spectrochemical series, we can propose the following electron configurations:

(a) $[Cu(OH_2)_6]^{2+}$ outer orbital (weak field) complex ion of Cu^{2+} (d^9)

paramagnetic (1 unpaired e^-)

outer electronic configuration:

$$\underbrace{\uparrow\downarrow\ \uparrow\downarrow\ \uparrow\downarrow\ \uparrow\downarrow\ \uparrow}_{3d} \quad \underbrace{\text{xx xx xx xx xx xx}}_{sp^3d^2} \quad \underbrace{\underline{\ }\ \underline{\ }\ \underline{\ }\ \underline{\ }}_{4d}$$

(b) $[MnF_6]^{3-}$ outer orbital (weak field) complex ion of Mn^{3+} (d^4)

paramagnetic (4 unpaired e^-)

outer electronic configuration:

$$\underbrace{\uparrow\ \uparrow\ \uparrow\ \uparrow\ \underline{\ }}_{3d} \quad \underbrace{\text{xx xx xx xx xx xx}}_{sp^3d^2} \quad \underbrace{\underline{\ }\ \underline{\ }\ \underline{\ }}_{4d}$$

(c) [Co(CN)₆]³⁻ $[Co(CN)_6]^{3-}$ inner orbital (strong field) complex ion of Co^{3+} (d^6)
diamagnetic (0 unpaired e^-)

outer electronic configuration:

$$\underset{3d}{\underline{\uparrow\downarrow}\,\underline{\uparrow\downarrow}\,\underline{\uparrow\downarrow}} \quad \underset{d^2sp^3}{\underline{xx}\,\underline{xx}\,\underline{xx}\,\underline{xx}\,\underline{xx}\,\underline{xx}} \quad \underset{4d}{\underline{}\,\underline{}\,\underline{}\,\underline{}\,\underline{}}$$

(d) [Cr(NH₃)₆]³⁺ inner orbital (strong field) complex ion of Cr^{3+} (d^3)
paramagnetic (3 unpaired e^-)

outer electronic configuration:

$$\underset{3d}{\underline{\uparrow}\,\underline{\uparrow}\,\underline{\uparrow}} \quad \underset{d^2sp^3}{\underline{xx}\,\underline{xx}\,\underline{xx}\,\underline{xx}\,\underline{xx}\,\underline{xx}} \quad \underset{4d}{\underline{}\,\underline{}\,\underline{}\,\underline{}\,\underline{}}$$

25-36. *Refer to Sections 25-8, 25-9 and 25-10, Tables 25-7 and 25-8, and Exercise 25-34 Solution.*

Using the spectrochemical series, we can propose the following electron configurations and hybridizations:

(a) [CrBr₂Cl₄]³⁻ inner orbital complex ion of Cr^{3+} (d^3); hybridization: d^2sp^3
paramagnetic (3 unpaired e^-)

outer electronic configuration:

$$\underset{3d}{\underline{\uparrow}\,\underline{\uparrow}\,\underline{\uparrow}} \quad \underset{d^2sp^3}{\underline{xx}\,\underline{xx}\,\underline{xx}\,\underline{xx}\,\underline{xx}\,\underline{xx}} \quad \underset{4d}{\underline{}\,\underline{}\,\underline{}\,\underline{}\,\underline{}}$$

(b) [Co(en)₃]³⁺ inner orbital (strong field) complex ion of Co^{3+} (d^6); hybridization: d^2sp^3
diamagnetic (0 unpaired e^-)

outer electronic configuration:

$$\underset{3d}{\underline{\uparrow\downarrow}\,\underline{\uparrow\downarrow}\,\underline{\uparrow\downarrow}} \quad \underset{d^2sp^3}{\underline{xx}\,\underline{xx}\,\underline{xx}\,\underline{xx}\,\underline{xx}\,\underline{xx}} \quad \underset{4d}{\underline{}\,\underline{}\,\underline{}\,\underline{}\,\underline{}}$$

(c) [Fe(OH₂)₆]³⁺ outer orbital (weak field) complex ion of Fe^{3+} (d^5); hybridization: sp^3d^2
paramagnetic (5 unpaired e^-)

outer electronic configuration:

$$\underset{3d}{\underline{\uparrow}\,\underline{\uparrow}\,\underline{\uparrow}\,\underline{\uparrow}\,\underline{\uparrow}} \quad \underset{sp^3d^2}{\underline{xx}\,\underline{xx}\,\underline{xx}\,\underline{xx}\,\underline{xx}\,\underline{xx}} \quad \underset{4d}{\underline{}\,\underline{}\,\underline{}}$$

(d) [Fe(NO₂)₆]³⁻ inner orbital (strong field) complex ion of Fe^{3+} (d^5); hybridization: d^2sp^3
paramagnetic (1 unpaired e^-)

outer electronic configuration:

$$\underset{3d}{\underline{\uparrow\downarrow}\,\underline{\uparrow\downarrow}\,\underline{\uparrow}} \quad \underset{d^2sp^3}{\underline{xx}\,\underline{xx}\,\underline{xx}\,\underline{xx}\,\underline{xx}\,\underline{xx}} \quad \underset{4d}{\underline{}\,\underline{}\,\underline{}\,\underline{}\,\underline{}}$$

25-38. *Refer to Section 25-8 and Table 25-7.*

	Complex Ion	Ox. No. of Metal	Electronic Configuration		Unpaired e^- in Complex
			Isolated Metal Ion	Metal Ion in Complex	
(a)	[Fe(CN)₆]³⁻	+3	[Ar] $\uparrow\ \uparrow\ \uparrow\ \uparrow\ \uparrow$	[Ar] $\uparrow\downarrow\,\uparrow\downarrow\,\uparrow\ \underline{}\ \underline{}$	1
(b)	[Fe(OH₂)₆]³⁺	+3	[Ar] $\uparrow\ \uparrow\ \uparrow\ \uparrow\ \uparrow$	[Ar] $\uparrow\ \uparrow\ \uparrow\ \uparrow\ \uparrow$	5
(c)	[Mn(OH₂)₆]²⁺	+2	[Ar] $\uparrow\ \uparrow\ \uparrow\ \uparrow\ \uparrow$	[Ar] $\uparrow\ \uparrow\ \uparrow\ \uparrow\ \uparrow$	5
(d)	[Co(NH₃)₆]³⁺	+3	[Ar] $\underset{3d}{\uparrow\downarrow\,\uparrow\ \uparrow\ \uparrow\ \uparrow}$	[Ar] $\underset{3d}{\uparrow\downarrow\,\uparrow\downarrow\,\uparrow\downarrow\ \underline{}\ \underline{}}$	0

Crystal Field Theory is a theory of bonding in transition metal complexes in which the bonds between metal ions and ligands are strictly electrostatic interactions. During bonding, the repulsions between ligand electrons and metal electrons in d orbitals create an electric field, i.e., the crystal field, which splits the d orbitals into two sets, the t_{2g} set at lower energy and the e_g set at higher energy. The energy separation between the two sets in an octahedral complex is named Δ_{oct} and is proportional to the crystal field strength of the ligands, that is, how strongly the ligand electrons repel the metal electrons.

When electrons undergo transitions from a lower energy t_{2g} orbital to a higher energy e_g orbital, an amount of energy equivalent to the wavelengths of visible light are absorbed, resulting in transition metal complexes with the complementary color of the light absorbed. Δ_{oct} can be determined experimentally from the wavelength of the light absorbed:

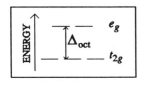

$$\Delta_{oct} = EN_A = \frac{hcN_A}{\lambda}$$

where

E = energy of the absorbed photon
N_A = Avogadro's Number
h = Planck's constant
c = speed of light
λ = wavelength

It is possible to arrange the common ligands in the order of increasing crystal field strengths, by interpreting the visible spectra of many complexes. This is the spectrochemical series, shown in Exercise 25-34. From the above equation, we can deduce that transition metal complexes that are colored, e.g. red or orange, are absorbing the complementary colors, green or blue. These absorbed wavelengths are at the shorter end of the visible light range and correspond to larger Δ_{oct} values. The ligands in these complexes have larger crystal field strengths and are located at the high end of the spectrochemical series.

A high spin complex is the Crystal Field designation for an outer orbital complex where all t_{2g} and e_g orbitals are singly occupied before pairing begins. A low spin complex, on the other hand, is the Crystal Field designation for an inner orbital complex. It contains electrons paired in t_{2g} orbitals before e_g orbitals are occupied. However, the low spin configuration exists only for octahedral complexes having metal ions with d^4 - d^7 configurations. For d^1 - d^3 and d^8 - d^{10} ions, only one possibility exists which is designated as high spin. In the case of d^4 - d^7 configurations: (1) weak ligand field strength is associated with high spin (outer orbital) complexes, whereas strong ligand field strength is associated with low spin (inner orbital) complexes, and (2) the spectrochemical series ranks ligands in order of increasing ligand field strength. The following are the predictions:

	Complex Ion	Metal Ion Configuration	Complex Configuration
(a)	$[Cu(OH_2)_6]^{2+}$	d^9	high spin
(b)	$[MnF_6]^{3-}$	d^4	high spin (weak field strength ligands)
(c)	$[Co(CN)_6]^{3-}$	d^6	low spin (strong field strength ligands)
(d)	$[Cr(NH_3)_6]^{3+}$	d^3	high spin (by convention)

Metal Ion	Ligand Field Strength	Example	
V^{2+}	weak	$[VF_6]^{4-}$	hexafluorovanadate(II) ion
		$[V(OH_2)_6]^{2+}$	hexaaquavanadium(II) ion
Mn^{2+}	strong	$[Mn(en)_3]^{2+}$	tris(ethylenediamine)manganese(II) ion
		$[Mn(NH_3)_6]^{2+}$	hexaamminemanganese(II) ion
Mn^{2+}	weak	$[MnF_6]^{4-}$	hexafluoromanganate(II) ion
		$[MnBr_6]^{4-}$	hexabromomanganate(II) ion
Ni^{2+}	weak	$[Ni(OH_2)_6]^{2+}$	hexaaquanickel(II) ion
		$[NiF_6]^{4-}$	hexafluoronickelate(II) ion
Cu^{2+}	weak	$[Cu(OH_2)_6]^{2+}$	hexaaquacopper(II) ion
		$[CuF_6]^{4-}$	hexafluorocuprate(II) ion
Fe^{3+}	strong	$[Fe(CN)_6]^{3-}$	hexacyanoferrate(III) ion
		$[Fe(NH_3)_6]^{3+}$	hexaammineiron(III) ion
Cu^{+}	weak	$[CuCl_6]^{5-}$	hexachlorocuprate(I) ion
		$[Cu(OH_2)_6]^{+}$	hexaaquacopper(I) ion
Ru^{3+}	strong	$[Ru(NH_3)_6]^{3+}$	hexaammineruthenium(III) ion
		$[Ru(en)_3]^{3+}$	tris(ethylenediamine)ruthenium(III) ion

25-46. *Refer to Section 25-11, Table 25-8 and the Key Terms for Chapter 25.*

Crystal field stabilization energy (CFSE) is a measure of the net energy of stabilization gained by a metal ion's nonbonding d electrons as a result of complex formation.

(a) $[Co(NH_3)_6]^{3+}$ is a low spin d^6 complex ion with a $t_{2g}^6 e_g^0$ configuration.

CFSE $= (6)(-2/5)(\Delta_{oct}) + (0)(+3/5)(\Delta_{oct}) = (6)(-2/5)(22{,}900 \text{ cm}^{-1} \times 0.01196 \text{ kJ/mol·cm}^{-1}) =$ **-657 kJ/mol**

(b) $[Ti(OH_2)_6]^{2+}$ is a high spin d^2 complex ion with a $t_{2g}^2 e_g^0$ configuration.

CFSE $= (2)(-2/5)(\Delta_{oct}) + (0)(+3/5)(\Delta_{oct}) = (2)(-2/5)(20{,}300 \text{ cm}^{-1} \times 0.01196 \text{ kJ/mol·cm}^{-1}) =$ **-194 kJ/mol**

(c) $[Cr(OH_2)]_6^{3+}$ is a high spin d^3 complex ion with a $t_{2g}^3 e_g^0$ configuration.

CFSE $= (3)(-2/5)(\Delta_{oct}) + (0)(+3/5)(\Delta_{oct}) = (3)(-2/5)(17{,}600 \text{ cm}^{-1} \times 0.01196 \text{ kJ/mol·cm}^{-1}) =$ **-253 kJ/mol**

(d) $[Co(CN)_6]^{3-}$ is a low spin d^6 complex ion with a $t_{2g}^6 e_g^0$ configuration.

CFSE $= (6)(-2/5)(\Delta_{oct}) + (0)(+3/5)(\Delta_{oct}) = (6)(-2/5)(33{,}500 \text{ cm}^{-1} \times 0.01196 \text{ kJ/mol·cm}^{-1}) =$ **-962 kJ/mol**

(e) $[Co(OH_2)_6]^{2+}$ is a high spin d^7 complex ion with a $t_{2g}^5 e_g^2$ configuration.

CFSE $= (5)(-2/5)(\Delta_{oct}) + (2)(+3/5)(\Delta_{oct})$
$= (5)(-2/5)(10{,}000 \text{ cm}^{-1} \times 0.01196 \text{ kJ/mol·cm}^{-1}) + (2)(+3/5)(10{,}000 \text{ cm}^{-1} \times 0.01196 \text{ kJ/mol·cm}^{-1})$
$=$ **-95.7 kJ/mol**

(f) $[Cr(en)_3]^{3+}$ is a high spin d^3 complex ion with a $t_{2g}^3 e_g^0$ configuration.

CFSE $= (3)(-2/5)(\Delta_{oct}) + (0)(+3/5)(\Delta_{oct}) = (3)(-2/5)(21{,}900 \text{ cm}^{-1} \times 0.01196 \text{ kJ/mol·cm}^{-1}) =$ **-314 kJ/mol**

(g) $[Cu(OH_2)_6]^{2+}$ is a high spin d^9 complex ion with a $t_{2g}^6 e_g^3$ configuration.

CFSE $= (6)(-2/5)(\Delta_{oct}) + (3)(+3/5)(\Delta_{oct})$
$= (6)(-2/5)(13,000 \text{ cm}^{-1} \times 0.01196 \text{ kJ/mol·cm}^{-1}) + (3)(+3/5)(13,000 \text{ cm}^{-1} \times 0.01196 \text{ kJ/mol·cm}^{-1})$
$= \textbf{-93.3 kJ/mol}$

(h) $[V(OH_2)_6]^{3+}$ is a high spin d^2 complex ion with a $t_{2g}^2 e_g^0$ configuration.

CFSE $= (2)(-2/5)(\Delta_{oct}) + (0)(+3/5)(\Delta_{oct}) = (2)(-2/5)(18,000 \text{ cm}^{-1} \times 0.01196 \text{ kJ/mol·cm}^{-1}) = \textbf{-172 kJ/mol}$

25-48. *Refer to Section 25-11 and Tables 25-8 and 25-11.*

(a) $[Mn(OH_2)_6]^{3+}$ is a high spin d^4 complex ion with a $t_{2g}^3 e_g^1$ configuration.

CFSE $= (3)(-2/5)(\Delta_{oct}) + (1)(+3/5)(\Delta_{oct}) = \textbf{-3/5 } \Delta_{\textbf{oct}}$

(b) $[Mn(CN)_6]^{3-}$ is a low spin d^4 complex ion with a $t_{2g}^4 e_g^0$ configuration.

CFSE $= (4)(-2/5)(\Delta_{oct}) + (0)(+3/5)(\Delta_{oct}) = \textbf{-8/5 } \Delta_{\textbf{oct}}$

25-50. *Refer to Section 15-13.*

Balanced equation: $Ag^+(aq) + 3I^-(aq) \rightarrow [AgI_3]^{2-}$

A *decrease* in entropy for the above reaction is expected since the reactant side has 4 moles of ions while the product side contains only 1 mole of ions.

$\Delta S^\circ_{rxn} = [S^\circ_{[AgI_3]^{2-}(aq)}] - [S^\circ_{Ag^+(aq)} + 3S^\circ_{I^-(aq)}]$
$= [(1 \text{ mol})(253.1 \text{ J/mol·K})] - [(1 \text{ mol})(72.68 \text{ J/mol·K}) + (3 \text{ mol})(111.3 \text{ J/mol·K})]$
$= \textbf{-153.5 J/(mol rxn)·K}$

Since $\Delta S^\circ_{rxn} < 0$, the entropy of this reaction is indeed decreasing, as predicted.

25-52. *Refer to Section 18-4 and Appendices G and I.*

Balanced equations: $[Cu(NH_3)_4]Cl_2 \rightarrow [Cu(NH_3)_4]^{2+} + 2Cl^-$ (to completion)
$[Cu(NH_3)_4]^{2+} \rightleftarrows Cu^{2+} + 4NH_3$ (reversible) $K_d = 8.5 \times 10^{-13}$
$NH_3 + H_2O \rightleftarrows NH_4^+ + OH^-$ (reversible) $K_b = 1.8 \times 10^{-5}$

Plan: (1) Calculate the concentration of NH_3 from the equilibrium expression for the dissociation of the complex ion. Note: this is possible only if we assume that the ionization of NH_3 does not appreciably alter the concentration of NH_3.

(2) Calculate the $[OH^-]$ and pH of the solution using the equilibrium expression for the ionization of NH_3. Note: we are ignoring the effect of the hydrolysis of Cu^{2+} ion on pH.

(1) Let x = $[[Cu(NH_3)_4]^{2+}]$ that dissociates. Then

	$[Cu(NH_3)_4]^{2+}$	$\rightleftarrows$	Cu^{2+}	+	$4 NH_3$
initial	0.25 M		0 M		0 M
change	- x M		+ x M		+ 4x M
at equilibrium	(0.25 - x) M		x M		4x M

$K_d = \dfrac{[Cu^{2+}][NH_3]^4}{[[Cu(NH_3)_4]^{2+}]} = \dfrac{(x)(4x)^4}{(0.25 - x)} = 8.5 \times 10^{-13} \approx \dfrac{(x)(4x)^4}{0.25} = \dfrac{256x^5}{0.25}$ Solving, x $= (8.3 \times 10^{-16})^{1/5}$
$= 9.6 \times 10^{-4}$

Therefore, $[NH_3] = 4x = 3.8 \times 10^{-3} M$

(2) Let $y = [NH_3]_{ionized}$. Then, $[NH_3] = 3.8 \times 10^{-3} - y$; $[NH_4^+] = [OH^-] = y$

$$K_b = \frac{[NH_4^+][OH^-]}{[NH_3]} = \frac{y^2}{(3.8 \times 10^{-3} - y)} = 1.8 \times 10^{-5}$$

Solving the quadratic equation: $y^2 + (1.8 \times 10^{-5})y - 6.8 \times 10^{-8} = 0$,
$$y = 2.5 \times 10^{-4} \text{ or } -2.7 \times 10^{-4} \text{ (discard)}$$

Therefore, $[OH^-] = 2.5 \times 10^{-4} M$; pOH = 3.60; **pH = 10.40**

25-54. *Refer to Sections 20-2, 20-3 and 20-6, Example 20-14, and Appendices H and I.*

(a) Balanced equation: $Zn(OH)_2(s) \rightleftarrows Zn^{2+}(aq) + 2OH^-(aq)$ $\qquad$ $K_{sp} = 4.5 \times 10^{-17}$

Let x = molar solubility of $Zn(OH)_2$. Then, $[Zn^{2+}] = x$; $[OH^-] = 2x$

$K_{sp} = [Zn^{2+}][OH^-]^2 = (x)(2x)^2 = 4x^3 = 4.5 \times 10^{-17}$ $\qquad$ Solving, $x = 2.2 \times 10^{-6}$

Therefore, molar solubility = **2.2×10^{-6} mol $Zn(OH)_2$/L**

(b) Balanced equations: $NaOH(aq) \rightarrow Na^+(aq) + OH^-(aq)$ $\qquad$ (to completion)

$\qquad\qquad\qquad\qquad$ $Zn(OH)_2(s) \rightleftarrows Zn^{2+}(aq) + 2OH^-(aq)$ $\qquad$ (reversible) $\qquad$ $K_{sp} = 4.5 \times 10^{-17}$

If we proceed with part (b) as if it were an ordinary example of a common ion effect problem with $Zn(OH)_2$ dissolving into a solution with a known concentration of OH^- ion, we would get the wrong answer. We must take into account the effect of the formation of the complex ion, $Zn(OH)_4^{2-}$, on the solubility of $Zn(OH)_2$.

$$Zn(OH)_4^{2-}(aq) \rightleftarrows Zn^{2+}(aq) + 4OH^-(aq) \qquad \text{(reversible)} \qquad K_d = 3.5 \times 10^{-16}$$

Much more $Zn(OH)_2$ will dissolve since essentially all of the released Zn^{2+} ions are incorporated into the soluble $Zn(OH)_4^{2-}$ complex because K_d is so small. When significant complex ion formation occurs as it does in this case, the molar solubility must include the concentrations of *all* the Zn^{2+} species. Also we cannot assume that $[OH^-]$ remains at 0.30 M throughout this process. The net concentration of OH^- ion decreases slightly: for every one formula unit of $Zn(OH)_2$ that dissolves producing 2 OH^- ions, one formula unit of $Zn(OH)_4^{2-}$ will form, removing 4 OH^- ions. The net result is that for every 1 mol/L of $Zn(OH)_2$ that dissolves (the molar solubility), approximately 1 mol/L of $Zn(OH)_4^{2-}$ is produced, but 2 mol/L of OH^- is lost. Therefore,

$$[OH^-] = 0.30 \, M - 2[Zn(OH)_4^{2-}]$$

Plan: $\quad$ The molar solubility of $Zn(OH)_2$ equals the sum of the concentrations of the 2 soluble Zn species, Zn^{2+} and $Zn(OH)_4^{2-}$. The concentrations can be calculated by solving the following 3 equations in 3 unknowns:

(1) $[OH^-] = 0.30 \, M - 2[Zn(OH)_4^{2-}]$

(2) $K_{sp} = 4.5 \times 10^{-17} = [Zn^{2+}][OH^-]^2$

(3) $K_d = 3.5 \times 10^{-16} = \dfrac{[Zn^{2+}][OH^-]^4}{[Zn(OH)_4^{2-}]}$

Step 1: $\quad$ Using equation (1), let $x = [OH^-]$. Then, $[Zn(OH)_4^{2-}] = \dfrac{0.30 - x}{2}$

Step 2: $\quad$ Divide equation (3) by equation (2) to remove the $[Zn^{2+}]$ term, substitute and solve for x.

$$\frac{K_d}{K_{sp}} = \frac{3.5 \times 10^{-16}}{4.5 \times 10^{-17}} = 7.8 = \frac{\left(\dfrac{[Zn^{2+}][OH^-]^4}{[Zn(OH)_4^{2-}]}\right)}{[Zn^{2+}][OH^-]^2} = \frac{[OH^-]^2}{[Zn(OH)_4^{2-}]} = \frac{x^2}{\left(\dfrac{0.30 - x}{2}\right)} = \frac{2x^2}{0.30 - x}$$

Step 3: Solving the quadratic equation: $2x^2 + 7.8x - 2.3 = 0$, we have: $x = 0.28$ or -4.2 (discard)

Therefore, $[OH^-] = x = 0.28\ M$ (a value slightly less than $0.30\ M$, as expected)

$$[Zn^{2+}] = \frac{K_{sp}}{[OH^-]^2} = \frac{4.5 \times 10^{-17}}{(0.28)^2} = 5.7 \times 10^{-16}\ M$$

$$[Zn(OH)_4{}^{2-}] = \frac{[Zn^{2+}][OH^-]^4}{K_d} = \frac{(5.7 \times 10^{-16})(0.28)^4}{3.5 \times 10^{-16}} = 0.010\ M$$

molar solubility = total number of moles of $Zn(OH)_2$ that dissolved to produce 1 L of saturated solution
$$= [Zn^{2+}] + [Zn(OH)_4{}^{2-}]$$
$$= (5.7 \times 10^{-16}\ M) + (0.010\ M)$$
$$\approx \textbf{0.010 mol } Zn(OH)_2\textbf{/L}$$

(c) $[Zn(OH)_4{}^{2-}] = \textbf{0.010}\ \textbf{\textit{M}}$ (see part (b))

Note: If you assume that $[OH^-]$ is constant at $0.30\ M$, then an approximate value for molar solubility is calculated.

Substituting into the rearranged K_{sp} expression: $[Zn^{2+}] = \dfrac{K_{sp}}{[OH^-]^2} = \dfrac{4.5 \times 10^{-17}}{(0.30)^2} = 5.0 \times 10^{-16}\ M$

Substituting into the rearranged K_d expression: $[Zn(OH)_4{}^{2-}] = \dfrac{[Zn^{2+}][OH^-]^4}{K_d} = \dfrac{(5.0 \times 10^{-16})(0.30)^4}{3.5 \times 10^{-16}} = 0.012\ M$

Therefore, molar solubility $= [Zn^{2+}] + [Zn(OH)_4{}^{2-}] = (5.0 \times 10^{-16}\ M) + (0.012\ M) \approx 0.012$ mol $Zn(OH)_2$/L

26 Nuclear Chemistry

26-2. *Refer to Sections 1-1, and 26-3.*

Einstein's equation relates matter and energy:

$$E = mc^2 \qquad \text{where } E = \text{amount of energy released}$$
$$m = \text{mass of matter transformed into energy}$$
$$c = \text{speed of light in a vacuum, } 3.00 \times 10^8 \text{ m/s}$$

If m is expressed in kg and c in m/s, the obtained E will be in units of J.

26-4. *Refer to Sections 26-1 and 5-4.*

Nucleons are the particles comprising the nucleus, i.e., it is a collective term for the protons and neutrons in a nucleus. The number of protons is the atomic number; the sum of the protons and the neutrons (the nucleons) is the mass number.

26-6. *Refer to Section 26-3 and Figure 26-10.*

The plot of binding energy per nucleon versus mass number for all the isotopes shows that binding energies/nucleon increase very rapidly with increasing mass number, reaching a maximum of 8.80 MeV per nucleon at mass number 56 for $^{56}_{26}$Fe, then decrease slowly.

26-8. *Refer to Section 26-3, Table 26-1, Examples 26-1 and 26-2, and Appendix C.*

(a) One neutral atom of $^{79}_{35}$Br contains 35 e^-, 35 p^+ and 44 n^0.

electrons:	35 × 0.00054858 amu	= 0.019 amu
protons:	35 × 1.0073 amu	= 35.256 amu
neutrons:	44 × 1.0087 amu	= 44.383 amu
	sum	= 79.658 amu

Δm = (sum of masses of e^-, p^+ and n^0) - (actual mass of a ^{79}Br atom)

= 79.658 amu - 78.91834 amu

= 0.740 amu

Therefore, the mass deficiency for ^{79}Br is **0.740 amu/atom** or **0.740 g/mol**.

(b) Note: 1 joule = 1 kg × (1 m/s)^2

The nuclear binding energy, $BE = (\Delta m)c^2$

$= (0.740 \times 10^{-3} \text{ kg/mol})(3.00 \times 10^8 \text{ m/s})^2$

$= 6.66 \times 10^{13} \text{ kg·m}^2/\text{mol·s}^2$

$= 6.66 \times 10^{13}$ J/mol or **6.66×10^{10} kJ/mol of ^{79}Br atoms**

26-10. Refer to Section 26-3, Table 26-1, Examples 26-1 and 26-2, and Appendix C.

(a) A neutral atom of $^{52}_{24}$Cr contains 24 e^-, 24 p^+ and 28 n^0.

electrons:	24 × 0.00054858 amu =	0.013 amu
protons:	24 × 1.0073 amu	= 24.175 amu
neutrons:	28 × 1.0087 amu	= 28.244 amu
	sum =	52.432 amu

Δm = (sum of masses of e^-, p^+ and n^0) - (actual mass of a ^{52}Cr atom)

$= 52.432$ amu - 51.94059 amu

$= 0.491$ amu

the mass deficiency, Δm, for ^{52}Cr is **0.491 amu/atom**

(b) This is equivalent to a mass deficiency of **0.491 g/mol**.

(c) Note: $1\ J = 1\ kg \times (1\ cm/s)^2$

The nuclear binding energy, $BE = (\Delta m)c^2$

$$= \left[\frac{0.491\ g}{1\ mol} \times \frac{1\ kg}{1000\ g} \times \frac{1\ mol}{6.02 \times 10^{23}\ atoms} \right] (3.00 \times 10^8\ m/s)^2$$

$$= 7.34 \times 10^{-11}\ kg \cdot m^2/atom \cdot s^2$$

$$= \mathbf{7.34 \times 10^{-11}\ J/atom}$$

(d) BE (kJ/mol) $= \dfrac{7.34 \times 10^{-11}\ J}{1\ atom} \times \dfrac{6.02 \times 10^{23}\ atoms}{1\ mol} \times \dfrac{1\ kJ}{1000\ J} = \mathbf{4.42 \times 10^{10}\ kJ/mol}$

(e) Since there are 52 nucleons in a ^{52}Cr atom,

BE (MeV/nucleon) for ^{52}Cr $= \dfrac{7.34 \times 10^{-11}\ J}{1\ atom} \times \dfrac{1\ atom}{52\ nucleons} \times \dfrac{1\ MeV}{1.60 \times 10^{-13}\ J} = \mathbf{8.82\ MeV/nucleon}$

26-12. Refer to Section 26-3 and Examples 26-1 and 26-2.

(a) One neutral atom of $^{20}_{10}$Ne contains 10 e^-, 10 p^+ and 10 n^0.

electrons:	10 × 0.00054858 amu =	0.005 amu
protons:	10 × 1.0073 amu	= 10.073 amu
neutrons:	10 × 1.0087 amu	= 10.087 amu
	sum =	20.165 amu

Δm = (sum of masses of e^-, p^+ and n^0) - (actual mass of a ^{20}Ne atom)

$= 20.165$ amu - 19.99244 amu

$= 0.173$ amu

Therefore, the mass deficiency for ^{20}Ne is 0.173 amu/atom or 0.173 g/mol.

Recall: 1 joule $= 1\ kg \times (1\ m/s)^2$

The nuclear binding energy, $BE = (\Delta m)c^2$

$$= (0.173 \times 10^{-3}\ kg/mol)(3.00 \times 10^8\ m/s)^2$$

$$= 1.56 \times 10^{13}\ kg \cdot m^2/mol \cdot s^2$$

$$= 1.56 \times 10^{13}\ J/mol\ or\ \mathbf{1.56 \times 10^{10}\ kJ/mol\ of\ {}^{20}Ne\ atoms}$$

(b) One neutral atom of $^{87}_{37}Rb$ contains 37 e^-, 37 p^+ and 50 n^0.

electrons: 37 × 0.00054858 amu = 0.020 amu

protons: 37 × 1.0073 amu   = 27.270 amu

neutrons: 50 × 1.0087 amu   = 50.435 amu

                               sum = 87.725 amu

Δm = 87.725 amu - 86.909187 amu = 0.816 amu/atom or 0.816 g/mol

$BE = (\Delta m)c^2 = (0.816 \times 10^{-3}$ kg/mol$)(3.00 \times 10^8$ m/s$)^2 = 7.34 \times 10^{13}$ kg·m^2/mol·s^2

$$= 7.34 \times 10^{13} \text{ J/mol}$$

$$\text{or } \mathbf{7.34 \times 10^{10} \text{ kJ/mol of } ^{87}Rb \text{ atoms}}$$

(c) One neutral atom of $^{106}_{46}Pd$ contains 46 e^-, 46 p^+ and 60 n^0.

electrons: 46 × 0.00054858 amu = 0.025 amu

protons: 46 × 1.0073 amu   = 46.336 amu

neutrons: 60 × 1.0087 amu   = 60.522 amu

                               sum =106.883 amu

Δm = 106.883 amu - 105.9032 amu = 0.980 amu/atom or 0.980 g/mol

$BE = (\Delta m)c^2 = (0.980 \times 10^{-3}$ kg/mol$)(3.00 \times 10^8$ m/s$)^2 = 8.82 \times 10^{13}$ kg·m^2/mol·s^2

$$= 8.82 \times 10^{13} \text{ J/mol}$$

$$\text{or } \mathbf{8.82 \times 10^{10} \text{ kJ/mol of } ^{106}Pd \text{ atoms}}$$

From the above calculations, we can see that ^{87}Rb, an isotope in the intermediate mass range, has the highest binding energy per nucleon of the three isotopes.

26-14.   *Refer to Section 26-4 and Table 26-3.*

(a) In an electric field, an alpha (α) particle (a helium nucleus with a +2 charge) will be drawn toward the negative electrode, while a beta (β^-) particle (an electron with a -1 charge) will be drawn toward the positive electrode. Gamma (γ) radiation (very high energy electromagnetic radiation) will be unaffected by the electric field.

(b) The α particle and β^- particle will be drawn in opposite directions in a magnetic field and the γ radiation will be unaffected by a magnetic field.

(c) A piece of paper will reduce the α radiation significantly, but not the β^- or γ radiation; a thick concrete slab will prevent the α particles and β^- particles from passing, and most of the γ radiation.

26-16.   *Refer to Sections 26-9 and 26-11, and Example 26-3.*

There are many radionuclides that have medical uses, including:

(1) cobalt-60: This is used to arrest certain types of cancer. The technique uses the gamma rays produced in the decay of ^{60}Co to destroy cancerous tissue.

(2) plutonium-238: The energy produced in its decay is converted to electrical energy which powers heart pacemakers. Its relatively long half-life allows the device to be used for ten years before replacement.

Several radioisotopes are used as radioactive tracers. These are injected into the body and allow physicians to study biological processes. These include

(3) sodium-24: This is used to follow the blood flow and locate stoppages in the circulatory system.

(4) thallium-201: This isotope helps locate healthy heart tissue.

(5) technicium-99: This isotope concentrates in abnormal heart tissue. Both this and thallium-201 can help to look at the damage caused by heart disease.

(6) iodine-131: This concentrates in the thyroid gland, liver and certain parts of the brain. It is used to monitor goiter and other thyroid problems as well as liver and brain tumors.

26-18. *Refer to Section 26-8.*

(1) Photographic Detection: Radioactive substances affect photographic plates. Although the intensity of the affected spot is related to the amount of radiation, precise measurement by this method is tedious.

(2) Detection by Fluorescence: Fluorescent substances can absorb radiation and subsequently emit visible light. This is the basis for scintillation counting and can be used for quantitative detection.

(3) Cloud Chambers: A chamber containing air saturated with vapor is used. Radioactive particles ionize air molecules in the chamber. Cooling the chamber causes droplets of liquid to condense on these ions, giving observable fog-like tracks.

(4) Gas Ionization Counters: A common gas ionization counter is the Geiger-Müller counter where the electronic pulses derived from the ionization process are registered as counts.

26-20. *Refer to Section 26-2.*

Scientists have known that nuclides which have certain "magic numbers" of protons and neutrons are especially stable. Nuclides with a number of protons or a number of neutrons or a sum of the two equal to 2, 8, 20, 28, 50, 82 or 126 have unusual stability. Examples of this are $_2^4He$, $_8^{16}O$, $_{20}^{42}Ca$, $_{38}^{88}Sr$, and $_{82}^{208}Pb$. This suggests an energy level (shell) model for the nucleus similar to the shell model of electron configurations.

26-22. *Refer to Section 26-6 and Figure 26-1.*

A nuclide below the band of stability, i.e., with a neutron/proton ratio which is smaller than that for a stable isotope of the element can increase its ratio by undergoing:

(1) positron emission $\qquad\qquad _1^1p \rightarrow\ _0^1n +\ _{+1}^0\beta$
(2) electron capture (K capture) $\qquad _1^1p +\ _{-1}^0e \rightarrow\ _0^1n$

The net result of both processes is the loss of one proton and the gain of one neutron, thereby increasing the n/p ratio. Also, a heavier nuclide can undergo alpha emission to increase its n/p ratio.

26-24. *Refer to Sections 26-5, 26-6 and 26-7, and Figure 26-1.*

	Protons	Neutrons	n/p Ratio	Stable Mass No.	Stable n/p Ratio
(a) $_{79}^{193}Au$	79	114	1.44	197	1.49
(b) $_{75}^{189}Re$	75	114	1.52	185, 187	1.47, 1.49
(c) $_{59}^{137}Pr$	59	78	1.32	141	1.39

Both (a) ^{193}Au and (c) ^{137}Pr have n/p ratios below that which is stable, so they will either undergo positron emission or electron capture. However, (b) ^{189}Re has an n/p ratio greater than that which is stable, and so it will undergo beta emission (neutron emission is less common).

26-26. *Refer to Sections 26-5, 26-6 and 26-7, and Figure 26-1.*

(a) $_{27}^{60}Co$ (n/p ratio too high)   beta emission (neutron emission is less common)

(b) $_{11}^{20}Na$ (n/p ratio too low)   positron emission or electron capture (K capture)

(c) $_{88}^{218}Rn$ alpha emission

(d) $_{29}^{67}Cu$ beta emission

(e) $^{238}_{92}U$ — alpha emission

(f) $^{11}_{6}C$ — positron emission or electron capture (K capture)

26-28. *Refer to Sections 26-5 and 26-12.*

In equations for nuclear reactions, the sums of the mass numbers and atomic numbers of the reactants must equal the sums for the products. Therefore,

(a) $^{23}_{11}Na + \boxed{^{1}_{1}H} \rightarrow ^{23}_{12}Mg + ^{1}_{0}n$

(b) $^{59}_{27}Co + ^{1}_{0}n \rightarrow ^{56}_{25}Mn + \boxed{^{4}_{2}He}$

(c) $^{232}_{90}Th + \boxed{^{12}_{6}C} \rightarrow ^{240}_{96}Cm + 4\,^{1}_{0}n$

(d) $\boxed{^{28}_{13}Al} + ^{1}_{1}H \rightarrow ^{29}_{14}Si + ^{0}_{0}\gamma$

(e) $^{26}_{12}Mg + \boxed{^{1}_{1}H} \rightarrow ^{26}_{13}Al + ^{1}_{0}n$

(f) $^{40}_{18}Ar + ^{4}_{2}He \rightarrow \boxed{^{43}_{19}K} + ^{1}_{1}H$

26-30. *Refer to Section 26-12.*

The equation for a nuclear reaction can be given in the following abbreviated form:

parent nucleus (bombarding particle, emitted particle) daughter nucleus.

(a) $^{14}_{7}N + ^{4}_{2}He \rightarrow ^{17}_{8}O + ^{1}_{1}H$

(b) $^{106}_{46}Pd + ^{1}_{0}n \rightarrow ^{106}_{45}Rh + ^{1}_{1}H$

(c) $^{23}_{11}Na + ^{1}_{0}n \rightarrow \boxed{^{24}_{12}Mg} + ^{0}_{-1}e$

26-32. *Refer to Section 26-12 and Exercise 26-30 Solution.*

(a) $^{6}_{3}Li\ (n, \alpha)\ ^{3}_{1}H$

(b) $^{31}_{15}P\ (d, p)\ ^{32}_{15}P$

(c) $^{238}_{92}U\ (n, \beta^-)\ ^{239}_{93}Np$

26-34. *Refer to Sections 26-5 and 26-12.*

(a) $^{63}_{28}Ni \rightarrow ^{63}_{29}Cu + ^{0}_{-1}e$

(b) $2\,^{2}_{1}H \rightarrow ^{3}_{2}He + ^{1}_{0}n$

(c) $\boxed{^{10}_{5}B} + ^{1}_{0}n \rightarrow ^{7}_{3}Li + ^{4}_{2}He$

(d) $^{14}_{7}N + ^{1}_{0}n \rightarrow ^{3}_{1}H + 3\,^{4}_{2}He$

26-36. *Refer to Section 26-7.*

Plan: Balance the nuclear reaction and identify the unknown "radioactinium."

$$^{235}_{92}U \rightarrow \boxed{^{227}_{90}Th} + 2\,^{4}_{2}He + 2\,^{0}_{-1}\beta$$

Therefore, "radioactinium" is the element thorium, **Th**, with atomic number **90** and mass number **227**.

26-38. *Refer to Sections 26-7 and 26-12.*

Balanced nuclear reactions: (1) $^{249}_{98}Cf + ^{12}_{6}C \rightarrow ^{257}_{104}Unq + 4\,^{1}_{0}n$

(2) $^{257}_{104}Unq \rightarrow \boxed{^{253}_{102}No} + ^{4}_{2}He$

The element **nobelium**, No, is formed.

26-40. Refer to Sections 26-5 and 26-12.

In equations for nuclear reactions, the sums of the mass numbers and atomic numbers of the reactants must equal the sums for the products. Therefore,

(a) $^{14}_{7}N + \boxed{^{4}_{2}He} \rightarrow \, ^{17}_{8}O + \, ^{1}_{1}p$ (b) $^{235}_{92}U + \, ^{1}_{0}n \rightarrow \, ^{137}_{54}Xe + 2 \, ^{1}_{0}n + \boxed{^{97}_{41}Nb}$ (c) $^{241}_{95}Am + \, ^{12}_{6}C \rightarrow 4 \, ^{1}_{0}n + \boxed{^{249}_{101}Md}$

26-42. Refer to Section 26-9.

The half-life of a radionuclide represents the amount of time required for half of the sample to decay. Relative stabilities of radionuclides are indicated by their half-life values. The shorter the half-life, the less stable is the radionuclide.

26-44. Refer to Section 26-11.

The radioisotope carbon-14 is produced continuously in the atmosphere as nitrogen atoms capture cosmic-ray neutrons:

$$^{14}_{7}N + \, ^{1}_{0}n \rightarrow \, ^{14}_{6}C + \, ^{1}_{1}H$$

The carbon-14 atoms react with O_2 to form $^{14}CO_2$. Like ordinary $^{12}CO_2$, it is removed from the atmosphere by living plants through the process of photosynthesis. As long as the cosmic-ray intensity remains constant, the amount of $^{14}CO_2$ and therefore its ratio to $^{12}CO_2$ in the atmosphere remains constant. Consequently, a certain fraction of carbon atoms in all living substances becomes carbon-14, a beta particle emitter with a half-life of 5730 years:

$$^{14}_{6}C \rightarrow \, ^{14}_{7}N + \, ^{0}_{-1}e$$

A steady state ratio of $^{14}C/^{12}C$ is maintained in living plants and organisms. After death the plant no longer carries out photosynthesis, so it no longer takes up $^{14}CO_2$. The radioactive emissions from the carbon-14 in dead tissue then decrease with the passage of time. The activity per gram of carbon in the sample in comparison with that in air gives a measure of the length of time elapsed since death. This is the basis of radiocarbon dating. This technique is useful only when dating objects that are less than 50,000 years old (roughly 10 times the half-life of carbon-14). Older objects have too little activity to be accurately dated. This technique depends on cosmic-ray intensity being constant or at least predictable in order to keep the $^{14}C/^{12}C$ known throughout the time interval. Also, the sample must not be contaminated with organic matter having a different $^{14}C/^{12}C$ ratio.

26-46. Refer to Section 26-9.

For first order kinetics, $\log\left(\dfrac{A_o}{A}\right) = \dfrac{kt}{2.303}$ and $t_{1/2} = \dfrac{0.693}{k}$ where A_o = initial amount of isotope
A = amount remaining after time, t
k = rate constant (units of time^{-1})
$t_{1/2}$ = half-life of isotope

Plan: (1) Calculate the rate constant, k, from the half-life of carbon-11.
        (2) Assume that the initial amount of carbon-11 is 100%. Calculate the time required for decay, t, using the first order rate equation.

(1) $k = \dfrac{0.693}{t_{1/2}} = \dfrac{0.693}{20.3 \text{ min}} = 0.0341 \text{ min}^{-1}$

(2)   When 95.0% of the sample has decayed away, 5.0% of the sample remains.
     Substituting into the first order rate equation,

$$\log\left(\frac{100.0\%}{5.0\%}\right) = \frac{(0.0341 \text{ min}^{-1})t}{2.303}$$

$$1.30 = 0.0148t$$

$$t = \textbf{87.9 min}$$

When 99.5% of the sample has decayed, 0.5% of the sample remains. Substituting,

$$\log\left(\frac{100.0\%}{0.5\%}\right) = \frac{(0.0341 \text{ min}^{-1})t}{2.303}$$

$$2.30 = 0.0148t$$

$$t = \textbf{155 min}$$

26-48. *Refer to Section 26-9 and Exercise 26-46 Solution.*

Balanced equation: $^{8}_{4}\text{Be} \rightarrow 2\,^{4}_{2}\text{He}$

Plan: (1) Determine the rate constant, k, from the half-life for ^{8}Be.
(2) Calculate the time required for 99.99% of ^{8}Be to decay.

(1) $k = \dfrac{0.693}{t_{1/2}} = \dfrac{0.693}{0.07 \text{ fs}} = 9.9 \text{ fs}^{-1}$ (Note: 1 fs = 10^{-15} s)

(2) The first order rate equation: $\log\left(\dfrac{A_o}{A}\right) = \dfrac{kt}{2.303}$

If 99.99% of ^{8}Be decayed away, then 0.01% remains. Substituting,

$$\log\left(\frac{100.00\%}{0.01\%}\right) = \frac{(9.9 \text{ fs}^{-1})t}{2.303}$$

$$4.0 = 4.3t$$

$$t = \textbf{0.93 fs}$$

26-50. *Refer to Sections 26-9 and 26-11, and Exercise 26-49.*

Plan: (1) Calculate the first order rate constant, k, from the half-life of carbon-14.
(2) Determine the age of the wood.

(1) $k = \dfrac{0.693}{t_{1/2}} = \dfrac{0.693}{5730 \text{ yr}} = 1.21 \times 10^{-4} \text{ yr}^{-1}$

(2) The first order rate equation: $\log\left(\dfrac{A_o}{A}\right) = \dfrac{kt}{2.303}$

Note: The ^{14}C activity is directly proportional to the amount of ^{14}C present. Therefore, we can substitute the activity values directly into the first order rate equation.

Substituting, $\log\left(\dfrac{15.3 \text{ min·gC}}{12.7 \text{ min·gC}}\right) = \dfrac{(1.21 \times 10^{-4} \text{ yr}^{-1})t}{2.303}$

$$8.09 \times 10^{-2} = (5.25 \times 10^{-5})t$$

$$t = \textbf{1540 yr}$$

26-52. *Refer to Sections 26-9 and 26-11.*

Plan: (1) Calculate the first order rate constant, k, from the half-life of carbon-14.
(2) Determine the age of the object.

(1) $k = \dfrac{0.693}{t_{1/2}} = \dfrac{0.693}{5730 \text{ yr}} = 1.21 \times 10^{-4} \text{ yr}^{-1}$

(2) The first order rate equation: $\log\left(\dfrac{A_o}{A}\right) = \dfrac{kt}{2.303}$

Substituting, $\log\left[\dfrac{7.50\ \mu g}{0.72\ \mu g}\right] = \dfrac{(1.21 \times 10^{-4}\ \text{yr}^{-1})t}{2.303}$

$$1.02 = (5.25 \times 10^{-5})t$$

$$t = \mathbf{1.94 \times 10^4\ yr}$$

26-54. Refer to Sections 26-13 and 26-14, and the Key Terms for Chapter 26.

A chain reaction is a reaction that sustains itself once it has begun and may even expand. Normally, the limiting reactant is regenerated as a product to maintain the progress of the chain. Nuclear fission processes are considered chain reactions because the number of neutrons produced in the reaction equals or is greater than the number of neutrons absorbed by the fissioning nucleus. For example:

$$^{235}_{92}\text{U} + ^{1}_{0}n \rightarrow [^{236}_{92}\text{U}] \rightarrow ^{140}_{56}\text{Ba} + ^{93}_{36}\text{Kr} + 3\,^{1}_{0}n + \text{energy}$$

The critical mass of a fissionable material is the minimum mass of a particular fissionable nuclide in a set volume that is necessary to sustain a nuclear chain reaction.

26-56. Refer to Sections 26-13 and 26-3.

(a) $^{7}_{3}\text{Li} + ^{1}_{1}\text{H} \rightarrow \boxed{^{4}_{2}\text{He}} + ^{4}_{2}\text{He}$

(b) Plan: (1) Determine the mass difference between the products and the reactants. The difference, Δm, is directly related to the energy involved in the reaction.
    (2) Calculate the amount of energy involved.

(1) Δm = mass of products - mass of reactants

$= (2 \times$ mass of $^{4}_{2}\text{He}) - ($mass of $^{7}_{3}\text{Li} + ^{1}_{1}\text{H})$

$= (2 \times 4.00260$ amu$) - (7.01600$ amu $+ 1.007825$ amu$)$

$= -0.01862$ amu or -0.01862 g/mol rxn

(2) $\Delta E = (\Delta m)c^2$

$= (-1.862 \times 10^{-5}\ \text{kg/mol rxn})(3.00 \times 10^8\ \text{m/s})^2$

$= -1.68 \times 10^{12}\ \text{kg} \cdot \text{(m/s)}^2/\text{mol rxn}$

$= -1.68 \times 10^{12}\ \text{J/mol rxn}$ or $\mathbf{-1.68 \times 10^9\ kJ/mol\ rxn}$

Since $\Delta E < 0$, energy is being released in this fusion reaction, as expected.

26-58. Refer to Section 26-15.

The primary advantage of nuclear energy is that enormous amounts of energy are liberated per unit mass of fuel. Also, the air pollution (oxides of S, N, C and particulate matter) caused by fossil fuel electric power plants is not a problem with nuclear energy plants. In European countries, where fossil fuel reserves are scarce, most of the electricity is generated by nuclear power plants for these reasons.

There are, however, some disadvantages associated with nuclear power from controlled fission reactions. The radionuclides must be properly shielded to protect the workers and the environment from radiation and contamination. Spent fuel, containing long-lived radioisotopes, must be disposed of carefully using special containers placed underground in geologically inactive areas. This is because the radiation from the fuel is biologically dangerous and must be contained until the fuel has decayed to the point when it is no longer dangerous. The problem is that the time involved could be several hundred thousand years. If there is inadequate cooling in the reactor, there is the possibility of overheating the fuel and causing a "meltdown." This cooling water can cause biological damage to aquatic life if it is returned to the natural water system while it is still too warm. Finally, it is possible that Pu-239 could be stolen and used for bomb production.

In the future, when nuclear fusion power plants are in operation, most of these disadvantages will not be a concern. For example, fusion reactions produce only short-lived isotopes and so there would be no long-term storage problems. An added advantage is that there is a virtually inexhaustible supply of deuterium fuel in the world's oceans.

26-60. *Refer to Section 26-15.*

Uranium ores contain only about 0.7% ^{235}U which is fissionable. Most of the rest is nonfissionable ^{238}U. To enrich ^{235}U for use in nuclear power plants, the oxide is converted to UF_4 with HF and then oxidized to UF_6 by fluorine. The vapor of $^{235}UF_6$ and $^{238}UF_6$ is then subjected to repeated diffusion through porous barriers to concentrate $^{235}UF_6$ (Graham's Law). Gas centrifuges are now used for the concentration process which is also based upon the difference in masses of the two U isotopes.

26-62. *Refer to Section 26-16.*

The major advantages of fusion as a potential energy source are three-fold:

(1) Fusion reactions are accompanied by much greater energy production per unit mass of reacting atoms than fission reactions.
(2) The deuterium fuel for fusion reactions is present in a virtually inexhaustible supply in the world oceans.
(3) Fusion reactions produce only short-lived radionuclides; there would be no long-term waste-disposal problem.

The only disadvantage of fusion is that extremely high temperatures are required to initiate the fusion process. A structural material that can withstand the high temperatures (4×10^7 K or more) and contain the fusion reaction, does not as yet exist.

26-64. *Refer to Sections 26-12 and 26-3.*

(a) $^{14}_{7}N + ^{4}_{2}He \rightarrow ^{17}_{8}O + ^{1}_{1}H$

(b) Plan: (1) Determine the mass difference between the products and the reactants. The difference, Δm, is directly related to the energy involved in the reaction.
 (2) Calculate the amount of energy involved.

 (1) Δm = mass of products - mass of reactants

 = (mass of $^{17}_{8}O + ^{1}_{1}H$) - (mass of $^{14}_{7}N$ + mass of $^{4}_{2}He$)

 = (16.99913 amu + 1.007825 amu) - (14.00307 amu + 4.00200 amu)

 = 0.00128 amu or 0.00128 g/mol rxn

 (2) $\Delta E = (\Delta m)c^2$

 = $(1.28 \times 10^{-6}$ kg/mol rxn$)(3.00 \times 10^8$ m/s$)^2$

 = 1.15×10^{11} kg·(m/s)^2/mol rxn

 = 1.15×10^{11} J/mol rxn or **1.15×10^8 kJ/mol rxn**

Since $\Delta E > 0$, energy is being absorbed in this reaction, as expected.

26-66. *Refer to Sections 26-9 and 26-11.*

Plan: (1) Calculate the first order rate constant, k, from the half-life of uranium-238.
  (2) Determine the age of the rock.

(1) $k = \dfrac{0.693}{t_{1/2}} = \dfrac{0.693}{4.51 \times 10^9 \text{ yr}} = 1.54 \times 10^{-10} \text{ yr}^{-1}$

(2) The first order rate equation: $\log\left(\dfrac{N_o}{N}\right) = \dfrac{kt}{2.303}$

where N = number of ^{238}U atoms remaining
  N_o = number of ^{238}U atoms originally present

Therefore, N_o = number of ^{238}U atoms remaining + number of ^{238}U atoms decayed
  = number of ^{238}U atoms remaining + number of ^{206}Pb atoms produced

So, $\left(\dfrac{N_o}{N}\right) = \dfrac{68.3 \ ^{238}\text{U atoms} + 31.7 \ ^{206}\text{Pb atoms}}{68.3 \ ^{238}\text{U atoms}} = \left(\dfrac{100.0}{68.3}\right)$

Substituting, $\log\left(\dfrac{100.0}{68.3}\right) = \dfrac{(1.54 \times 10^{-10} \text{ yr}^{-1})t}{2.303}$

$0.166 = (6.69 \times 10^{-11})t$

$t = \mathbf{2.48 \times 10^9 \text{ yr}}$

26-68. *Refer to Sections 26-13 and 26-3.*

Balanced equations: fission $^{235}_{92}\text{U} + ^1_0n \rightarrow ^{94}_{40}\text{Zr} + ^{140}_{58}\text{Ce} + 6\ ^{0}_{-1}e + 2\ ^1_0n$
  fusion $2\ ^2_1\text{H} \rightarrow ^3_1\text{H} + ^1_1\text{H}$

Plan: (1) Determine the mass difference between the products and the reactants. The difference, Δm, is directly related to the energy involved in the reaction.
  (2) Calculate the amount of energy involved.

(1) fission: Δm = mass of products - mass of reactants
  = [mass of $^{94}_{40}$Zr + mass of $^{140}_{58}$Ce + (6 × mass of $^{0}_{-1}e$) + (2 × mass of ^1_0n)]
   - [mass of $^{235}_{92}$U + mass of ^1_0n]
  = [93.9061 amu + 139.9053 amu + (6 × 0.000549 amu) + (2 × 1.0087 amu)]
   - [235.0439 amu + 1.0087 amu)
  = -0.2205 amu or -0.2205 g/mol rxn

  fusion: Δm = [mass of $^3_1$H + mass of $^1_1$H] - [2 × mass of $^2_1$H]
  = [3.01605 amu + 1.007825 amu] - [2 × 2.0140 amu]
  = -0.00412 amu or -0.00412 g/mol rxn

(2) fission: $\Delta E = (\Delta m)c^2 = (-2.205 \times 10^{-4} \text{ kg/mol rxn})(3.00 \times 10^8 \text{ m/s})^2 = -1.98 \times 10^{13} \text{ kg·(m/s)}^2/\text{mol rxn}$
  $= -1.98 \times 10^{13} \text{ J/mol rxn}$

$\Delta E \text{ (J/amu } ^{235}\text{U)} = \dfrac{-1.94 \times 10^{13} \text{ J}}{1 \text{ mol rxn}} \times \dfrac{1 \text{ mol rxn}}{1 \text{ mol } ^{235}\text{U}} \times \dfrac{1 \text{ mol } ^{235}\text{U}}{6.02 \times 10^{23} \text{ atoms } ^{235}\text{U}} \times \dfrac{1 \text{ atom } ^{235}\text{U}}{235.0439 \text{ amu}}$

$= -1.40 \times 10^{-13} \text{ J/amu } ^{235}\text{U}$

fusion: $\Delta E = (\Delta m)c^2 = $ (-4.12 × 10^{-6} kg/mol rxn)(3.00 × 10^8 m/s)^2 = -3.71 × 10^{11} kg·(m/s)^2/mol rxn

$$= -3.71 \times 10^{11} \text{ J/mol rxn}$$

$$\Delta E \text{ (J/amu }^2\text{H)} = \frac{-3.71 \times 10^{11} \text{ J}}{1 \text{ mol rxn}} \times \frac{1 \text{ mol rxn}}{2 \text{ mol }^2\text{H}} \times \frac{1 \text{ mol }^2\text{H}}{6.02 \times 10^{23} \text{ atoms }^2\text{H}} \times \frac{1 \text{ atom }^2\text{H}}{2.0140 \text{ amu}}$$

$$= -1.53 \times 10^{-13} \text{ J/amu }^2\text{H}$$

Therefore, the above **fusion** process produces about 10% more energy per amu of material than fission. However, the above fusion reaction is not a typical one because it involves the production of two particles from two particles of similar size. In general, fusion processes produce much more energy than fission processes on a per unit mass basis.

27 Organic Chemistry I: Compounds

27-2. *Refer to the Introduction to Chapter 27.*

(a) Carbon atoms bond to each other to a much greater extent than any other element. They form long chains, branched chains and rings which may also contain chains attached to them. Millions of such compounds are known which constitutes the study of organic chemistry.

(b) Catenation means "chain-making" which describes the ability of an element to bond to itself.

27-4. *Refer to the Introduction to Chapter 27.*

(a) Most synthetic organic materials are derived from petroleum, coal and natural gas.

(b) Most geochemists believe that petroleum, natural gas and coal are derived from plant matter, buried millions of years ago. Since the source of carbon for plants is CO_2, we can say that the ultimate source of many naturally occurring organic compounds which are based on carbon, is CO_2.

27-6. *Refer to Section 27-1 and Table 27-1.*

In alkanes such as (a) methane, CH_4, (b) ethane, C_2H_6, (c) propane, C_3H_8, and (d) n-butane, C_4H_{10} (Figures 27-2 to 27-5), the geometry about each C atom is tetrahedral. All the carbon atoms are connected to each other to form chains, and in the case of n-C_4H_{10}, a "straight" chain of 4 carbon atoms without branching is formed. Each of the C atoms undergoes sp^3 hybridization, and forms σ bonds with each other by using the sp^3 hybrid orbitals. The C atoms at the end of each chain are in the form of CH_3, each bonded to 3 H atoms by overlapping with their $1s$ orbitals to give σ bonds. The C atoms in the interior of each chain are in the form of CH_2, each bonded to 2 H atoms in the same fashion.

These four molecules are the first four members of the alkanes, a homologous series of saturated hydrocarbons with the general formula, C_nH_{2n+2}. The difference between them is in the number of C atoms in the compound; the formula of each alkane differs from the next by one CH_2 group.

27-8. *Refer to Section 27-1.*

(a) A homologous series is a series of compounds in which each member differs from the next by a specific number and kind of atoms.

(b) The alkane series contains saturated hydrocarbons such as CH_4, C_2H_6, C_3H_8 and C_4H_{10}. Each member differs from the next by CH_2 and are therefore examples of compounds that are members of a homologous series. Refer to Table 27-2 for the names and formulas of more members of this homologous series.

(c) A methylene group is a CH_2 group.

(d) The structures of homologous series members differ by a CH_2 unit from one member to the next. The properties of the members of a homologous series are closely related. For example, the boiling point of a compound in the homologous series given in (b) is higher than the compounds before it in the series, but less than the compounds after it due to increasing London dispersion forces.

(e) Homologous series that are also aliphatic hydrocarbons are the alkanes, C_nH_{2n+2}, the alkenes, C_nH_{2n}, and the alkynes, C_nH_{2n-2}; they all differ by a CH_2 unit from one member to the next.

Cycloalkanes are cyclic saturated hydrocarbons with the general formula C_nH_{2n}. Therefore, a substance with the formula C_3H_8 could *not* be a cycloalkane, since C_3H_8 conforms to the general formula, C_nH_{2n+2}, the molecular formula for an alkane. It is, however, too small to be a branched alkane with a methyl group attached to the longest chain. In fact, C_3H_8 is propane.

(a) $CH_3CH_2CH_2CH(CH_3)CH_2CH_3$ 3-methylhexane

(b) $CH_3CH_2CH(CH_2CH_3)CH_2CH_3$ 3-ethylpentane

(c) $CH_3CH(CH_2CH_2CH_3)CH_2CH_3$ 3-methylhexane (Hint: draw it and look for the longest carbon chain.)

(d) $CH_3CH_2CH_2CH(CH_3)CH_3$ 2-methylpentane

(e) $CH_3CH(CH_3)CH_2CH_2CH_3$ 2-methylpentane

1-cyclopropylbutane   1-cyclopropyl-2-methylpropane

2-cyclopropylbutane   2-cyclopropyl-2-methylpropane

  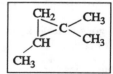

methylcyclopropane   1,1-dimethylcyclopropane   1,1,2-trimethylcyclopropane

(a) Alkenes contain C=C double bonds formed at the expense of two hydrogen atoms. Therefore, the general formula for alkenes is C_nH_{2n}, while that for alkanes is C_nH_{2n+2}.

(b) When an alkane loses two H atoms, the resulting species could undergo either ring-enclosure to give a cycloalkane, or H-shifting and π bond formation to give an alkene. Therefore, cycloalkanes and alkenes are isomers, both having the general formula, C_nH_{2n}.

27-20. *Refer to Section 27-3 and Table 27-1.*

(a) ethene

$$CH_2=CH_2$$

Both carbon atoms in ethene undergo sp^2 hybridization. The C-H bonds involve overlap of sp^2 carbon orbitals with $1s$ orbitals of the H atoms. The carbon-carbon double bond involves the overlap of sp^2 orbitals from each carbon to give the σ bond and the side-on overlap of a p orbital from each carbon atom to give the π bond.

(b) propene

$$\overset{1}{CH_2}=\overset{2}{CH}-\overset{3}{CH_3}$$

Carbon atom (3) uses sp^3 hybrid orbitals to form four sigma bonds, three by overlap with the hydrogen $1s$ orbitals and one by overlap with an sp^2 orbital from the central carbon (2). The two carbon atoms involved in the double bond undergo sp^2 hybridization. They form C-H bonds by overlapping with $1s$ orbitals of the H atoms. The C=C double bond is formed similarly to that described in (a).

(c) 1-butene

$$\overset{1}{CH_2}=\overset{2}{CH}-\overset{3}{CH_2}-\overset{4}{CH_3}$$

Carbon atoms (1) and (2) undergo sp^2 hybridization, while carbon atoms (3) and (4) undergo sp^3 hybridization. The overlap of the hybrid orbitals with the $1s$ orbitals of H atoms gives the C-H bonds. The C=C double bond is formed similarly to that described in (a). The C-C single bonds involve sp^2-sp^3 overlap for the C(2)-C(3) single bond and sp^3-sp^3 overlap for the C(3)-C(4) single bond.

(d) 2-methyl-2-butene

$$\overset{5}{CH_3}$$
$$\underset{1}{CH_3}-\underset{2}{C}=\underset{3}{CH}-\underset{4}{CH_3}$$

Carbon atoms (2) and (3) undergo sp^2 hybridization, while carbon atoms (1), (4) and (5) undergo sp^3 hybridization. The orbital overlaps are similar to those in 1-butene except carbon atoms (2) and (3) are involved in the double bond and carbon atoms (2) and (3) are also bonded to another carbon atom by sp^2-sp^3 overlaps. The overlap of the hybrid orbitals with the $1s$ orbitals of H atoms gives the C-H bonds.

27-22. *Refer to Section 27-3.*

Carbon-carbon double bonds (1.34 Å) are shorter than carbon-carbon single bonds (1.54 Å). This is due to the additional side-on overlap of the p orbitals creating the π bond. Two shared electron pairs draw the atoms closer together than a single electron pair does.

27-24. *Refer to Section 27-4 and Figure 27-10.*

The C≡C triple bond consists of one σ and two π bonds. The σ bond is formed by overlapping head-on the sp hybridized orbitals of the corresponding carbon atoms. The π bonds are formed by overlapping side-on the two remaining sets of p orbitals. The triply bonded atoms and their adjacent atoms lie on a straight line.

27-26. *Refer to Sections 27-2, 27-3 and 27-4, and Example 27-8.*

(a)

$$CH_3-CH_2-C\equiv C-CH_2-CH_3$$

3-hexyne

(b)

$$CH_2=CH-CH=CH-CH_3$$

1,3-pentadiene

(c)

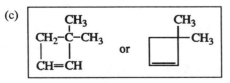

3,3-dimethylcyclobutene

(d)

$$\begin{array}{c} CH_2CH_3 \\ | \\ HC\equiv C-CH-CH-CH_2-CH_3 \\ | \\ CH_2CH_3 \end{array}$$

3,4-diethyl-1-hexyne

27-28. *Refer to Sections 27-2, 27-3 and 27-4, and Example 27-9.*

(a) 1-methylcyclopentene

(c) 2-butyne

(e) 3-methylpentane

(b) 2-methylbutane

(d) 2-methyl-2-butene

(f) 3-ethylhexane

27-30. *Refer to Section 27-6.*

A phenyl group, C_6H_5-, results when an H atom is removed from a benzene ring. It could take the place of any H in an organic compound. In particular, when a phenyl group replaces a hydrogen atom on a naphthalene molecule, **two** isomers of monophenylnaphthalene are possible:

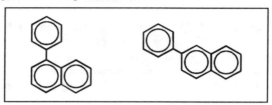

27-32. *Refer to Section 27-6.*

A total of 3 isomers of dibromobenzene are possible:

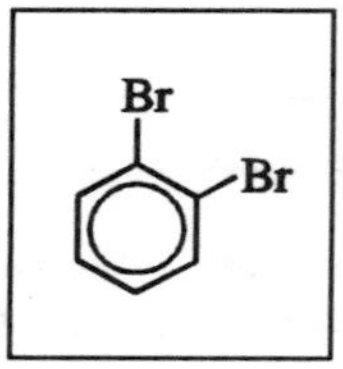

1,2-dibromobenzene
(*o*-dibromobenzene)

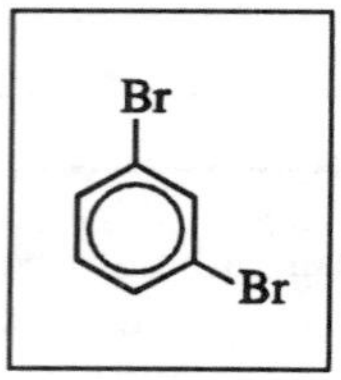

1,3-dibromobenzene
(*m*-dibromobenzene)

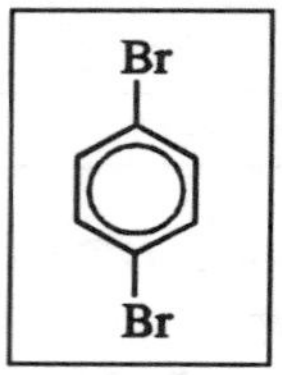

1,4-dibromobenzene
(*p*-dibromobenzene)

27-34. *Refer to Section 27-6.*

(a) 1,2-dimethylbenzene (*o*-xylene)

(b) 1,2,4-trimethylbenzene

(c) 1,5-diethyl-2,4-dimethylbenzene

(d) 1,2,3,4,5-pentamethylbenzene

27-36. *Refer to Section 27-8.*

(a) chlorotriphenylmethane

(b) 1-chloro-2-methylpropane

(c) 1,2-dichloroethane

(d) 1,1,2-trichloroethene

27-38. *Refer to Sections 27-6 and 27-8.*

(a) 1,2,3-trichlorobenzene

(b) 4-chloro-1-methylbenzene (*p*-chlorotoluene)

(c) 2,5-dibromo-1,4-diiodobenzene

(d) 2,4-dibromo-1,3,5-trichlorobenzene

(a) Alcohols and phenols are hydrocarbon derivatives which contain the hydroxyl group (-OH) as their functional group.

(b) Alcohols are derived from aliphatic hydrocarbons by replacing at least one hydrogen atom with a hydroxyl (-OH) group. On the other hand, in phenols, the -OH group must attach directly to an aromatic ring. Phenols are weak acids, while alcohols are neutral.

(c) Alcohols and phenols can be viewed as derivatives of hydrocarbons in which a hydrogen atom is replaced by an -OH group. On the other hand, they can also be viewed as derivatives of water in which a hydrogen atom is replaced by an organic group.

27-42. *Refer to Section 27-9.*

(a) The four saturated alcohols that contain four carbon atoms and one -OH group per molecule are:

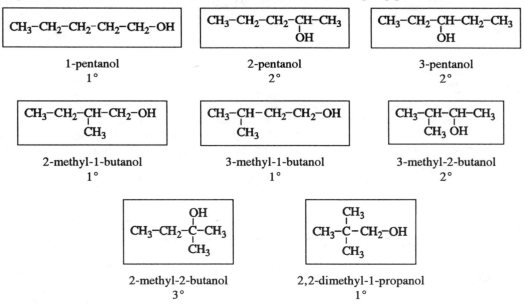

(b) The eight saturated alcohols that contain five carbon atoms and one -OH group per molecule are:

27-44. *Refer to Section 27-9 and Table 27-8.*

Data in Table 27-8 show that the boiling points of normal primary alcohols increase and their solubilities in water decrease with increasing molecular weight. The boiling point increases because the London dispersion forces increase with the size of the molecules. The solubility decreases because the alcohols become less polar down the list. The alcohols, ROH, have a polar hydroxyl group end and a nonpolar alkyl group end. Due to the principle, "like dissolves like," as the nonpolar end of the molecules becomes larger and larger, their solubilities in water decrease rapidly because H_2O is a very polar solvent. In fact, the C_1 - C_3 alcohols are miscible with H_2O in all proportions. Beginning with the butyl alcohols, solubility in H_2O decreases rapidly with increasing molecular weight.

31-46. *Refer to Section 27-9.*

Most phenols are relatively high-molecular weight compounds with large nonpolar portions and therefore, they exhibit low solubilities in water.

27-48. *Refer to Section 27-9.*

(a) 2-methyl-1-butanol

(b) 3,3-dimethyl-1-butanol

(c) 1,2-propanediol (propylene glycol)

(d) 2-methyl-2-propanol

27-50. *Refer to Section 27-9.*

(a) *p*-iodophenol

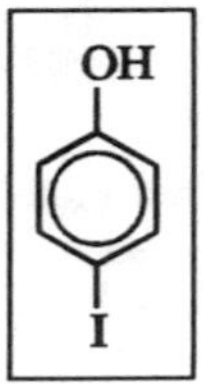

(b) 4-nitrophenol

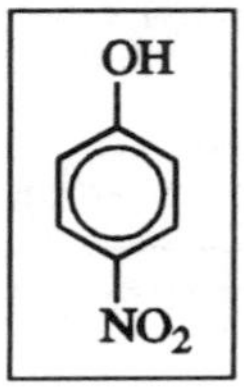

(c) *m*-nitrophenol

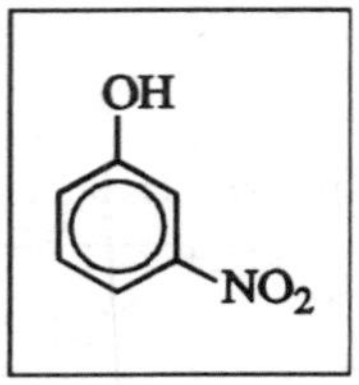

27-52. *Refer to Section 27-10.*

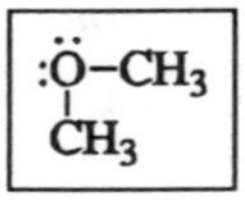

In dimethyl ether, the oxygen atom is sp^3 hybridized. Two single bonds are formed by the overlap of two of its sp^3 hybrid orbitals with two of the sp^3 hybrid orbitals on the adjacent carbon atoms. In each of the remaining two hybrid orbitals on the oxygen atom are contained a lone pair of electrons. The resulting molecule is polar. The intermolecular forces found operating between molecules of dimethyl ether are therefore dipole-dipole interactions and London forces.

27-54. *Refer to Section 27-10.*

(a)

$$H_3C-O-CH_3$$

methoxymethane

(b)

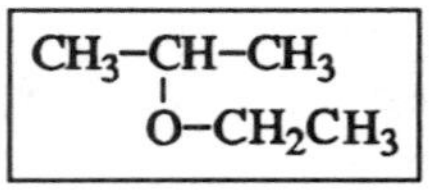

$$CH_3-CH-CH_3$$
$$\quad\quad |$$
$$\quad\; O-CH_2CH_3$$

2-ethoxypropane

(c)

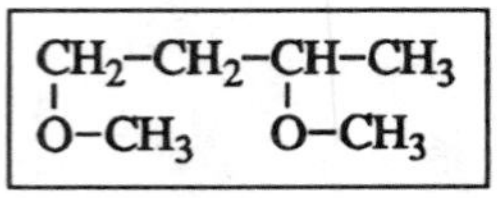

$$CH_2-CH_2-CH-CH_3$$
$$\;\; |\quad\quad\quad\;\; |$$
$$\; O-CH_3 \quad O-CH_3$$

1,3-dimethoxybutane

(d)

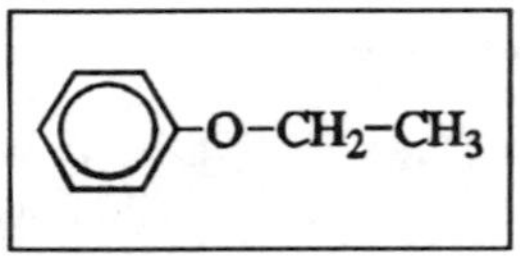

$$\text{⬡}-O-CH_2-CH_3$$

ethoxybenzene

(e)

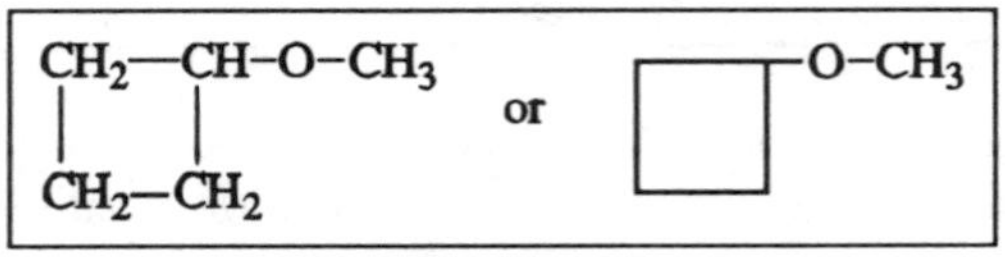

$$CH_2-CH-O-CH_3$$
$$\; |\quad\quad |$$
$$CH_2-CH_2$$

or

$$\square-O-CH_3$$

methoxycyclobutane

(a) The amines are derivatives of ammonia, NH_3, in which one or more H atoms have been replaced by organic groups. They have the general formula: RNH_2, R_2NH or R_3N, where R is any alkyl or aryl group. Amines are basic; their basicity is derived from the lone pair of electrons on the N atoms.

(b) Amines are considered to be derivatives of ammonia. The structures of NH_3, primary, secondary and tertiary amines are shown below. From the comparison, it is obvious that amines can be treated as if one, two or three hydrogen atoms of ammonia have been replaced by organic groups.

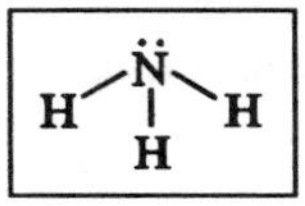

| ammonia | primary amine | secondary amine | tertiary amine |

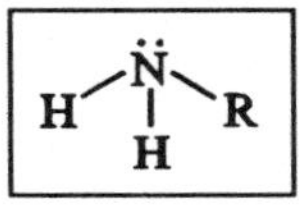

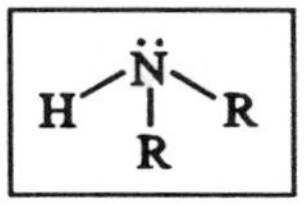

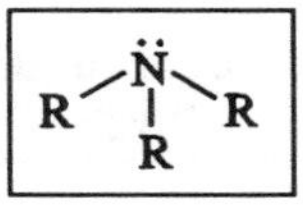

27-58. *Refer to Section 27-11.*

(a) diethylamine

(b) 4-nitroaniline (*p*-nitroaniline)

(c) methylaminocyclopentane (cyclopentylmethylamine)

(d) tributylamine

27-60. *Refer to Section 27-12 and Table 27-10.*

(a) A carboxylic acid is an acidic organic compound containing the carboxyl group, $-\overset{\overset{\displaystyle O}{\|}}{C}-OH$.

(b)

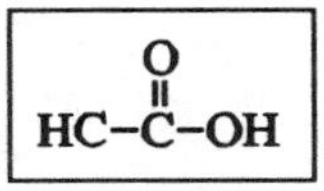

| methanoic acid (formic acid) | ethanoic acid (acetic acid) | propanoic acid (propionic acid) |

| butanoic acid (*n*-butyric acid) | 2-methylbutanoic acid (α-methylbutyric acid) |

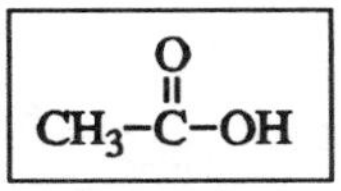

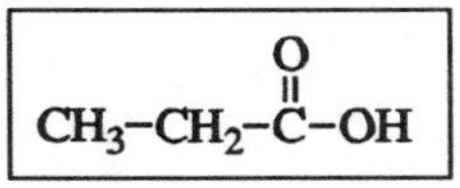

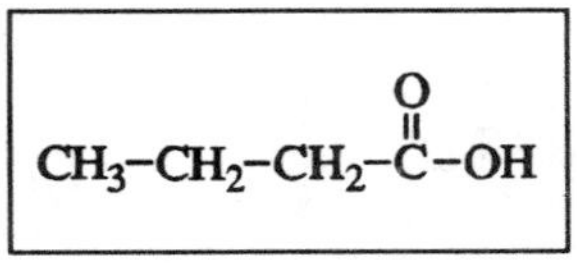

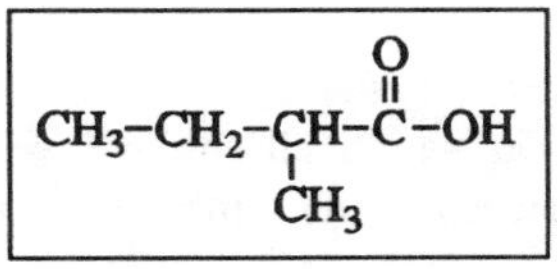

27-62. *Refer to Section 27-13.*

(a) An acyl chloride, sometimes called an acid chloride, is a compound that is a derivative of a carboxylic acid, made by replacing the -OH with a Cl atom. It has the general formula: $R-\overset{\overset{\displaystyle O}{\|}}{C}-Cl$

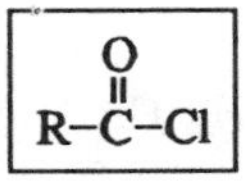

(b)

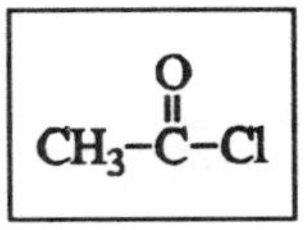

| acetyl chloride | propionyl chloride | butyryl chloride | benzoyl chloride |

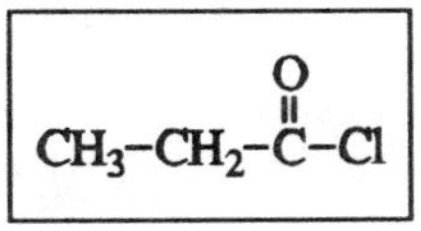

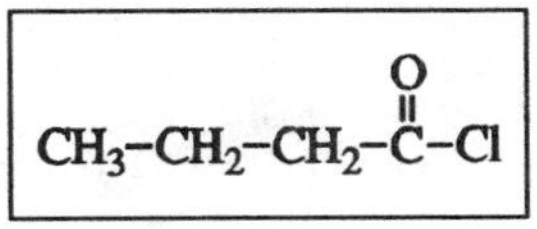

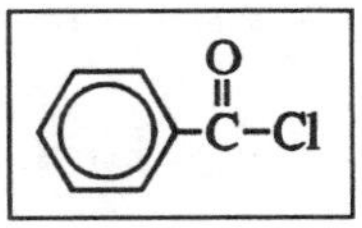

(a) The general formula for an acid anhydride is:

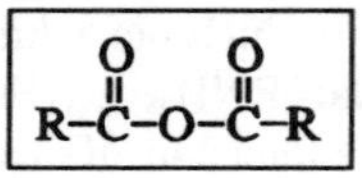

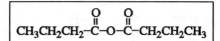

(b)

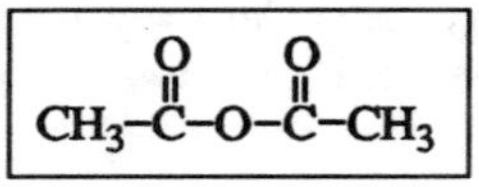

$CH_3-C(=O)-O-C(=O)-CH_3$	$CH_3CH_2-C(=O)-O-C(=O)-CH_2CH_3$	$CH_3CH_2CH_2-C(=O)-O-C(=O)-CH_2CH_2CH_3$
acetic anhydride	propionic anhydride	butyric anhydride

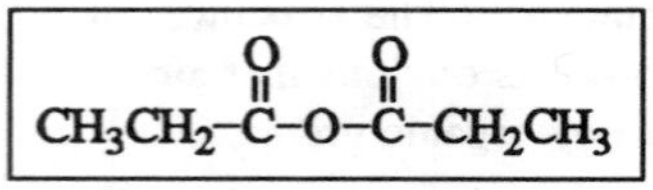

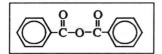

benzoic anhydride

Ester	Formula	Odor or Origin
isoamyl acetate	$CH_3COOC_5H_{11}$	bananas
ethyl butyrate	$C_3H_7COOC_2H_5$	pineapples
amyl butyrate	$C_3H_7COOC_5H_{11}$	apricots
octyl acetate	$CH_3COOC_8H_{17}$	oranges
isoamyl isovalerate	$C_4H_9COOC_5H_{11}$	apples
methyl anthranilate	$C_6H_4(NH_2)(COOCH_3)$	grapes

(a) As shown, glycerides are the triesters of glycerol. Simple glycerides are esters in which all three R groups are identical whereas mixed glycerides contain a mixture of various R groups.

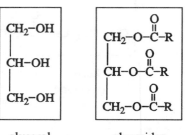

glycerol   glycerides

(b)

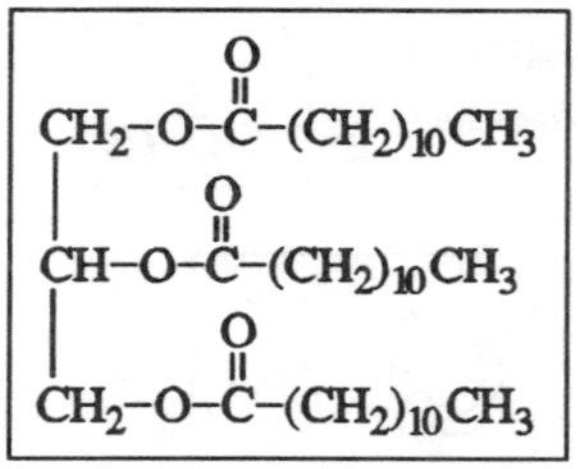

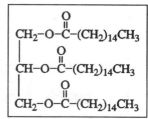

glyceryl trilaurate   glyceryl tripalmitate

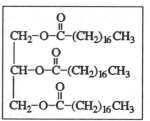

glyceryl tristearate

Waxes are esters of fatty acids with alcohols other than glycerol. Most of them are derived from long-chain fatty acids and long-chain monohydric alcohols, both usually having an even number of carbon atoms. Therefore, a wax usually contains an even number of carbon atoms also.

(a) phenyl benzoate

(b) *n*-propyl butanoate

(a), (b)

Classification	Examples	Sources
Aldehyde	benzaldehyde	almonds
	cinnamaldehyde	cinnamon
	vanillin	vanilla bean
Ketone	muscone	musk deer
	testosterone	male sex hormone
	camphor	camphor tree

(c) The aldehydes listed in (a) have many uses. The three aldehydes can be used to add flavor to food. Muscone is the compound that gives the scent to musk perfumes, deodorants, cologne and aftershave lotions. Testosterone is used to regulate male sexual and reproductive functions. Camphor is used in medicine as a diaphoretic, stimulant and sedative.

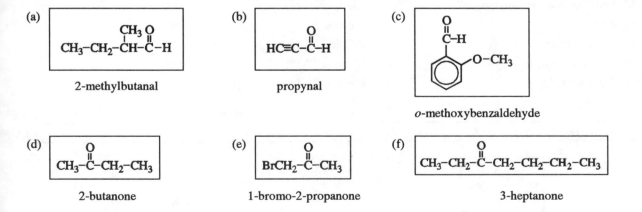

(a)

$$CH_3-CH_2-\overset{\overset{\displaystyle CH_3}{|}}{CH}-\overset{\overset{\displaystyle O}{||}}{C}-H$$

2-methylbutanal

(b)

$$HC\equiv C-\overset{\overset{\displaystyle O}{||}}{C}-H$$

propynal

(c)

o-methoxybenzaldehyde

(d)

$$CH_3-\overset{\overset{\displaystyle O}{||}}{C}-CH_2-CH_3$$

2-butanone

(e)

$$BrCH_2-\overset{\overset{\displaystyle O}{||}}{C}-CH_3$$

1-bromo-2-propanone

(f)

$$CH_3-CH_2-\overset{\overset{\displaystyle O}{||}}{C}-CH_2-CH_2-CH_2-CH_3$$

3-heptanone

(a) carboxylic acid

(b) ester

(c) amide

(d) ether (epoxide)

(e) acyl chloride (acid chloride)

27-80. *Refer to Section 27-15 and Figure 27-21.*

(a) $R-\overset{\overset{O}{\|}}{C}-R$ ketone

(b) Ar-OH   phenol
 R-O-R   ether
 R_2-CH-OH   secondary alcohol
 R_2-NH   secondary amine

(c) R-O-R   ether

(d) Ar-OH   phenol
 R-O-R   ether
 R_2-CH-OH   secondary alcohol
 R_3-N   tertiary amine
 C=C   alkene (double bond)

(e) Ar-OH   phenol
 R_2-CH-OH   secondary alcohol
 R_2-NH   secondary amine

27-82. *Refer to the Sections as stated.*

(a) 1-pentanol   (Section 27-9)
(b) 2-methylcyclopentanol   (Section 27-9)
(c) 2-aminopropane (isopropylamine)   (Section 27-11)
(d) 2-chloropropene   (Section 27-3)
(e) 1,4-dibromobenzene (*p*-dibromobenzene) (Section 27-6)
(f) triethylamine   (Section 27-11)
(g) diphenyl ether   (Section 27-10)
(h) 2,4,6-tribromoaniline   (Section 27-11)

27-84. *Refer to the Sections as stated.*

(a) 2-methyl-1-butanol   (Section 27-9)
(b) 1-aminobutane (*n*-butylamine)   (Section 27-11)
(c) pentanal   (Section 27-14)
(d) cyclopentanone   (Section 27-14)
(e) 2-methoxybutane   (Section 27-10)
(f) 2,2-dimethylpropanoic acid   (Section 27-12)

27-86. *Refer to Section 27-14.*

2-butanone

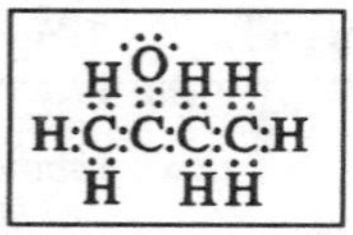

27-88. *Refer to Section 27-11.*

Lidocaine has replaced novocain as the favored anesthetic in dentistry. Both compounds have a tertiary amine group (a diethylamino group) in common.

28 Organic Chemistry II: Molecular Geometry and Reactions

28-2. *Refer to Section 28-2 and Figures 28-1, 28-2, 28-3, 28-4, 28-5 and 28-6.*

There are two types of stereoisomerism. (1) Geometrical isomers differ only in the spatial orientation of groups about a plane or direction, i.e., they differ in orientation either (i)around a double bond (see 2-butene) or (ii) across the ring in a cyclic compound (see 1,2-dichlorocyclobutane). Both *cis* and *trans* isomers exist.

  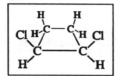 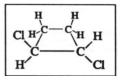

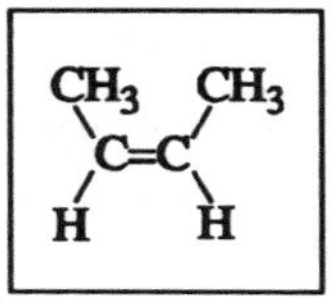

  cis-2-butene 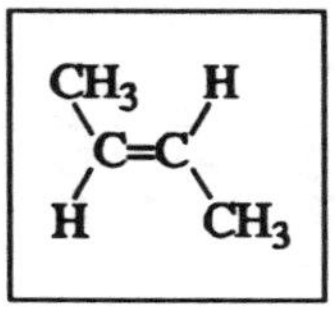   *trans*-2-butene   *cis*-1,2-dichlorocyclobutane   *trans*-1,2-dichlorocyclobutane

(2) The second type of stereoisomerism is optical isomerism, in which two molecules that are mirror images of each other are not superimposable on each other. Consider the compound 2-butanol, $CH_3CH(OH)CH_2CH_3$. It has two optical isomers, because it is not superimposable on its mirror image.

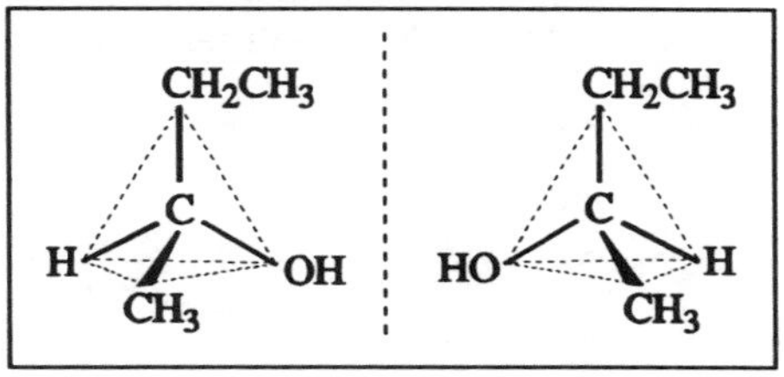

28-4. *Refer to Section 28-2, and Figures 28-1 and 28-2.*

The compounds that exist as *cis* and *trans* isomers are (c) 2-bromo-2-butene and (d) 1,2-dichlorocyclopentane.

 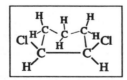

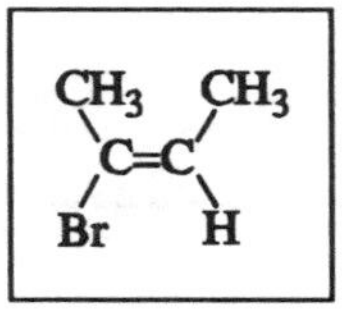

2-bromo-*cis*-2-butene 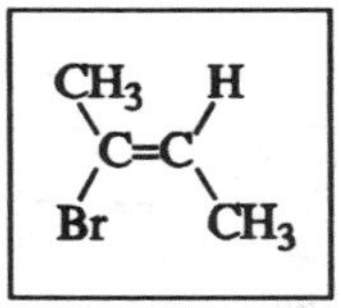   2-bromo-*trans*-2-butene   *cis*-1,2-dichlorocyclopentane 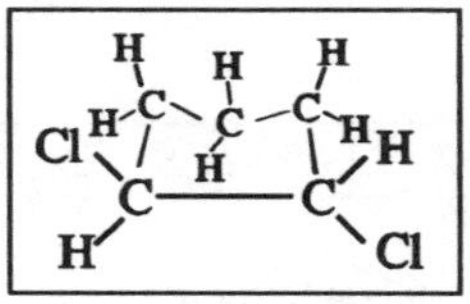   *trans*-1,2-dichlorocyclopentane

---

28-6. *Refer to Section 28-2, the Key Terms for Chapter 28 and Figures 28-3, 28-4, 28-5 and 28-6.*

Optical isomerism is exhibited by compounds that are chiral, i.e., are not superimposable on their mirror images. Such a compound and its mirror image are called optical isomers or enantiomers. They have identical physical and chemical properties except when they interact with other chiral molecules. A solution of one of the pair is capable of rotating a plane of polarized light to the right and is called dextrorotatory. An equimolar solution of its optical isomer, called levorotatory, will rotate the plane of polarized light by the same amount, but to the left.

28-8. *Refer to Section 28-2 and Exercise 28-7.*

The compounds exhibiting optical isomerism in Exercise 28-7 are (a) and (c):

(a)

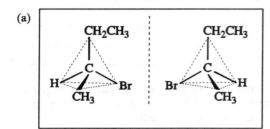

(c)

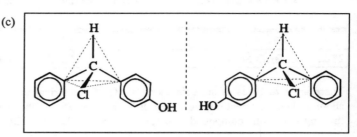

28-10. *Refer to Sections 28-4 and 28-5, and the Key Terms for Chapter 28.*

(a) A substitution reaction is a reaction in which an atom (or group of atoms) replaces another atom (or group of atoms) on a carbon in an organic reaction. No change occurs in the degree of saturation at the reactive carbon atom.

(b) A halogenation reaction is a substitution reaction in which one or more hydrogen atoms of a hydrocarbon is replaced by the corresponding number of halogen atoms.

(c) A free-radical chain reaction is one with a free radical or atom as a chain carrier, which perpetuates the reaction. It normally consists of three steps. For example, for the overall reaction: $A_2 + B_2 \rightarrow 2AB$

   (1) Chain Initiation: $A_2 \rightarrow 2A$

   (2) Propagation: $A + B_2 \rightarrow AB + B$
                                 $B + A_2 \rightarrow AB + A$

   (3) Termination: $2A \rightarrow A_2$
                                 $A + B \rightarrow AB$
                                 $2B \rightarrow B_2$

(d) An addition reaction is a reaction in which there is an increase in the number of groups attached to carbon. Two atoms or groups of atoms are added to the molecule, one on each side of a double or a triple bond. The molecule becomes more nearly saturated.

28-12. *Refer to Section 28-4.*

(a) The chlorination of ethane in ultraviolet light is a free radical chain reaction. It begins when the chlorine molecule is split into two very reactive Cl atoms, which can attack ethane, extracting one of its H atoms to form HCl and a C_2H_5 radical. This, in turn, extracts a Cl atom from Cl_2 to form a monosubstituted chloroethane. When a second hydrogen atom is replaced, a mixture of two disubstituted ethanes are produced as shown below. Subsequent substitution will eventually give C_2Cl_6 as the highest chlorinated product.

(b), (c)

$$
Cl\text{--}Cl \quad + \quad \underset{\substack{\\ H\ H}}{\overset{\substack{H\ H \\ |\ \ | \\ H\text{--}C\text{--}C\text{--}H \\ |\ \ |}}{}} \quad \xrightarrow{uv} \quad \underset{\substack{\\ H\ H}}{\overset{\substack{H\ H \\ |\ \ | \\ H\text{--}C\text{--}C\text{--}Cl \\ |\ \ |}}{}} \quad + \quad H\text{--}Cl
$$

        chlorine   ethane   chloroethane   hydrogen
                                                                                        chloride

$$\text{Cl-Cl} \quad + \quad \underset{\text{chloroethane}}{\text{H-}\overset{\overset{\displaystyle H}{|}}{\underset{\underset{\displaystyle H}{|}}{C}}\text{-}\overset{\overset{\displaystyle H}{|}}{\underset{\underset{\displaystyle H}{|}}{C}}\text{-Cl}} \quad \xrightarrow{uv} \quad \underset{\text{1,1-dichloroethane}}{\text{H-}\overset{\overset{\displaystyle H}{|}}{\underset{\underset{\displaystyle H}{|}}{C}}\text{-}\overset{\overset{\displaystyle Cl}{|}}{\underset{\underset{\displaystyle H}{|}}{C}}\text{-Cl}} \quad + \quad \underset{\substack{\text{hydrogen} \\ \text{chloride}}}{\text{H-Cl}}$$

chlorine

$$\underset{uv}{\searrow}$$

$$\underset{\text{1,2-dichloroethane}}{\text{H-}\overset{\overset{\displaystyle Cl}{|}}{\underset{\underset{\displaystyle H}{|}}{C}}\text{-}\overset{\overset{\displaystyle H}{|}}{\underset{\underset{\displaystyle H}{|}}{C}}\text{-Cl}} \quad + \quad \underset{\substack{\text{hydrogen} \\ \text{chloride}}}{\text{H-Cl}}$$

The substitution reaction will continue in the presence of excess Cl_2 to give the following chlorinated compounds:

$$\underset{\text{1,1,1-trichloroethane}}{\text{H-}\overset{\overset{\displaystyle H}{|}}{\underset{\underset{\displaystyle H}{|}}{C}}\text{-}\overset{\overset{\displaystyle Cl}{|}}{\underset{\underset{\displaystyle Cl}{|}}{C}}\text{-Cl}} \qquad \underset{\text{1,1,2-trichloroethane}}{\text{H-}\overset{\overset{\displaystyle Cl}{|}}{\underset{\underset{\displaystyle H}{|}}{C}}\text{-}\overset{\overset{\displaystyle Cl}{|}}{\underset{\underset{\displaystyle H}{|}}{C}}\text{-Cl}} \qquad \underset{\text{1,1,1,2-tetrachloroethane}}{\text{H-}\overset{\overset{\displaystyle Cl}{|}}{\underset{\underset{\displaystyle H}{|}}{C}}\text{-}\overset{\overset{\displaystyle Cl}{|}}{\underset{\underset{\displaystyle Cl}{|}}{C}}\text{-Cl}}$$

$$\underset{\text{1,1,2,2-tetrachloroethane}}{\text{H-}\overset{\overset{\displaystyle Cl}{|}}{\underset{\underset{\displaystyle Cl}{|}}{C}}\text{-}\overset{\overset{\displaystyle Cl}{|}}{\underset{\underset{\displaystyle Cl}{|}}{C}}\text{-H}} \qquad \underset{\text{pentachloroethane}}{\text{H-}\overset{\overset{\displaystyle Cl}{|}}{\underset{\underset{\displaystyle Cl}{|}}{C}}\text{-}\overset{\overset{\displaystyle Cl}{|}}{\underset{\underset{\displaystyle Cl}{|}}{C}}\text{-Cl}} \qquad \underset{\text{hexachloroethane}}{\text{Cl-}\overset{\overset{\displaystyle Cl}{|}}{\underset{\underset{\displaystyle Cl}{|}}{C}}\text{-}\overset{\overset{\displaystyle Cl}{|}}{\underset{\underset{\displaystyle Cl}{|}}{C}}\text{-Cl}}$$

28-14. *Refer to Sections 28-4 and 28-5.*

The characteristic reaction of the relatively unreactive alkanes is the substitution reaction which involves the replacement of one σ bonded atom for another and requires heat or light. The more reactive alkenes are characterized by addition reactions to the double bond, many of which occur easily at room temperature. The carbon-carbon double bond is a reaction site and is classified as a functional group. The π portion of the double bond can be utilized to accommodate two incoming atoms, converting the double bond into one single σ bond between the carbon atoms and the π portion into two single σ bonds between each carbon and the two incoming atoms.

28-16. *Refer to Section 28-5.*

(1) $\quad$ Cl-Cl $\quad + \quad$ $CH_2{=}CH{-}CH_3$ $\quad \rightarrow \quad$ $\underset{\substack{| \quad | \\ Cl \quad Cl}}{CH_2{-}CH{-}CH_3}$

$\quad$ chlorine $\qquad\qquad$ propene $\qquad\qquad\qquad$ 1,2-dichloropropane

(2) $\quad$ Br-Br $\quad + \quad$ $CH_3{-}CH{=}CH{-}CH_3$ $\quad \rightarrow \quad$ $\underset{\substack{| \quad | \\ Br \quad Br}}{CH_3{-}CH{-}CH{-}CH_3}$

$\quad$ bromine $\qquad\qquad$ 2-butene $\qquad\qquad\qquad$ 2,3-dibromobutane

---

28-18. *Refer to Section 28-5 and the Key Terms for Chapter 28.*

(a) Hydrogenation refers to the reaction in which molecular hydrogen, H_2, adds across a double or triple bond. This reaction requires elevated temperatures, high pressure and the presence of an appropriate heterogeneous catalyst (finely divided Pt, Pd or Ni).

(b) Hydrogenation is an important industrial process in many areas. For example, unsaturated hydrocarbons can be converted to saturated hydrocarbons by hydrogenation to manufacture high octane gasoline and aviation fuels. It is also employed to convert unsaturated vegetable oils to solid cooking fats.

(c),(d) $CH_2=CH_2$ + H_2 $\xrightarrow[\Delta]{\text{catalyst}}$ $CH_3\text{-}CH_3$
           ethene   ethane

$CH_3\text{-}CH=CH\text{-}CH_3$ + H_2 $\xrightarrow[\Delta]{\text{catalyst}}$ $CH_3\text{-}CH_2\text{-}CH_2\text{-}CH_3$
       2-butene   butane

28-20. Refer to Section 28-5.

(a) The unsaturated π bonds are very susceptible to addition reactions because they are sources of electrons. Alkynes contain two π bonds, while alkenes contain only one π bond. Therefore, alkynes are more reactive.

(b) The most common kind of reaction that alkynes undergo is addition of atoms or groups across the triple bond.

(c), (d) (1) $CH_3\text{-}CH_2\text{-}C\equiv CH$ $\xrightarrow[\text{Pt}]{H_2}$ $CH_3\text{-}CH_2\text{-}CH=CH_2$ $\xrightarrow[\text{Pt}]{H_2}$ $CH_3\text{-}CH_2\text{-}CH_2\text{-}CH_3$
                   1-butyne   1-butene   butane

(2) $CH_3\text{-}C\equiv CH$ $\xrightarrow{Br_2}$ $CH_3\text{-}CBr=CHBr$ $\xrightarrow{Br_2}$ $CH_3\text{-}CBr_2\text{-}CHBr_2$
          propyne   1,2-dibromopropene   1,1,2,2-tetrabromopropane

(3) $HC\equiv CH$ $\xrightarrow{Cl_2}$ $ClHC=CHCl$ $\xrightarrow{Cl_2}$ $Cl_2HC\text{-}CHCl_2$
         ethyne   1,2-dichloroethene   1,1,2,2-tetrachloroethane

---

28-22. Refer to Section 28-4.

(a) aliphatic substitution:

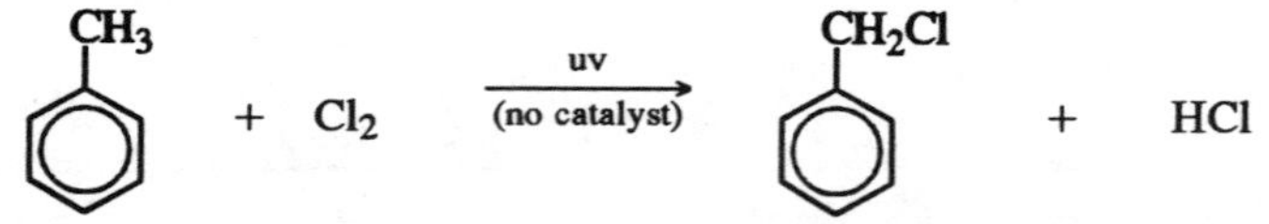

toluene   + Cl_2 $\xrightarrow[\text{(no catalyst)}]{\text{uv}}$ benzyl chloride + HCl

aromatic substitution:

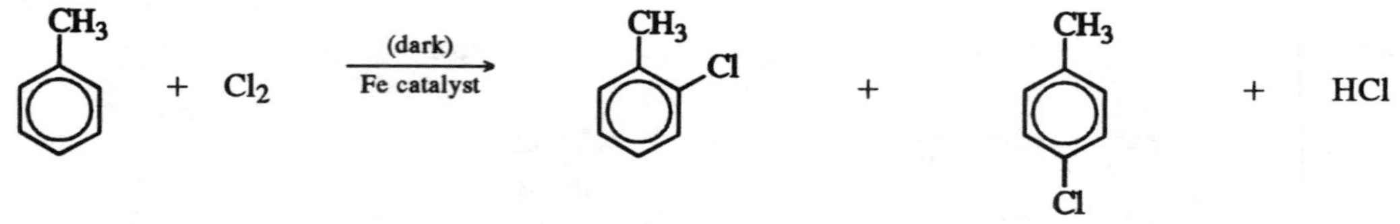

toluene + Cl_2 $\xrightarrow[\text{Fe catalyst}]{\text{(dark)}}$ o-chlorotoluene (60%) + p-chlorotoluene (40%) + HCl

(b) aliphatic substitution:

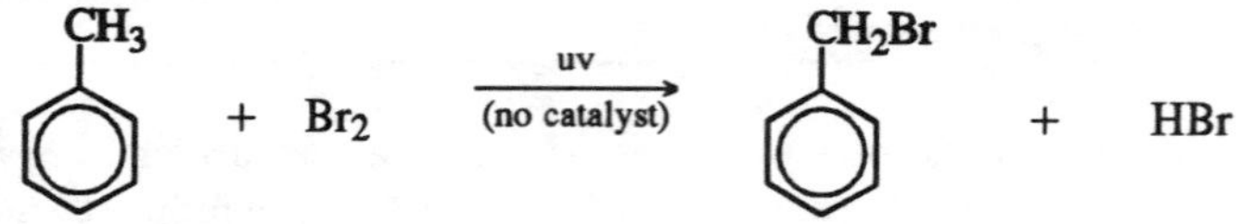

toluene + Br_2 $\xrightarrow[\text{(no catalyst)}]{\text{uv}}$ benzyl bromide + HBr

aromatic substitution:

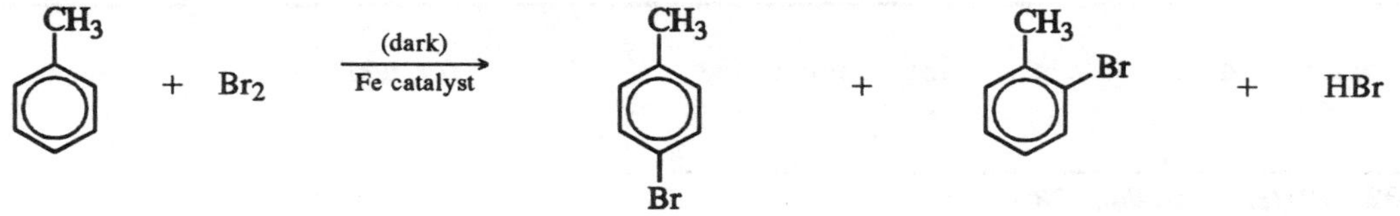

toluene $+$ Br_2 $\xrightarrow[\text{Fe catalyst}]{\text{(dark)}}$ p-bromotoluene (67%) $+$ o-bromotoluene (33%) $+$ HBr

(c) aliphatic substitution: (not found)

aromatic substitution:

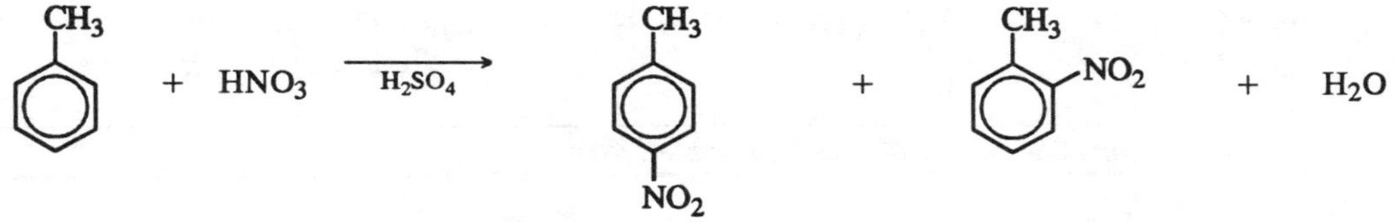

toluene $+$ HNO_3 $\xrightarrow{H_2SO_4}$ p-nitrotoluene (58%) $+$ o-nitrotoluene (38%) $+$ H_2O

28-24. *Refer to the Sections as stated.*

(a) $O{=}C{=}O$ (Section 28-8)

(b) 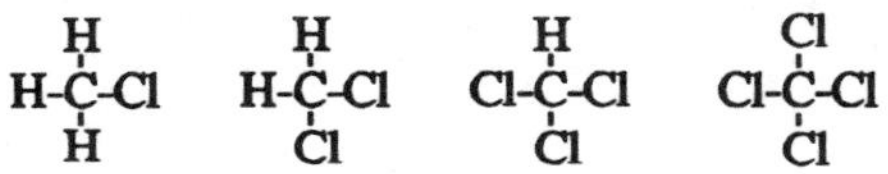
$$\begin{array}{cccc}
H & H & H & Cl \\
H{-}\overset{\displaystyle H}{\underset{\displaystyle H}{C}}{-}Cl & H{-}\overset{\displaystyle H}{\underset{\displaystyle Cl}{C}}{-}Cl & Cl{-}\overset{\displaystyle H}{\underset{\displaystyle Cl}{C}}{-}Cl & Cl{-}\overset{\displaystyle Cl}{\underset{\displaystyle Cl}{C}}{-}Cl
\end{array}$$
     (Section 28-4)

(c) $CH_3{-}CH_2{-}OH$ (Section 28-5)

28-26. *Refer to Section 28-5.*

The presence or absence of unsaturation as found in alkenes and alkynes can be easily determined by a simple, qualitative test using a solution of bromine. Br_2, a dark red liquid, is dissolved in a nonpolar solvent. When added to an unknown organic compound, the solution of Br_2 will lose its red color as the Br_2 reacts with the multiple bond and becomes incorporated into the molecule via an addition reaction. The Br_2 solution will react only with 2-hexene (test tube 1), styrene (test tube 2) and cyclohexene (test tube 3).

28-28. *Refer to Section 28-4.*

(a) An inorganic ester may be thought of as a compound that contains one or more alkyl groups covalently bonded to the anion of a ternary inorganic acid.

(b), (c) (1) $CH_3OH + HONO_2 \rightarrow CH_3ONO_2 + H_2O$
                         methyl nitrate

         (2) $CH_3CH_2OH + HONO_2 \rightarrow CH_3CH_2ONO_2 + H_2O$
                                  ethyl nitrate

         (3) $CH_3CH_2CH_2OH + HONO_2 \rightarrow CH_3CH_2CH_2ONO_2 + H_2O$
                                 n-propyl nitrate

28-30. *Refer to Sections 32-4 and 32-5.*

(a) substitution   (b) addition   (c) addition

28-32. *Refer to Sections 28-7 and 18-4.*

Amines are Brönsted-Lowry bases due to the presence of a lone pair of electrons on N to accommodate incoming protons. In aqueous solution, an amine will hydrolyze in an equilibrium to produce hydroxide ions.

ammonia $NH_3(aq) + H_2O(\ell) \rightleftarrows NH_4^+ + OH^-(aq)$

amine (e.g., 1°) $RNH_2(aq) + H_2O(\ell) \rightleftarrows RNH_3^+(aq) + OH^-(aq)$

28-34. *Refer to Sections 28-7 and 18-4, and Example 18-14.*

Balanced equation: $C_6H_5NH_2(aq) + H_2O(\ell) \rightleftarrows C_6H_5NH_3^+(aq) + OH^-(aq)$ $K_b = 4.2 \times 10^{-10}$

Let $x = [C_6H_5NH_2]_{ionized}$.   Then, $x = [C_6H_5NH_3^+] = [OH^-]$

	$C_6H_5NH_2$	+	H_2O	$\rightleftarrows$	$C_6H_5NH_3^+$	+	OH^-
initial	0.100 M				0 M		$\approx$ 0 M
change	- x M				+ x M		+ x M
at equilibrium	(0.100 - x) M				x M		x M

$$K_b = \frac{[C_6H_5NH_3^+][OH^-]}{[C_6H_5NH_2]} = 4.2 \times 10^{-10} = \frac{x^2}{0.100 - x} \approx \frac{x^2}{0.100}$$ Solving, $x = 6.5 \times 10^{-6}$

Therefore, $[C_6H_5NH_2] = \mathbf{0.100 \ M}$
            $[C_6H_5NH_3^+] = \mathbf{6.5 \times 10^{-6} \ M}$
            $[OH^-] = \mathbf{6.5 \times 10^{-6} \ M}$
            $[H_3O^+] = K_w/[OH^-] = \mathbf{1.5 \times 10^{-9} \ M}$

28-36. *Refer to Section 28-7.*

The compounds that are the stronger acids are:

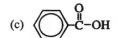

(a) 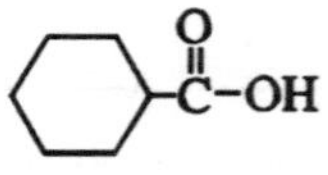   (b) 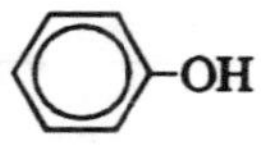   (c)   (d)

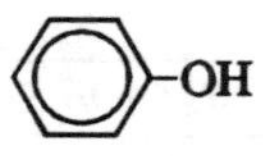

28-38. *Refer to Section 28-7.*

(a), (b)   (1) $2CH_3OH + 2Na \rightarrow 2[Na^+ + CH_3O^-] + H_2$
              methanol   sodium methoxide

          (2) $2CH_3CH_2OH + 2Na \rightarrow 2[Na^+ + CH_3CH_2O^-] + H_2$
               ethanol   sodium ethoxide

          (3) $2CH_3CH_2C_2OH + 2Na \rightarrow 2[Na^+ + CH_3CH_2CH_2O^-] + H_2$
               1-propanol   sodium propoxide

(c) The reactions of alcohols with sodium are similar to the reaction of metallic sodium with water. Both types of reactions are oxidation-reduction reactions involving the displacement of hydrogen from an O-H bond by sodium and the production of H_2 gas.

28-40. *Refer to Section 28-5 and Exercise 28-26 Solution.*

A simple test to distinguish between 1-pentene and cyclopentane is to add a few drops of a red Br_2 solution to the unknown liquid. The reddish color will disappear if the liquid is an alkene or alkyne e.g., 1-pentene, due to the addition of Br_2 to the multiple bond. No such addition reaction occurs between Br_2 and cyclopentane.

28-42. *Refer to Section 28-8.*

(a) oxidation   (b) oxidation   (c) reduction   (d) reduction

---

28-44. *Refer to Section 28-8.*

(a) Elemental carbon particles (soot) are produced in burning if there is incomplete combustion. Aromatic hydrocarbons, such as benzene, are very stable due to resonance and therefore when combusted, they release less energy to the combustion process than expected. This in turn causes the carbon atoms to be less efficiently oxidized. Carbon atoms then are oxidized to an oxidation state of zero, producing soot, rather than to an oxidation state of +4, the oxidation state of carbon in CO_2.

(b) The flames would be expected to be yellow (a reducing flame), a sign of incomplete combustion, rather than blue (an oxidizing flame), a sign of complete combustion.

28-46. *Refer to Section 28-8.*

(1)
$$\underset{\text{2-propanol}}{CH_3\text{-}\underset{\underset{\text{OH}}{|}}{CH}\text{-}CH_3} \xrightarrow[\text{OH}^-]{\text{KMnO}_4} \underset{\substack{\text{propanone} \\ \text{(dimethyl ketone or acetone)}}}{CH_3\text{-}\overset{\overset{\text{O}}{\|}}{C}\text{-}CH_3}$$

(2)
$$\underset{\text{2-butanol}}{CH_3\text{-}CH_2\text{-}\underset{\underset{\text{OH}}{|}}{CH}\text{-}CH_3} \xrightarrow[\text{OH}^-]{\text{KMnO}_4} \underset{\substack{\text{2-butanone} \\ \text{(methyl ethyl ketone)}}}{CH_3\text{-}CH\text{-}\overset{\overset{\text{O}}{\|}}{C}\text{-}CH_3}$$

(3)

cyclohexanol $\xrightarrow[\text{dil. H}_2\text{SO}_4]{\text{K}_2\text{CrO}_7}$ cyclohexanone

---

28-48. *Refer to Section 28-9.*

(1) CH_3COOH + CH_3CH_2OH → $CH_3COOCH_2CH_3$ + H_2O
   acetic acid   ethanol   ethyl acetate

(2) CH_3CH_2COOH + CH_3OH → $CH_3CH_2COOCH_3$ + H_2O
   propanoic acid   methanol   methyl propanoate
   (propionic acid)   (methyl propionate)

(3) C_6H_5COOH + CH_3CH_2OH → $C_6H_5COOCH_2CH_3$ + H_2O
   benzoic acid   ethanol   ethyl benzoate

28-50. *Refer to Section 28-10.*

(a) CH_3COOCH_3 + Na^+OH^- $\xrightarrow{\Delta}$ $CH_3COO^-Na^+$ + CH_3OH
   methyl acetate   sodium acetate   methanol

(b) $HCOOCH_2CH_3$ + Na^+OH^- $\xrightarrow{\Delta}$ $HCOO^-Na^+$ + CH_3CH_2OH
   ethyl formate   sodium formate   ethanol

(c) $CH_3COOCH_2CH_2CH_2CH_3$ + Na^+OH^- $\xrightarrow{\Delta}$ $CH_3COO^-Na^+$ + $CH_3CH_2CH_2CH_2OH$
   n-butyl acetate   sodium acetate   1-butanol

(d) $CH_3COO(CH_2)_7CH_3$ + Na^+OH^- $\xrightarrow{\Delta}$ $CH_3COO^-Na^+$ + $CH_3(CH_2)_7OH$
   n-octyl acetate   sodium acetate   1-octanol

28-52. *Refer to Section 28-11.*

(a) Polymerization is the combination of many monomers, usually small molecules, to form large molecules called polymers which contain repetitive units of the monomers.

(b) Examples of polymerization reactions:

(1) $nCH_2{=}CH_2$ $\xrightarrow{\text{catalyst}}$ $(CH_2{-}CH_2)_n$
              ethylene   polyethylene

(2) $nCF_2{=}CF_2$ $\xrightarrow[\Delta]{\text{catalyst}}$ $(CF_2{-}CF_2)_n$
          tetrafluoroethene   "Teflon"

(3) $\overset{\displaystyle Cl}{|}$ $\overset{\displaystyle Cl}{|}$
       $nCH_2{=}CH{-}C{=}CH_2$ $\longrightarrow$ $(CH_2{-}CH{=}C{-}CH_2)_n$
          chloroprene   neoprene

28-54. *Refer to Section 28-11.*

It is definitely possible for a single monomer to polymerize to form a condensation homopolymer. An example is the formation of a polypeptide from the polymerization of a single amino acid. The individual amino acids are joined together by peptide bonds, which are formed by the elimination of a molecule of water between the amino group of one amino acid and the carboxylic acid group of another.

28-56. Refer to Section 28-11.

Changes in the polymer structure that can increase its rigidity and raise the melting point include introducing (1) cross-linking between polymer chains, (2) bulky substituents or branches on the chains, and (3) groups that can interact with strong intermolecular forces such as hydrogen bonding.

28-58. Refer to Section 28-11 and Exercise 28-56 Solution.

(a) Natural rubber is an elastic hydrocarbon polymer obtained from the sap of the rubber tree, called latex. A molecule of rubber (MW $\approx$ 136,000 g/mol) is composed of approximately 2000 units of 2-methyl-1,3-butadiene, also named isoprene.

$$2n \ CH_2=\overset{\overset{\displaystyle CH_3}{|}}{C}-CH=CH_2 \longrightarrow \ (CH_2-\overset{\overset{\displaystyle CH_3}{|}}{C}=CH-CH_2-CH_2-\overset{\overset{\displaystyle CH_3}{|}}{C}=CH-CH-CH_2)_n$$

$$\text{isoprene} \qquad\qquad\qquad \text{natural rubber}$$

(b) Vulcanization is a process in which sulfur is first added to rubber, then the system is heated to about 140°.

(c) Vulcanization causes cross-linking between the long rubber polymer chains. The result is a stronger, more elastic rubber, which is more resistant to cold and heat.

(d) At the same time the sulfur is mixed into the rubber, fillers, such as zinc oxide, barium sulfate, titanium dioxide and antimony(V) sulfate, and reinforcing agent, such as carbon black, are added.

(e) The purpose of fillers and reinforcing agents is to increase the durability of rubber and alter its color.

28-60. Refer to Section 32-11.

Polyamides are polymeric amides, a class of condensation polymer. Nylon, a very important fiber product, is the best known polymeric amide. Polyamides can be formed by the reactions (1) between a dicarboxylic acid with a diamine, or (2) between amino acids with the elimination of H_2O molecules..

28-62. Refer to Section 28-11.

Consider the following unbalanced polymerization reaction producing a polyester:

two repeating units of the polymer

28-64. Refer to Section 28-11.

Copolymers are formed when two different monomers are mixed and then polymerized. It is possible to produce a copolymer by addition polymerization. An example is SBR, the most important rubber produced in the United States. It is a copolymer of styrene and butadiene in a 1:3 ratio.

(a) Common Nylon is called Nylon 66 because the parent diamine and dicarboxylic acid of Nylon 66 each contain six carbon atoms.

(b) The parent diamine and dicarboxylic acid of Nylon xy would contain x and y carbon atoms, respectively.

Nylon 45:

$$-NH\left[\overset{\overset{O}{\|}}{C}-(CH_2)_3-\overset{\overset{O}{\|}}{C}-NH-(CH_2)_4-NH\right]_n\overset{\overset{O}{\|}}{C}-$$

Nylon 64:

$$-NH\left[\overset{\overset{O}{\|}}{C}-(CH_2)_2-\overset{\overset{O}{\|}}{C}-NH-(CH_2)_6-NH\right]_n\overset{\overset{O}{\|}}{C}-$$

In order to participate in polymer formation, a molecule must be able to react at both ends (difunctional) so that the polymer chain can grow in length. Three types of molecules that can polymerize are (1) alkenes (e.g., ethene molecules reacting to form polyethylene), (2) molecules with two identical functional groups (e.g., dicarboxylic acid reacting with a diamine to produce Nylon), and (3) molecules containing two different functional groups (e.g., amino acids containing an amine group and a carboxylic acid group reacting to form proteins).

The link between adjacent units or monomers in a polypeptide is formed in a condensation reaction between the amine group of one amino acid and the carboxylic acid group of another with the elimination of H_2O molecules. These links are called peptide bonds.

A total of 9 dipeptides can be formed from the three amino acids, A, B and C:

A-A	B-A	C-A
A-B	B-B	C-B
A-C	B-C	C-C

Note: The dipeptide A-B is different from B-A. For example, consider the condensation reaction between $NH_2CHRCOOH$ and $NH_2CHR'COOH$. The two products that will form are:

$$NH_2\underset{R}{C}H\overset{\overset{O}{\|}}{C}NH\underset{R'}{C}H\overset{\overset{O}{\|}}{C}OH \quad \text{and} \quad NH_2\underset{R'}{C}H\overset{\overset{O}{\|}}{C}NH\underset{R}{C}H\overset{\overset{O}{\|}}{C}OH$$

The dipeptide precursor of aspartame is made from the two amino acids given here. Aspartame is the methyl ester of the dipeptide.

phenylalanine   aspartic acid

(a)    2 [benzene]—CH$_2$OH   +   2 Na $\longrightarrow$ 2 [benzene]—CH$_2$O$^-$ Na$^+$ +   H$_2$ (Section 28-7)

an alkoxide

(b)   CH$_3$CH$_2$COCH$_3$ $\xrightarrow{\text{KOH}(aq)}{\Delta}$ CH$_3$CH$_2$CO$^-$ K$^+$ +   CH$_3$OH   (Section 28-10)

potassium propanoate   methanol
(potassium salt of carboxylic acid)

(c) 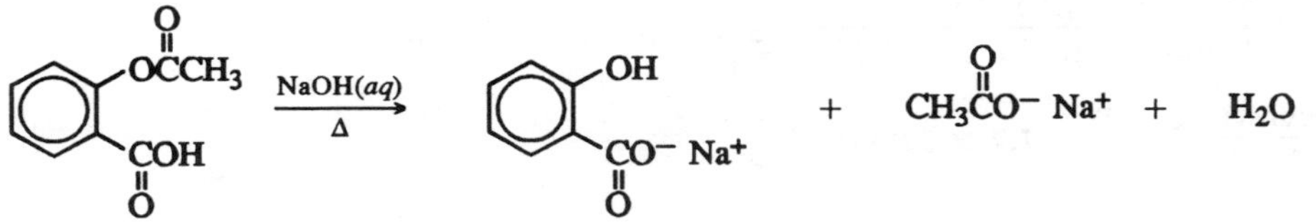   [2-(acetyloxy)benzoic acid structure with OCCH$_3$ and COH groups] $\xrightarrow{\text{NaOH}(aq)}{\Delta}$ [salicylate structure with OH and CO$^-$ Na$^+$]   +   CH$_3$CO$^-$ Na$^+$ +   H$_2$O   (Sections 28-7 and 28-10)

sodium 2-hydroxybenzoate   sodium acetate

The heat of combustion of ethanol (ethyl alcohol) is significantly lower than that of the saturated alkanes on a per gram basis. On a per mole basis, ethanol's heat of combustion is lower than those of all the saturated alkanes except methane.

Balanced equation: C(s) + H$_2$O(g) $\rightleftarrows$ CO(g) + H$_2$(g)

Plan: When carbon is treated with steam, equimolar amounts of CO and H$_2$ are produced. Therefore when the above reaction reaches equilibrium, the partial pressures of CO and H$_2$ are equal. Solve directly for P_{CO} and P_{H_2} using the K_p expression.

Let x = P_{CO} = P_{H_2} at equilibrium

$$K_p = \frac{(P_{CO})(P_{H_2})}{(P_{H_2O})} = \frac{x^2}{15.6} = 3.2 \qquad \text{Solving, x = 7.1}$$

Therefore, P_{CO} = P_{H_2} = **7.1 atm**

---

(1) Balanced equations: NaC$_6$H$_5$COO $\rightarrow$ Na$^+$ + C$_6$H$_5$COO$^-$ (to completion)
                C$_6$H$_5$COO$^-$ + H$_2$O $\rightleftarrows$ C$_6$H$_5$COOH + OH$^-$ (reversible)

Let x = [C$_6$H$_5$COO$^-$]$_{hydrolyzed}$.   Then, 0.10 - x = [C$_6$H$_5$COO$^-$];   x = [C$_6$H$_5$COOH] = [OH$^-$]

$$K_b = \frac{K_w}{K_{a(C_6H_5COOH)}} = \frac{1.0 \times 10^{-14}}{6.3 \times 10^{-5}} = 1.6 \times 10^{-10} = \frac{[C_6H_5COOH][OH^-]}{[C_6H_5COO^-]} = \frac{x^2}{0.10 - x} \approx \frac{x^2}{0.10}$$

Solving, x = 4.0 $\times$ 10^{-6}

Therefore, [OH$^-$] = 4.0 $\times$ 10^{-6} M; pOH = 5.40; pH = **8.60**

(2) Acetic acid is a weaker acid than benzoic acid since the K_a for acetic acid is less than the K_a for benzoic acid. Therefore, when the relative base strengths of their conjugate bases are compared, the acetate ion, CH_3COO^-, is a stronger base than the benzoate ion, $C_6H_5COO^-$. In other words, a 0.10 M solution of the benzoate ion is **more acidic** (less basic) than a 0.10 M solution of the acetate ion.

28-84. *Refer to Section 14-15 and Example 14-14.*

Plan: (1) Determine the moles of poly(vinyl alcohol), a nonelectrolyte, from its osmotic pressure, using $\pi = iMRT$ or $\pi = i(n/V)RT$.
(2) Calculate the molecular weight.

(1) $n = \dfrac{\pi V}{iRT} = \dfrac{(55/760 \text{ atm})(0.101 \text{ L})}{(1)(0.0821 \text{ L·atm/mol·K})(25°C + 273°)} = 3.0 \times 10^{-4}$ mol poly(vinyl alcohol)

(2) $MW = \dfrac{2.30 \text{ g poly(vinyl alcohol)}}{3.0 \times 10^{-4} \text{ mol}} = 7.7 \times 10^3$ **g/mol**